London Mathematical Society Lecture Note Series

All the titles listed below can be obtained from good booksellers or from Cambridge University Press. For a complete series listing visit
`http://publishing.cambridge.org/stm/mathematics/lmsn/`

285. Rational points on curves over finite fields, H. Niederreiter & C. Xing
286. Clifford algebras and spinors, 2nd edn, P. Lounesto
287. Topics on Riemann surfaces and Fuchsian groups, E. Bujalance, A.F. Costa & E. Martinez (eds)
288. Surveys in combinatorics, 2001, J.W.P. Hirschfeld (ed)
289. Aspects of Sobolev-type inequalities, L. Saloffe-Coste
290. Quantum groups and Lie theory, A. Pressley
291. Tits buildings and the model theory of groups, K. Tent
292. A quantum groups primer, S. Majid
293. Second order partial differential equations in Hilbert spaces, G. da Prato & J. Zabczyk
294. Introduction to operator space theory, G. Pisier
295. Geometry and integrability, L. Mason & Y. Nutku (eds)
296. Lectures on invariant theory, I. Dolgachev
297. The homotopy theory of simply-connected 4-manifolds, H.J. Baues
298. Higher operads, higher categories, T. Leinster
299. Kleinian groups and hyperbolic 3-manifolds, Y. Komori, V. Markovic & C. Series (eds)
300. Introduction to Möbius differential geometry, U. Hertrich-Jeromin
301. Stable modules and the D(2)-problem, F.A.E. Johnson
302. Discrete and continuous nonlinear Schrödinger systems, M. Ablowitz, B. Prinari & D. Trubatch
303. Number theory and algebraic geometry M. Reid & A. Skorobogatov
304. Groups St Andrews 2001 in Oxford vol. 1, C.M. Campbell, E.F. Robertson & G.C. Smith (eds)
305. Groups St Andrews 2001 in Oxford vol. 2, C.M. Campbell, E.F. Robertson & G.C. Smith (eds)
306. Geometric mechanics and symmetry: the Peyresq lectures, J. Montaldi & T. Ratiu (eds)
307. Surveys in combinatorics, 2003, C.D. Wensley (ed)
308. Topology, geometry and quantum field theory, U.L. Tillmann (ed)
309. Corings and comodules, T. Brzezinski & R. Wisbauer
310. Topics in dynamics and ergodic theory, S. Bezuglyi & S. Kolyada (eds)
311. Groups: topological, combinatorial and arithmetic aspects, T.W. Müller (ed)
312. Foundations of computational mathematics: Minneapolis, 2002, F. Cucker *et al.* (eds)
313. Transcendental aspects of algebraic cycles, S. Müller-Stach & C. Peters (eds)
314. Linear logic in computer science, T. Ehrhard *et al.* (eds)

London Mathematical Society Lecture Note Series: 306

Geometric Mechanics and Symmetry

The Peyresq Lectures

Edited by

James Montaldi
University of Manchester

Tudor Ratiu
Ecole Polytechnique Fédérale de Lausanne

CAMBRIDGE
UNIVERSITY PRESS

CAMBRIDGE UNIVERSITY PRESS
Cambridge, New York, Melbourne, Madrid, Cape Town,
Singapore, São Paulo, Delhi, Tokyo, Mexico City

Cambridge University Press
The Edinburgh Building, Cambridge CB2 8RU, UK

Published in the United States of America by Cambridge University Press, New York

www.cambridge.org
Information on this title: www.cambridge.org/9780521539579

First published 2005

A catalogue record for this publication is available from the British Library

ISBN 978-0-521-53957-9 Paperback

Additional resources for this publication at www.cambridge.org/9780521539579

Contents

Each set of lectures begins with its own table of contents.

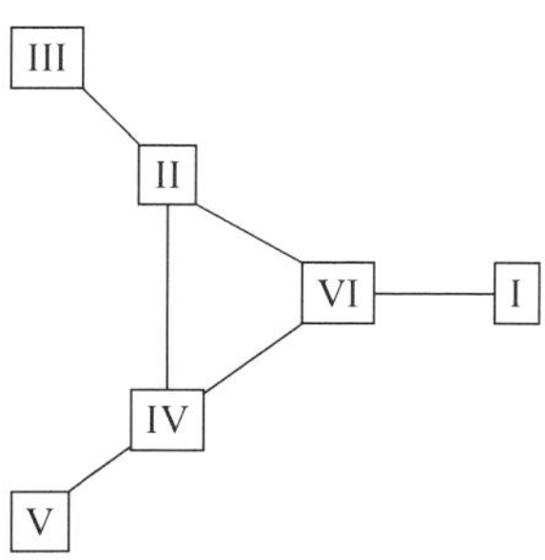

Interrelationships between the courses.

Contributors

Henk Broer, Department of Mathematics, University of Groningen, P.O. Box 800, 9700 AV Groningen, The Netherlands.

Pietro-Luciano Buono, Faculty of Science, University of Ontario Institute of Technology, 2000 Simcoe St. North, Oshawa, ON L1H 7K4, Canada.

Richard Cushman, Mathematics Institute, University of Utrecht, Budapestlaan 6, 3508TA Utrecht, The Netherlands.

Maria-Cristina Ciocci, Department of Pure Mathematics and Computer Algebra, Ghent University, Krijgslaan 281, 9000 Gent, The Netherlands.

Konstantinos Efstathiou, Department of Physics, University of Athens, Panepistimiopolis Zografos, Athens, Greece.

Frédéric Laurent-Polz, Institut Nonlinéaire de Nice, 1361 route des Lucioles, 06560 Valbonne, France.

Anna Litvak-Hinenzon, Mathematics Institute, University of Warwick, Coventry CV4 7AL, UK.

Ken Meyer, Department of Mathematical Sciences, University of Cincinnati, PO Box 210025, Cincinnati, OH 45221-0025, USA.

James Montaldi, School of Mathematics, University of Manchester, Manchester, UK.

Tudor Ratiu, Départment de Mathématiques, Ecole Polytechnique Fédérale de Lausanne, 1015 Lausanne, Switzerland.

Esmeralda Sousa Dias, Departmento de Matemática, Instituto Superior Técnico, Av. Rovisco Pais 1049-001, Lisbon, Portugal.

Bob Rink, Department of Mathematics, Imperial College of Science Technology and Medicine, London SW7 2AZ, UK.

Dmitrií Sadovskií, Université du Littoral, MREID, 145 av. Maurice Schumann, 59140 Dunkerque, France.

Luca Sbano, Mathematics Institute, University of Warwick, Coventry CV4 7AL, UK.

Glaucio Terra, Instituto de Matematica e Estatistica da Universidade de Sao Paulo, Rua do Matão 1010, Cidade Universitária, CEP 05508-090, São Paulo, SP, Brazil.

Răzvan Tudoran, Departamentul de Matematică, Universitatea de Vest, RO1900 Timişoara, Romania.

Johan Matheus Tuwankotta, Department of Mathematics, Bandung Institute of Technology Jl. Ganesa no. 10, Bandung, 40132 Jawa Barat, Indonesia.

Preface

In the summers of 2000 and 2001, we organized two European Summer Schools in Geometric Mechanics. They were both held in the wonderful environment provided by the village-cum-international conference centre at Peyresq in the Alpes de Haute Provence in France, about 100km North of Nice. Each school consisted of 6 short lecture courses, as well as numerous short talks given by participants, of whom there were about 40 at each school. The majority of participants were from Europe with a few coming from West of the Atlantic or East of the Urals, and we were pleased to see a number of participants from the first year returning in the second. Several of the courses and short talks led to collaborations between participants and/or lecturers.

The summer schools were funded principally by the European Commission under the High-Level Scientific Conferences section of the Fifth Framework Programme. Additional funding was very kindly provided by the *Fondation Peiresc.* The principal aim of the two schools was to provide young scientists with a quick introduction to the geometry and dynamics involved in geometric mechanics and to bring them to a level of understanding where they could begin work on research problems. The schools were also closely linked to the Mechanics and Symmetry in Europe (MASIE) research training network, organized by Mark Roberts, and several of the participants went on to become successful PhD students or postdocs in MASIE.

Of the lecture courses, seven have been written up for this book—mostly by the participants themselves with varying degrees of collaboration from the lecturers. The book is divided into 6 chapters, each representing a course of 5 or 6 lectures, with the exception of Ratiu's which are taken from two courses. The notes on Stability in Hamiltonian systems by Rink and Tuwankotta based on Meyer's lectures on N-body problems have been placed first as they require the least background knowledge. They cover not only Lyapounov's and Dirichlet's stability theorems but also the *in*stability theorem of Chetaev, with applications to the restricted 3-body problem. Second are the notes from Ratiu's courses which give an introduction to the mathematical formalism of geometric mechanics, beginning with the Hamiltonian, Lagrangian and Poisson formalisms, and continuing with aspects of reduction and reconstruction, the whole being laced with numerous examples, and including some meterial on Euler-Poincaré equations. This last topic is the basis of the third set of lecture notes: Holm's course on the Euler-Poincaré approach to fluid dynamics, showing especially how this approach helps to model the multiscale physics involved.

The fourth chapter contains Cushman's lectures on the global geometry of integrable systems, describing particularly the monodromy in such systems, which has recently proved to be so important in explaining some features of molecular spectra. When integrability breaks down, one requires KAM theory which is described in Broer's lectures presented in the following chapter. The theory is described there for dissipative systems, showing how quasiperiodic attractors persist and bifurcate in families of systems, but applies also to conservative systems as is described in the appendix to that chapter.

The final chapter consists of (a slightly expanded version of) Montaldi's lecture course on Hamiltonian bifurcations in symmetric systems. These deal firstly with bifurcations near equilibria including Hamiltonian-Hopf bifurcation, and then with bifurcations of relative equilibria.

We believe all the participants and lecturers would like to join us in thanking Mme. Mady Smets and the staff of the *Peyresq Foyer d'Humanisme* for their warmth, generosity and hospitality, and for the smooth running of the centre without which the Schools would not have had the academic success they did.

I

Stability in Hamiltonian Systems:

Applications to the restricted three-body problem

Bob Rink & Theo Tuwankotta

Based on lectures by Ken Meyer

1 Introduction

As participants in the MASIE-project, we attended the summer school *Mechanics and Symmetry* in Peyresq, France, during the first two weeks of September 2000. These lecture notes are based on the notes we took there from Professor Meyer's lecture series *"N-Body Problems"*.

The N-body problem is a famous classical problem. It consists in describing the motion of N planets that interact with a gravitational force. Already in 1772, Euler described the three-body problem in his effort to study the motion of the moon. In 1836 Jacobi brought forward an even more specific part of the three body problem, namely that in which one of the planets has a very small mass. This system is the topic of this paper and is nowadays called the *restricted three-body problem.* It is a conservative system with two degrees of freedom, which gained extensive study in mechanics.

The N-body problem has always been a major topic in mathematics and physics. In 1858, Dirichlet claimed to have found a general method to treat any problem in mechanics. In particular, he said to have proven the stability of the planetary system. This statement is still questionable because he passed away without leaving any proof. Nevertheless, it initiated Weierstrass and his students Kovalevski and Mittag-Leffler to try and rediscover the method mentioned by Dirichlet. Mittag-Leffler even managed to convince the King of Sweden and Norway to establish a prize for finding a series expansion for coordinates of the N-body problem valid for all time, as indicated by Dirichlet's statement. In 1889, this prize was awarded to Poincaré, although he did not solve the problem. His essay, however, produced a lot of original ideas which later turned out to be very important for mechanics. Moreover, some of them even stimulated other branches of mathematics, for instance topology, to be born and later on gain extensive study. Despite of all this effort, the N-body problem is still unsolved for N>2[1].

This paper focuses on the relatively simple restricted three-body problem. This describes the motion of a test particle in the combined gravitational field of two planets and it could serve for instance as a model for the motion of a satellite in the Earth-Moon system or a comet in the Sun-Jupiter system. The restricted three-body problem has a number of relative equilibria, which we compute. The remaining text will mainly be concerned with general Hamiltonian equilibria. Stability criteria for these equilibria will be derived, as well as detection methods for bifurcations of periodic solutions. Classical and more advanced mathematical techniques are used, such as spectral analysis, Liapunov functions, Birkhoff-Gustavson normal forms, Poincaré sections, and Kolmogorov twist stability. All help to study the motion of the test particle near the relative equilibria of the restricted problem.

2 The restricted three-body problem

Before introducing the restricted three-body problem, let us study the two-body problem, the motion of two planets interacting via gravitation. Denote by $\boldsymbol{X_1}, \boldsymbol{X_2} \in \mathbb{R}^3$ the positions of the planets 1 and 2 respectively. Let us assume that planet 1 has mass $0 < \mu < 1$, planet 2 has mass $1 - \mu$ and the gravitational constant is equal to 1. These assumptions are not very restrictive, because they can always be arranged by a rescaling of time. The equations of

[1] Summarized from [10], [11] and [8]

motion for the two-body problem then read:

$$\begin{aligned} \frac{d^2\boldsymbol{X_1}}{dt^2} &= -\tfrac{(1-\mu)}{||\boldsymbol{X_1}-\boldsymbol{X_2}||^3}(\boldsymbol{X_1}-\boldsymbol{X_2}) \\ \frac{d^2\boldsymbol{X_2}}{dt^2} &= -\tfrac{\mu}{||\boldsymbol{X_1}-\boldsymbol{X_2}||^3}(\boldsymbol{X_2}-\boldsymbol{X_1})\,. \end{aligned} \tag{2.1}$$

Let us denote the center of mass

$$\boldsymbol{Z} := \mu\boldsymbol{X_1} + (1-\mu)\boldsymbol{X_2}\,. \tag{2.2}$$

Then we derive from (2.1) and (2.2) that $\frac{d^2\boldsymbol{Z}}{dt^2} = 0$, expressing that the center of mass moves with constant speed. Now we transform to co-moving coordinates

$$\boldsymbol{Y_i} = \boldsymbol{X_i} - \boldsymbol{Z} \text{ for } i = 1, 2, \tag{2.3}$$

and we write down the equations of motions in these new variables:

$$\frac{d^2\boldsymbol{Y_1}}{dt^2} = -\frac{(1-\mu)^3}{||\boldsymbol{Y_1}||^3}\boldsymbol{Y_1}\,, \qquad \frac{d^2\boldsymbol{Y_2}}{dt^2} = -\frac{\mu^3}{||\boldsymbol{Y_2}||^3}\boldsymbol{Y_2}\,. \tag{2.4}$$

Let us analyze these equations a bit more. First of all, we see from the definitions (2.2) and (2.3) that $\mu\boldsymbol{Y_1} + (1-\mu)\boldsymbol{Y_2} = 0$, so $\boldsymbol{Y_1}$ and $\boldsymbol{Y_2}$ lie on a line through the origin of $\mathbb{R}^3$, both at another side of the origin, and their length ratio $\frac{||\boldsymbol{Y_1}||}{||\boldsymbol{Y_2}||}$ is fixed to the value $\frac{1-\mu}{\mu}$. The line connecting $\boldsymbol{Y_1}, \boldsymbol{Y_2}$ and the origin is called the *line of syzygy*. Because $\boldsymbol{Y}_2 = -\frac{\mu}{1-\mu}\boldsymbol{Y}_1$, we in fact only need to study the first equation of (2.4). The motion of the second planet then follows automatically.

Secondly, by differentiation one finds that the angular momentum $\boldsymbol{Y_1} \times \frac{d\boldsymbol{Y_1}}{dt}$ is independent of time. Indeed, $\frac{d}{dt}(\boldsymbol{Y_1} \times \frac{d\boldsymbol{Y_1}}{dt}) = \frac{d\boldsymbol{Y_1}}{dt} \times \frac{d\boldsymbol{Y_1}}{dt} + \boldsymbol{Y_1} \times \frac{d^2\boldsymbol{Y_1}}{dt^2} = 0$, because both terms are the cross-products of collinear vectors.

In the case that $\boldsymbol{Y_1} \times \frac{d\boldsymbol{Y_1}}{dt} = 0$, and assuming that $\boldsymbol{Y_1}(0) \neq 0$, we have that $\frac{d\boldsymbol{Y_1}}{dt}$ has the same direction as $\boldsymbol{Y_1}$, so the motion takes place in a one-dimensional subspace: $\boldsymbol{Y_1}, \frac{d\boldsymbol{Y_1}}{dt}, \boldsymbol{Y_2}, \frac{d\boldsymbol{Y_2}}{dt} \in \boldsymbol{Y_1}(0)\mathbb{R} = \boldsymbol{Y_2}(0)\mathbb{R}$. It is not difficult to derive the following scalar second order differential equation for the motion in this subspace: $\frac{d^2}{dt^2}||\boldsymbol{Y_1}|| = -(1-\mu)^3/||\boldsymbol{Y_1}||^2$. It turns out that in this case $\boldsymbol{Y_1}$ and $\boldsymbol{Y_2}$ fall into the origin in a finite time.

In the case that $\boldsymbol{Y_1} \times \frac{d\boldsymbol{Y_1}}{dt} \neq 0$, the motion takes place in the plane perpendicular to $\boldsymbol{Y_1} \times \frac{d\boldsymbol{Y_1}}{dt}$, because both $\boldsymbol{Y_1}$ and $\frac{d\boldsymbol{Y_1}}{dt}$ are perpendicular to the constant vector $\boldsymbol{Y_1} \times \frac{d\boldsymbol{Y_1}}{dt}$. By rotating our coordinate frame, we can arrange that $\boldsymbol{Y_1} \times \frac{d\boldsymbol{Y_1}}{dt}$ is some multiple of the third basis vector. Thus we can consider the equations (2.4) as two second order planar equations. It is well-known that the planar solutions of $\frac{d^2\boldsymbol{Y_1}}{dt^2} = -\frac{(1-\mu)^3}{||\boldsymbol{Y_1}||^3}\boldsymbol{Y_1}$ with $\boldsymbol{Y_1} \times \frac{d\boldsymbol{Y_1}}{dt} \neq 0$ describe one of the conic sections: a circle, an ellipse, a parabola or a hyperbola. $\boldsymbol{Y_2}$ clearly describes a similar conic section.

Let us now assume that a certain solution of the two-body problem is given to us. We want to study the motion of a *test particle* in the gravitational field of the two main bodies, which we call *primaries*. The test particle is assumed to have zero mass. Therefore it does not affect the primaries, but it does feel the gravitational force of the primaries acting on it. The resulting problem is called the restricted three-body problem. It could serve as a model for a satellite in the Earth-Moon system or a comet in the Sun-Jupiter system. Let $\boldsymbol{X} \in \mathbb{R}^3$ denote the position of the test particle. Then the restricted three-body problem is given by

$$\frac{d^2\boldsymbol{X}}{dt^2} = -\frac{\mu}{||\boldsymbol{X} - \boldsymbol{X_1}||^3}(\boldsymbol{X} - \boldsymbol{X_1}) - \frac{(1-\mu)}{||\boldsymbol{X} - \boldsymbol{X_2}||^3}(\boldsymbol{X} - \boldsymbol{X_2}) , \qquad (2.5)$$

in which $(\boldsymbol{X_1}, \boldsymbol{X_2})$ is the given solution of the two-body problem. One can again transform to co-moving coordinates, setting $\boldsymbol{Y} = \boldsymbol{X} - \boldsymbol{Z}$, which results in the system

$$\frac{d^2\boldsymbol{Y}}{dt^2} = -\frac{\mu}{||\boldsymbol{Y} - \boldsymbol{Y_1}||^3}(\boldsymbol{Y} - \boldsymbol{Y_1}) - \frac{(1-\mu)}{||\boldsymbol{Y} - \boldsymbol{Y_2}||^3}(\boldsymbol{Y} - \boldsymbol{Y_2}) . \qquad (2.6)$$

At this point we start making assumptions. Let us assume that the primaries move in a circular orbit around their center of mass with constant angular velocity . This is approximately true for the Earth-Moon system and the Sun-Jupiter system. We set the angular velocity equal to 1. Without loss of generality, we can assume that the motion of the primaries takes place in the plane perpendicular to the third basis-vector. Thus, after translating time if necessary,

$$\boldsymbol{Y_1} = R(t)\begin{pmatrix} 1-\mu \\ 0 \\ 0 \end{pmatrix} , \quad \boldsymbol{Y_2} = R(t)\begin{pmatrix} -\mu \\ 0 \\ 0 \end{pmatrix} , \qquad (2.7)$$

in which $R(t)$ is the rotation matrix:

$$R(t) := \begin{pmatrix} \cos t & -\sin t & 0 \\ \sin t & \cos t & 0 \\ 0 & 0 & 1 \end{pmatrix} . \qquad (2.8)$$

Note that we have introduced a rotating coordinate frame in which the motion of the primaries has become stationary. At this point we put in our test particle and again we make an assumption, namely that it moves in the same plane as the primaries do. So we set

$$\boldsymbol{Y} = R(t)\begin{pmatrix} x_1 \\ x_2 \\ 0 \end{pmatrix}. \tag{2.9}$$

Let $(\boldsymbol{x}, 0)^T = (x_1, x_2, 0)^T$ be the coordinates of the test particle in the rotating coordinate frame. By inserting (2.7), (2.8) and (2.9) into (2.6), multiplying the resulting equation from the left by $R(t)^{-1}$ and using two following identities

$$\frac{d^2}{dt^2}\begin{pmatrix} \cos t & -\sin t \\ \sin t & \cos t \end{pmatrix} = -\begin{pmatrix} \cos t & -\sin t \\ \sin t & \cos t \end{pmatrix},$$

$$\begin{pmatrix} \cos t & -\sin t \\ \sin t & \cos t \end{pmatrix}^{-1}\frac{d}{dt}\begin{pmatrix} \cos t & -\sin t \\ \sin t & \cos t \end{pmatrix} = \begin{pmatrix} 0 & -1 \\ 1 & 0 \end{pmatrix},$$

we deduce the planar equations of motion for $\boldsymbol{x} \in \mathbb{R}^2$:

$$\frac{d^2\boldsymbol{x}}{dt^2} - \boldsymbol{x} + \begin{pmatrix} 0 & -2 \\ 2 & 0 \end{pmatrix}\frac{d\boldsymbol{x}}{dt} =$$
$$-\frac{\mu}{\|\boldsymbol{x} - \left(\begin{smallmatrix} 1-\mu \\ 0 \end{smallmatrix}\right)\|^3}\left(\boldsymbol{x} - \left(\begin{smallmatrix} 1-\mu \\ 0 \end{smallmatrix}\right)\right) - \frac{1-\mu}{\|\boldsymbol{x} - \left(\begin{smallmatrix} -\mu \\ 0 \end{smallmatrix}\right)\|^3}\left(\boldsymbol{x} - \left(\begin{smallmatrix} -\mu \\ 0 \end{smallmatrix}\right)\right).$$

Finally, setting $\boldsymbol{y} = \frac{d\boldsymbol{x}}{dt} + \left(\begin{smallmatrix} 0 & -1 \\ 1 & 0 \end{smallmatrix}\right)\boldsymbol{x}$, we find that these are Hamiltonian equations of motion on $\mathbb{R}^4 \setminus \{\boldsymbol{x} = \left(\begin{smallmatrix} 1-\mu \\ 0 \end{smallmatrix}\right), \left(\begin{smallmatrix} -\mu \\ 0 \end{smallmatrix}\right)\}$ with Hamiltonian

$$H = \frac{1}{2}(y_1^2 + y_2^2) - (x_1y_2 - x_2y_1) - \frac{\mu}{\|\boldsymbol{x} - \left(\begin{smallmatrix} 1-\mu \\ 0 \end{smallmatrix}\right)\|} - \frac{1-\mu}{\|\boldsymbol{x} - \left(\begin{smallmatrix} -\mu \\ 0 \end{smallmatrix}\right)\|}, \tag{2.10}$$

where we have equipped $\mathbb{R}^4$ with the canonical symplectic form $dx_1 \wedge dy_1 + dx_2 \wedge dy_2$, i.e. the equations of motion are given by $\frac{dx_i}{dt} = \frac{\partial H}{\partial y_i}$, $\frac{dy_i}{dt} = -\frac{\partial H}{\partial x_i}$.

3 Relative equilibria

Let us look for equilibrium solutions of the Hamiltonian vector field induced by (2.10). These correspond to stationary motion of the test particle relative to the rotating coordinate frame and are therefore called *relative equilibria.* In the original coordinates they correspond to the test particle rotating around the center of mass of the primaries with angular velocity 1.

First of all, to facilitate notation, we introduce the potential energy function

$$V(\boldsymbol{x}) := -\frac{\mu}{\|\boldsymbol{x} - \left(\begin{smallmatrix} 1-\mu \\ 0 \end{smallmatrix}\right)\|} - \frac{1-\mu}{\|\boldsymbol{x} - \left(\begin{smallmatrix} -\mu \\ 0 \end{smallmatrix}\right)\|}.$$

To find the equilibrium solutions of (2.10) we set all the partial derivatives of H equal to zero and find

$$y_1 + x_2 = 0 \;,\; y_2 - x_1 = 0 \;,\; -y_2 + \frac{\partial V}{\partial x_1}(\boldsymbol{x}) = 0 \;,\; y_1 + \frac{\partial V}{\partial x_2}(\boldsymbol{x}) = 0 \;,$$

or equivalently,

$$\frac{\partial V}{\partial x_1}(\boldsymbol{x}) = x_1 \;,\; \frac{\partial V}{\partial x_2}(\boldsymbol{x}) = x_2 \;, \tag{3.1}$$

where $\boldsymbol{y}$ at the equilibrium point can easily be found once we solved (3.1) for $\boldsymbol{x}$ at the equilibrium point. Note that $\boldsymbol{x}$ solves (3.1) if and only if $\boldsymbol{x}$ is a stationary point of the function

$$U(\boldsymbol{x}) := \frac{1}{2}(x_1^2 + x_2^2) - V(\boldsymbol{x}) \;,$$

called the *amended potential.*

Let us first look for equilibrium points of the amended potential that lie on the line of syzygy, i.e. for which $x_2 = 0$. Note that $\frac{\partial U}{\partial x_2}(\boldsymbol{x}) = 0$ is automatically satisfied in this case since $\frac{\partial V}{\partial x_2}\,|_{x_2=0} \equiv 0$. $\frac{\partial U}{\partial x_1}(\boldsymbol{x}) = 0$ reduces to

$$\frac{d}{dx_1}U(x_1, 0) = \frac{d}{dx_1}\left(\frac{1}{2}x_1^2 + \frac{\mu}{|x_1 + \mu - 1|} + \frac{1-\mu}{|x_1 + \mu|}\right) = 0 \,. \tag{3.2}$$

Clearly, $U(x_1, 0)$ goes to infinity if x_1 approaches $-\infty, -\mu, 1-\mu$ or ∞, so $U(x_1, 0)$ has at least one critical point on each of the intervals $(-\infty, -\mu)$, $(-\mu, 1-\mu)$ and $(1-\mu, \infty)$. But we also calculate that $\frac{d^2}{dx_1^2}U(x_1, 0) = 1 + 2\frac{\mu}{|x_1+\mu-1|^3} + 2\frac{1-\mu}{|x_1+\mu|^3} > 0$. So $U(x_1, 0)$ is convex on each of these intervals and we conclude that there is exactly one critical point in each of the intervals. The three relative equilibria on the line of syzygy are called the *Eulerian* equilibria. They are denoted by $\mathcal{L}_1$, $\mathcal{L}_2$ and $\mathcal{L}_3$, where $\mathcal{L}_1 \in (-\infty, -\mu) \times \{0\}$, $\mathcal{L}_2 \in (-\mu, 1-\mu) \times \{0\}$ and $\mathcal{L}_3 \in (1-\mu, \infty) \times \{0\}$.

Now we shall look for equilibrium points that do not lie on the line of syzygy. Let us use $d_1 = ||\boldsymbol{x} - \binom{1-\mu}{0}|| = \sqrt{(x_1 + \mu - 1)^2 + x_2^2}$ and $d_2 = ||\boldsymbol{x} - \binom{-\mu}{0}|| = \sqrt{(x_1 + \mu)^2 + x_2^2}$ as coordinates in each of the half-planes $\{x_2 > 0\}$ and $\{x_2 < 0\}$. Then U can be written as $U = \frac{\mu}{2}d_1^2 + \frac{1-\mu}{2}d_2^2 - \frac{\mu(1-\mu)}{2} + \frac{\mu}{d_1} + \frac{1-\mu}{d_2}$. So the critical points of U are given by $d_i = d_i^{-2}$ i.e. $d_1 = d_2 = 1$. This gives us the two *Lagrangean* equilibria which lie at the third vertex of the equilateral triangle with the primaries at its base-points: $\mathcal{L}_4 = (\frac{1}{2} - \mu, \frac{1}{2}\sqrt{3})^T$ and $\mathcal{L}_5 = (\frac{1}{2} - \mu, -\frac{1}{2}\sqrt{3})^T$.

This paper discusses some useful tools for the study of the flow of Hamiltonian vector fields near equilibrium points. We will for instance establish stability criteria for Hamiltonian equilibria and study bifurcations of periodic solutions near Hamiltonian equilibria. The Eulerian and Lagrangean equilibria of the restricted three-body problem will serve as an instructive and inspiring example.

4 Linear Hamiltonian systems

One of the techniques to prove stability for an equilibrium of a system of differential equations, is to analyze the linearized system around that equilibrium. Stability or instability then may follow from the eigenvalues of the matrix of the linearized system. In Hamiltonian systems, these eigenvalues have a special structure which implies that the linear theory can only be used to prove instability, not stability. We will start by giving a brief introduction to linear Hamiltonian systems. We then conclude this section with a lemma which shows why one can not conclude stability from the linear analysis.

Consider a symplectic vector space $\mathbb{R}^{2n}$ with coordinates $\boldsymbol{z} = (\boldsymbol{x}, \boldsymbol{y})^T$ and the symplectic form is $d\boldsymbol{x} \wedge d\boldsymbol{y} := \sum_{j=1}^{n} dx_j \wedge dy_j$. Then every continuously differentiable function $H : \mathbb{R}^{2n} \to \mathbb{R}$ induces the Hamiltonian vector field X_H on $\mathbb{R}^{2n}$ defined by $X_H(z) = J(\nabla H(z))^T$, in which the $2n \times 2n$ matrix

$$J = \begin{pmatrix} 0 & I_n \\ -I_n & 0 \end{pmatrix}$$

is called the *standard symplectic matrix*. Note that X_H gives rise to the Hamiltonian system of differential equations $\frac{dx_i}{dt} = \frac{\partial H}{\partial y_i}$, $\frac{dy_i}{dt} = -\frac{\partial H}{\partial x_i}$. The function H is called the Hamiltonian function of the vector field X_H.

Suppose that for $\boldsymbol{z}_\circ \in \mathbb{R}^{2n}$ we have $\nabla H(\boldsymbol{z}_\circ) = 0$, then $\boldsymbol{z}_\circ$ is called a *rest point*, *equilibrium point*, *fixed point*, or *critical point* of H. Note that $X_H(\boldsymbol{z}_\circ) = 0$ so $\boldsymbol{z}_\circ$ is fixed by the flow of X_H. By translating our coordinate frame, we can arrange that $\boldsymbol{z}_\circ = 0$. We will assume that H is a sufficiently smooth function in a neighborhood of its equilibrium 0, so that we can write $H(\boldsymbol{z}) = H_2(\boldsymbol{z}) + \mathcal{O}(||\boldsymbol{z}||^3)$ as $\boldsymbol{z} \to 0$, where H_2 is a quadratic form on $\mathbb{R}^{2n}$. The linearized vector field of X_H at 0 is the Hamiltonian vector field X_{H_2} induced by the quadratic Hamiltonian H_2. This encourages us to study quadratic Hamiltonians and their induced linear Hamiltonian vector fields.

Let $H_2 : \mathbb{R}^{2n} \to \mathbb{R}$ be a quadratic form, determined by the symmetric $2n \times 2n$ matrix Q, i.e. $H_2(\boldsymbol{z}) = \frac{1}{2}\boldsymbol{z}^T Q \boldsymbol{z}$ with $Q^T = Q$. H_2 generates a linear Hamiltonian vector field:

$$X_{H_2}(\boldsymbol{z}) = J(\nabla H_2(\boldsymbol{z}))^T = JQ\boldsymbol{z} \, . \tag{4.1}$$

Matrices S of the form $S = JQ$ for some symmetric matrix Q are called *infinitesimally symplectic* or *Hamiltonian*. The set of all infinitesimally symplectic matrices is denoted by

$$\begin{aligned}\mathfrak{sp}(n) &:= \{S \in \mathbb{R}^{2n\times 2n} \mid S = JQ \text{ for some } Q = Q^T\}\\ &= \{S \in \mathbb{R}^{2n\times 2n} \mid S^T J + JS = 0\}\,.\end{aligned}$$

Note that the standard symplectic matrix J satisfies $J^{-1} = J^T = -J$. Now take any infinitesimally symplectic matrix S of the form $S = JQ$, with Q symmetric. Then the simple calculation

$$J^{-1}(-S^T)J = J^{-1}(-JQ)^T J = -J^{-1}(QJ^T)J = -J^{-1}Q = JQ = S,$$

shows that S and $-S^T$ are similar. But similar matrices have equal eigenvalues. And because S has real coefficients, this observation leads to the following lemma:

Lemma 4.1 *If $S \in \mathfrak{sp}(n)$ and λ is an eigenvalue of S, then also $-\lambda, \overline{\lambda}$ and $-\overline{\lambda}$ are eigenvalues of S.*

Now let us consider the exponential of an infinitesimally symplectic matrix, $\exp(S) = \exp(JQ)$, which is the fundamental matrix for the time-1 flow of the linear Hamiltonian vector field $\boldsymbol{z} \mapsto S\boldsymbol{z} = JQ\boldsymbol{z}$. It is a nice exercise to show that it satisfies $(\exp(S))^T J \exp(S) = J$. In general, a matrix $P \in \mathbb{R}^{2n\times 2n}$ satisfying $P^T JP = J$ is called *symplectic*. The set of symplectic matrices is denoted

$$\mathrm{Sp}(n) := \{P \in \mathbb{R}^{2n\times 2n} \mid P^T JP = J\}.$$

For a symplectic matrix P one easily derives that $J^{-1}P^{-T}J = P$, so P^{-T} and P are similar. This leads to:

Lemma 4.2 *If $P \in \mathrm{Sp}(n)$ and λ is an eigenvalue of P, then so too are $\lambda^{-1}, \overline{\lambda}$ and $\overline{\lambda}^{-1}$.*

We remark here that $\mathrm{Sp}(n)$ is a Lie group with matrix multiplication. Its Lie algebra is exactly $\mathfrak{sp}(n)$.

Remember that we studied linear Hamiltonian systems to determine stability or instability of an equilibrium from the spectrum of its linearized vector field. From lemma 4.1 we see if one eigenvalue has a nonzero real part, then there must be an eigenvalue with positive real part. In this case the equilibrium is unstable. The other possibility is that all eigenvalues are purely imaginary. In this case, adding nonlinear terms could destabilize the equilibrium. So lemma 4.1 states that for Hamiltonian systems, the linear theory can only be useful to prove instability of an equilibrium.

Lemma 4.2 states a similar thing for symplectic maps: if $\Psi : \mathbb{R}^{2n} \to \mathbb{R}^{2n}$ is a symplectic diffeomorphism with a fixed point, then the linearization of Ψ at that fixed point can only be used to prove instability of the fixed point, not stability.

The reader should be convinced now that we need more sophisticated mathematical techniques if we want to have stability results. Some of them will be explained in the following section.

5 Liapunov's and Chetaev's theorems

We will now describe a direct method to determine stability of an equilibrium. We will give references for the proofs and explain the interpretation of the theory instead. In section 6 we shall apply the obtained results to the relative equilibria of the restricted three-body problem.

Consider a general system of differential equations,

$$\dot{\boldsymbol{v}} = f(\boldsymbol{v}) \,, \tag{5.1}$$

where f is a C^r vector field on $\mathbb{R}^m$ and $f(0) = 0$. Let $V : U \to \mathbb{R}$ be a positive definite C^1 function on a neighborhood U of the origin, i.e. $V(0) = 0$ and $V(\boldsymbol{z}) > 0, \ \ \forall \boldsymbol{z} \in U \backslash \{0\}$. If $\boldsymbol{u}$ is a solution of (5.1), then the derivative of V along $\boldsymbol{u}$ is $\frac{d}{dt} V(\boldsymbol{u}(t)) = \nabla V(\boldsymbol{u}(t)) \cdot \dot{\boldsymbol{u}}(t) = \nabla V(\boldsymbol{u}(t)) \cdot f(\boldsymbol{u}(t))$. So let us define the *orbital derivative* $\dot{V} : U \to \mathbb{R}$ of V as

$$\dot{V}(\boldsymbol{v}) := \nabla V(\boldsymbol{v}) \cdot f(\boldsymbol{v}) \,.$$

Theorem 5.1 (Liapunov's theorem) *Given such a function V for the system of equations (5.1), we have:*

(i) *If* $\dot{V}(\boldsymbol{v}) \leq 0, \ \forall \boldsymbol{v} \in U \backslash \{0\}$ *then the origin is stable.*
(ii) *If* $\dot{V}(\boldsymbol{v}) < 0, \ \forall \boldsymbol{v} \in U \backslash \{0\}$ *then the origin is asymptotically stable.*
(iii) *If* $\dot{V}(\boldsymbol{v}) > 0, \ \forall \boldsymbol{v} \in U \backslash \{0\}$ *then the origin is unstable.*

The function V is called a *Liapunov function.*

Let us see what this means for $m = 2$. Since V is a positive definite function, 0 is a local minimum of V. This implies that there exists a small neighborhood U' of 0 such that the level sets of V lying in U' are closed curves. Recall that $\nabla V(\boldsymbol{u}_c)$ is a normal vector to the level set C of V at $\boldsymbol{u}_c$ pointing outward. If an orbit $\boldsymbol{u}(t)$ crosses this level curve C at $\boldsymbol{u}_c$, then the velocity vector of the orbit and the gradient $\nabla V(\boldsymbol{u}_c)$ will form an angle θ for which

$$\cos(\theta) = \frac{\dot{V}(\boldsymbol{u}_c)}{||\nabla V(\boldsymbol{u}_c)|| \, ||f(\boldsymbol{u}_c)||} \,.$$

$\dot{V}(\boldsymbol{u}) < 0$ implies that $\pi/2 < \theta < 3\pi/2$. It follows that the orbit is moving inwards the level curve C in this case. If $\dot{V}(\boldsymbol{u}) = 0$, the orbit follows C. If $\dot{V}(\boldsymbol{u}) > 0$ we see the orbit moving outwards of C, that is away from the origin. See [7] for proof of Liapunov's theorem.

An immediate implication of Liapunov's theorem is the following. Consider a Hamiltonian system

$$\dot{\boldsymbol{z}} = J(\nabla H(\boldsymbol{z}))^T . \tag{5.2}$$

A good candidate for the Liapunov function in this Hamiltonian system would be the Hamiltonian function itself, because the orbits of a Hamiltonian system lie in the level set of the Hamiltonian. So $\dot{V} = \dot{H} = 0$. Thus, if H is locally positive definite then Liapunov's theorem applies. And if H is negative definite, one can choose $-H$ as a Liapunov function. We have:

Theorem 5.2 (Dirichlet's Theorem) *The origin is a stable equilibrium of (5.2), if it is an isolated local maximum or local minimum of the Hamiltonian H.*

The condition for instability in Liapunov's theorem is very strong since it requires the orbital derivative to be positive everywhere in U. The following theorem is a way to conclude instability under somewhat weaker conditions.

Theorem 5.3 (Chetaev's theorem) *Let U be a small neighborhood of the origin where the C^1 Chetaev function $V : U \to \mathbb{R}$ is defined. Let Ω be an open subset of U such that*

(i) $0 \in \partial\Omega$,
(ii) $V(\boldsymbol{v}) = 0, \forall\, \boldsymbol{v} \in \partial\Omega \cap U$,
(iii) $V(\boldsymbol{v}) > 0$ *and* $\dot{V}(\boldsymbol{v}) > 0, \forall\, \boldsymbol{v} \in \Omega \cap U$.

Then the origin is an unstable equilibrium of (5.2).

The interpretation of this theorem is the following. An orbit $\boldsymbol{u}(t; \boldsymbol{u}_\circ)$ starting in $\Omega \cap U$, will never cross $\partial\Omega$ due to the properties (2) and (3) of the Chetaev function. From the second part of property (3) it now follows that $V(\boldsymbol{u}(t; \boldsymbol{u}_\circ))$ is increasing whenever $\boldsymbol{u}(t; \boldsymbol{u}_\circ)$ lies in $\Omega \cap U$. This orbit can not stay in $\partial\Omega \cap U$ due to the fact that U is open. Thus, $\boldsymbol{u}(t)$ moves away from the origin. Hence the origin is unstable.

6 Applications to the restricted problem

In this section we apply the theory of the previous sections to the relative equilibria of the restricted three-body problem.

Consider a Hamiltonian system with Hamiltonian

$$H = \frac{1}{2}\omega({x_1}^2 + {y_1}^2) + \lambda x_2 y_2 + \mathcal{O}(||\boldsymbol{z}||^3), \tag{6.1}$$

where $\omega, \lambda \neq 0$ are reals. One can calculate that the Hamiltonians of the restricted three-body problem at the Eulerian equilibria $\mathcal{L}_1, \mathcal{L}_2$ and $\mathcal{L}_3$ can be written in this form for all values of the parameter μ. The eigenvalues of the linearized system are $\pm i\omega$ and $\pm\lambda$, so the origin is an unstable equilibrium for the system induced by (6.1). But we can say more about the flow near this equilibrium.

We will first make a little excursion to a theorem on the existence of periodic solutions, known as *Liapunov's center theorem.*

Theorem 6.1 (Liapunov's center theorem) *Consider a Hamiltonian system of differential equations on $\mathbb{R}^{2m}$, $\dot{\boldsymbol{u}} = f(\boldsymbol{u})$ with $f(0) = 0$. Suppose that the eigenvalues of the linearized system around 0 are nonzero and given as $\pm i\omega, \lambda_3, \ldots, \lambda_{2m}$, where $\omega \in \mathbb{R}$ and $\lambda_j \in \mathbb{C}$. If $\lambda_j / i\omega \notin \mathbb{Z}$ for all j, then there is a smooth $2-$dimensional surface through the origin, tangent to the eigenspace corresponding to $\pm i\omega$, filled with periodic solutions with period close to $2\pi/\omega$ (as $\boldsymbol{u} \to 0$).*

This surface of periodic solutions is called the *Liapunov center.* Consider now the Hamiltonian system (6.1), for which $m = 4$. The eigenvalues of the linearized system are $\pm i\omega$, and $\pm\,\lambda$ where λ is real. Therefore Liapunov's Center Theorem holds: there exists such a Liapunov center through the origin of the system (6.1). In fact, we have the following result.

Proposition 6.2 *The equilibrium at the origin for the Hamiltonian system with Hamiltonian (6.1) is unstable. There is a Liapunov Center through the origin. Furthermore, there is a neighborhood of the origin such that every solution which begins at an initial position away from the Liapunov center, leaves this neighborhood in either positive or negative time.*

It remains to prove the last statement. First of all, let us write $H = H_2 + H_r$, where H_r represents the higher order terms of H near 0. H_r starts with third order terms in $\boldsymbol{z}$. Secondly, to make life easier, let us assume that the Liapunov center is located at $x_2 = 0, y_2 = 0$. This implies that

$$\frac{\partial H_r}{\partial x_2}(x_1, 0, y_1, 0) = \frac{\partial H_r}{\partial y_2}(x_1, 0, y_1, 0) = 0\,. \tag{6.2}$$

Define $V(\boldsymbol{z}) = ({x_2}^2 - {y_2}^2)/2$. The orbital derivative of V is

$$\dot{V} = \lambda({x_2}^2 + {y_2}^2) + W(\boldsymbol{z})$$

where

$$W(\boldsymbol{z}) := x_2 \frac{\partial H_r}{\partial y_2} - y_2 \frac{\partial H_r}{\partial x_2}.$$

From (6.2) we have that $W(\boldsymbol{z})$ is at least quadratic in (x_2, y_2). As a consequence we can choose a neighborhood U of 0 such that $|W(\boldsymbol{z})| \leq \lambda({x_2}^2 + {y_2}^2)/2$ on U. Taking $\Omega = \{\boldsymbol{z} \mid {x_2}^2 > {y_2}^2\}$ and applying Chetaev's theorem, we conclude that every solution starting in $(U \backslash \{x_2 = y_2 = 0\}) \cap \Omega$ will leave U in positive time. Reversing the time, we conclude that taking $\Omega = \{\boldsymbol{z} \mid {x_2}^2 < {y_2}^2\}$, every solution starting in $(U \backslash \{x_2 = y_2 = 0\}) \cap \Omega$ will leave U in negative time.

Modulo small modifications if the Liapunov center is not flat, this concludes the proof of proposition 6.2. Recall that proposition 6.2 also completely describes the flow of the restricted three-body problem near the Eulerian equilibria $\mathcal{L}_1, \mathcal{L}_2$ and $\mathcal{L}_3$.

Secondly, consider the Hamiltonian

$$H = \alpha(x_1 y_1 + x_2 y_2) + \beta(y_1 x_2 - x_1 y_2) + \mathcal{O}(||\boldsymbol{z}||^3), \tag{6.3}$$

where $\alpha, \beta \neq 0$ are real. The Hamiltonian of the restricted three-body problem at $\mathcal{L}_4$ and $\mathcal{L}_5$ is of this type for the parameter values $\mu_1 < \mu < 1 - \mu_1$. The eigenvalues of the linearized system are $\pm\alpha \pm i\beta$, so the origin is unstable. Moreover, choosing $V(\boldsymbol{z}) = ({x_1}^2 + {x_2}^2 - {y_1}^2 - {y_2}^2)/2$ as a Liapunov function we can verify the following result:

Proposition 6.3 *The Lagrangean equilibria $\mathcal{L}_4$ and $\mathcal{L}_5$ of the restricted three-body problem are unstable for $\mu_1 < \mu < 1 - \mu_1$. Furthermore there is a neighborhood of these points with the property that every nonzero solution starting in this neighborhood, will eventually leave it in positive time.*

By now we determined the stability of the equilibria of the restricted three-body problem except for the Lagrangean points at the parameter values $0 < \mu \leq \mu_1$ and $1 - \mu_1 \leq \mu < 1$. In the cases that $0 < \mu < \mu_1$ and $1 - \mu_1 < \mu < 1$, the Hamiltonian can be expanded around the Lagrangean points as

$$H = \frac{1}{2}\omega_1({x_1}^2 + {y_1}^2) + \frac{1}{2}\omega_2({x_2}^2 + {y_2}^2) + \mathcal{O}(||\boldsymbol{z}||^3),$$

for certain nonzero reals ω_1 and ω_2. The eigenvalues of the linearized vector field are $\pm i\omega_1, \pm i\omega_2$, so we can not conclude stability or instability from the eigenvalues. An extra problem arises because ω_1 and ω_2 turn out to have different signs, whatever the value of μ. So unfortunately, Dirichlet's theorem is not applicable. More sophisticated tools are needed here.

The solution is to take into account also the nonlinear terms in the expansion of the system around its equilibrium. That is to take a closer look at the $\mathcal{O}(||\boldsymbol{z}||^3)$-terms of the Hamiltonian. A common way to do that is using the theory of *normal forms*.

7 Normal forms

The idea behind normal forms is to construct a transformation of phase-space that brings a given system of differential equations into the 'simplest possible' form up to a certain order of accuracy. This idea will be made more precise in this section.

Let P_k be the space of homogeneous polynomials of degree k in the canonical variables $(x_1, \ldots, x_n, y_1, \ldots, y_n)$, so

$$P_k := \operatorname{span}_{\mathbb{R}} \left\{ x_1^{k_1} \cdots x_n^{k_n} y_1^{k_{n+1}} \cdots y_n^{k_{2n}} \mid \sum_{j=1}^{2n} k_j = k \right\}.$$

The space of all convergent power series without linear part, $P \subset \bigoplus_{k \geq 2} P_k$, is a Lie algebra with the Poisson bracket

$$\{\cdot, \cdot\} : P \times P \to P, \ (f, g) \mapsto \{f, g\},$$

where

$$\{f, g\} := d\boldsymbol{x} \wedge d\boldsymbol{y}(X_f, X_g) = \sum_{j=1}^{n} \left(\frac{\partial f}{\partial x_j} \frac{\partial g}{\partial y_j} - \frac{\partial f}{\partial y_j} \frac{\partial g}{\partial x_j} \right).$$

For each $h \in P$, its adjoint $\mathrm{ad}_h : P \to P$ is the linear operator defined by $\mathrm{ad}_h(H) = \{h, H\}$. Note that whenever $h \in P_k$, then $\mathrm{ad}_h : P_l \to P_{k+l-2}$.

Let us take an $h \in P$. It can be shown that for this h there is an open neighborhood U of the origin such that for every $|t| \leq 1$ each time-t flow $e^{tX_h} : U \to \mathbb{R}^{2n}$ of the Hamiltonian vector field X_h induced by h is a symplectic diffeomorphism on its image. These time-t flows define a family of mappings $(e^{tX_h})^* : P \to P$ by sending $H \in P$ to $(e^{tX_h})^* H := H \circ e^{tX_h}$. Differentiating the curve $t \mapsto (e^{tX_h})^* H$ with respect to t we find that it satisfies the linear differential equation $\frac{d}{dt}(e^{tX_h})^* H = dH \cdot X_h = -\mathrm{ad}_h(H)$ with initial condition $(e^{0X_h})^* H = H$. The solution reads $(e^{tX_h})^* H = e^{-t\mathrm{ad}_h} H$. In particular the symplectic transformation e^{-X_h} transforms H into

$$H' := (e^{-X_h})^* H = e^{\mathrm{ad}_h} H = H + \{h, H\} + \frac{1}{2!}\{h, \{h, H\}\} + \ldots \ . \quad (7.1)$$

The diffeomorphism e^{-X_h} sends 0 to 0 (because $X_h(0) = 0$). If $h \in \bigoplus_{k \geq 3} P_k$, then $De^{-X_h}(0) = Id$. A diffeomorphism with these two properties is called a *near-identity transformation*.

An element $H \in P$ can be written as $H = \sum_{k=2}^{\infty} H_k$, where $H_k \in P_k$. Assume now, as will usually be the case for the problems we consider in this paper, that $\mathrm{ad}_{H_2} : P_k \longrightarrow P_k$ is semisimple (i.e. complex-diagonalizable) for every $k \geq 3$. Then $P_k = \ker \mathrm{ad}_{H_2} \oplus \mathrm{im}\, \mathrm{ad}_{H_2}$, as is clear from the diagonalizability. In particular H_3 is uniquely decomposed as $H_3 = f_3 + g_3$, with $f_3 \in \ker \mathrm{ad}_{H_2}$, $g_3 \in \mathrm{im}\, \mathrm{ad}_{H_2}$. Now choose an $h_3 \in P_3$ such that $\mathrm{ad}_{H_2}(h_3) = g_3$. One could for example choose $h_3 = \tilde{g}_3 := (\mathrm{ad}_{H_2}|_{\mathrm{im}\, \mathrm{ad}_{H_2}})^{-1}(g_3)$. But clearly the choice $h_3 = \tilde{g}_3 + p_3$ suffices for any $p_3 \in \ker \mathrm{ad}_{H_2} \cap P_3$. For the transformed Hamiltonian $H' = (e^{-X_{h_3}})^* H$ we calculate from (7.1) that $H_2' = H_2$, $H_3' = f_3 \in \ker \mathrm{ad}_{H_2}$, $H_4' = H_4 + \{h_3, H_3 - \frac{1}{2} g_3\}$, etc. The reader should verify this! But now we can again write $H_4' = f_4 + g_4$ with $f_4 \in \ker \mathrm{ad}_{H_2}$, $g_4 \in \mathrm{im}\, \mathrm{ad}_{H_2}$ and it is clear that by a suitable choice of $h_4 \in P_4$ the Lie-transformation $e^{-X_{h_4}}$ transforms our H' into H'' for which $H_2'' = H_2$, $H_3'' = f_3 \in \ker \mathrm{ad}_{H_2}$ and $H_4'' = f_4 \in \ker \mathrm{ad}_{H_2}$. Continuing in this way, we can for any finite $r \geq 3$ find a sequence of symplectic near-identity transformations $e^{-X_{h_3}}, \ldots, e^{-X_{h_r}}$ with the property that $e^{-X_{h_k}}$ only changes the H_l with $l \geq k$, whereas the composition $e^{-X_{h_3}} \circ \ldots \circ e^{-X_{h_r}}$ transforms H into $\overline{H}$ with the property that $\overline{H}_k$ Poisson commutes with H_2 for every $2 \leq k \leq r$. The previous analysis culminates in the following

Theorem 7.1 (Birkhoff-Gustavson) *Let $r > 2$ be a given natural number. If $H = \sum_{k=2}^{\infty} H_k \in P$ is such that $\mathrm{ad}_{H_2} : P_k \rightarrow P_k$ is semisimple for each $k \geq 3$, then there is an open neighborhood $U \in \mathbb{R}^{2n}$ of the origin and an analytic symplectic diffeomorphism $\Psi : U \rightarrow \Psi(U) \subset \mathbb{R}^{2n}$ such that $\Psi(0) = 0$, $D\Psi(0) = Id$ and $\overline{H} := H \circ \Psi = \sum_{k=2}^{\infty} \overline{H}_k \in P$ has the properties that $\overline{H}_2 = H_2$ and $\{\overline{H}_2, \overline{H}_k\} = 0$ for all $2 \leq k \leq r$.*

The near-identity transformation Ψ is the composition of $r-2$ time-1 flows of Hamiltonian vector fields, which can subsequently be determined. Note that it need not be unique.

The transformed Hamiltonian $\overline{H}$ is called a normal form of H of order r. It can explicitly be determined following the procedure of the paragraph that precedes theorem 7.1 and using formula (7.1). The study of $\overline{H}$ can give us useful information on solutions of the original Hamiltonian H near its equilibrium point 0. It helps for instance to detect bifurcations and to construct approximations of solutions. More on normalization by Lie-transformations can be found in [3].

It is very common to study the truncated Hamiltonian system induced by

$H_2 + \overline{H}_3 + \ldots \overline{H}_r$. Its solutions approximate the solutions of the original system induced by $\overline{H}$. But the truncated system has an advantage: it admits at least two integrals. Not only the truncated Hamiltonian itself, but also H_2 is an integral of motion. Therefore the truncated normal form has an S^1-symmetry which allows us to make a reduction to a lower-dimensional Hamiltonian system. We will not treat these techniques.

Remark 7.2 Near-identity transformations Ψ with the properties of theorem 7.1 can be found in various ways. Lie-transformations, i.e. compositions of time-1 flows of Hamiltonian vector fields, are just one method. Other methods use power series expansions or averaging techniques. The method of Lie-transformations has the big advantage that the formula for the transformed Hamiltonian, (7.1), is fairly simple.

Normal form techniques also exist for critical points of non-Hamiltonian vector fields. Nothing changes dramatically, except that the near-identity transformations are of course no longer symplectic.

Remark 7.3 Let $S : \mathbb{R}^{2n} \to \mathbb{R}^{2n}$ be a *linear symmetry* of the Hamiltonian $H \in P$, that is S is a linear symplectic transformation keeping H invariant: $S^*(d\boldsymbol{x} \wedge d\boldsymbol{y}) = d\boldsymbol{x} \wedge d\boldsymbol{y}$ and $S^*H = H \circ S = H$. It is not hard to show that this implies that the Hamiltonian vector field X_H induced by H is equivariant under S : $S \cdot X_H = X_H \circ S$. In other words: if $\gamma : \mathbb{R} \to \mathbb{R}^{2n}$ is an integral curve of X_H, then $S \circ \gamma : \mathbb{R} \to \mathbb{R}^{2n}$ is also an integral curve of X_H. This explains the name symmetry.

Similarly, let $R : \mathbb{R}^{2n} \to \mathbb{R}^{2n}$ be a *linear reversing symmetry* of the Hamiltonian H, i.e. R is a linear anti-symplectic transformation that keeps H invariant: $R^*(d\boldsymbol{x} \wedge d\boldsymbol{y}) = -d\boldsymbol{x} \wedge d\boldsymbol{y}$ and $R^*H = H \circ R = H$. One now shows that X_H is anti-equivariant under R : $R \cdot X_H = -X_H \circ R$. Thus, if $\gamma : \mathbb{R} \to \mathbb{R}^{2n}$ is an integral curve of X_H, then $R \circ \gamma \circ -Id : \mathbb{R} \to \mathbb{R}^{2n}$ is also an integral curve of X_H. This explains the name reversing symmetry.

The group generated by the linear symmetries and linear reversing symmetries of the Hamiltonian $H \in P$, is called the *reversing symmetry group* of H. It can be shown (cf. [3]) that the near-identity transformation Ψ in theorem 7.1 can always be chosen in such a way that $\overline{H} = H \circ \Psi$ is again invariant under the elements of the reversing symmetry group of H. Alternatively stated: one can construct normal forms $\overline{H}$ of H which have the same linear symmetries and linear reversing symmetries as H.

8 The Poincaré section

In this section we summarize the most important properties of the so-called Poincaré map. Although the Poincaré map is a very useful tool for the study of any ordinary differential equation, we will introduce it here for Hamiltonian systems only. More extensive information can be found in [1], ch. 7-8.

Theorem 8.1 *Let H be a Hamiltonian on a $2n$-dimensional symplectic manifold M with symplectic form ω. Suppose that $\gamma : \mathbb{R} \to M$ is a periodic solution of the Hamiltonian vector field X_H induced by H and that γ lies in a regular energy-surface of H, i.e. $H^{-1}(\{\gamma(0)\})$ is a manifold. Then there is a codimension 1 submanifold $S \subset H^{-1}(\{\gamma(0)\})$ and open submanifolds S_1 and S_2 of S with the following properties:*

- *$X_H(m) \notin T_m S$ for all $m \in S$.*
- *$\gamma(0) \in S_1 \cap S_2$*
- *S_1 and S_2 are codimension 2 symplectic submanifolds of M. If $\iota_i : S_i \to M$ are the inclusions, then the symplectic forms ω_i of S_i are given by $\omega_i = \iota_i^* \omega$.*
- *For every $m \in S_1$ there is a time $t(m) > 0$ such that m is mapped to S_2 by the time-$t(m)$ flow of X_H, i.e. $e^{t(m)X_H}(m) \in S_2$. There exists a unique smallest positive number $d(m)$ with this property. d is a smooth function on S_1.*
- *The flow of X_H defines a unique symplectic diffeomorphism $\mathcal{P} : S_1 \to S_2$. $\mathcal{P}$ is given by sending $m \in S_1$ to $e^{d(m)X_H}(m) \in S_2$.*

The proof is highly based on the implicit function theorem, cf. [1]. The property $X_H(m) \notin T_m S$ implies that $T_m(H^{-1}(\{m\})) = X_H(m) \oplus T_m S$. This explains why S is sometimes called a *local transversal section* to the flow of X_H at γ. But usually we call S a *Poincaré section* at γ. The mapping $\mathcal{P}$ is called a *Poincaré* map or *first return map*. We remark that any two Poincaré maps $\mathcal{P}^1 : S_1^1 \to S_2^1$ and $\mathcal{P}^2 : S_1^2 \to S_2^2$ at γ are locally conjugate, i.e. there is an open neighborhood U of $\gamma(0)$ in S_1^1 and a symplectic diffeomorphism $\Phi : U \to S_1^2$ such that $\Phi \circ \mathcal{P}^1 = \mathcal{P}^2 \circ \Phi$. For Φ one could take the mapping that takes $m \in U$ and let it follow X_H until it hits S_1^2 at $\Phi(m)$.

It is clear that a study of the Poincaré map could provide us with very useful information on the flow of X_H in a neighborhood of the periodic solution γ, like stability and instability. First of all, let us study the derivative of the Poincaré map, $T_{\gamma(0)}\mathcal{P} : T_{\gamma(0)}S \to T_{\gamma(0)}S$. Note that, since any two Poincaré maps are locally conjugate, their derivatives are similar linear mappings. Hence they have the same eigenvalues. This allows us to make the following definition:

Definition 8.2 *Let $\gamma : \mathbb{R} \to M$ be a periodic solution of a Hamiltonian vector field on a symplectic manifold. The* characteristic multipliers *of γ are the eigenvalues of $T_{\gamma(0)}\mathcal{P}$, where $\mathcal{P}$ is any Poincaré map at γ.*

In local Darboux coordinates, $T_{\gamma(0)}\mathcal{P}$ can be represented by a symplectic matrix. Thus, the characteristic multipliers of γ come in quadruples: if λ is a multiplier, then so are $\lambda^{-1}, \bar{\lambda}$ and $\bar{\lambda}^{-1}$. Whenever one of the multipliers does not lie on the unit circle in $\mathbb{C}$, there must be a multiplier outside the unit circle. It is not very surprising that this can be used to prove that γ is an unstable periodic orbit. So γ can only be stable if all its multipliers have complex modulus 1. As usual, this is not sufficient to prove the stability of γ. Stability can sometimes be proved using variants of Liapunov's theorem. We will not go into this idea here.

Instead, we will focus on two-degrees of freedom Hamiltonian systems. In local Darboux coordinates, a Poincaré map near a periodic orbit γ in this case is an area-preserving planar map leaving the origin fixed. There are only two multipliers. γ is unstable if one of them does not lie on the unit circle. So suppose the multipliers are of the form $e^{\pm i\omega}$, with $\omega \in \mathbb{R}$. This expresses that the Poincaré map is, up to linear approximation, simply a rotation around 0 over an angle ω. This doesn't say anything yet about the stability of γ. But we shall see that under certain assumptions on the higher order approximations of $\mathcal{P}$ around 0, one can indeed prove that γ is stable. It turns out that $\mathcal{P}$ has to be a so-called *twist map*. The resulting type of stability goes under the name *Moser twist stability*.

9 The twist map and Arnold's stability theorem

Let $F : \mathbb{R}^2 \to \mathbb{R}^2$ be a diffeomorphism of $\mathbb{R}^2$. Note that it defines a discrete dynamical system. As an example, we have seen the Poincaré map of a two-degrees of freedom system. In this section we will be concerned with a special type of diffeomorphism, so called *twist maps*.

For $\alpha, \omega \in \mathbb{R}$ with $\alpha \neq 0$, consider the 2-dimensional diffeomorphism given by

$$\begin{pmatrix} I \\ \theta \end{pmatrix} \mapsto \begin{pmatrix} I \\ \theta + \omega + \alpha I, \end{pmatrix} \tag{9.1}$$

where we have used the polar coordinates notation $x = \sqrt{2I}\cos(\theta), y = \sqrt{2I}\sin(\theta)$, so $I \in \mathbb{R}_{\geq 0}$, $\theta \in \mathbb{R}/2\pi\mathbb{Z}$. It is easy to see that (9.1) rotates every circle $x^2 + y^2 = 2I_\circ$ over an angle $\delta := (\omega + \alpha I_\circ)$ that depends on the radius of the circle. Such a map is called a *twist map*. Note that if $\delta/2\pi$ is rational, then the motion on this circle is periodic. If $\delta/2\pi \notin \mathbb{Q}$, then the orbit of any point on the circle $x^2 + y^2 = 2I_\circ$ is dense in the circle. The latter type of dynamics is called *quasi-periodic*. So we see that $\mathbb{R}^2$ is densely filled with periodic and quasiperiodic orbits of the twist map.

Let us now look at perturbations of twist maps. Let $F : \mathbb{R}^2 \to \mathbb{R}^2$ be given by

$$\begin{pmatrix} I \\ \theta \end{pmatrix} \mapsto \begin{pmatrix} I + \varepsilon^{r+s} f_1(I, \theta, \varepsilon) \\ \theta + \omega + \varepsilon^s g(I) + \varepsilon^{s+r} f_2(I, \theta, \varepsilon), \end{pmatrix} \tag{9.2}$$

where $I \in \mathbb{R}_{\geq 0}$, $\theta \in \mathbb{R}/2\pi\mathbb{Z}$, $\omega \in \mathbb{R}$. We require the following properties:

(i) f_1 and f_2 are smooth functions for $0 \leq a \leq I < b < \infty$, $0 \leq \varepsilon \leq \varepsilon_\circ$, and $\theta \in \mathbb{R}/2\pi\mathbb{Z}$.
(ii) $r \geq 1$ and $s \geq 0$ are two integers.
(iii) g is a smooth function on $0 \leq a \leq I < b < \infty$.
(iv) $dg(I)/dI \neq 0$ for $0 \leq a \leq I < b < \infty$.

Theorem 9.1 (Moser Twist Stability) *Given such a map F with the following additional property. If Ξ is any closed curve of the form*

$$\Xi = \{(I, \theta) \mid I = \Theta(\theta), \Theta : \mathbb{R}/2\pi\mathbb{Z} \to [a, b] \textit{ continuous } \}$$

then $\Xi \cap F(\Xi) \neq \emptyset$. Then, for sufficiently small ε, there is a continuous F-invariant curve Γ of the form $\Gamma = \{(I, \theta) \mid I = \Phi(\theta), \Phi : \mathbb{R}/2\pi\mathbb{Z} \to [a, b]$ continuous$\}$.

This theorem was proposed by Kolmogorov and proved by Moser [9]. Note that the unperturbed map is just a rotation, its eigenvalues are $e^{\pm i\omega}$. Up to order ε^s, we have a pure twist map, according to assumption 4. So we are looking here at a perturbation of a twist map restricted to an annulus. Another important remark is about the additional condition in the theorem. This condition excludes the situation where a closed curve of the prescribed form is mapped completely inside or outside itself. This is an important restriction and it prevents the perturbation from being arbitrary. The condition is satisfied for area-preserving maps.

We may now ask the question of stability of the fixed point 0 of the perturbed map F. Theorem 9.1 states now that if the restriction of F to any small annulus of the form $a \leq I < b$ satisfies the conditions of theorem 9.1, then there is an invariant curve in that annulus. In particular, we can choose this annulus as small as we like, so 0 is a stable fixed point of F.

We want to apply this to Poincaré maps in order to prove the stability of periodic solutions of two-degrees of freedom Hamiltonian systems. As was explained in the previous section, one can construct such a Poincaré map around a periodic solution and it is represented by an area-preserving map in $\mathbb{R}^2$ with a fixed point. If the two multipliers of γ lie on the unit circle, then γ could be stable. Once we can show that the Poincaré map is in fact a perturbed twist map (in the sense of the previous theorem), this stability is indeed proved.

We will use normal form theory to view the Poincaré map as a perturbed twist map. Let us assume that around 0 our Hamiltonian can be expanded as

$$H = \frac{1}{2}\omega_1(x_1^2 + y_1^2) - \frac{1}{2}\omega_2(x_2^2 + y_2^2) + H_3 + H_4 + \dots ,$$

with $\omega_j \neq 0$ real numbers. The following result is pure 'algebra of normal forms':

Theorem 9.2 (Birkhoff) *Let H be of the above form and suppose that $\frac{\omega_1}{\omega_2} = \frac{p}{q}$ where p and q are relatively prime. Then any normal form of H of order smaller then or equal to $p + q - 1$ is of the following simple form:*

$$\begin{aligned} \overline{H}(\boldsymbol{x}, \boldsymbol{y}) &= H_2(\boldsymbol{x}, \boldsymbol{y}) + \overline{H}_4(x_1^2 + y_1^2, x_2^2 + y_2^2) + \cdots \\ &\quad \cdots + \overline{H}_{2m}(x_1^2 + y_1^2, x_2^2 + y_2^2) + \mathcal{O}(||\boldsymbol{z}||^{2m+1}), \end{aligned} \tag{9.3}$$

with $m < (p + q)/2$.

Proof: We want to investigate the eigenvalues of ad_{H_2}. For this purpose we diagonalize it by transforming to complex coordinates: $z_j = x_j + iy_j$, $j = 1, 2$. The symplectic form in these coordinates reads $\frac{i}{2}(dz_1 \wedge d\overline{z}_1 + dz_2 \wedge d\overline{z}_2)$ and the corresponding Poisson bracket is

$$\{f, g\} = 4\sum_{j=1}^{2} \left(\frac{\partial f}{\partial \overline{z}_j}\frac{\partial g}{\partial z_j} - \frac{\partial f}{\partial z_j}\frac{\partial g}{\partial \overline{z}_j} \right),$$

It is easy to check that $H_2 = \frac{\omega_1}{2} z_1\overline{z}_1 - \frac{\omega_2}{2} z_2\overline{z}_2$ and thus ad_{H_2} acts diagonally on monomials as follows:

$$\begin{aligned} \mathrm{ad}_{H_2} : {z_1}^{\alpha_1}{z_2}^{\alpha_2}{\overline{z}_1}^{\beta_1}{\overline{z}_2}^{\beta_2} \quad &\mapsto \quad 2\left((\alpha_1 - \beta_1)\,\omega_1 - (\alpha_2 - \beta_2)\,\omega_2\right) \times \\ &\qquad\qquad {z_1}^{\alpha_1}{z_2}^{\alpha_2}{\overline{z}_1}^{\beta_1}{\overline{z}_2}^{\beta_2}. \end{aligned}$$

So a monomial ${z_1}^{\alpha_1}{z_2}^{\alpha_2}{\overline{z}_1}^{\beta_1}{\overline{z}_2}^{\beta_2}$ can only occur in $\overline{H}_{\alpha_1+\alpha_2+\beta_1+\beta_2}$ if

$$(\alpha_1 - \beta_1)\,\omega_1 - (\alpha_2 - \beta_2)\,\omega_2 = 0.$$

There are two possibilities. First of all it can happen that $\alpha_1 = \beta_1$, $\alpha_2 = \beta_2$.

In this case, ${z_1}^{\alpha_1}{z_2}^{\alpha_2}{\overline{z}_1}^{\beta_1}{\overline{z}_2}^{\beta_2} = (z_1\overline{z}_1)^{\alpha_1}(z_2\overline{z}_2)^{\alpha_2} = (x_1^2+y_1^2)^{\alpha_1}(x_2^2+y_2^2)^{\alpha_2}$, so $\overline{H}_{\alpha_1+\alpha_2+\beta_1+\beta_2}$ is of the form prescribed in the theorem. The second possibility is that $\frac{p}{q} = \frac{\omega_1}{\omega_2} = \frac{\alpha_2-\beta_2}{\alpha_1-\beta_1}$. But p and q are relatively prime, so this is impossible if $|\alpha_2-\beta_2| < p$ or $|\alpha_1-\beta_1| < q$. In particular. this is impossible if $\alpha_1+\alpha_2+\beta_1+\beta_2 < p+q$. □

A Hamiltonian of the form in Birkhoff's theorem is said to be in *Birkhoff normal form*. Birkhoff's theorem has the following consequence: if ω_1/ω_2 is irrational, then H can be brought into Birkhoff normal form up to any desired order. Birkhoff wanted to use this observation to construct a coordinate transformation that brought H into an integrable form, the $x_j^2+y_j^2$ being the integrals. Unfortunately, one can not expect that the involved infinite sequence of normalization transformations is convergent. So in general, if $\omega_1/\omega_2 \notin \mathbb{Q}$, H still need not be integrable.

Now let us describe how the Birkhoff normal form helps us constructing a Poincaré map. First of all, we introduce the so-called *symplectic polar coordinates* by transforming

$$x_j = \sqrt{2I_j}\cos(\varphi_j) \text{ and } y_j = \sqrt{2I_j}\sin(\varphi_j)\,.$$

For $I_j > 0, \varphi_j \in \mathbb{R}/2\pi\mathbb{Z}$ this is a symplectic transformation, i.e. $dx \wedge dy = d\varphi \wedge dI$. Note that $2I_j = x_j^2+y_j^2$ so up to high order, Hamiltonians in Birkhoff normal form depend only on I, that is they are integrable up to this order. By an appropriate rescaling of the variables, one can also introduce a small parameter ε in the system. The resulting Hamiltonian system (9.3) then reads:

$$\dot{I}_j = \mathcal{O}(\varepsilon^{2m-1})\,,\ \dot{\varphi}_j = \omega_j + \varepsilon^2\frac{\partial\overline{H}_4(I)}{\partial I_j} + \ldots + \varepsilon^{2m-2}\frac{\partial\overline{H}_{2m}(I)}{\partial I_j} + \mathcal{O}(\varepsilon^{2m-1})\,. \tag{9.4}$$

From these simple equations, an approximation for the Poincaré map is easily constructed. Briefly, this runs as follows. First of all one restricts to a level set of H, which is approximately defined by $H_2(I) + \ldots \varepsilon^{2m-2}\overline{H}_{2m}(I) = h$. In this surface one chooses the set $\{\varphi_1 = 0\}$ as a transversal section to the flow. It is possible to use (I_2,φ_2) as coordinates for this section. With the help of equations (9.4) one can now approximate the return time to the section and finally also the Poincaré map.

So a combination of normal form theory and Moser's stability theorem can be used to prove stability of periodic solutions in a neighborhood of an equilibrium point. The surprise is now, that the theory can be extended in order to actually prove the stability of the equilibrium itself:

Theorem 9.3 (Arnold's Stability Theorem) *Consider the Hamiltonian system with Hamiltonian (9.3). If there exists a* $2 \le k \le m$ *such that* $D_{2k} :=$

$H_{2k}(\omega_2, \omega_1) \neq 0$ then the origin is stable. Moreover, arbitrarily close to the origin in $\mathbb{R}^4$ there are invariant tori filled with quasi-periodic solutions.

The proof is based on normal form theory and the idea of Poincaré maps. It is rather hard though and we refer the reader to [2] or [7].

For the restricted three-body problem with parameter values $0 < \mu < \mu_1$ (with $\mu \neq \mu_2 := 0.0242938\ldots$, $\mu_3 := 0.0135116\ldots$ [1]) Deprit and Deprit-Bartholomé in 1967 calculated the normal form of the Hamiltonian at $\mathcal{L}_4$ and $\mathcal{L}_5$. They found that $D_4 \neq 0$ except for $\mu \approx 0.010$. Nowadays we know that $D_6 \neq 0$ at this parameter value. Thus, by Arnold's theorem we have the following result.

Proposition 9.4 *In the restricted three-body problem, the Lagrangean equilibria $\mathcal{L}_4$ and $\mathcal{L}_5$ are stable for $0 < \mu < \mu_1$ and $1 - \mu_1 < \mu < 1$ with $\mu \notin \{\mu_2, \mu_3, 1 - \mu_2, 1 - \mu_3\}$.*

Thus, the stability of the equilibria of the restricted three-body problem has been established except for $\mathcal{L}_4$ and $\mathcal{L}_5$ if $\mu \in \{\mu_1, \mu_2, \mu_3, 1 - \mu_1, 1 - \mu_2, 1 - \mu_3\}$. We refer to [6] for the analysis of these cases.

In the Sun-Jupiter system, the result of proposition 9.4 can really be observed: if we draw the equilateral triangles with the sun and Jupiter at its base points, then we find two groups of asteroids at the third vertex. They are called the Trojans and the Greeks.

Acknowledgments

We wish to thank Ken Meyer for his inspiring lectures and for sending us the many articles that helped us writing these lecture notes.

With pleasure we think back of our time in Peyresq, the friends we met there, the Génépi and the beautiful mountains. We are especially indebted to James Montaldi, for being so helpful all the time, and the Fondation Peiresc for having hosted the summer school.

We also acknowledge our colleagues at the Mathematisch Instituut, Universiteit Utrecht, for supporting us. Especially Ferdinand Verhulst who read the manuscript. Part of this article has been presented in the Ph.D.-student colloquium of the Universiteit Utrecht by Bob Rink. Theo Tuwankotta thanks CICAT TU Delft for the financial support that made it possible for him to come to Peyresq. He also thanks Santi Goenarso for her contributions in many ways.

[1] μ_2 and μ_3 are real numbers which produce the relations $\omega_1 : \omega_2 = 1 : 2$ and $\omega_1 : \omega_2 = 1 : 3$, respectively. This causes the theorems 9.2 and 9.3 not to be applicable.

Bibliography

[1] Abraham, R., J.E. Marsden, *Foundations of Mechanics*. The Benjamin/Cummings Publ. Co., Reading, Mass., 1987.

[2] Arnold, V.I., Proof of a theorem of A.N. Kolmogorov on the invariance of quasi-periodic motions under small perturbations of the Hamiltonian, *Russian Math. surveys*, **18** (1963), 9–36.

[3] Churchill, R.C., M. Kummer, D.L. Rod, On averaging, reduction, and symmetry in hamiltonian systems. *J. Diff. Eq.* **49** (1983), 359–414.

[4] Hale, J. and H. Koçak, *Dynamics and Bifurcations*, Text on Applied Mathematics 3, Springer-Verlag, 1991.

[5] Meyer, K.R., *Periodic Solutions of the N-Body-Problem*, Lecture Notes in Mathematics, Springer-Verlag, 1999.

[6] Meyer, K.R. and G.R. Hall, *Introduction to Hamiltonian dynamical systems and the N-body problem*, Applied Math. Sciences 90, Springer-Verlag, 1992.

[7] Meyer, K.R. and D.S. Schmidt, The stability of the Lagrange triangular point and a theorem of Arnol'd, *J. Diff. Eq.* **62** (1986), 222–236.

[8] Henrard, J., K.R. Meyer and D. Schmidt, *The Trojan Problem: A Study in Stability and Bifurcation*, in preparation.

[9] Moser, J., On invariant curves of area-preserving mappings of an annulus, *Nachr. Akad. Wiss., Göttingen Math. Phys.* **K1** (1962), 1–20.

[10] Siegel, C.L. and J.K. Moser, *Lectures on Celestial Mechanics*. Classics in Mathematics, Springer-verlag, 1970.

[11] Stewart, I., *Does God play dice?* Penguin Books, 1990.

II

A Crash Course in Geometric Mechanics

T.S. Ratiu, R. Tudoran, L. Sbano, E. Sousa Dias & G. Terra

Notes of the courses given by Tudor Ratiu

1 Introduction

These lecture notes are the direct result of presentations held in two consecutive years (2000 and 2001) at the Peyresq Conference Center, in the Alpes de Haute Provence, North of Nice. These summer schools were organized in conjunction with the Research Training Network MASIE (Mechanics and Symmetry in Europe) of the Fifth Framework Program of the European Commission and were intended for graduate students and postdocs who needed a crash course in geometric mechanics. They were also tailored to link with the other lectures and provide the necessary background for them. There are already many books on this subject and its links to symplectic and Poisson geometry (e.g. [AbMa78], [Arnold79], [GuSt84], [JoSa98], [LiMa87], [MaRa94], [McDSal95]) and the literature on this subject is overwhelming. So the goal of these two one-week intensive lecture courses was to find a quick way through this subject and give the young researchers enough tools to be able to sift and sort through the books and papers necessary for their own work. This is why these lectures present occasionally detailed proofs and sometimes only quick surveys of more extensive subjects that are, however, explained with care. The examples, on the other hand, are all carried out with detailed computations in order to show how one applies the theory in concrete cases. There are, essentially, four main examples that reappear throughout these lectures: particle dynamics, the free and heavy tops, the motion of a charged particle in a magnetic field, and ideal incompressible fluid flow as well as related systems such as the Korteweg-de Vries and the Camassa-Holm equations. Each example illustrates several different constructions prevalent in geometric mechanics and is at the root of many developments and generalizations.

There is nothing original in these lectures and they are entirely based on three main sources: [MaRa94], the yet unfinished book [MaRa03], and some unpublished notes [MaRa95] on the geometric theory of fluid dynamics. When carrying out reduction, several strong regularity hypotheses will be made. The singular case is considerably more involved and we refer to [OR04] for an in-depth analysis of this case. All the missing proofs of results quoted here can be found in these works as well as the books referred to before.

The reader is assumed to be familiar with calculus on manifolds and the elementary theory of Lie groups and Lie algebras, as found in e.g. [AbMa78], [AMR88], [DFN95], [Jost], [Lang], [MaRa94], [Sp79], [Serre], or [W83]. [Bou71, Bou89] is always helpful when a quick recall of the statement of a theorem is needed.

The authors would like to take this opportunity to thank Olivier Brahic for allowing them to use his notes from Ratiu's second lecture course.

1.1 Lagrangian and Hamiltonian Formalism

Let us start with Newton's equations for N particles $\mathbf{q} := (\mathbf{q}_1, \ldots, \mathbf{q}_N) \in \mathbb{R}^{3N}$ with masses $m_1, \ldots, m_N \in \mathbb{R}$. If $\mathbf{F} = (F_1, ..., F_N)$ are the forces acting on these particles then ***Newton's equations*** are

$$m\mathbf{a} = \mathbf{F}, \tag{1.1}$$

where $\mathbf{a} = \ddot{\mathbf{q}}$ is the acceleration of the system. Assuming that the forces are induced by a potential $V : \mathbb{R}^{3N} \to \mathbb{R}$, that is,

$$\mathbf{F}(\mathbf{q}) = -\nabla V(\mathbf{q}), \tag{1.2}$$

equations (1.1) become

$$m_i \ddot{\mathbf{q}}_i = -\frac{\partial V}{\partial \mathbf{q}_i}, \quad i = 1, \ldots, N, \tag{1.3}$$

where $\partial V/\partial \mathbf{q}_i$ denotes the gradient relative to the variable $\mathbf{q}_i$.

A straightforward verification shows that if one defines the ***Lagrangian*** function $L : \mathbb{R}^{6N} = \{(\mathbf{q}, \dot{\mathbf{q}}) \mid \mathbf{q}, \dot{\mathbf{q}} \in \mathbb{R}^{3N}\} \to \mathbb{R}$ by

$$L(\mathbf{q}, \dot{\mathbf{q}}) := \frac{1}{2} \sum_{i=1}^{N} m_i \|\dot{\mathbf{q}}_i\|^2 - V(\mathbf{q}). \tag{1.4}$$

and assumes that $\dot{\mathbf{q}} = d\mathbf{q}/dt$, then Newton's equations (1.2) are equivalent to ***Lagrange's equations***

$$\frac{d}{dt}\left(\frac{\partial L}{\partial \dot{\mathbf{q}}_i}\right) - \frac{\partial L}{\partial \mathbf{q}_i} = 0, \qquad i = 1, \ldots, N, \tag{1.5}$$

where $\partial L/\partial \dot{\mathbf{q}}_i, \partial L/\partial \mathbf{q}_i \in \mathbb{R}^3$ denote the gradients in $\mathbb{R}^3$ of L relative to $\dot{\mathbf{q}}_i, \mathbf{q}_i \in \mathbb{R}^3$.

On other hand, the Lagrange equations (1.5) are equivalent with the variational principle of Hamilton: *the solutions of* (1.5) *are critical points of the action functional defined on the space of smooth paths with fixed endpoints.* More precisely, let $\Lambda([a, b], \mathbb{R}^{3N})$ be the space of all possible smooth trajectories $\mathbf{q} : [a, b] \to \mathbb{R}^{3N}$ with fixed endpoints $\mathbf{q}_a = \gamma(a)$, $\mathbf{q}_b = \gamma(b)$. The ***action functional*** is defined by:

$$\mathcal{A}[\mathbf{q}(\cdot)] := \int_a^b L(\mathbf{q}(t), \dot{\mathbf{q}}(t))\, dt, \tag{1.6}$$

where $\dot{\mathbf{q}} = d\mathbf{q}(t)/dt$. In $\Lambda([a, b], \mathbb{R}^{3N})$ consider a deformation $\mathbf{q}(t, s)$, $s \in (-\epsilon, \epsilon)$, $\epsilon > 0$, with fixed endpoints $\mathbf{q}_a$, $\mathbf{q}_b$, of a curve $\mathbf{q}_0(t)$, that is, $\mathbf{q}(t, 0) = \mathbf{q}_0(t)$ for all $t \in [a, b]$ and $\mathbf{q}(a, s) = \mathbf{q}_0(a) = \mathbf{q}_a$, $\mathbf{q}(b, s) = \mathbf{q}_0(b) = \mathbf{q}_b$ for all $s \in (-\epsilon, \epsilon)$. Define a ***variation*** of the curve $\mathbf{q}_0(\cdot)$ in $\Lambda([a, b], \mathbb{R}^{3N})$ by

$$\delta\mathbf{q}(\cdot) := \left.\frac{d}{ds}\right|_{s=0} \mathbf{q}(\cdot, s) \in T_{\mathbf{q}_0(\cdot)}\Lambda([a, b], \mathbb{R}^{3N}),$$

and the ***first variation*** of $\mathcal{A}$ at $\mathbf{q}_0(t)$ to be the following derivative:

$$\mathbf{D}\mathcal{A}[\mathbf{q}_0(\cdot)](\delta\mathbf{q}(\cdot)) := \left.\frac{d}{ds}\right|_{s=0} \mathcal{A}[\mathbf{q}(\cdot, s)]. \tag{1.7}$$

Note that $\delta\mathbf{q}(a) = \delta\mathbf{q}(b) = \mathbf{0}$. With these notations, the ***variational principle of Hamilton*** states that the curve $\mathbf{q}_0(t)$ satisfies the Lagrange equations (1.5) if and only if $\mathbf{q}_0(\cdot)$ is a critical point of the action functional, that is, $\mathbf{D}\mathcal{A}[\mathbf{q}_0(\cdot)] = 0$. Indeed, using the equality of mixed partials, integrating by parts, and taking into account that $\delta\mathbf{q}(a) = \delta\mathbf{q}(b) = 0$, we get

$$\begin{aligned}
\mathbf{D}\mathcal{A}[\mathbf{q}_0(\cdot)](\delta\mathbf{q}(\cdot)) &= \left.\frac{d}{ds}\right|_{s=0} \mathcal{A}[\mathbf{q}(\cdot, s)] = \left.\frac{d}{ds}\right|_{s=0} \int_a^b L(\mathbf{q}(t,s), \dot{\mathbf{q}}(t,s))\, dt \\
&= \int_a^b \left[\frac{\partial L}{\partial \mathbf{q}_i}\delta\mathbf{q}(t,s) + \frac{\partial L}{\partial \dot{\mathbf{q}}_i}\delta\dot{\mathbf{q}}_i\right] dt \\
&= -\int_a^b \left[\frac{d}{dt}\left(\frac{\partial L}{\partial \dot{\mathbf{q}}_i}\right) - \frac{\partial L}{\partial \mathbf{q}_i}\right]\delta\mathbf{q}_i dt = 0
\end{aligned}$$

for all smooth $\delta\mathbf{q}_i(t)$ satisfying $\delta\mathbf{q}_i(a) = \delta\mathbf{q}_i(b) = 0$, which proves the claim.

Next, introduce the ***conjugate momenta***

$$\mathbf{p}^i := \frac{\partial L}{\partial \dot{\mathbf{q}}_i} = m_i \dot{\mathbf{q}}_i \in \mathbb{R}^3, \quad i = 1, \ldots, N. \tag{1.8}$$

Define the change of variables $(\mathbf{q}, \dot{\mathbf{q}}) \mapsto (\mathbf{q}, \mathbf{p})$, called the ***Legendre transform***, and the ***Hamiltonian***

$$\begin{aligned}
H(\mathbf{q}, \mathbf{p}) :&= \mathbf{p} \cdot \dot{\mathbf{q}}(\mathbf{q}, \mathbf{p}) - L(\mathbf{q}, \dot{\mathbf{q}}(\mathbf{q}, \mathbf{p})) \\
&= \frac{1}{2}\sum_{i=1}^N m_i \|\dot{\mathbf{q}}_i\|^2 + V(\mathbf{q}) \\
&= \frac{1}{2}\sum_{i=1}^N \frac{1}{m_i}\|\mathbf{p}^i\|^2 + V(\mathbf{q})
\end{aligned} \tag{1.9}$$

which is the total energy of the system, expressed in the variables $(\mathbf{q}, \mathbf{p})$. Then one has

$$\frac{\partial H}{\partial \mathbf{p}^i} = \frac{1}{m_i}\mathbf{p}^i = \dot{\mathbf{q}}_i = \frac{d\mathbf{q}_i}{dt}$$

and

$$\frac{\partial H}{\partial \mathbf{q}_i} = \frac{\partial V}{\partial \mathbf{q}_i} = -\frac{\partial L}{\partial \mathbf{q}_i}.$$

Therefore, by the Lagrange equations (1.5) we have

$$\dot{\mathbf{p}}^i = \frac{d\mathbf{p}^i}{dt} = \frac{d}{dt}\left(\frac{\partial L}{\partial \dot{\mathbf{q}}_i}\right) = \frac{\partial L}{\partial \mathbf{q}_i} = -\frac{\partial H}{\partial \mathbf{q}_i}.$$

This shows that Lagrange's equations (1.5) are equivalent to ***Hamilton's equations***

$$\dot{\mathbf{q}}_i = \frac{\partial H}{\partial \mathbf{p}^i} \qquad \dot{\mathbf{p}}^i = -\frac{\partial H}{\partial \mathbf{q}_i}, \tag{1.10}$$

where, as before, $\partial H/\partial \mathbf{q}_i, \partial H/\partial \mathbf{p}^i \in \mathbb{R}^3$ are the gradients of H relative to $\mathbf{q}_i, \mathbf{p}^i \in \mathbb{R}^3$.

Note that, whereas Lagrange's equations are of second order and concern curves in the configuration space (the space of $\mathbf{q}$'s), Hamilton's equations are of first order and describe the dynamics of curves belonging to ***phase space***, a space of dimension twice the dimension of the configuration space whose points are pairs formed by configurations $\mathbf{q}$ and conjugate momenta $\mathbf{p}$.

An easy verification shows that Hamilton's equations (1.10) can be equivalently written as

$$\dot{F} = \{F, H\} \quad \text{for all } F \in \mathcal{F}(P), \tag{1.11}$$

where $\mathcal{F}(P)$ denotes the smooth functions on the phase space $P := \mathbb{R}^{6N}$, and the ***Poisson bracket*** is defined by

$$\{G, K\} := \sum_{i=1}^{N} \left(\frac{\partial G}{\partial \mathbf{q}_i} \cdot \frac{\partial K}{\partial \mathbf{p}^i} - \frac{\partial G}{\partial \mathbf{p}^i} \cdot \frac{\partial K}{\partial \mathbf{q}_i} \right) \quad \text{for all} \quad G, K \in \mathcal{F}(P). \tag{1.12}$$

Indeed, from (1.12) and (1.10) we have, for any $F \in \mathcal{F}(P)$,

$$\begin{aligned} \sum_{i=1}^{N} \left(\frac{\partial F}{\partial \mathbf{q}_i} \cdot \dot{\mathbf{q}}_i + \frac{\partial F}{\partial \mathbf{p}^i} \cdot \dot{\mathbf{p}}^i \right) &= \frac{dF}{dt} \\ &= \{F, H\} = \sum_{i=1}^{N} \left(\frac{\partial F}{\partial \mathbf{q}_i} \cdot \frac{\partial H}{\partial \mathbf{p}^i} - \frac{\partial F}{\partial \mathbf{p}^i} \cdot \frac{\partial H}{\partial \mathbf{q}_i} \right) \end{aligned}$$

which is equivalent to (1.10) since $F \in \mathcal{F}(P)$ is arbitrary.

Summarizing, for classical mechanical systems in Euclidean space describing particle motion, whose total energy is given by kinetic plus potential energy, we have shown that Newton's equations are equivalent to:

- Lagrange's equations
- Hamilton's variational principle
- Hamilton's equations of motion
- Hamilton's equations in Poisson bracket formulation.

In the course of these lectures we shall focus on each one of these four pictures and shall explain the geometric structure underlying them when the configuration space is a general manifold. It turns out that, in general, they are not equivalent and, moreover, some of these formulations have very useful generalizations, particularly appropriate for systems with symmetry, the case we shall consider next by means of an example.

1.2 The Heavy Top

In these lectures we shall discuss the equivalences just described in the context of systems with symmetry when one can eliminate variables. To see what is involved in this case, let us consider in detail an example, namely the motion of a heavy top moving about a fixed point; the exposition below is mostly based on [MaRa94, MaRaWe84a, MaRaWe84b, HMR98]. In this case, all computations can be easily carried out explicitly. We shall describe this system both in the Lagrangian and Hamiltonian picture and shall find two additional equivalent formulations that take into account the symmetries of this system. This example will serve then as model for the reduction theory presented in these lectures.

The Lie algebra $\mathfrak{so}(3)$ and its dual. To be efficient in the computations that follow we briefly recall the main formulas regarding the special orthogonal group $SO(3) := \{A \mid A \text{ a } 3 \times 3 \text{ orthogonal matrix }, \det(A) = 1\}$, its Lie algebra $\mathfrak{so}(3)$ formed by 3×3 skew symmetric matrices, and its dual $\mathfrak{so}(3)^*$. All these formulas will be proved in §5.1 and §6.3. The Lie algebra $(\mathfrak{so}(3), [\cdot,\cdot])$, where $[\cdot,\cdot]$ is the commutator bracket of matrices, is isomorphic to the Lie algebra $(\mathbb{R}^3, \times)$, where $\times$ denotes the vector product in $\mathbb{R}^3$, by the isomorphism

$$\mathbf{u} := (u^1, u^2, u^3) \in \mathbb{R}^3 \mapsto \hat{\mathbf{u}} := \begin{bmatrix} 0 & -u^3 & u^2 \\ u^3 & 0 & -u^1 \\ -u^2 & u^1 & 0 \end{bmatrix} \in \mathfrak{so}(3). \tag{1.13}$$

Equivalently, this isomorphism is given by

$$\hat{\mathbf{u}}\mathbf{v} = \mathbf{u} \times \mathbf{v} \quad \text{for all} \quad \mathbf{u}, \mathbf{v} \in \mathbb{R}^3. \tag{1.14}$$

The following properties for $\mathbf{u}, \mathbf{v}, \mathbf{w} \in \mathbb{R}^3$ are easily checked:

$$(\mathbf{u} \times \mathbf{v})\hat{} = [\hat{\mathbf{u}}, \hat{\mathbf{v}}] \tag{1.15}$$

$$[\hat{\mathbf{u}}, \hat{\mathbf{v}}]\mathbf{w} = (\mathbf{u} \times \mathbf{v}) \times \mathbf{w} \tag{1.16}$$

$$\mathbf{u} \cdot \mathbf{v} = -\frac{1}{2}\operatorname{trace}(\hat{\mathbf{u}}\hat{\mathbf{v}}). \tag{1.17}$$

If $A \in SO(3)$ and $\hat{\mathbf{u}} \in \mathfrak{so}(3)$ denote, as usual, by $\operatorname{Ad}_A \hat{\mathbf{u}} := A\hat{\mathbf{u}}A^{-1}$ the adjoint action of $SO(3)$ on its Lie algebra $\mathfrak{so}(3)$. Then

$$(A\mathbf{u})\hat{} = \operatorname{Ad}_A \hat{\mathbf{u}} := A\hat{\mathbf{u}}A^T \tag{1.18}$$

since $A^{-1} = A^T$, the transpose of A. Also

$$A(\mathbf{u} \times \mathbf{v}) = A\mathbf{u} \times A\mathbf{v} \tag{1.19}$$

for any $\mathbf{u}, \mathbf{v} \in \mathbb{R}^3$ and $A \in SO(3)$. It should be noted that this relation is *not* valid if A is just an orthogonal matrix; if A is not in the component of the identity matrix, then one gets a minus sign on the right hand side.

The dual $\mathfrak{so}(3)^*$ is identified with $\mathbb{R}^3$ by the isomorphism $\mathbf{\Pi} \in \mathbb{R}^3 \mapsto \tilde{\mathbf{\Pi}} \in \mathfrak{so}(3)^*$ given by $\tilde{\mathbf{\Pi}}(\hat{\mathbf{u}}) := \mathbf{\Pi} \cdot \mathbf{u}$ for any $\mathbf{u} \in \mathbb{R}^3$. Then the coadjoint action of $SO(3)$ on $\mathfrak{so}(3)^*$ is given by (see §5.1 for the explicit computation)

$$\mathrm{Ad}^*_{A^{-1}} \tilde{\mathbf{\Pi}} = (A\mathbf{\Pi})\tilde{} \,. \tag{1.20}$$

The coadjoint action of $\mathfrak{so}(3)$ on $\mathfrak{so}(3)^*$ is given by (see §6.3 for the detailed computation)

$$\mathrm{ad}^*_{\hat{\mathbf{u}}} \tilde{\mathbf{\Pi}} = (\mathbf{\Pi} \times \mathbf{u})\tilde{} \,. \tag{1.21}$$

Euler angles. The Lie group $SO(3)$ is diffeomorphic to the real three dimensional projective space $\mathbb{RP}(3)$. The Euler angles that we shall review below provide a very convenient chart for $SO(3)$.

Let $\mathbf{E}_1, \mathbf{E}_2, \mathbf{E}_3$ be an orthonormal basis of $\mathbb{R}^3$ thought of as the ***reference configuration***. Points in the reference configuration, called ***material*** or ***Lagrangian points***, are denoted by $\mathbf{X}$ and their components, called ***material*** or ***Lagrangian coordinates*** by (X^1, X^2, X^3). Another copy of $\mathbb{R}^3$ is thought of as the ***spatial*** or ***Eulerian configuration***; its points, called ***spatial*** or ***Eulerian points*** are denoted by $\mathbf{x}$ whose components (x^1, x^2, x^3) relative to an orthonormal basis $\mathbf{e}_1, \mathbf{e}_2, \mathbf{e}_3$ are called ***spatial*** or ***Eulerian coordinates***. A ***configuration*** is a map from the reference to the spatial configuration that will be assumed to be an orientation preserving diffeomorphism. If the configuration is defined only on a subset of $\mathbb{R}^3$ with certain good properties such as being a submanifold, as will be the case for the heavy top, then it is assumed that the configuration is a diffeomorphism onto its image. A ***motion*** $\mathbf{x}(\mathbf{X}, t)$ is a time dependent family of configurations. In what follows we shall only consider motions that are given by rotations, that is, we shall assume that $\mathbf{x}(\mathbf{X}, t) = A(t)\mathbf{X}$ with $A(t)$ an orthogonal matrix. Since the motion is assumed to be smooth and equal to the identity at $t = 0$, it follows that $A(t) \in SO(3)$.

Define the time dependent orthonormal basis $\boldsymbol{\xi}_1, \boldsymbol{\xi}_2, \boldsymbol{\xi}_3$ by $\boldsymbol{\xi}_i := A(t)\mathbf{E}_i$, for $i = 1, 2, 3$. This basis is anchored in the body and moves together with it. The ***body*** or ***convected coordinates*** are the coordinates of a point relative to the basis $\boldsymbol{\xi}_1, \boldsymbol{\xi}_2, \boldsymbol{\xi}_3$. Note that the components of a vector $\mathbf{V}$ relative to the basis $\mathbf{E}_1, \mathbf{E}_2, \mathbf{E}_3$ are the same as the components of the vector $A(t)\mathbf{V}$ relative to the basis $\boldsymbol{\xi}_1, \boldsymbol{\xi}_2, \boldsymbol{\xi}_3$. In particular, the body coordinates of $\mathbf{x}(\mathbf{X}, t) = A(t)\mathbf{X}$ are X^1, X^2, X^3.

The Euler angles encode the passage from the spatial basis $\mathbf{e}_1, \mathbf{e}_2, \mathbf{e}_3$ to the body basis $\boldsymbol{\xi}_1, \boldsymbol{\xi}_2, \boldsymbol{\xi}_3$ by means of three consecutive counterclockwise rotations performed in a specific order: first rotate around the axis $\mathbf{e}_3$ by the angle φ and denote the resulting position of $\mathbf{e}_1$ by ON (line of nodes), then rotate about ON by the angle θ and denote the resulting position of $\mathbf{e}_3$ by $\boldsymbol{\xi}_3$, and finally rotate about $\boldsymbol{\xi}_3$ by the angle ψ. Note that, by construction, $0 \leq \varphi, \psi < 2\pi$ and $0 \leq \theta < \pi$ and that the method just described provides a bijective map between (φ, ψ, θ) variables and the group $SO(3)$. However, this bijective map is not a chart since its differential vanishes at $\varphi = \psi = \theta = 0$. So for $0 < \varphi, \psi < 2\pi$, $0 < \theta < \pi$ the ***Euler angles*** (φ, ψ, θ) form a chart. If one carries out explicitly the rotation just described the resulting linear map performing the motion $\mathbf{x}(\mathbf{X}, t) = A(t)\mathbf{X}$ has the matrix relative to the bases $\boldsymbol{\xi}_1, \boldsymbol{\xi}_2, \boldsymbol{\xi}_3$ and $\mathbf{e}_1, \mathbf{e}_2, \mathbf{e}_3$ equal to

$$A = \begin{bmatrix} \cos\psi\cos\varphi - \cos\theta\sin\varphi\sin\psi & \cos\psi\sin\varphi + \cos\theta\cos\varphi\sin\psi & \sin\theta\sin\psi \\ -\sin\psi\cos\varphi - \cos\theta\sin\varphi\cos\psi & -\sin\psi\sin\varphi + \cos\theta\cos\varphi\cos\psi & \sin\theta\cos\psi \\ \sin\theta\sin\varphi & -\sin\theta\cos\varphi & \cos\theta \end{bmatrix}; \tag{1.22}$$

this computation is carried out in practically any mechanics book such as [Arnold79] or [MaRa94].

The total energy of the heavy top. A heavy top is by definition a rigid body moving about a fixed point in $\mathbb{R}^3$. Let $\mathcal{B}$ be an open bounded set whose closure is a reference configuration. Points on the reference configuration are denoted, as before, by $\mathbf{X} = (X^1, X^2, X^3)$, with X^1, X^2, X^3 the material coordinates relative to a fixed orthonormal frame $\mathbf{E}_1, \mathbf{E}_2, \mathbf{E}_3$. The map $\eta : \mathcal{B} \to \mathbb{R}^3$, with enough smoothness properties so that all computations below make sense, which is, in addition, orientation preserving and invertible on its image, is a configuration of the top. The spatial points $\mathbf{x} := \eta(\mathbf{X}) \in \eta(\mathcal{B})$ have coordinates x^1, x^2, x^3 relative to an orthonormal basis $\mathbf{e}_1, \mathbf{e}_2, \mathbf{e}_3$. Since the body is rigid and has a fixed point, its motion $\eta_t : \mathcal{B} \to \mathbb{R}^3$ is necessarily of the form

$$\eta_t(\mathbf{X}) := \mathbf{x}(\mathbf{X}, t) = A(t)\mathbf{X}$$

with $A(t) \in SO(3)$; this is a 1932 theorem of Mazur and Ulam which states that any isometry of $\mathbb{R}^3$ that leaves the origin fixed is necessarily a rotation. If $\boldsymbol{\xi}_1, \boldsymbol{\xi}_2, \boldsymbol{\xi}_3$ is the orthonormal basis of $\mathbb{R}^3$ defined by $\boldsymbol{\xi}_i := A(t)\mathbf{E}_i$, for $i = 1, 2, 3$, then the body coordinates of a vector are its components relative to this basis anchored in the body an moving together with it.

The ***material*** or ***Lagrangian velocity*** is defined by

$$\mathbf{V}(\mathbf{X}, t) := \frac{\partial \mathbf{x}(\mathbf{X}, t)}{\partial t} = \dot{A}(t)\mathbf{X}. \tag{1.23}$$

The ***spatial*** or ***Eulerian velocity*** is defined by

$$\mathbf{v}(\mathbf{x}, t) := \mathbf{V}(\mathbf{X}, t) = \dot{A}(t)\mathbf{X} = \dot{A}(t)A(t)^{-1}\mathbf{x}. \tag{1.24}$$

The ***body*** or ***convective velocity*** is defined by

$$\begin{aligned} \mathcal{V}(\mathbf{X}, t) : &= -\frac{\partial \mathbf{X}(\mathbf{x}, t)}{\partial t} = A(t)^{-1}\dot{A}(t)A(t)^{-1}\mathbf{x} \\ &= A(t)^{-1}\mathbf{V}(\mathbf{X}, t) = A(t)^{-1}\mathbf{v}(\mathbf{x}, t). \end{aligned} \tag{1.25}$$

Denote by ρ_0 the density of the top in the reference configuration. Then the kinetic energy at time t in material, spatial, and convective representation is given by

$$K(t) = \frac{1}{2}\int_{\mathcal{B}} \rho_0(\mathbf{X})\|\mathbf{V}(\mathbf{X}, t)\|^2 d^3\mathbf{X} \tag{1.26}$$

$$= \frac{1}{2}\int_{A(t)\mathcal{B}} \rho_0(A(t)^{-1}\mathbf{x})\|\mathbf{v}(\mathbf{x}, t)\|^2 d^3\mathbf{x} \tag{1.27}$$

$$= \frac{1}{2}\int_{\mathcal{B}} \rho_0(\mathbf{X})\|\mathcal{V}(\mathbf{X}, t)\|^2 d^3\mathbf{X} \tag{1.28}$$

If we denote

$$\hat{\omega}_S(t) := \dot{A}(t)A(t)^{-1} \tag{1.29}$$

$$\hat{\omega}_B(t) := A(t)^{-1}\dot{A}(t) \tag{1.30}$$

and take into account (1.24), (1.25), and (1.14), we conclude that

$$\begin{aligned} \mathbf{v}(\mathbf{x}, t) &= \omega_S(t) \times \mathbf{x} \\ \mathcal{V}(\mathbf{X}, t) &= \omega_B(t) \times \mathbf{X} \end{aligned}$$

which shows that ω_S and ω_B are the ***spatial*** and ***body angular velocities*** respectively. Using the Euler angles representation (1.22), the expressions for ω_S and ω_B are

$$\omega_S = \begin{bmatrix} \dot{\theta}\cos\varphi + \dot{\psi}\sin\varphi\sin\theta \\ \dot{\theta}\sin\varphi - \dot{\psi}\cos\varphi\sin\theta \\ \dot{\varphi} + \dot{\psi}\cos\theta \end{bmatrix} \tag{1.31}$$

$$\omega_B = \begin{bmatrix} \dot{\theta}\cos\psi + \dot{\varphi}\sin\psi\sin\theta \\ -\dot{\theta}\sin\psi + \dot{\varphi}\cos\psi\sin\theta \\ \dot{\varphi}\cos\theta + \dot{\psi} \end{bmatrix}. \tag{1.32}$$

Thus, by (1.28), the kinetic energy in convective representation has the expression

$$K(t) = \frac{1}{2}\int_{\mathcal{B}} \rho_0(\mathbf{X})\|\omega_B(t) \times \mathbf{X}\|^2 d^3\mathbf{X} =: \frac{1}{2}\langle\langle \omega_B(t), \omega_B(t)\rangle\rangle. \tag{1.33}$$

This is the quadratic form associated to the bilinear symmetric form on $\mathbb{R}^3$

$$\langle\langle \mathbf{a}, \mathbf{b}\rangle\rangle := \int_{\mathcal{B}} \rho_0(\mathbf{X})(\mathbf{a} \times \mathbf{X}) \cdot (\mathbf{b} \times \mathbf{X}) d^3\mathbf{X} = \mathbb{I}\mathbf{a} \cdot \mathbf{b}, \tag{1.34}$$

where $\mathbb{I} : \mathbb{R}^3 \to \mathbb{R}^3$ is the symmetric isomorphism (relative to the dot product) whose components are given by $\mathbb{I}_{ij} := \mathbb{I}\mathbf{E}_j \cdot \mathbf{E}_i = \langle\langle \mathbf{E}_j, \mathbf{E}_i\rangle\rangle$, that is,

$$\mathbb{I}_{ij} = -\int_{\mathcal{B}} \rho_0(\mathbf{X}) X^i X^j d^3\mathbf{X} \quad \text{if} \quad i \neq j$$

and

$$\mathbb{I}_{ii} = \int_{\mathcal{B}} \rho_0(\mathbf{X}) \left(\|\mathbf{X}\|^2 - (X^i)^2\right) d^3\mathbf{X}.$$

These are the expressions of the moment of inertia tensor in classical mechanics, that is, $\mathbb{I}$ is the ***moment of inertia tensor***. Since $\mathbb{I}$ is symmetric, it can be diagonalized. The basis in which it is diagonal is called in classical mechanics the ***principal axis body frame*** and the diagonal elements I_1, I_2, I_3 of $\mathbb{I}$ in this basis are called the ***principal moments of inertia*** of the top. From now on, we choose the basis $\mathbf{E}_1, \mathbf{E}_2, \mathbf{E}_3$ to be a principal axis body frame.

Identify in what follows the linear functional $\langle\langle \omega_B, \cdot\rangle\rangle$ on $\mathbb{R}^3$ with the vector $\mathbf{\Pi} := \mathbb{I}\omega_B \in \mathbb{R}^3$. In Euler angles this equals

$$\mathbf{\Pi} = \begin{bmatrix} I_1(\dot{\varphi}\sin\psi\sin\theta + \dot{\theta}\cos\psi) \\ I_2(\dot{\varphi}\cos\psi\sin\theta - \dot{\theta}\sin\psi) \\ I_3(\dot{\varphi}\cos\theta + \dot{\psi}) \end{bmatrix}. \tag{1.35}$$

Using (1.33) and (1.34), and noting that $\omega_B = \mathbb{I}^{-1}\mathbf{\Pi}$, the expression of the kinetic energy on the dual of $\mathfrak{so}(3)^*$ identified with $\mathbb{R}^3$, is

$$K(\mathbf{\Pi}) = \frac{1}{2}\mathbf{\Pi} \cdot \mathbb{I}^{-1}\mathbf{\Pi} = \frac{1}{2}\left(\frac{\Pi_1^2}{I_1} + \frac{\Pi_2^2}{I_2} + \frac{\Pi_3^2}{I_3}\right). \tag{1.36}$$

The kinetic energy on $\mathbb{R}^3$ given by (1.33) can be expressed as a function on $\mathfrak{so}(3)$ using (1.17), namely

$$\begin{aligned} K(\omega_B) &= \frac{1}{2}\omega_B \cdot \mathbb{I}\omega_B = -\frac{1}{4}\operatorname{trace}\left(\hat{\omega}_B(\mathbb{I}\omega_B)\hat{\ }\right) \\ &= -\frac{1}{4}\operatorname{trace}\left(\hat{\omega}_B(\hat{\omega}_B J + J\hat{\omega}_B)\right), \end{aligned} \tag{1.37}$$

where J is a diagonal matrix whose entries are given by the relations $I_1 = J_2 + J_3$, $I_2 = J_3 + J_1$, and $I_3 = J_1 + J_2$, that is, $J_1 = (-I_1 + I_2 + I_3)/2$, $J_2 = (I_1 - I_2 + I_3)/2$, and $J_3 = (I_1 + I_2 - I_3)/2$. The last equality in (1.37) follows from the identity $(\mathbb{I}\omega_B)\hat{} = \hat{\omega}_B J + J\hat{\omega}_B$, proved by a direct verification. Formulas (1.37) and (1.30) immediately yield the expression of the kinetic energy on the tangent bundle $TSO(3)$:

$$K(A, \dot{A}) = -\frac{1}{4}\,\mathrm{trace}((JA^{-1}\dot{A} + A^{-1}\dot{A}J)A^{-1}\dot{A}). \tag{1.38}$$

Since left translation of $SO(3)$ on itself lifts to the left action $B \cdot (A, \dot{A}) := (BA, B\dot{A})$ on $TSO(3)$, the expression (1.38) of $K(A, \dot{A})$ immediately implies that K is invariant relative to this action. Thus, *the kinetic energy of the heavy top is left invariant.*

Left translating the inner product $\langle\langle \cdot, \cdot \rangle\rangle$ from the tangent space to the identity to the tangent space at an arbitrary point of $SO(3)$, defines a left invariant Riemannian metric on $SO(3)$ whose kinetic energy is (1.38). Relative to this metric, the Legendre transformation gives the canonically conjugate variables

$$p_\varphi := \frac{\partial K}{\partial \dot{\varphi}}, \quad p_\psi := \frac{\partial K}{\partial \dot{\psi}}, \quad p_\theta := \frac{\partial K}{\partial \dot{\theta}}.$$

We shall summarize at the end of this discussion various formulas in terms of the Euler angles, including this one. Expressing now the kinetic energy in the variables $(\varphi, \psi, \theta, p_\varphi, p_\psi, p_\theta)$ will give thus a left invariant function on $T^*SO(3)$.

Next we turn to the expression of the potential energy. It is given by the height of the center of mass over the horizontal plane perpendicular to the direction of gravity. Let ℓ denote the length of the segment between the fixed point and the center of mass and let $\boldsymbol{\chi}$ be the unit vector supported by this line segment. Let $M = \int_{\mathcal{B}} \rho_0(\mathbf{X}) d^3\mathbf{X}$ be the total mass of the top, g the value of the gravitational acceleration, and $\mathbf{k}$ the spatial unit vector pointing in opposite direction to gravity. Then the potential energy at time t equals

$$V(t) = Mg\mathbf{k} \cdot A\ell\boldsymbol{\chi} = Mg\ell\mathbf{k} \cdot \boldsymbol{\lambda} = Mg\ell\boldsymbol{\Gamma} \cdot \boldsymbol{\chi} \tag{1.39}$$

where $\boldsymbol{\lambda} := A\boldsymbol{\chi}$ and $\boldsymbol{\Gamma} := A^{-1}\mathbf{k}$. The three expressions represent the potential energy in material, spatial, and body representation, respectively. It is clear that the potential energy is invariant only with respect to rotations about the axis of gravity, which shows that the total energy

$$H(A, \dot{A}) = -\frac{1}{4}\,\mathrm{trace}((JA^{-1}\dot{A} + A^{-1}\dot{A}J)A^{-1}\dot{A}) + Mg\ell\mathbf{k} \cdot A\boldsymbol{\chi} \tag{1.40}$$

is also invariant only under this circle subgroup of $SO(3)$. In Euler angles and their conjugate momenta, (1.40) becomes

$$H = \frac{1}{2}\left[\frac{[(p_\varphi - p_\psi \cos\theta)\sin\psi + p_\theta \sin\theta\cos\psi]^2}{I_1 \sin^2\theta}\right.$$
$$\left. + \frac{[(p_\varphi - p_\psi \cos\theta)\cos\psi - p_\theta \sin\theta\sin\psi]^2}{I_2 \sin^2\theta} + \frac{p_\psi^2}{I_3}\right]$$
$$+ Mg\ell\cos\theta, \tag{1.41}$$

where, without loss of generality, we assumed that χ in body coordinates is equal to $(0, 0, 1)$.

Since $\dot{p}_\varphi = -\partial H/\partial\varphi = 0$, it follows that $p_\varphi = \mathbf{\Pi}\cdot\mathbf{\Gamma}$ is conserved.

On $\mathbb{R}^3 \times \mathbb{R}^3$ the expression of the total energy is hence

$$H(\mathbf{\Pi}, \mathbf{\Gamma}) = \frac{1}{2}\mathbf{\Pi}\cdot\mathbb{I}^{-1}\mathbf{\Pi} + Mg\ell\mathbf{\Gamma}\cdot\chi. \tag{1.42}$$

In addition to the conservation of $\mathbf{\Pi}\cdot\mathbf{\Gamma}$, we also have $\|\mathbf{\Gamma}\| = 1$. The significance of these two conserved quantities in body representation will become clear only after understanding the Poisson geometry underlying the motion given by (1.42).

For completeness we summarize in Table 1.1 the relationship between the variables introduced till now.

The equations of motion of the heavy top. In a chart on $T^*SO(3)$ given by the Euler angles and their conjugate momenta, the equations of motion are

$$\dot{\varphi} = \frac{\partial H}{\partial p_\varphi}, \qquad \dot{\psi} = \frac{\partial H}{\partial p_\psi}, \qquad \dot{\theta} = \frac{\partial H}{\partial p_\theta}$$
$$\dot{p}_\varphi = -\frac{\partial H}{\partial \varphi}, \qquad \dot{p}_\psi = -\frac{\partial H}{\partial \psi}, \qquad \dot{p}_\theta = -\frac{\partial H}{\partial \theta}$$

with H given by (1.41).

Consider now the map

$$\mathbf{J} : (\varphi, \psi, \theta, p_\varphi, p_\psi, p_\theta) \mapsto (\mathbf{\Pi}, \mathbf{\Gamma}) \tag{1.43}$$

given by the formulas above. This is *not* a change of variables because $\|\mathbf{\Gamma}\| = 1$. A lengthy direct computation, using the formulas above, shows that these equations imply the ***Euler-Poisson equations***

$$\dot{\mathbf{\Pi}} = \mathbf{\Pi}\times\mathbf{\Omega} + Mg\ell\mathbf{\Gamma}\times\chi, \qquad \dot{\mathbf{\Gamma}} = \mathbf{\Gamma}\times\mathbf{\Omega} \tag{1.44}$$

where $\mathbf{\Omega} := \omega_B = \mathbb{I}^{-1}\mathbf{\Pi}$.

These equations can be obtained in two ways.

$$
\begin{aligned}
\Pi_1 &= [(p_\varphi - p_\psi \cos\theta)\sin\psi + p_\theta \sin\theta\cos\psi]/\sin\theta \\
&= I_1(\dot\varphi\sin\theta\sin\psi + \dot\theta\cos\psi) \\
\Pi_2 &= [(p_\varphi - p_\psi \cos\theta)\cos\psi - p_\theta \sin\theta\sin\psi]/\sin\theta \\
&= I_2(\dot\varphi\sin\theta\cos\psi - \dot\theta\sin\psi) \\
\Pi_3 &= p_\psi = I_3(\dot\varphi\cos\theta + \dot\psi) \\
\Gamma_1 &= \sin\theta\sin\psi \\
\Gamma_2 &= \sin\theta\cos\psi \\
\Gamma_3 &= \cos\theta \\
p_\varphi &= \mathbf{\Pi}\cdot\mathbf{\Gamma} \\
&= I_1(\dot\varphi\sin\theta\sin\psi + \dot\theta\cos\psi)\sin\theta\sin\psi \\
&\quad + I_2(\dot\varphi\sin\theta\cos\psi - \dot\theta\sin\psi)\sin\theta\cos\psi \\
&\quad + I_3(\dot\varphi\cos\theta + \dot\psi)\cos\theta \\
p_\psi &= \Pi_3 = I_3(\dot\varphi\cos\theta + \dot\psi) \\
p_\theta &= (\Gamma_2\Pi_1 - \Gamma_1\Pi_2)/\sqrt{1-\Gamma_3^2} \\
&= I_1(\dot\varphi\sin\theta\sin\psi + \dot\theta\cos\psi)\cos\psi \\
&\quad - I_2(\dot\varphi\sin\theta\cos\psi - \dot\theta\sin\psi)\sin\psi \\
\dot\varphi &= \frac{1}{I_1}\frac{\Pi_1\Gamma_1}{1-\Gamma_3^2} + \frac{1}{I_2}\frac{\Pi_2\Gamma_2}{1-\Gamma_3^2} \\
\dot\psi &= \frac{\Pi_3}{I_3} - \frac{\Pi_1\Gamma_1\Gamma_3}{I_1(1-\Gamma_3^2)} - \frac{\Pi_2\Gamma_2\Gamma_3}{I_2(1-\Gamma_3^2)} \\
\dot\theta &= \frac{\Pi_1\Gamma_2}{I_1\sqrt{1-\Gamma_3^2}} - \frac{\Pi_2\Gamma_1}{I_2\sqrt{1-\Gamma_3^2}}
\end{aligned}
$$

Table 1.1. *Summary of the variables for the heavy top*

(i) The canonical Poisson bracket of two functions $f, h : T^*SO(3) \to \mathbb{R}$ in a chart given by the Euler angles and their conjugate momenta is

$$
\{f,h\} = \frac{\partial f}{\partial\varphi}\frac{\partial h}{\partial p_\varphi} - \frac{\partial f}{\partial p_\varphi}\frac{\partial h}{\partial\varphi} + \frac{\partial f}{\partial\psi}\frac{\partial h}{\partial p_\psi} - \frac{\partial f}{\partial p_\psi}\frac{\partial h}{\partial\psi} + \frac{\partial f}{\partial\theta}\frac{\partial h}{\partial p_\theta} - \frac{\partial f}{\partial p_\theta}\frac{\partial h}{\partial\theta}.
$$

A direct long computation shows that if $F, H : \mathbb{R}^3 \times \mathbb{R}^3 \to \mathbb{R}$, then

$$
\{F\circ\mathbf{J}, H\circ\mathbf{J}\} = \{F,H\}_-\circ\mathbf{J},
$$

where $\mathbf{J}$ is given by (1.43) and

$$
\begin{aligned}
\{F,H\}_-(\mathbf{\Pi},\mathbf{\Gamma}) = &-\mathbf{\Pi}\cdot(\nabla_{\mathbf{\Pi}}F \times \nabla_{\mathbf{\Pi}}H) \\
&-\mathbf{\Gamma}\cdot(\nabla_{\mathbf{\Pi}}F \times \nabla_{\mathbf{\Gamma}}H + \nabla_{\mathbf{\Gamma}}F \times \nabla_{\mathbf{\Pi}}H);
\end{aligned}
\tag{1.45}
$$

$\nabla_{\mathbf{\Pi}} F$ and $\nabla_{\mathbf{\Gamma}} F$ denote the partial gradients relative to the variables $\mathbf{\Pi}$ and $\mathbf{\Gamma}$ respectively. An additional long computation shows that this defines a Poisson bracket, that is, it is bilinear, skew symmetric, and satisfies both the Jacobi and the Leibniz identities. Finally, if H is given by (1.42), it is easy to see that the equation $\dot{F} = \{F, H\}$ for any $F : \mathbb{R}^3 \times \mathbb{R}^3 \to \mathbb{R}$ is equivalent to the Euler-Poisson equations (1.44).

Note that the bracket of any function with an arbitrary function of $\|\mathbf{\Gamma}\|^2$ and $\mathbf{\Pi} \cdot \mathbf{\Gamma}$ is zero. The functions $\|\mathbf{\Gamma}\|^2$ and $\mathbf{\Pi} \cdot \mathbf{\Gamma}$ are the ***Casimir functions*** of the bracket (1.45).

(ii) Equations (1.44) can also be obtained from a variational principle. Given is the Lagrangian

$$L(\mathbf{\Omega}, \mathbf{\Gamma}) := \frac{1}{2} \mathbb{I}\mathbf{\Omega} \cdot \mathbf{\Omega} - Mg\ell \mathbf{\Gamma} \cdot \boldsymbol{\chi} \tag{1.46}$$

and the second Euler-Poisson equation $\dot{\mathbf{\Gamma}} = \mathbf{\Gamma} \times \mathbf{\Omega}$ whose solution is $\mathbf{\Gamma}(t) = A(t)^{-1}\mathbf{k}$, where $\mathbf{\Omega}(t) = A(t)^{-1}\dot{A}(t)$. Consider the variational principle for L

$$\delta \int_a^b L(\mathbf{\Omega}, \mathbf{\Gamma}) dt = 0$$

but only subject to the restricted variations of the form

$$\delta\mathbf{\Omega} := \dot{\mathbf{\Sigma}} + \mathbf{\Omega} \times \mathbf{\Sigma} \qquad \delta\mathbf{\Gamma} := \mathbf{\Gamma} \times \mathbf{\Sigma} \tag{1.47}$$

where $\mathbf{\Sigma}(t)$ is an arbitrary curve vanishing at the endpoints a and b, i.e

$$\mathbf{\Sigma}(a) = \mathbf{\Sigma}(b) = 0. \tag{1.48}$$

Using integration by parts together with the vanishing conditions at the endpoints, $\nabla_{\mathbf{\Omega}} L(\mathbf{\Omega}, \mathbf{\Gamma}) = \mathbb{I}\mathbf{\Omega} = \mathbf{\Pi}$, and $\nabla_{\mathbf{\Gamma}} L(\mathbf{\Omega}, \mathbf{\Gamma}) = -Mg\ell\boldsymbol{\chi}$, we get

$$\begin{aligned}
0 = \delta \int_a^b L(\mathbf{\Omega}, \mathbf{\Gamma}) dt &= \int_a^b \nabla_{\mathbf{\Omega}} L(\mathbf{\Omega}, \mathbf{\Gamma}) \cdot \delta\mathbf{\Omega} dt + \int_a^b \nabla_{\mathbf{\Gamma}} L(\mathbf{\Omega}, \mathbf{\Gamma}) \cdot \delta\mathbf{\Gamma} dt \\
&= \int_a^b \mathbf{\Pi} \cdot \delta\mathbf{\Omega} dt - Mg\ell \int_a^b \boldsymbol{\chi} \cdot \delta\mathbf{\Gamma} dt \\
&= \int_a^b \mathbf{\Pi} \cdot (\dot{\mathbf{\Sigma}} + \mathbf{\Omega} \times \mathbf{\Sigma}) dt - Mg\ell \int_a^b \boldsymbol{\chi} \cdot (\mathbf{\Gamma} \times \mathbf{\Sigma}) dt \\
&= -\int_a^b \dot{\mathbf{\Pi}} \cdot \mathbf{\Sigma} dt + \int_a^b \mathbf{\Pi} \cdot (\mathbf{\Omega} \times \mathbf{\Sigma}) dt - Mg\ell \int_a^b \mathbf{\Sigma} \cdot (\boldsymbol{\chi} \times \mathbf{\Gamma}) dt \\
&= \int_a^b \left(-\dot{\mathbf{\Pi}} + \mathbf{\Pi} \times \mathbf{\Omega} + Mg\ell \mathbf{\Gamma} \times \boldsymbol{\chi} \right) \cdot \mathbf{\Sigma} dt.
\end{aligned}$$

The arbitrariness of $\mathbf{\Sigma}$ yields the first of the Euler-Poisson equations (1.44).

Remark. If $g\ell = 0$, then the heavy top becomes a free rigid body. In this case there is only one equation, namely the Euler equation $\dot{\mathbf{\Pi}} = \mathbf{\Pi} \times \mathbf{\Omega}$, the Poisson bracket of two smooth functions $F, H : \mathbb{R}^3 \to \mathbb{R}$ is given by $\{F, H\}(\mathbf{\Pi}) = -\mathbf{\Pi} \cdot (\nabla F \times \nabla H)$, and the variational principle is $\delta \int_a^b L(\mathbf{\Omega}) dt = 0$ for variations $\delta\mathbf{\Omega} = \dot{\mathbf{\Sigma}} + \mathbf{\Omega} \times \mathbf{\Sigma}$ where $\mathbf{\Sigma}(a) = \mathbf{\Sigma}(b) = 0$. The Poisson bracket of any function with an arbitrary smooth function of $\|\mathbf{\Pi}\|^2$ vanishes and thus the motion of the rigid body takes place on $\mathbf{\Pi}$-spheres of constant radius. For the free rigid body, both the Lagrangian and the total energy coincide with the kinetic energy $L(\mathbf{\Omega}) = H(\mathbf{\Pi}) = \mathbf{\Pi} \cdot \mathbf{\Omega}/2$, where $\mathbf{\Pi} = \mathbb{I}\mathbf{\Omega}$, and thus the solutions of the free rigid body motion are geodesics on $SO(3)$ relative to the left invariant metric whose value at the identity is (1.34). The solutions of the Euler equation $\dot{\mathbf{\Pi}} = \mathbf{\Pi} \times \mathbf{\Omega}$ are therefore obtained by intersecting concentric spheres $\{\mathbf{\Pi} \mid \|\mathbf{\Pi}\| = R\}$ with the family of ellipsoids $\{\mathbf{\Pi} \mid \mathbf{\Pi} \cdot \mathbb{I}^{-1}\mathbf{\Pi} = C\}$ for any constants $R, C \geq 0$. In this way one immediately sees that there are six equilibria, four of them stable and two of them unstable. The stable ones correspond to rotations about the short and long axes of the moment of inertia and the unstable one corresponds to rotations about the middle axis.

One of the goals of these lectures is to explain the geometry behind all these phenomena that have appeared here as computational accidents. As we shall see, none of them are arbitrary occurrences and they all have a symplectic geometrical underpinning.

2 Hamiltonian Formalism

In this lecture we recall the fundamental concepts on symplectic manifolds and canonical transformations, as found, for example in, [Arnold79], [AbMa78], [MaRa94], [LiMa87], or [McDSal95].

2.1 Symplectic Manifolds

Definition 2.1 A ***symplectic manifold*** is a pair (P, Ω) where P is a Banach manifold and Ω is a closed (weakly) nondegenerate two-form on P. If Ω is strongly nondegenerate, (P, Ω) is called a ***strong symplectic manifold***.

Recall that Ω is ***weakly***, respectively ***strongly***, ***nondegenerate*** if the smooth vector bundle map covering the identity $\flat : TP \to T^*P$ given by $v \mapsto v^\flat := \Omega(v, \cdot)$ is injective, respectively bijective, on each fiber. If P is finite dimensional, there is no distinction between these concepts and nondegeneracy implies that P is even dimensional.

If the manifold is a Banach space V and the two-form is constant on V, then (V, Ω) is called a ***symplectic Banach space***. For example, $\mathbb{R}^{2n}$ endowed with the canonical symplectic structure

$$\Omega((\mathbf{u}, \mathbf{v}), (\mathbf{u}', \mathbf{v}')) = \mathbf{u} \cdot \mathbf{v}' - \mathbf{u}' \cdot \mathbf{v},$$

for $\mathbf{u}, \mathbf{v}, \mathbf{u}', \mathbf{v}' \in \mathbb{R}^n$, is a symplectic finite dimensional vector space. Another standard example is a complex Hilbert space with the symplectic form given by the negative of the imaginary part of the Hermitian inner product. For example, $\mathbb{C}^n$ has Hermitian inner product given by $\mathbf{z} \cdot \mathbf{w} := \sum_{j=1}^{n} z_j \overline{w}_j$, where $\mathbf{z} = (z_1, \ldots, z_n), \mathbf{w} = (w_1, \ldots, w_n) \in \mathbb{C}^n$. The symplectic form is thus given by $\Omega(\mathbf{z}, \mathbf{w}) := -\operatorname{Im}(\mathbf{z} \cdot \mathbf{w})$ and it is identical to the one given before on $\mathbb{R}^{2n}$ by identifying $\mathbf{z} = \mathbf{u} + i\mathbf{v} \in \mathbb{C}^n$ with $(\mathbf{u}, \mathbf{v}) \in \mathbb{R}^{2n}$ and $\mathbf{w} = \mathbf{u}' + i\mathbf{v}' \in \mathbb{C}^n$ with $(\mathbf{u}', \mathbf{v}') \in \mathbb{R}^{2n}$.

The local structure of strong symplectic manifolds is given by the following basic result.

Theorem 2.2 (Darboux) *If (P, Ω) is a strong symplectic manifold then each point of P admits a chart in which Ω is constant. If P is finite dimensional then around each point there are coordinates $(q^1, \ldots, q^n, p_1, \ldots, p_n)$, where $2n = \dim P$, such that $\Omega = \mathbf{d}q^i \wedge \mathbf{d}p_i$ (with the usual summation convention).*

The Darboux theorem for $2n$-dimensional symplectic manifolds hence states that, locally, the two-form Ω is given by the matrix

$$\mathbb{J} = \begin{bmatrix} 0 & \mathrm{I} \\ -\mathrm{I} & 0 \end{bmatrix}$$

where I denotes the $n \times n$ identity matrix, that is, locally the symplectic form is the one given above on $\mathbb{R}^{2n}$.

Remark 2.3 (i) The Darboux theorem is false for weak symplectic manifolds. For a counterexample, see Exercise 5.1-3 in [MaRa94].

(ii) There is a relative version of the Darboux theorem, namely, if S is a submanifold of P and Ω_0, Ω_1 are two strong symplectic forms on P that coincide when evaluated at points of S, then there is an open neighborhood V of S and a diffeomorphism $\varphi : V \to \varphi(V)$ such that $\varphi|_S = id$ and $\varphi^*\Omega_1 = \Omega_0$.

(iii) There is a generalization of the Darboux theorem for G-equivariant forms if G is a compact Lie group. It is still true that every point admits a G-invariant neighborhood on which the symplectic form is constant. However, the constant symplectic forms are no longer equivalent under G-equivariant changes of coordinates and the number of inequivalent symplectic forms depends on the representation type of the compact Lie group G, i.e., if it is real, complex, or quaternionic; see [DeMe93] and [MD93] for details.

Denote by $\mathfrak{X}(P)$ the set of vector fields on P and by $\Omega^k(P)$ the set of k-forms on P. The map $\flat$ induces a similar map between $\mathfrak{X}(P)$ and $\Omega^1(P)$, namely

$$\mathfrak{X}(P) \ni X \mapsto \mathbf{i}_X\Omega := \Omega(X, \cdot) =: X^\flat \in \Omega^1(P).$$

If Ω is strongly nondegenerate, this map is an isomorphism, so for any smooth map $H : P \to \mathbb{R}$ there exists a vector field X_H, called the ***Hamiltonian vector field*** defined by H, such that

$$\mathbf{i}_{X_H}\Omega = \mathbf{d}H,$$

where $\mathbf{d} : \Omega^k(P) \to \Omega^{k+1}(P)$ denotes the exterior differential. If P is only weakly nondegenerate, the uniqueness but not the existence of the Hamiltonian vector field is guaranteed. So for weak symplectic manifolds, not every function defines a Hamiltonian vector field. See [ChMa74] for an approach to infinite dimensional Hamiltonian systems.

If P is $2n$-dimensional, the local expression of X_H in Darboux coordinates is

$$X_H = \frac{\partial H}{\partial p_i}\frac{\partial}{\partial q^i} - \frac{\partial H}{\partial q^i}\frac{\partial}{\partial p_i}$$

and ***Hamilton's equations*** $\dot{z} = X_H(z)$ are given by

$$\dot{q}^i = \frac{\partial H}{\partial p_i}, \quad \dot{p}_i = -\frac{\partial H}{\partial q^i} \quad i = 1, \ldots, n.$$

The two-form Ω induces the volume ***Liouville form***

$$\Lambda := \frac{(-1)^{n(n-1)/2}}{n!}\underbrace{(\Omega \wedge \cdots \wedge \Omega)}_{n \text{ times}}, \tag{2.1}$$

where $2n = \dim P$, which has the local expression

$$\Lambda = \mathbf{d}q^1 \wedge \cdots \wedge \mathbf{d}q^n \wedge \mathbf{d}p_1 \wedge \cdots \wedge \mathbf{d}p_n.$$

In particular, any symplectic manifold is oriented.

2.2 Symplectic Transformations

Given two symplectic manifolds (P_1, Ω_1), (P_2, Ω_2), a C^∞ mapping

$$\phi : (P_1, \Omega_1) \to (P_2, \Omega_2)$$

is said to be a ***symplectic transformation*** if $\phi^*\Omega_2 = \Omega_1$, that is,

$$\Omega_2(\phi(z))(T_z\phi(v_1), T_z\phi(v_2)) = \Omega_1(z)(v_1, v_2)$$

for all $z \in P_1$ and $v_1, v_2 \in T_zP_1$, where $T_z\phi : T_zP_1 \to T_zP_2$ is the tangent map of ϕ at $z \in P_1$. Any symplectic map has injective derivative due to the weak nondegeneracy of the symplectic forms.

Proposition 2.4 *Let (P, Ω) be a symplectic manifold. The flow ϕ_t of $X \in \mathfrak{X}(P)$ consists of symplectic transformations if and only if it is **locally Hamiltonian**, that is, $\pounds_X\Omega = 0$, where $\pounds_X$ denotes the Lie derivative operator defined by X.*

Remark 2.5 There is a general formula relating the Lie derivative and the dynamics of a vector field. If $\mathcal{T}$ is an arbitrary (time independent) tensor field on P and F_t the flow of the (time independent) vector field X, then

$$\frac{d}{dt}F_t^*\mathcal{T} = F_t^*\pounds_X\mathcal{T} \tag{2.2}$$

where $\pounds_X\mathcal{T}$ denotes the Lie derivative of $\mathcal{T}$ along X. We shall use this formula in the proof below.

Proof Note that

$$\phi_t^*(\Omega) = \Omega \Longleftrightarrow \frac{d}{dt}\phi_t^*(\Omega) = 0,$$

which by (2.2) is equivalent to

$$\phi_t^*\pounds_X\Omega = 0 \Longleftrightarrow \pounds_X\Omega = 0$$

since ϕ_t^* is an isomorphism. ■

The name "locally Hamiltonian vector field" is justified by the following fact. Since $\pounds_X\Omega = \mathbf{i}_X\mathbf{d}\Omega + \mathbf{d}\mathbf{i}_X\Omega$ and $\mathbf{d}\Omega = 0$, the condition $\pounds_X\Omega = 0$ is equivalent to $\mathbf{d}\mathbf{i}_X\Omega = 0$, that is, $\mathbf{i}_X\Omega$ is closed. The Poincaré lemma implies the existence of a *local* function H such that $\mathbf{i}_X\Omega = \mathbf{d}H$.

Proposition 2.6 (Energy conservation) *Let ϕ_t be the flow of the Hamiltonian vector field X_H. Then $H \circ \phi_t = H$.*

Proof Since

$$\frac{d}{dt}\phi_t(z) = X_H(\phi_t(z))$$

and

$$\Omega(z)(X_H(z), u) = \langle \mathbf{d}H(z), u \rangle \quad \text{for all} \quad z \in P \quad \text{and} \quad u \in T_z P,$$

where $\langle \cdot, \cdot \rangle$ is the pairing between one-forms and vectors, we have by the chain rule

$$\begin{aligned}\frac{d}{dt} H(\phi_t(z)) &= \left\langle \mathbf{d}H(\phi_t(z)), \frac{d\phi_t(z)}{dt} \right\rangle = \langle \mathbf{d}H(\phi_t(z)), X_H(\phi_t(z)) \rangle \\ &= \Omega\left(\phi_t(z)\right)(X_H(\phi_t(z)), X_H(\phi_t(z))) = 0,\end{aligned}$$

which means that $H \circ \phi_t$ is constant relative to t. Since $\phi_0(z) = z$ for any $z \in P$, the result follows. ■

Proposition 2.7 *A diffeomorphism $\phi : P_1 \to P_2$ between strong symplectic manifolds is symplectic if and only if $\phi^* X_H = X_{H \circ \phi}$ for any $H : U \to \mathbb{R}$, where U is an arbitrary open subset of P_2.*

Proof For a strong symplectic manifold (P, Ω), the tangent space $T_z P$ equals the collection of all vectors of the form $X_K(z)$, for all smooth functions $K : U \to \mathbb{R}$ and all open neighborhoods U of z.

With this observation in mind and using the fact that ϕ is a diffeomorphism and the symplectic form Ω_2 strong, we have for all V open in P_1, $z \in V$, $v \in T_z P_1$, and all $H : \phi(V) \to \mathbb{R}$ the following equivalence

$$\begin{aligned}\phi^* \Omega_2 = \Omega_1 \Longleftrightarrow & \Omega_2(\phi(z)) \left(T_z\phi(X_{H \circ \phi}(z)), T_z\phi(v)\right) \\ &= \Omega_1(z) \left(X_{H \circ \phi}(z), v\right)\end{aligned} \tag{2.3}$$

However,

$$\begin{aligned}\Omega_1(z) \left(X_{H \circ \phi}(z), v\right) &= \langle \mathbf{d}(H \circ \phi)(z), v \rangle \\ &= \langle \mathbf{d}H(\phi(z)), T_z\phi(v) \rangle \\ &= \Omega_2(\phi(z)) \left(X_H(\phi(z)), T_z\phi(v)\right).\end{aligned} \tag{2.4}$$

By (2.3) and (2.4) the relation $\phi^* \Omega_2 = \Omega_1$ is thus equivalent to

$$\Omega_2(\phi(z)) \left(T_z\phi(X_{H \circ \phi}(z)), T_z\phi(v)\right) = \Omega_2(\phi(z)) \left(X_H(\phi(z)), T_z\phi(v)\right)$$

for any smooth function $H : U \to \mathbb{R}$, $U \subset P_2$ open, $\phi(z) \in U$, and $v \in T_z P_1$. By nondegeneracy of Ω_2 this is equivalent to

$$T_z\phi \circ X_{H \circ \phi} = X_H \circ \phi \Longleftrightarrow \phi^* X_H = X_{H \circ \phi}$$

which proves the statement. ■

2.3 Poisson Brackets

Given a strong symplectic manifold (P, Ω), define the ***Poisson bracket*** of two smooth functions $F, G : P \to \mathbb{R}$ by

$$\{F, G\} := \Omega(X_F, X_G). \tag{2.5}$$

In this section all symplectic manifolds are assumed to be strong. As a consequence of Proposition 2.7 and of the Poisson bracket definition we have the following.

Proposition 2.8 *A diffeomorphism $\phi : P_1 \to P_2$ is symplectic if and only if*

$$\{F, G\}_2 \circ \phi = \{F \circ \phi, G \circ \phi\}_1$$

for all smooth functions $F, G : U \to \mathbb{R}$, where U is any open subset of P_2.

Combined with Proposition 2.4 this yields the following statement.

Corollary 2.9 *The flow ϕ_t is that of a locally Hamiltonian vector field if and only if*

$$\phi_t^*\{F, G\} = \{\phi_t^* F, \phi_t^* G\}$$

for all smooth functions $F, G : U \to \mathbb{R}$, where U is an arbitrary open subset of P_2.

Proposition 2.10 *The vector space of smooth functions on the strong symplectic manifold (P, Ω) is a Lie algebra under the Poisson bracket.*

Proof The definition of $\{F, G\}$ immediately implies that it is bilinear, skew-symmetric, and it satisfies the Leibniz identity in each factor. To prove the Jacobi identity, note that

$$\{F, G\} = \langle \mathbf{i}_{X_F}\Omega, X_G\rangle = \langle \mathbf{d}F, X_G\rangle = X_G[F]$$

and so

$$\{\{F, G\}\,, H\} = X_H\left[\{F, G\}\right].$$

Thus the Jacobi identity is equivalent to

$$X_H\left[\{F, G\}\right] = \{X_H[F]\,, G\} + \{F\,, X_H[G]\} \tag{2.6}$$

which is obtained by differentiating in t at $t = 0$ the identity

$$\phi_t^*\{F, G\} = \{\phi_t^* F, \phi_t^* G\},$$

where ϕ_t is the flow of X_H (see Corollary 2.9). Indeed,

$$\begin{aligned}
X_H\left[\{F,G\}\right] &= \left.\frac{d}{dt}\right|_{t=0} \phi_t^*\{F,G\} = \left.\frac{d}{dt}\right|_{t=0} \{\phi_t^*F,\phi_t^*G\} \\
&= \left.\frac{d}{dt}\right|_{t=0} \Omega\left(X_{\phi_t^*F}\,,\, X_{\phi_t^*G}\right) \\
&= \Omega\left(X_{X_H[F]}\,,\, X_G\right) + \Omega\left(X_F\,,\, X_{X_H[G]}\right) \\
&= \{X_H[F]\,,\, G\} + \{F\,,\, X_H[G]\}
\end{aligned}$$

which proves the derivation identity (2.6). ■

Corollary 2.11 *The set of Hamiltonian vector fields on P is a Lie subalgebra of the set of the vector fields on P because*

$$[X_F, X_G] = X_{-\{F,G\}}.$$

Proof As derivations,

$$\begin{aligned}
[X_F, X_G][H] &= X_F[X_G[H]] - X_G[X_F[H]] = X_F\left[\{H,G\}\right] - X_G\left[\{H,F\}\right] \\
&= \{\{H,G\},F\} - \{\{H,F\},G\} = -\{H,\{F,G\}\} \\
&= -X_{\{F,G\}}[H],
\end{aligned}$$

where we have applied the Jacobi identity in the fourth equality. ■

The next corollary gives Hamilton's equations in Poisson bracket form.

Corollary 2.12 *If ϕ_t is the flow of X_H and $F : U \to \mathbb{R}$ is an arbitrary smooth function defined on the open subset U of P then*

$$\frac{d}{dt}(F\circ\phi_t) = \{F\circ\phi_t, H\} = \{F,H\}\circ\phi_t. \tag{2.7}$$

Proof We have

$$\begin{aligned}
\frac{d}{dt}(F\circ\phi_t)(z) &= \left\langle \mathbf{d}F(\phi_t(z)), \frac{d\phi_t(z)}{dt}\right\rangle = \langle \mathbf{d}F(\phi_t(z)), X_H(\phi_t(z))\rangle \\
&= \{F,H\}(\phi_t(z)).
\end{aligned}$$

As ϕ_t is symplectic, this is just $\{F\circ\phi_t, H\circ\phi_t\}(z)$. Conservation of energy gives $\{F\circ\phi_t, H\circ\phi_t\}(z) = \{F\circ\phi_t, H\}(z)$ which proves (2.7). ■

Equation (2.7) is often written in the compact form

$$\dot{F} = \{F,H\}.$$

Corollary 2.13 *The smooth function F is a constant of motion for X_H if and only if $\{F, H\} = 0$.*

2.4 Cotangent Bundles

In mechanics, the phase space is very often the cotangent bundle T^*Q of a configuration space Q. A cotangent bundle has a weak canonical symplectic structure and so any cotangent bundle is symplectic. The ***canonical two-form*** Ω on T^*Q is constructed from the ***canonical one-form*** Θ in the following way. Define $\Omega = -\mathbf{d}\Theta$ where Θ is the one-form given by

$$\Theta_{\alpha_q}\left(w_{\alpha_q}\right) := \langle \alpha_q, T_{\alpha_q}\pi(w_{\alpha_q})\rangle, \tag{2.8}$$

where $\alpha_q \in T_q^*Q$, $w_{\alpha_q} \in T_{\alpha_q}(T^*Q)$, $\pi : T^*Q \to Q$ is the projection, and $T\pi : T(T^*Q) \to TQ$ is the tangent map of π. The structure is called "canonical" because all the elements occurring in the definition are naturally associated to any cotangent bundle.

In local coordinates (q^i, p_i) on T^*Q the expressions of the canonical one- and two-forms are

$$\Theta = p_i \mathbf{d}q^i, \qquad \Omega = -\mathbf{d}\Theta = \mathbf{d}q^i \wedge \mathbf{d}p_i.$$

Thus the natural cotangent bundle coordinates are Darboux coordinates for the canonical symplectic form.

Given a diffeomorphism $f : Q \to S$ define the ***cotangent lift*** of f as the map $T^*f : T^*S \to T^*Q$ given by

$$T^*f(\alpha_s)\left(v_{f^{-1}(s)}\right) := \alpha_s\left(Tf(v_{f^{-1}(s)})\right), \tag{2.9}$$

where $\alpha_s \in T_s^*S$ and $v_{f^{-1}(s)} \in T_{f^{-1}(s)}Q$.

Theorem 2.14 *The cotangent lift T^*f of a diffeomorphism $f : Q \to S$ preserves the canonical one-form Θ.*

Proof For simplicity, we give the proof in finite dimensions. Denote by $(q^1, \ldots, q^n, p_1, \ldots, p_n)$ and $(s^1, \ldots, s^n, r_1, \ldots, r_n)$ canonical coordinates on T^*Q and T^*S, respectively. If $(s^1, ..., s^n) = f(q^1, ..., q^n)$, the cotangent lift T^*f is given by

$$(s^1, \ldots, s^n, r_1, \ldots, r_n) \mapsto (q^1, \ldots, q^n, p_1, \ldots, p_n), \tag{2.10}$$

where

$$p_j = \frac{\partial s^i}{\partial q^j} r_i. \tag{2.11}$$

To see that the lift (2.10) of f preserves the one-form is only a matter of using the chain rule and (2.11), that is,

$$r_i ds^i = r_i \frac{\partial s^i}{\partial q^j} dq^j = p_j dq^j.$$

Therefore $(T^* f)^* \Theta_Q = \Theta_S$ where Θ_Q and Θ_S are the canonical one-forms on Q and S respectively. ■

The converse of theorem 2.14 is also true, that is, if $\varphi : T^*S \to T^*Q$ is a diffeomorphism such that $\varphi^* \Theta_Q = \Theta_S$ then there exists a diffeomorphism $f : Q \to S$ such that $\phi = T^* f$. The proof of this fact is more involved; see Proposition 6.3.2 in [MaRa94].

2.5 Magnetic Terms

If $A \in \Omega^1(Q)$ define the ***fiber translation map*** $t_A : T^*Q \to T^*Q$ by

$$t_A(\alpha_q) := \alpha_q + A(q) \qquad \alpha_q \in T_q^*Q.$$

Proposition 2.15 *Let Θ be the canonical one-form on T^*Q and $t_A : T^*Q \to T^*Q$ the fiber translation by $A \in \Omega^1(Q)$. Then*

$$t_A^*(\Theta) = \Theta + \pi^* A$$

where $\pi : T^*Q \to Q$ is the canonical projection. Hence

$$t_A^* \Omega = \Omega - \pi^* \mathbf{d}A,$$

where $\Omega = -\mathbf{d}\Theta$ is the canonical symplectic form on T^*Q. Thus t_A is a symplectic transformation if and only if $\mathbf{d}A = 0$.

Proof We do the proof in finite dimensions. In local coordinates we have

$$t_A(q^i, p_j) = (q^i, p_j + A_j(q))$$

so

$$t_A^* \Theta = t_A^*(p_i \mathbf{d}q^i) = (p_i + A_i(q))\mathbf{d}q^i = p_i \mathbf{d}q^i + A_i(q)\mathbf{d}q^i,$$

which is the coordinate expression of $\Theta + \pi^* A$. ■

Let Ω be the canonical two-form on T^*Q. If $B \in \Omega^2(Q)$ is closed (i.e $\mathbf{d}B = 0$) then

$$\Omega_B := \Omega - \pi^* B$$

is a (weak) symplectic form on T^*Q. Indeed, the matrix form of Ω_B is

$$\begin{bmatrix} -B & \mathrm{I} \\ -\mathrm{I} & 0 \end{bmatrix}, \tag{2.12}$$

which shows that Ω_B is nondegenerate since (2.12) is nonsingular. The extra term $-B$ in the symplectic form Ω_B is called a ***magnetic term***.

Proposition 2.16 *Let B and $B^{'}$ be closed two-forms on Q such that $B - B^{'} = \mathbf{d}A$. Then the map*

$$t_A : (T^*Q, \Omega_B) \to (T^*Q, \Omega_{B'})$$

is a symplectic diffeomorphism of (T^*Q, Ω_B) with $(T^*Q, \Omega_{B'})$.

Proof As

$$t_A^*\Omega = \Omega - \pi^*\mathbf{d}A = \Omega - \pi^*B + \pi^*B^{'}$$

and $\pi \circ t_A = \pi$ we have

$$t_A^*(\Omega - \pi^*B^{'}) = \Omega - \pi^*B$$

which proves the statement. ■

Example: A particle in a magnetic field. Consider the motion of a charged particle in a given time independent magnetic field $\mathbf{B} := B_x\mathbf{i} + B_y\mathbf{j} + B_z\mathbf{k}$, where $\mathbf{i}, \mathbf{j}, \mathbf{k}$ is the usual orthonormal basis of $\mathbb{R}^3$. As customary in electromagnetism, we assume that $\operatorname{div}\mathbf{B} = 0$. The divergence free vector field $\mathbf{B}$ uniquely defines a closed two-form B on $\mathbb{R}^3$ by

$$B := \mathbf{i}_{\mathbf{B}}(\mathbf{d}x \wedge \mathbf{d}y \wedge \mathbf{d}z) = B_x\mathbf{d}y \wedge \mathbf{d}z + B_y\mathbf{d}z \wedge \mathbf{d}x + B_z\mathbf{d}x \wedge \mathbf{d}y.$$

The Lorentz force law for a particle with charge e and mass m is

$$m\frac{d\mathbf{v}}{dt} = \frac{e}{c}\mathbf{v} \times \mathbf{B}, \tag{2.13}$$

where $\mathbf{q} = (x, y, z)$ is the position of the particle, $\mathbf{v} = (\dot{x}, \dot{y}, \dot{z})$ its velocity, e its charge, m its mass, and c the speed of light. What is the Hamiltonian formulation of these equations? As we shall see, there are two different answers.

First, let $(p_x, p_y, p_z) := \mathbf{p} := m\mathbf{v}$, consider on $(\mathbf{q}, \mathbf{p})$-space, that is, $T^*\mathbb{R}^3$, the magnetic symplectic form

$$\begin{aligned} \Omega_B &= \mathbf{d}x \wedge \mathbf{d}p_x + \mathbf{d}y \wedge \mathbf{d}p_y + \mathbf{d}z \wedge \mathbf{d}p_z - \frac{e}{c}B \\ &= m(\mathbf{d}x \wedge \mathbf{d}\dot{x} + \mathbf{d}y \wedge \mathbf{d}\dot{y} + \mathbf{d}z \wedge \mathbf{d}\dot{z}) - \frac{e}{c}B, \end{aligned}$$

and the Hamiltonian given by the kinetic energy of the particle

$$H = \frac{1}{2m}\|\mathbf{p}\|^2 = \frac{m}{2}(\dot{x}^2 + \dot{z}^2 + \dot{y}^2).$$

Let us show that (2.13) is given by the Hamiltonian vector field X_H determined by the equation

$$\mathbf{d}H = \mathbf{i}_{X_H}\Omega_B. \tag{2.14}$$

If $X_H(\mathbf{q},\mathbf{p}) = (u, v, w, \dot{u}, \dot{v}, \dot{w})$, then (2.14) is equivalent to

$$m(\dot{x}\mathbf{d}\dot{x} + \dot{y}\mathbf{d}\dot{y} + \dot{z}\mathbf{d}\dot{z}) = m(u\mathbf{d}\dot{x} - \dot{u}\mathbf{d}x + v\mathbf{d}\dot{y} - \dot{v}\mathbf{d}y + w\mathbf{d}\dot{z} - \dot{w}\mathbf{d}z)$$
$$-\frac{e}{c}[B_x v\mathbf{d}z - B_x w\mathbf{d}y - B_y u\mathbf{d}z + B_y w\mathbf{d}x + B_z u\mathbf{d}y - B_z v\mathbf{d}x].$$

Comparing coefficients we get

$$\begin{aligned} u &= \dot{x} \\ v &= \dot{y} \\ w &= \dot{z} \\ m\dot{u} &= \tfrac{e}{c}(B_z v - B_y w) \\ m\dot{v} &= \tfrac{e}{c}(B_x w - B_z u) \\ m\dot{w} &= \tfrac{e}{c}(B_y u - B_x v), \end{aligned}$$

which are equivalent to (2.13). *Thus Hamilton's equations for the Lorentz force law need to be taken relative to the magnetic symplectic form and the kinetic energy of the particle.*

Second, write $B = \mathbf{d}A$ (always possible on $\mathbb{R}^3$), or equivalently, $\mathbf{B} = \nabla \times \mathbf{A}$, where $\mathbf{A}^\flat = A$, that is $\mathbf{A} = A_x\mathbf{i} + A_y\mathbf{j} + A_z\mathbf{k}$ and $A = A_x\mathbf{d}x + A_y\mathbf{d}y + A_z\mathbf{d}z$. Then the map $t_A : (\mathbf{x},\mathbf{p}) \mapsto (\mathbf{x}, \mathbf{p} + \frac{e}{c}\mathbf{A})$, pulls back the canonical two-form Ω of $T^*\mathbb{R}^3$ to Ω_B by Proposition 2.15. So, *the Lorentz equations are also Hamiltonian relative to the canonical two-form Ω and Hamilton's function*

$$H_A(\mathbf{x},\mathbf{p}) = \frac{1}{2m}\left\|\mathbf{p} - \frac{e}{c}A\right\|^2.$$

(See also [MaRa94] and [Jost]). It is not always possible to write $B = \mathbf{d}A$ if the domain on which one considers these equations is not simply connected, the magnetic monopole being such a case. We shall return to this example in the context of the Lagrangian formalism in §3.3.

3 Lagrangian Formalism

This lecture describes the local geometry of the Lagrangian formalism and ends with some remarks on global Lagrangian variational principles. The references for this material continue to be [Arnold79], [AbMa78], [MaRa94], [LiMa87].

3.1 Lagrangian Systems

The Legendre transformation. A smooth function $L : TQ \to \mathbb{R}$ is called a ***Lagrangian***. The ***Legendre transformation*** $\mathbb{F}L : TQ \to T^*Q$ is the smooth vector bundle map covering the identity defined by

$$\langle \mathbb{F}L(v_q), w_q \rangle := \left. \frac{d}{d\epsilon} \right|_{\epsilon=0} L(v_q + \epsilon w_q).$$

In the finite dimensional case, the local expression of $\mathbb{F}L$ is

$$\mathbb{F}L(q^i, \dot{q}^i) = \left(q^i, \frac{\partial L}{\partial \dot{q}^i} \right) = (q^i, p_i).$$

Given a Lagrangian L, the ***action*** of L is the map $A : TQ \to \mathbb{R}$ given by

$$A(v) := \langle \mathbb{F}L(v), v \rangle, \tag{3.1}$$

and the ***energy*** of L is

$$E(v) := A(v) - L(v). \tag{3.2}$$

Let Θ and Ω be the canonical one and two-forms on T^*Q, respectively. The Legendre transformation $\mathbb{F}L$ induces a one-form Θ_L and a closed two-form Ω_L on TQ by

$$\begin{aligned} \Theta_L &= (\mathbb{F}L)^* \Theta \\ \Omega_L &= -\mathbf{d}\Theta_L = (\mathbb{F}L)^* \Omega. \end{aligned}$$

In finite dimensions, the local expressions are

$$\begin{aligned} \Theta_L &:= \frac{\partial L}{\partial \dot{q}^i} \mathbf{d}q^i \\ \Omega_L &:= \frac{\partial^2 L}{\partial \dot{q}^i \partial q^j} \mathbf{d}q^i \wedge \mathbf{d}q^j + \frac{\partial^2 L}{\partial \dot{q}^i \partial \dot{q}^j} \mathbf{d}q^i \wedge \mathbf{d}\dot{q}^j. \end{aligned}$$

The closed two-form Ω_L can be written as the $2n \times 2n$ skew-symmetric matrix

$$\Omega_L = \begin{pmatrix} \mathcal{A} & \frac{\partial^2 L}{\partial \dot{q}^i \partial \dot{q}^j} \\ -\frac{\partial^2 L}{\partial \dot{q}^i \partial \dot{q}^j} & 0 \end{pmatrix} \tag{3.3}$$

where $\mathcal{A}$ is the skew-symmetrization of the $n \times n$ matrix $\left(\frac{\partial^2 L}{\partial \dot{q}^i \partial q^j} \right)$. The non-degeneracy of Ω_L is equivalent to the invertibility of the matrix $\left(\frac{\partial^2 L}{\partial \dot{q}^i \partial \dot{q}^j} \right)$. If

(3.3) is invertible, the Lagrangian L is said to be ***regular***. In this case, by the implicit function theorem, $\mathbb{F}L$ is locally invertible. If $\mathbb{F}L$ is a diffeomorphism, L is called ***hyperregular***.

Lagrangian systems. Lagrangian systems are special vector fields on the tangent bundle that are naturally associated to a function $L : TQ \to \mathbb{R}$.

Definition 3.1 A vector field Z on TQ is called a ***Lagrangian vector field*** if

$$\Omega_L(v)(Z(v), w) = \langle \mathbf{d}E(v), w \rangle,$$

for all $v \in T_qQ$, $w \in T_v(TQ)$.

Proposition 3.2 *The energy is conserved along the flow of a Lagrangian vector field Z.*

Proof Let $v(t) \in TQ$ be an integral curve of Z. Skew-symmetry of Ω_L implies

$$\begin{aligned}\frac{d}{dt}E(v(t)) &= \langle \mathbf{d}E(v(t)), \dot{v}(t) \rangle = \langle \mathbf{d}E(v(t)), Z(v(t)) \rangle \\ &= \Omega_L(v(t))\,(Z(v(t)), Z(v(t))) = 0.\end{aligned}$$

Thus $E(v(t))$ is constant in t. ■

Second order equations. Lagrangian vector fields are intimately tied to second order equations.

Definition 3.3 Let $\tau_Q : TQ \to Q$ be the canonical projection. A vector field $Z \in \mathfrak{X}(TQ)$ is called a ***second order equation*** if

$$T\tau_Q \circ Z = \mathrm{id}_{TQ}\,.$$

If $c(t)$ is an integral curve of Z, the curve $\tau_Q(c(t))$ on Q is called a ***base integral curve*** of Z.

Let $\mathbf{V}$ be the model of Q. In a canonical tangent bundle chart $U \times \mathbf{V}$ of TQ, U open in $\mathbf{V}$, the vector field Z can be written as

$$Z(\mathbf{q}, \dot{\mathbf{q}}) = (\mathbf{q}, \dot{\mathbf{q}}, Z_1(\mathbf{q}, \dot{\mathbf{q}}), Z_2(\mathbf{q}, \dot{\mathbf{q}})),$$

for some smooth maps $Z_1, Z_2 : U \times \mathbf{V} \to \mathbf{V}$. Thus,

$$T\tau_Q(Z(\mathbf{q}, \dot{\mathbf{q}})) = (\mathbf{q}, Z_1(\mathbf{q}, \dot{\mathbf{q}}))).$$

Since Z is a second order equation, this implies that $Z_1(\mathbf{q}, \dot{\mathbf{q}}) = \dot{\mathbf{q}}$. That is, in a local chart, the vector field is given by

$$Z(\mathbf{q}, \dot{\mathbf{q}}) = (\mathbf{q}, \dot{\mathbf{q}}, \dot{\mathbf{q}}, Z_2(\mathbf{q}, \dot{\mathbf{q}})).$$

So the dynamics is determined by

$$\begin{aligned} \frac{d\mathbf{q}}{dt} &= Z_1(\mathbf{q}, \dot{\mathbf{q}}) = \dot{\mathbf{q}} \\ \frac{d\dot{\mathbf{q}}}{dt} &= Z_2(\mathbf{q}, \dot{\mathbf{q}}) \end{aligned}$$

which is equivalent to the usual second order equation

$$\frac{d^2\mathbf{q}}{dt^2} = Z_2(\mathbf{q}, \dot{\mathbf{q}}).$$

The relationship between second order equations and Lagrangian vector fields is given by the following statement, whose proof is a straightforward (but somewhat lengthy) computation.

Theorem 3.4 *Let Z be the Lagrangian vector field on Q defined by L and suppose that Z is a second order equation. Then, in a canonical chart $U \times \mathbf{V}$ of TQ, an integral curve $(\mathbf{q}(t), \dot{\mathbf{q}}(t))$ of Z satisfies the* ***Euler-Lagrange equations***

$$\begin{aligned} \frac{dq^i}{dt} &= \dot{q}^i \\ \frac{d}{dt}\frac{\partial L}{\partial \dot{q}^i} &= \frac{\partial L}{\partial q^i}, \qquad i = 1, \ldots, n. \end{aligned}$$

Furthermore, if L is regular then Z is always second order.

3.2 Geodesics

An important example of a Lagrangian vector field is the geodesic spray of a pseudo-Riemannian metric. Recall that a ***pseudo-Riemannian manifold*** is a smooth manifold Q endowed with a symmetric nondegenerate covariant tensor g. Thus, on each tangent space T_qQ there is a nondegenerate (but indefinite, in general) inner product $g(q)$. If g is positive definite, then the pair (Q, g) is called a ***Riemannian manifold***.

If (Q, g) is a pseudo-Riemannian manifold, there is a natural Lagrangian on it given by the ***kinetic energy*** K of the metric g, namely,

$$K(v) := \frac{1}{2} g(q)(v_q, v_q),$$

for $q \in Q$ and $v_q \in T_qQ$. In finite dimensions, in a local chart,

$$K(q^k, \dot{q}^l) = \frac{1}{2} g_{ij}(q^k) \dot{q}^i \dot{q}^j.$$

The Legendre transformation is in this case $\mathbb{F}K(v_q) = g(q)(v_q, \cdot)$, for $v_q \in T_qQ$. The Euler-Lagrange equations become the ***geodesic equations*** for the metric g, given (for finite dimensional Q in a local chart) by

$$\ddot{q}^i + \Gamma^i_{jk} \dot{q}^j \dot{q}^k = 0, \quad i = 1, \ldots n,$$

where

$$\Gamma^h_{jk} = \frac{1}{2} g^{hl} \left(\frac{\partial g_{jl}}{\partial q^k} + \frac{\partial g_{kl}}{\partial q^j} - \frac{\partial g_{jk}}{\partial q^l} \right)$$

are the ***Christoffel symbols***. The Lagrangian vector field associated to K is called the ***geodesic spray***. Since the Legendre transformation is a diffeomorphism (in finite dimensions or in infinite dimensions if the metric is assumed to be strong), the geodesic spray is always a second order equation.

Let us link this approach to geodesics to the classical formulation using covariant derivatives. The covariant derivative $\nabla : \mathfrak{X}(Q) \times \mathfrak{X}(Q) \rightarrow \mathfrak{X}(Q)$, $(X, Y) \mapsto \nabla_X(Y)$, of the Levi-Civita connection on (Q, g) is given in local charts by

$$\nabla_X(Y) = X^i Y^j \Gamma^k_{ij} \frac{\partial}{\partial q^k} + X^i \frac{\partial Y^j}{\partial q^i} \frac{\partial}{\partial q^j}.$$

If $c(t)$ is a curve on Q and $X \in \mathfrak{X}(Q)$, the covariant derivative of X along $c(t)$ is defined by

$$\frac{DX}{Dt} := \nabla_{\dot{c}} X,$$

or locally,

$$\left(\frac{DX}{Dt} \right)^i = \Gamma^i_{jk}(c(t)) \dot{c}^j(t) X^k(c(t)) + \frac{d}{dt} X^i(c(t)).$$

A vector field is said to be ***parallel transported*** along $c(t)$ if $\frac{DX}{Dt} = 0$. Thus $\dot{c}(t)$ is parallel transported along $c(t)$ if and only if

$$\ddot{c}^i + \Gamma^i_{jk} \dot{c}^j \dot{c}^k = 0.$$

In classical differential geometry a ***geodesic*** is defined to be a curve $c(t)$ in Q whose tangent vector $\dot{c}(t)$ is parallel transported along $c(t)$. As the expression above shows, geodesics are base integral curves of the Lagrangian vector field defined by the kinetic energy of g.

A ***classical mechanical system*** is given by a Lagrangian of the form $L(v_q) =$

$K(v_q) - V(q)$, for $v_q \in T_qQ$. The smooth function $V : Q \to \mathbb{R}$ is called the ***potential energy***. The total energy of this system is given by $E = K + V$ and the Euler-Lagrange equations (which are always second order) are

$$\ddot{q}^i + \Gamma^i_{jk}\dot{q}^j\dot{q}^k + g^{il}\frac{\partial V}{\partial q^l} = 0, \quad i = 1, \ldots n,$$

where g^{ij} are the entries of the inverse matrix of (g_{ij}). If $Q = \mathbb{R}^3$ and the metric is given by $g_{ij} = \delta_{ij}$, these equations are Newton's equations of motion (1.3) of a particle in a potential field with which we began our discussion in the Introduction.

3.3 Hyperregular Lagrangians

We shall summarize here the precise equivalence between the Lagrangian and Hamiltonian formulation for hyperregular Lagrangians and Hamiltonians. The proofs are easy lengthy verifications; see [AbMa78] or [MaRa94].

(a) Let L be a hyperregular Lagrangian on TQ and $H = E \circ (\mathbb{F}L)^{-1}$, where E is the energy of L. Then the Lagrangian vector field Z on TQ and the Hamiltonian vector field X_H on T^*Q are related by the identity

$$(\mathbb{F}L)^* X_H = Z.$$

Furthermore, if $c(t)$ is an integral curve of Z and $d(t)$ an integral curve of X_H with $\mathbb{F}L(c(0)) = d(0)$, then $\mathbb{F}L(c(t)) = d(t)$ and their base integral curves coincide, that is, $\tau_Q(c(t)) = \pi_Q(d(t)) = \gamma(t)$, where $\tau_Q : TQ \to Q$ and $\pi_Q : T^*Q \to Q$ are the canonical bundle projections.

(b) A Hamiltonian $H : T^*Q \to \mathbb{R}$ is said to be ***hyperregular*** if the smooth vector bundle map covering the identity, $\mathbb{F}H : T^*Q \to TQ$, defined by

$$\langle \mathbb{F}H(\alpha_q), \beta_q \rangle := \left.\frac{d}{d\epsilon}\right|_{\epsilon=0} H(\alpha_q + \epsilon\beta_q), \qquad \alpha_q, \beta_q \in T^*_qQ,$$

is a diffeomorphism. Define the ***action*** of H by $G := \langle \Theta\, , X_H \rangle$. If H is a hyperregular Hamiltonian then the energies of L and H and the actions of L and H are related by

$$E = H \circ (\mathbb{F}H)^{-1}, \qquad A = G \circ (\mathbb{F}H)^{-1}.$$

In addition, the Lagrangian $L = A - E$ is hyperregular and $\mathbb{F}L = \mathbb{F}H^{-1}$.

(c) The constructions above define a bijective correspondence between hyperregular Lagrangians and Hamiltonians.

Example. Let us return to the example in §2.5 of a particle of charge e and mass m moving in a magnetic field $\mathbf{B}$. Recall that if $\mathbf{B} = \nabla \times \mathbf{A}$ is a given magnetic field on $\mathbb{R}^3$, then with respect to the canonical variables $(\mathbf{q}, \mathbf{p})$ on $T^*\mathbb{R}^3$, the canonical symplectic form $\mathbf{dq} \wedge \mathbf{dp}$, and the Hamiltonian

$$H(\mathbf{q}, \mathbf{p}) = \frac{1}{2m} \left\| \mathbf{p} - \frac{e}{c}\mathbf{A} \right\|^2, \tag{3.4}$$

Hamilton's equations coincide with Newton's equations for the Lorentz force law (2.13). It is obvious that H is hyperregular since

$$\dot{\mathbf{q}} = \mathbb{F}H(\mathbf{q}, \mathbf{p}) = \frac{1}{m}\left(\mathbf{p} - \frac{e}{c}\mathbf{A}\right)$$

has the inverse

$$\mathbf{p} = m\dot{\mathbf{q}} + \frac{e}{c}\mathbf{A}.$$

The Lagrangian is given by

$$\begin{aligned} L(\mathbf{q}, \dot{\mathbf{q}}) &= \mathbf{p} \cdot \dot{\mathbf{q}} - H(\mathbf{q}, \mathbf{p}) \\ &= \left(m\dot{\mathbf{q}} + \frac{e}{c}\mathbf{A}\right) \cdot \dot{\mathbf{q}} - \frac{1}{2m}\left\| \mathbf{p} - \frac{e}{c}\mathbf{A} \right\|^2 \\ &= m\|\dot{\mathbf{q}}\|^2 + \frac{e}{c}\mathbf{A} \cdot \dot{\mathbf{q}} - \frac{1}{2m}\|m\dot{\mathbf{q}}\|^2 \\ &= \frac{m}{2}\|\dot{\mathbf{q}}\|^2 + \frac{e}{c}\mathbf{A} \cdot \dot{\mathbf{q}}. \end{aligned} \tag{3.5}$$

Lagrange's equations for this L can be directly verified to yield $m\ddot{\mathbf{q}} = \frac{e}{c}\dot{\mathbf{q}} \times \mathbf{B}$, that is, Newton's equations for the Lorentz force (2.13).

These equations are not geodesic because the Lagrangian is not given by the kinetic energy of a metric. Can one enlarge the space such that these equations are induced by some geodesic equations on a higher dimensional space? The answer is positive and is given by the ***Kaluza-Klein construction***. We shall return once more to this example after we have presented the basic elements of reduction theory. The direct construction proceeds as follows. Define the manifold $Q_{KK} := \mathbb{R}^3 \times S^1$ with variables $(\mathbf{q}, \theta)$. On Q_{KK} introduce the one-form $A + \mathbf{d}\theta$ (a connection one-form on the trivial circle bundle $\mathbb{R}^3 \times S^1 \to \mathbb{R}^3$) and define the ***Kaluza-Klein Lagrangian***

$$\begin{aligned} L_{KK}(\mathbf{q}, \theta, \dot{\mathbf{q}}, \dot{\theta}) &= \frac{1}{2}m\|\dot{\mathbf{q}}\|^2 + \frac{1}{2}\left\|\left\langle A + \mathbf{d}\theta, (\mathbf{q}, \dot{\mathbf{q}}, \theta, \dot{\theta})\right\rangle\right\|^2 \\ &= \frac{1}{2}m\|\dot{\mathbf{q}}\|^2 + \frac{1}{2}\left(\mathbf{A} \cdot \dot{\mathbf{q}} + \dot{\theta}\right)^2. \end{aligned} \tag{3.6}$$

Note that L_{KK} is positive definite in $(\dot{\mathbf{q}}, \dot{\theta})$ so it is the kinetic energy of a metric, the ***Kaluza-Klein metric*** on Q_{KK}. Thus the Euler-Lagrange equations for L_{KK} are the geodesic equations of this metric on $\mathbb{R}^3 \times S^1$. (For the readers who know a little about connections and Lie algebras, it is obvious that this construction can be generalized to a principal bundle with compact structure group endowed with a connection; one gets in this way Yang-Mills theory.) The Legendre transformation for L_{KK} gives the momenta

$$\mathbf{p} = m\dot{\mathbf{q}} + (\mathbf{A} \cdot \dot{\mathbf{q}} + \dot{\theta})\mathbf{A} \qquad \text{and} \qquad \pi = \mathbf{A} \cdot \dot{\mathbf{q}} + \dot{\theta}. \tag{3.7}$$

Since L_{KK} does not depend on θ, the Euler-Lagrange equation

$$\frac{d}{dt}\frac{\partial L_{KK}}{\partial \dot{\theta}} = \frac{\partial L_{KK}}{\partial \theta} = 0$$

shows that $\pi = \partial L_{KK}/\partial \dot{\theta}$ is conserved. The ***charge*** is now defined by $e := c\pi$. The Hamiltonian H_{KK} associated to L_{KK} by the Legendre transformation (3.7) is

$$\begin{aligned}
H_{KK}(\mathbf{q}, \theta, \mathbf{p}, \pi) &= \mathbf{p} \cdot \dot{\mathbf{q}} + \pi\dot{\theta} - L_{KK}(\mathbf{q}, \dot{\mathbf{q}}, \theta, \dot{\theta}) \\
&= \mathbf{p} \cdot \frac{1}{m}(\mathbf{p} - \pi\mathbf{A}) + \pi(\pi - \mathbf{A} \cdot \dot{\mathbf{q}}) \\
&\quad - \frac{1}{2}m\|\dot{\mathbf{q}}\|^2 - \frac{1}{2}\pi^2 \\
&= \mathbf{p} \cdot \frac{1}{m}(\mathbf{p} - \pi\mathbf{A}) + \frac{1}{2}\pi^2 \\
&\quad - \pi\mathbf{A} \cdot \frac{1}{m}(\mathbf{p} - \pi\mathbf{A}) - \frac{1}{2m}\|\mathbf{p} - \pi\mathbf{A}\|^2 \\
&= \frac{1}{2m}\|\mathbf{p} - \pi\mathbf{A}\|^2 + \frac{1}{2}\pi^2.
\end{aligned} \tag{3.8}$$

Since $\pi = e/c$ is a constant, this Hamiltonian, regarded as a function of only the variables $(\mathbf{q}, \mathbf{p})$, is up to a constant equal to the Lorentz force Hamiltonian (3.4). This fact is not an accident and is due to the geometry of reduction. This example will be redone in §7.3 from the point of view of reduction theory.

3.4 Variational Principles

We now sketch the variational approach to mechanics.

Theorem 3.5 (Variational Principle of Hamilton) *Let L be a Lagrangian on TQ. A C^2 curve $c : [a, b] \to Q$ joining $q_1 = c(a)$ to $q_2 = c(b)$ satisfies the*

Euler-Lagrange equations if and only if

$$\delta \int_a^b L(c(t), \dot{c}(t))dt = 0.$$

Proof The meaning of the variational derivative in the statement is the following. Consider a family of C^2 curves $c_\lambda(t)$ for $|\lambda| < \varepsilon$ satisfying $c_0(t) = c(t)$, $c_\lambda(a) = q_1$, and $c_\lambda(b) = q_2$ for all $\lambda \in (-\varepsilon, \varepsilon)$. Then

$$\delta \int_a^b L(c(t), \dot{c}(t))dt := \left.\frac{d}{d\lambda}\right|_{\lambda=0} \int_a^b L(c_\lambda(t), \dot{c}_\lambda(t))dt.$$

Differentiating under the integral sign, working in local coordinates (covering the curve $c(t)$ by a finite number of coordinate charts), integrating by parts, denoting

$$v(t) := \left.\frac{d}{d\lambda}\right|_{\lambda=0} c_\lambda(t),$$

and taking into account that $v(a) = v(b) = 0$, yields

$$\int_a^b \left(\frac{\partial L}{\partial q^i} v^i + \frac{\partial L}{\partial \dot{q}^i} \dot{v}^i \right) dt = \int_a^b \left(\frac{\partial L}{\partial q^i} - \frac{d}{dt}\frac{\partial L}{\partial \dot{q}^i} \right) v^i dt.$$

This vanishes for any C^1 function $v(t)$ if and only if the Euler-Lagrange equations hold. ■

The integral appearing in this theorem

$$\mathcal{A}(c(\cdot)) := \int_a^b L(c(t), \dot{c}(t))dt$$

is called the ***action integral***. It is defined on C^2 curves $c : [a, b] \to Q$ with fixed endpoints, $c(a) = q_1$ and $c(b) = q_2$.

The next theorem emphasizes the role of the Lagrangian one- and two-forms in the variational principle. It is a direct corollary of the previous theorem.

Theorem 3.6 *Given a C^k Lagrangian $L : TQ \to \mathbb{R}$ for $k \geq 2$, there exists a unique C^{k-2} map $\mathcal{EL}(L) : \ddot{Q} \to T^*Q$, where*

$$\ddot{Q} := \left\{ \frac{d^2q}{dt^2}(0) \in T(TQ) \,\middle|\, q(t) \text{ is a } C^2 \text{ curve in } Q \right\}$$

is a submanifold of $T(TQ)$ (second order submanifold), and a unique C^{k-1}

one-form $\Theta_L \in \Omega^1(TQ)$*, such that for all* C^2 *variations* $q_\varepsilon(t)$ *(defined on a fixed t-interval) of* $q_0(t) := q(t)$*, we have*

$$\mathbf{d}\mathcal{A}[q(\cdot)] \cdot \delta q(\cdot) = \int_a^b \mathcal{EL}(L)\left(\frac{d^2q}{dt^2}\right) \cdot \delta q \, dt + \Theta_L\left(\frac{dq}{dt}\right)\widehat{\delta q}\Big|_a^b$$

where

$$\delta q(t) = \left.\frac{d}{d\varepsilon}\right|_{\varepsilon=0} q_\varepsilon(t), \qquad \widehat{\delta q}(t) = \left.\frac{d}{d\varepsilon}\right|_{\varepsilon=0} \frac{dq_\varepsilon(t)}{dt}.$$

The map $\mathcal{EL} : \ddot{Q} \to T^*Q$ is called the ***Euler-Lagrange operator*** and its expression in local coordinates is

$$\mathcal{EL}(q^j, \dot{q}^j, \ddot{q}^j)_i = \frac{\partial L}{\partial q^i} - \frac{d}{dt}\frac{\partial L}{\partial \dot{q}^i},$$

where it is understood that the formal time derivative is taken in the second summand and everything is expressed as a function of $(q^j, \dot{q}^j, \ddot{q}^j)$. The one-form Θ_L, whose existence and uniqueness is guaranteed by this theorem, appears as the boundary term of the derivative of the action integral if the end-points of the curves on the configuration manifold are free; it coincides with $(\mathbb{F}L)^*\Theta$, where Θ is the canonical one-form on T^*Q, defined in §3.1 as an easy verification in coordinates shows.

Remark. From the variational principle one can recover well-known results for regular Lagrangians; for proofs of these statements see e.g. [MaRa94], §8.2.

(i) If F_t is the flow of the Lagrangian vector field, then $F_t^*\Omega_L = \Omega_L$, where $\Omega_L = -\mathbf{d}\Theta_L$.

(ii) If L is time dependent, for $|t - t_0|$ small, and $q^i(s)$ the solution of the Euler-Lagrange equation subject to the condition $q^i(t_0) = \bar{q}^i$, the convex neighborhood theorem guarantees that the action integral

$$S(q^i, \bar{q}^i, t) = \int_{t_0}^t L(q^i(s), \dot{q}^i(s), s)\, ds,$$

satisfies the Hamilton-Jacobi equation

$$\frac{\partial S}{\partial t} + H\left(q, \frac{\partial S}{\partial q}, t\right) = 0. \tag{3.9}$$

There is another classical variational principle involving the Hamiltonian, known under the name of ***Hamilton's phase space principle***. Denote in what follows by $\pi_Q : T^*Q \to Q$ the cotangent bundle projection.

Theorem 3.7 *Let $H : T^*Q \to \mathbb{R}$ a smooth Hamiltonian. A C^1 curve $z : [a, b] \to T^*Q$ joining $z(a)$ to $z(b)$ satisfies Hamilton's equations if and only if*

$$\delta \int_a^b \big(\langle \Theta, \dot{z}(t) \rangle - H(z(t)) \big) dt = 0,$$

where the variations δz satisfy $T\pi_Q(\delta\alpha(a)) = T\pi_Q(\delta\alpha(b)) = 0$.

Proof One follows the method of proof of Theorem 3.5. The meaning of the variational derivative is the following. Consider a family of C^1 curves $z_\lambda(t)$ for $|\lambda| < \varepsilon$ satisfying $z_0(t) = z(t)$, $z_\lambda(a) = z_1$, and $z_\lambda(b) = z_2$ for all $\lambda \in (-\varepsilon, \varepsilon)$. Then

$$\delta \int_a^b \big(\langle \Theta, \dot{z}(t) \rangle - H(z(t)) \big) dt := \left.\frac{d}{d\lambda}\right|_{\lambda=0} \int_a^b \big(\langle \Theta, \dot{z}_\lambda(t) \rangle - H(z_\lambda(t)) \big) dt.$$

Differentiating under the integral sign, working in local coordinates (covering the curve $z(t)$ by a finite number of coordinate charts), integrating by parts, denoting

$$\delta z(t) := \left.\frac{d}{d\lambda}\right|_{\lambda=0} z_\lambda(t),$$

writing in coordinates $\delta z(t) = (\delta q^i(t), \delta p_i(t))$, and taking into account that $\delta q^i(a) = \delta q^i(b) = 0$, yields

$$\begin{aligned}
&\left.\frac{d}{d\lambda}\right|_{\lambda=0} \int_a^b \big((p_\lambda)_i(t)(\dot{q}_\lambda)^i - H((q_\lambda)^i(t), (p_\lambda)_i(t)) \big) dt \\
&\quad = \int_a^b \bigg((\delta p)_i(t)\dot{q}^i(t) + p_i(t)(\delta\dot{q})^i(t) - \frac{\partial H(q^i(t), p_i(t))}{\partial q^i}(\delta q)^i(t) \\
&\qquad\qquad - \frac{\partial H(q^i(t), p_i(t))}{\partial p_i}(\delta p)_i(t) \bigg) dt \\
&\quad = \int_a^b \bigg[\left(\dot{q}^i(t) - \frac{\partial H(q^i(t), p_i(t))}{\partial p_i} \right) (\delta p)_i(t) \\
&\qquad\qquad - \left(\dot{p}_i + \frac{\partial H(q^i(t), p_i(t))}{\partial q^i} \right) (\delta q)^i(t) \bigg] dt.
\end{aligned}$$

This vanishes for any functions $\delta q^i, \delta p_i$ if and only if Hamilton's equations hold. ■

Critical point theory. We have seen that under appropriate regularity assumptions, the vanishing of the first variation of the action integral is equivalent to the Euler-Lagrange equations. That is, critical points of the action integral are solutions of the equations of motion. Now, critical points of a function are connected to the topology of its level sets and are described by Morse and Ljusternik-Schnirelman theory. For functions on finite dimensional spaces see e.g. [DFN95] or [Milnor] for an exposition of this theory. For the action integral, which is defined on the infinite dimensional manifold of admissible motions, see e.g. [MawWil989] for the development of the relevant Morse and Ljusternik-Schnirelman theory.

Here is an example of a strategy for the search of periodic orbits in Lagrangian systems. Let Q be a n-dimensional configuration manifold. We search a trajectory $\gamma(t) \in Q$, such that $\gamma(t) = \gamma(T+t)$ for a certain number $T > 0$ (the period), where $\gamma(t)$ is a solution of the Euler-Lagrange equations. To do this, we study the action integral $\mathcal{A}$, where

$$\mathcal{A}[q] = \int_0^T L(\dot{q}(t), q(t))\, dt,$$

is defined on a space of T-periodic trajectories lying in Q of a certain differentiability class.

Since we want to consider continuous paths, by the Sobolev Embedding Theorem, we take

$$H^1([0,T];Q) := \{q(\cdot) : [0,T] \to Q \mid q(\cdot) \text{ of class } H^1, q(0) = q(T)\}$$

which is an infinite dimensional Hilbert manifold; it is a simple example of a manifold of maps from a compact manifold with boundary to an arbitrary smooth paracompact manifold (see [Palais68] or [EbMa70] for example). However, in this case, there is simpler more direct construction (see e.g. [Ben86]). Embed Q into some $\mathbb{R}^N$ and consider the space of loops $H^1([0,T];\mathbb{R}^N)$ of period T with values in $\mathbb{R}^N$, obtained as the completion of the space of smooth loops $C^\infty([0,T];\mathbb{R}^n)$ relative to the norm

$$\|q\|_1^2 := \int_0^T \|\dot{q}(t)\|^2 dt + \int_0^T \|q(t)\|^2 dt.$$

Therefore, $q(\cdot) \in H^1([0,T];\mathbb{R}^n)$ is an absolutely continuous loop with L^2 derivative; $H^1([0,T];\mathbb{R}^N)$ is compactly immersed in $C^0([0,T];\mathbb{R}^N)$ (see e.g. [BlBr92]). Then $H^1([0,T];Q)$ is the submanifold of $H^1([0,T];\mathbb{R}^N)$ consisting of all loops with values in Q.

Critical points of the action integral $\mathcal{A}$ are trajectories where the homology of the level sets of $\mathcal{A}$ changes. For the sake of simplicity we can consider them to be minimal. Therefore, given a subset $\mathcal{M} \subset H^1([0,T];Q)$ we could look for $\gamma^*(\cdot) \in \mathcal{M}$ subject to the condition

$$\gamma^* = \min_{\gamma \in \mathcal{M}} \mathcal{A}[\gamma].$$

Typical conditions that guarantee the existence of a minimum γ^* for $\mathcal{A}$ are:

(i) $\mathcal{M}$ is weakly closed in $H^1([0,T];Q)$,
(ii) $\mathcal{A}$ is C^1 on $\mathcal{M}$ and bounded from below,
(iii) the set $\{\gamma \in H^1([0,T];Q) \mid \mathcal{A}[\gamma] < \infty, \mathbf{d}\mathcal{A}[\gamma] = 0\}$ is compact in $H^1([0,T];Q)$.

Points (i) and (iii) imply that γ^* solves the equations of motion, that is,

$$\mathbf{d}\mathcal{A}[\gamma^*](u) = 0,$$

for all $u(\cdot) \in T_{\gamma^*}H^1([0,T];Q)$. We refer, as general references, to [BlBr92], [Jost], and [MawWil989], and for applications of this theory to the N-body problem to [AmCZ96].

The judicious choice of the domain $\mathcal{M}$, where one searches for critical points of $\mathcal{A}$, involves a lot of geometrical, analytical, and physical information, intimately tied to the problem under consideration. In fact, the form of the potential provides severe restrictions; for example, in the N-body problem one has to take into account all the trajectories leading to collisions of two or more bodies. Also, when a system is symmetric then the symmetry group gives additional structure to $\mathcal{M}$ and leads to topological constraints; for example, restrictions on the homotopy type for the trajectories enter in this case.

Regarding the connection between global variational methods and geometry we want to mention an interesting problem. We have seen that the Euler equations $\dot{\mathbf{\Pi}} = \mathbf{\Pi} \times \mathbf{\Omega}$ for the free rigid body motion can be formulated in a variational setting assuming a particular class of variations. This construction can be generalized to any Lie group, as we shall see in Theorem 6.6. It would be very interesting to exploit the geometrical structure and to extend the powerful methods of the calculus of variations to formulate results about the existence of critical points of such types of restricted variational principles.

4 Poisson Manifolds

This lecture quickly reviews the basic theory of Poisson manifolds. Very little will be proved here and we refer to standard books (such as [GuSt84, LiMa87, Marsden92, MaRa94, McDSal95, V96]) and [W] for the detailed discussion of the topics below. Unless otherwise specified, all manifolds in this chapter are finite dimensional. Whenever infinite dimensional manifolds will be used the results presented here are formal. A theory for infinite dimensional Poisson manifolds generalizing that of strong symplectic manifolds can be found in [OdzRa03] (for more examples see [OdzRa004] and [BelRa04]). No general satisfactory theory for infinite dimensional general Poisson manifolds has been developed yet, even though there are many significant examples, some of which will be presented in these lectures. Remarks regarding the infinite dimensional situation will be made throughout the text.

4.1 Fundamental Concepts

Definition 4.1 A ***Poisson bracket*** on a manifold P is a bilinear operation $\{\,,\}$ on the space $\mathcal{F}(P) := \{F : P \to \mathbb{R} \mid F \text{ is smooth}\}$ verifying the following conditions:

(i) $(\mathcal{F}(P), \{\,,\})$ is a Lie algebra, and
(ii) $\{\,,\}$ satisfies the Leibniz identity on each factor.

A manifold endowed with a Poisson bracket is called a ***Poisson manifold*** and is denoted by $(P, \{\,,\})$. The elements of the center $\mathcal{C}(P)$ of the Poisson algebra are called ***Casimir functions***.

A smooth map $\varphi : (P_1, \{\,,\}_1) \to (P_2, \{\,,\}_2)$ between the Poisson manifolds $(P_1, \{\,,\}_1)$ and $(P_2, \{\,,\}_2)$ is called a ***canonical*** or ***Poisson map***, if $\varphi^*\{F, H\}_2 = \{\varphi^* F, \varphi^* H\}_1$ for any $F, H \in \mathcal{F}(P_2)$.

Note that (i) is equivalent to the statement that the Poisson bracket is real, bilinear, antisymmetric, and satisfies the Jacobi identity. Furthermore, (ii) states that the linear map $F \mapsto \{F, H\}$ (and $H \mapsto \{F, H\}$) is a ***derivation***, that is,

$$\{FK, H\} = \{F, H\}\, K + F\, \{K, H\}$$

for all $F, H, K \in \mathcal{F}(P)$.

As in the case of strong symplectic manifolds, we can define Hamiltonian vector fields on a finite dimensional Poisson manifold.

Definition 4.2 Let $(P, \{\,,\})$ be a finite dimensional Poisson manifold and $H \in \mathcal{F}(P)$. The unique vector field X_H on P that satisfies $X_H[F] = \{F, H\}$ for all $F \in \mathcal{F}(P)$, is called the ***Hamiltonian vector field*** associated to the ***Hamiltonian function*** H.

Note that the Jacobi identity is equivalent to

$$[X_F, X_H] = -X_{\{F,H\}}$$

for all $F, H \in \mathcal{F}(P)$.

Any symplectic manifold (P, Ω) is Poisson. First, recall that there is a Poisson bracket naturally defined on P, namely, $\{F, H\} = \Omega(X_F, X_H)$, where X_H is defined by the identity $\mathbf{d}H = \Omega(X_H, \cdot)$. Second, the relation

$X_H[F] = \langle \mathbf{d}F, X_H \rangle = \Omega(X_F, X_H)$, shows that the Hamiltonian vector field defined via the symplectic form coincides with the Hamiltonian vector field defined using the Poisson bracket.

If P is a Poisson manifold, note that $F \in \mathcal{C}(P)$ if and only if $X_F = 0$.

If ϕ_t is the flow of X_H, then $H \circ \phi_t = H$, that is, H is conserved. Indeed,

$$\frac{d}{dt}(H \circ \phi_t) = \left\langle \mathbf{d}H, \frac{d}{dt}\phi_t \right\rangle = \langle \mathbf{d}H, X_H \rangle = \{H, H\} = 0.$$

Thus $H \circ \phi_t$ is constant in t and since ϕ_0 is the identity, it follows that $H \circ \phi_t = H$, for all t.

Hamilton's equations $\dot{z} = X_H(z)$ for the function $H \in \mathcal{F}(P)$ can be equivalently written as

$$\frac{d}{dt}(F \circ \phi_t) = \{F, H\} \circ \phi_t \quad \text{or, in shorthand notation,} \quad \dot{F} = \{F, H\}$$

for any $F \in \mathcal{F}(P)$. To see this, note first that if ϕ_t is the flow of X_H, we have $\phi_t^* X_H = X_H$, or, equivalently, $T\phi_t \circ X_H = X_H \circ \phi_t$. Thus, for any $z \in P$ we have

$$\begin{aligned}\frac{d}{dt}(F \circ \phi_t)(z) &= \left\langle \mathbf{d}F(\phi_t(z)), \frac{d}{dt}\phi_t(z) \right\rangle = \langle \mathbf{d}F(\phi_t(z)), X_H(\phi_t(z)) \rangle \\ &= \langle \mathbf{d}F(\phi_t(z)), T_z\phi_t(X_H(z)) \rangle = \langle \mathbf{d}(F \circ \phi_t)(z), X_H(z) \rangle \\ &= \{F \circ \phi_t, H\}(z).\end{aligned}$$

Since $H \circ \phi_t = H$ we conclude that Hamilton's equations on a Poisson manifold can be written as

$$\frac{d}{dt}(F \circ \phi_t) = \{F, H\} \circ \phi_t = \{F \circ \phi_t, H\} = \{F \circ \phi_t, H \circ \phi_t\}. \tag{4.1}$$

Proposition 4.3 *The flows of Hamiltonian vector fields are Poisson diffeomorphisms.*

Proof Let ϕ_t be the flow of the Hamiltonian vector field X_H, that is, $\frac{d}{dt}\phi_t = X_H \circ \phi_t$. We need to prove $\{F, K\} \circ \phi_t = \{F \circ \phi_t, K \circ \phi_t\}$ for any $F, K \in \mathcal{F}(P)$. To see this, note that for any $z \in P$,

$$\begin{aligned}\frac{d}{dt}\{F, K\}(\phi_t(z)) &= \left\langle \mathbf{d}\{F, K\}(\phi_t(z)), \frac{d}{dt}\phi_t(z) \right\rangle \\ &= \langle \mathbf{d}\{F, K\}(\phi_t(z)), X_H(\phi_t(z)) \rangle = \{\{F, K\}, H\}(\phi_t(z)),\end{aligned}$$

that is,

$$\frac{d}{dt}(\{F, K\} \circ \phi_t) = \{\{F, K\}, H\} \circ \phi_t = \{\{F, K\} \circ \phi_t, H\} \tag{4.2}$$

by (4.1). On the other hand, the bilinearity of the Poisson bracket gives

$$\frac{d}{dt}\{F \circ \phi_t, K \circ \phi_t\} = \left\{\frac{d}{dt}(F \circ \phi_t), K \circ \phi_t\right\} + \left\{F \circ \phi_t, \frac{d}{dt}(K \circ \phi_t)\right\}.$$

This equality, the Jacobi identity, and (4.1) imply that

$$\begin{aligned}\frac{d}{dt}\{F \circ \phi_t, K \circ \phi_t\} &= \{\{F \circ \phi_t, H\}, K \circ \phi_t\} + \{F \circ \phi_t, \{K \circ \phi_t, H\}\} \\ &= \{\{F \circ \phi_t, K \circ \phi_t\}, H\}. \end{aligned} \tag{4.3}$$

Comparing (4.2) and (4.3) one sees that both $\{F, K\} \circ \phi_t$ and $\{F \circ \phi_t, K \circ \phi_t\}$ satisfy the same equation, namely, $\dot{L} = \{L, H\}$. Since for $t = 0$, the functions $\{F, K\} \circ \phi_t$ and $\{F \circ \phi_t, K \circ \phi_t\}$ are both equal to $\{F, K\}$, it follows that $\{F, K\} \circ \phi_t = \{F \circ \phi_t, K \circ \phi_t\}$ for all t. ■

The same strategy of proof is used to show the following statement.

Proposition 4.4 *A smooth map* $\phi : (P_1, \{\,,\}_1) \to (P_2, \{\,,\}_2)$ *is Poisson if and only if for any* $H \in \mathcal{F}(V)$, V *open in* P_2, *we have* $T\phi \circ X_{H\circ\phi} = X_H \circ \phi$ *on* $\phi^{-1}(V)$.

Canonical maps are the key ingredient in the definition of the notion of a Poisson submanifold.

Definition 4.5 Let $(P_1, \{\cdot,\cdot\}_1)$ and $(P_2, \{\cdot,\cdot\}_2)$ be two Poisson manifolds, $P_1 \subset P_2$, such that the inclusion $i : P_1 \hookrightarrow P_2$ is an immersion. The Poisson manifold $(P_1, \{\cdot,\cdot\}_1)$ is called a ***Poisson submanifold*** of $(P_2, \{\cdot,\cdot\}_2)$ if i is a canonical map.

An immersed submanifold P_1 of P_2 is called a ***quasi Poisson submanifold*** of $(P_2, \{\cdot,\cdot\}_2)$ if for any $p \in P_1$, any open neighborhood U of p in P_2, and any $F \in C^\infty(U)$ we have

$$X_F(i(p)) \in T_p i(T_p P_1),$$

where X_F is the Hamiltonian vector field of F on U with respect to the Poisson bracket $\{\cdot,\cdot\}_2$ restricted to smooth functions defined on U.

The proofs of the following statements can be found in, e.g., [OR04], §4.1.

- If $(P_1, \{\cdot,\cdot\}_1)$ is a Poisson submanifold of $(P_2, \{\cdot,\cdot\}_2)$ then there is no other bracket $\{\cdot,\cdot\}'$ on P_1 making the inclusion $i : P_1 \hookrightarrow P_2$ into a canonical map.
- If P_1 is a quasi Poisson submanifold of $(P_2, \{\cdot,\cdot\}_2)$ then there exists a unique Poisson bracket on P_1 making it into a Poisson submanifold of P_2.

- Proposition 4.4 implies that any Poisson submanifold is quasi Poisson. However, a quasi Poisson submanifold P_1 of $(P_2, \{\cdot, \cdot\}_2)$ could carry a Poisson structure that has nothing to do with the one induced from P_2, so it won't be a Poisson submanifold of P_2 relative to this a priori given structure.
- If (P_1, Ω_1) and (P_2, Ω_2) are two symplectic manifolds such that $P_1 \subset P_2$, then P_1 is said to be a ***symplectic submanifold*** of P_2 if $i^*\Omega_2 = \Omega_1$, where $i : P_1 \hookrightarrow P_2$ is the inclusion. In this case, the inclusion is necessarily an immersion (P_1 and P_2 are assumed finite dimensional).
- Symplectic submanifolds of a symplectic manifold are, in general, neither Poisson nor quasi Poisson submanifolds.
- The only quasi Poisson submanifolds of a symplectic manifold are its open sets which are, in fact, Poisson submanifolds.

4.2 Structure Theorems

The derivation property of a Poisson bracket $\{\, , \}$ implies that the value $X_F(z)$ of the Hamiltonian vector field of F at $z \in P$ depends on F only through $\mathbf{d}F(z)$. Thus there is a contravariant antisymmetric two-tensor Λ on P, called the ***Poisson tensor***, defined by

$$\Lambda(z)(\mathbf{d}F(z), \mathbf{d}H(z)) = \{F, H\}(z)$$

for any $F, H \in \mathcal{F}(U)$, U open in P, $z \in U$.

In finite dimensions, if $(z^1, z^2, \ldots, z^n)$ are local coordinates, Λ is determined by the matrix $\left[\Lambda^{ij}\right]$, where $\Lambda^{ij} = \{z^i, z^j\}$, and hence the expression of the Poisson bracket of the two functions $F, H \in \mathcal{F}(P)$ is given in terms of Λ by

$$\{F, H\} = \Lambda^{ij} \frac{\partial F}{\partial z^i} \frac{\partial H}{\partial z^j}.$$

The ***rank*** of the Poisson structure at a point $z \in P$ is defined to be the rank of the matrix $\left[\Lambda^{ij}\right]$.

The Poisson tensor defines a vector bundle map $\Lambda^\sharp : T^*P \to TP$ by

$$\Lambda(z)(\alpha_z, \beta_z) = \langle \alpha_z, \Lambda^\sharp(\beta_z) \rangle, \qquad \alpha_z, \beta_z \in T_z^*P.$$

Since

$$\{F, H\}(z) = \Lambda(z)(\mathbf{d}F(z), \mathbf{d}H(z)) = \langle \mathbf{d}F(z), \Lambda_z^\sharp(\mathbf{d}H(z)) \rangle,$$

it follows that the Hamiltonian vector field X_H is given by

$$X_H(z) = \Lambda_z^\sharp(\mathbf{d}H(z)),$$

that is, $\Lambda_z^\sharp : \mathbf{d}H(z) \mapsto X_H(z)$. Thus the image of $\Lambda_z^\sharp$ is the set of all Hamiltonian vector fields evaluated at z.

Note that if the Poisson tensor is nondegenerate, that is $\Lambda^\sharp : T_z^*P \to T_zP$ is a isomorphism for all $z \in P$, then P is symplectic with the symplectic form $\Omega(X_F, X_H) := \{F, H\}$ for all locally defined Hamiltonian vector fields X_F, X_H (the closeness of Ω is equivalent to the Jacobi identity of the Poisson bracket).

The image $\mathrm{Im}(\Lambda^\sharp) \subset TP$ of $\Lambda^\sharp$ defines a ***smooth generalized distribution*** on P, i.e., $\mathrm{Im}(\Lambda_z^\sharp) \subset T_zP$ is a vector subspace for each $z \in P$ and for every point $z_0 \in P$ and every vector $v \in \mathrm{Im}(\Lambda_{z_0}^\sharp)$, there exists an open neighborhood U of z_0 and smooth vector field $X \in \mathfrak{X}(U)$ such that $X(u) \in \mathrm{Im}(\Lambda_u^\sharp)$ for all $u \in U$ and $X(z_0) = v$.

If the rank of the Poisson tensor is constant then $\mathrm{Im}(\Lambda^\sharp)$ is a smooth vector subbundle of TP. Furthermore, the Jacobi identity gives $[X_F, X_H] = X_{-\{F,H\}}$ which shows that the distribution $\mathrm{Im}(\Lambda^\sharp)$ is involutive. The Frobenius theorem guarantees then its integrability.

If $\mathrm{Im}(\Lambda^\sharp)$ is not a subbundle then integrability and involutivity (as used in the Frobenius theorem) are not equivalent. Recall that if $D \subset TP$ is a smooth generalized distribution, an immersed connected submanifold S of P, $S \subset P$, is said to be an ***integral manifold*** of D if for every $z \in S$, $T_zi(T_zS) \subset D(z)$, where $i : S \to P$ is the inclusion. The integral submanifold S is said to be of ***maximal dimension*** at a point $z \in S$ if $T_zi(T_zS) = D(z)$. The smooth generalized distribution D is ***integrable*** if for every point $z \in P$ there is an integral manifold of D everywhere of maximal dimension containing z. The smooth generalized distribution D is ***involutive*** if it is invariant under the (local) flows associated to differentiable sections of D. Note that this definition of involutivity is weaker than the one used in the Frobenius theorem and that it only coincides with it when the dimension of $D(z)$ is the same for any $z \in P$.

Theorem 4.6 (Stefan-Sussmann) *The smooth generalized distribution D on a finite dimensional manifold P is integrable if and only if it is involutive.*

Thus P carries a generalized foliation, all of whose integral manifolds are injectively immersed.

If P is a Poisson manifold, $\mathrm{Im}(\Lambda^\sharp)$ is an involutive smooth generalized distribution so each of its integral submanifolds has its tangent space at every point equal to the vector space of all Hamiltonian vector fields evaluated at that point. Thus, these integral submanifolds are symplectic and the Poisson bracket defined by their symplectic structure coincides with the original Poisson bracket on P. These integral manifolds are called the ***symplectic leaves*** of P. It turns out that the symplectic leaves are the equivalence classes of the following equivalence relation: $z_1 \mathfrak{R} z_2$ if and only if there is a piecewise smooth curve in P joining z_1 and z_2 each segment of which is an integral curve of some Hamiltonian vector field. The following theorem summarizes this discussion.

Theorem 4.7 (Symplectic Stratification Theorem) *Let P be a finite dimensional Poisson manifold. Then P is the disjoint union of its symplectic leaves. Each symplectic leaf in P is an injectively immersed Poisson submanifold and the induced Poisson structure on the leaf is symplectic. The dimension of a leaf through a point z equals the rank of the Poisson structure at that point.*

Note that if C is a Casimir function, then $\Lambda^\sharp(\mathbf{d}C) = 0$, which shows that Casimir functions are constant on symplectic leaves. However, one should not conclude that the symplectic leaves are level sets of Casimir functions. This is not even true for the maximal dimensional ones, which are generic. For example, symplectic leaves may be open or they may all have a common accumulation point. Worse, there are Poisson manifolds with no global Casimir functions. Locally, Casimir functions always exist generically as the next local structure theorem shows.

Theorem 4.8 (Weinstein) *Let P be a finite dimensional Poisson manifold and $z \in P$. There exists a neighborhood U of z and an isomorphism $\phi = \phi_S \times \phi_N : U \to S \times N$ where S is symplectic, N is Poisson and the rank of N at $\phi_N(z)$ is zero. The factors S, N are unique up to a local isomorphism.*

In this theorem S can be chosen to be an open set in the symplectic leaf through z and N any submanifold of P transverse to it such that $S \cap N = \{z\}$. While there is no canonical choice of N in general, the Poisson structure on it is uniquely determined up to a Poisson isomorphism. This Poisson structure is called the ***transverse Poisson structure*** at z.

Assume that the Poisson structure has rank $0 \leq 2k \leq \dim(P) = 2k + l$ at the point $z \in P$. Then there are coordinates $(q^1, \ldots, q^k, p_1, \ldots, p_k, y^1, \ldots, y^l)$ in a chart around z such that

$$\{q^i, q^j\} = \{p_i, p_j\} = \{q^i, y^j\} = \{p_i, y^j\} = 0, \; \{q^i, p_j\} = \delta^i_j$$

and the brackets $\{y^i, y^j\}$ depend only on $y^1, \ldots, y^l$ and vanish at the point z. In this chart, the transverse Poisson structure is given by the subspace defining these coordinates $y^1, \ldots, y^l$. If, in addition, there is a neighborhood of z such that the rank of the Poisson structure is constant on it then, shrinking if necessary the above chart such that it lies in this neighborhood, the coordinates

$y^1, \ldots, y^l$ can be chosen such that $\{y^i, y^j\} = 0$ for all $i, j = 1, \ldots, l$. In this case, the y^i are the local Casimir functions in a chart about z.

Comments on Banach Poisson manifolds. Definition 4.1 presents several problems in the infinite dimensional case that will be briefly reviewed here. Let $(P, \{\,,\})$ be a Banach Poisson manifold. The Leibniz property insures, as in finite dimensions, that the value of the Poisson bracket at $z \in P$ depends only on the differentials $\mathbf{d}F(z), \mathbf{d}H(z) \in T_z^*P$ which implies that there is a smooth section Λ of the vector bundle $\bigwedge^2 T^{**}P$ satisfying $\{F, H\} = \Lambda(\mathbf{d}F\mathbf{d}H)$.

If the Poisson tensor is strongly nondegenerate, that is $\Lambda^\sharp : T_z^*P \to T_zP$ is a isomorphism of Banach spaces for all $z \in P$, then P is strong symplectic with the symplectic form $\Omega(X_F, X_H) := \{F, H\}$ for all locally defined Hamiltonian vector fields X_F, X_H (the closedness of Ω is equivalent to the Jacobi identity of the Poisson bracket). However if $\Lambda^\sharp$ is one-to-one but not surjective, that is, the Poisson tensor is weakly nondegenerate, then P is not, in general, symplectic (see example of page 344 of [MaRa94]). Worse, a weak symplectic manifold is not a Poisson manifold since not every locally defined function defines a Hamiltonian vector field.

On a Banach Poisson manifold the rule $X_F := \Lambda^\sharp(\mathbf{d}F)$ defines a smooth section of $T^{**}P$ and hence is not, in general, a vector field on P. In analogy with the finite dimensional and the strong symplectic case, we need to require that X_F be a Hamiltonian vector field. In order to achieve this, we are forced to make the assumption that the Poisson bracket on P satisfies the condition $\Lambda^\sharp(T^*P) \subset TP \subset T^{**}P$. The study of such Banach Poisson manifolds was begun in [OdzRa03] with special emphasis on the Lie-Poisson case. See [OdzRa004] and [BelRa04] for further examples of such Banach Poisson manifolds.

While these manifolds are important in quantum mechanics, most infinite dimensional examples in classical continuum mechanics do not satisfy this hypothesis on $\Lambda^\sharp$; in fact, most of them have weak symplectic phase spaces. In this case, the beginning of a systematic theory of weak symplectic manifolds and the associated Hamiltonian dynamics can be found in [ChMa74]. For (weak) Poisson manifolds, not even a proposal of a theory is available today and the rigorous study of several examples coming from fluid dynamics, elasticity theory, and plasma physics should shed light on the general abstract case.

4.3 Examples of Poisson Brackets

1. **Symplectic Bracket.** As we mentioned before, a strong symplectic form Ω

on a manifold P gives the Poisson bracket

$$\{F, H\} := \Omega(X_F, X_H).$$

If C is a Casimir function on a connected strong symplectic manifold P, i.e., $\{C, F\} = 0$ for all $F \in \mathcal{F}(P)$, then $X_C = 0$. Strong nondegeneracy of Ω implies then that $\mathbf{d}C = 0$, which in turn shows that C is constant on the connected manifold P. Thus, on a connected strong symplectic manifold the Casimir functions are the constants, i.e., the center of $\mathcal{F}(P)$ is $\mathbb{R}$.

2. **Lie-Poisson Bracket.** Let $\mathfrak{g}$ be a Lie algebra with Lie bracket $[\,,]$ and $\mathfrak{g}^*$ its dual. Define the functional derivative of the smooth function $F : \mathfrak{g}^* \longrightarrow \mathbb{R}$ at $\mu \in \mathfrak{g}^*$ to be the unique element $\frac{\delta F}{\delta \mu} \in \mathfrak{g}$ given by

$$\mathbf{D}F(\mu) \cdot \delta\mu = \lim_{\epsilon \to 0} \frac{F(\mu + \epsilon\delta\mu) - F(\mu)}{\epsilon} = \left\langle \delta\mu, \frac{\delta F}{\delta \mu} \right\rangle$$

where $\langle\,,\rangle$ denotes the duality pairing between $\mathfrak{g}^*$ and $\mathfrak{g}$. Note that $\mathbf{D}F(\mu)$ is the usual Fréchet derivative , i.e., $\mathbf{D}F(\mu) \in L(\mathfrak{g}^*, \mathbb{R}) = \mathfrak{g}^{**}$. If $\mathfrak{g}$ is finite dimensional, then $\mathfrak{g}^{**} \cong \mathfrak{g}$ naturally and $\delta F/\delta\mu \in \mathfrak{g}$ is the element of $\mathfrak{g}$ representing the functional $\mathbf{D}F(\mu) \in \mathfrak{g}^{**}$ on $\mathfrak{g}^*$.

The Banach space $\mathfrak{g}^*$ is a Poisson manifold for each of the Lie-Poisson brackets $\{\,,\}_+$ and $\{\,,\}_-$ defined by

$$\{F, H\}_\pm (\mu) = \pm \left\langle \mu, \left[\frac{\delta F}{\delta\mu}, \frac{\delta H}{\delta\mu}\right]\right\rangle \tag{4.4}$$

for all $\mu \in \mathfrak{g}^*$ and $F, H \in \mathcal{F}(\mathfrak{g}^*)$. The bilinearity and skew-symmetry are obvious from the definition. The derivation property follows from the Leibniz rule for functional derivatives. For the direct proof of the Jacobi identity see [MaRa94], pg. 329.

In general, if $\mathfrak{g}$ is a Banach Lie algebra, $\mathfrak{g}^*$ is not a Banach Poisson manifold that satisfies the condition $\Lambda^\sharp(T^*\mathfrak{g}^*) \subset T\mathfrak{g}^*$ discussed at the end of §4.2. It was shown in [OdzRa03] that a Banach space $\mathfrak{b}$ is a Banach Lie-Poisson space $(\mathfrak{b}, \{\,,\})$ if and only if its dual $\mathfrak{b}^*$ is a Banach Lie algebra $(\mathfrak{b}^*, [\,,])$ satisfying

$$\mathrm{ad}_x^* \mathfrak{b} \subset \mathfrak{b} \subset \mathfrak{b}^{**} \quad \text{for all} \quad x \in \mathfrak{b}^*, \tag{4.5}$$

where $\mathrm{ad}_x : \mathfrak{b}^* \longrightarrow \mathfrak{b}^*$ is the adjoint representation $\mathrm{ad}_x y := [x, y]$ for any $x, y \in \mathfrak{b}^*$. Of course, the Poisson bracket of $F, H \in \mathcal{F}(\mathfrak{b})$ is given in this case by

$$\{F, H\}(b) = \langle [\mathbf{D}F(b), \mathbf{D}H(b)], b\rangle, \tag{4.6}$$

where $b \in \mathfrak{b}$ and $\mathbf{D}$ denotes the Fréchet derivative. If H is a smooth function on $\mathfrak{b}$, the associated Hamiltonian vector field is given by

$$X_H(b) = -\operatorname{ad}^*_{\mathbf{D}H(b)} b. \tag{4.7}$$

Note that if the condition $\operatorname{ad}^*_x \mathfrak{b} \subset \mathfrak{b}$ does not hold for every $x \in \mathfrak{b}^*$, then (4.7) does not make sense, in general. We shall encounter below such situations, in which case formulas (4.6) and (4.7) will be taken formally, or subject to the condition that only functions for which (4.7) makes sense will be used.

3. **Rigid body bracket.** As a particular case of the previous example, consider the Lie algebra $\mathfrak{g} = \mathfrak{so}(3)$ of the rotation group consisting of 3×3 skew-symmetric matrices. This Lie algebra is isomorphic to $(\mathbb{R}^3, \times)$, where $\times$ is the cross product of vectors, via the isomorphism (1.13), that is $(\mathbf{u} \times \mathbf{v})\hat{} = [\hat{\mathbf{u}}, \hat{\mathbf{v}}]$ for any $\mathbf{u}, \mathbf{v} \in \mathbb{R}^3$.

   We identify $\mathfrak{so}(3)^*$ with $\mathbb{R}^3$ using as pairing the Euclidean inner product. The Fréchet derivative and the functional derivative of a function defined on $\mathbb{R}^3$ coincide and are both equal to the usual gradient of the function. Thus the Lie-Poisson bracket (4.4) or (4.6) take the form

$$\{F, H\}_{\pm}(\mu) = \pm \mu \cdot (\nabla F \times \nabla H).$$

   Let us show that $C(\mu) = \Phi\left(\frac{1}{2}\|\mu\|^2\right)$ is a Casimir function, where $\Phi : \mathbb{R} \to \mathbb{R}$ is an arbitrary differentiable function. Indeed, since $\nabla C(\mu) = \Phi'\left(\frac{1}{2}\|\mu\|^2\right)\mu$, for any $F \in \mathcal{F}(\mathbb{R}^3)$ we have

$$\{C, F\}_{\pm}(\mu) = \pm \mu \cdot (\nabla C \times \nabla F) = \pm \Phi'\left(\frac{1}{2}\|\mu\|^2\right) \mu \cdot (\mu \times \nabla F) = 0.$$

4. **Frozen Lie-Poisson Bracket.** Let $\mathfrak{g}$ be a finite dimensional Lie algebra. For $\nu \in \mathfrak{g}^*$ define for any $F, H \in \mathcal{F}(\mathfrak{g}^*)$ the brackets

$$\{F, H\}^{\nu}_{\mathfrak{g}^*}(\mu) = \pm \left\langle \nu, \left[\frac{\delta F}{\delta \mu}, \frac{\delta H}{\delta \mu}\right] \right\rangle. \tag{4.8}$$

   A computation almost identical to the one needed to prove the Jacobi identity for the Lie-Poisson bracket shows that (4.8) also satisfies the Jacobi identity. This Poisson bracket on $\mathfrak{g}^*$ is called the ***frozen Lie-Poisson bracket*** at $\nu \in \mathfrak{g}^*$. A discussion similar to that at the end of the second example leads to a condition on (4.8) (like (4.5)) that makes it into a rigorous functional analytic Poisson bracket for any smooth functions F and H.

   It is worth noting that $\{\,,\} + s\{\,,\}^{\nu}$ is also a Poisson bracket on $\mathfrak{g}^*$ for any $\nu \in \mathfrak{g}^*$ and any $s \in \mathbb{R}$. One says that these two Poisson brackets are ***compatible***. The verification of this statement is direct, using the previously alluded proofs of the Jacobi identity for the Lie-Poisson and frozen Lie-Poisson brackets.

5. **Ideal Fluid Bracket in Velocity Representation.** In this infinite dimensional example we shall work formally or with the understanding that we restrict the class of functions to those admitting functional derivatives.

   Let $\mathfrak{X}_{\mathrm{div}}(D)$ be the Lie algebra of smooth divergence free vector fields tangent to the boundary ∂D defined on an oriented Riemannian manifold D with Riemannian metric g and Riemannian volume form μ. Consider the weakly nondegenerate pairing $\langle\, ,\rangle : \mathfrak{X}_{\mathrm{div}}(D) \times \mathfrak{X}_{\mathrm{div}}(D) \to \mathbb{R}$ given by the L^2 pairing

$$\langle u, v\rangle = \int_D g(u, v)\, \mu.$$

   Thus, we formally regard $\mathfrak{X}_{\mathrm{div}}(D)^*$ as being $\mathfrak{X}_{\mathrm{div}}(D)$ and apply the formulas from the general theory. The plus Lie-Poisson bracket is

$$\{F, H\}(v) = -\int_D g\left(v, \left[\frac{\delta F}{\delta v}, \frac{\delta H}{\delta v}\right]\right) \mu, \tag{4.9}$$

   where the functional derivative $\frac{\delta F}{\delta v}$ is the element of $\mathfrak{X}_{\mathrm{div}}(D)$ defined by

$$\lim_{\epsilon \to 0} \frac{F(v + \epsilon \delta v) - F(v)}{\epsilon} = \int_D g\left(\frac{\delta F}{\delta v}, v\right) \mu.$$

   The minus sign in front of the right hand side of (4.9) appears because the usual Lie bracket of vector fields is the *negative* of the left Lie algebra bracket of the diffeomorphism group, whose Lie algebra has underlying vector space $\mathfrak{X}(D)$; this will be shown explicitly in Section 5.1, Example 2.

6. **Ideal Fluid Bracket in Vorticity Representation.** We continue the previous example realizing the dual of $\mathfrak{X}_{\mathrm{div}}(D)$ in a different manner; this approach is due to [MaWei83]. The natural dual of $\mathfrak{X}(D)$ is the space of one-forms $\Omega^1(D)$ via the weak pairing $(u, \alpha) \in \mathfrak{X}(D) \times \Omega^1(D) \mapsto \int_D \alpha(u)\mu \in \mathbb{R}$. However, restricted to $\mathfrak{X}_{\mathrm{div}}(D)$ this pairing is degenerate. By the Hodge decomposition theorem, the kernel of the linear map sending $\alpha \in \Omega^1(D)$ to the element of $\mathfrak{X}(D)^*$ given by $u \in \mathfrak{X}(D) \mapsto \int_D \alpha(u)\mu \in \mathbb{R}$ is $\mathbf{d}\mathcal{F}(D)$ and thus we can identify, formally, $\mathfrak{X}(D)^*$ with $\Omega^1(D)/\mathbf{d}\mathcal{F}(D)$.

   Next, note that the linear map $[\alpha] \in \Omega^1(D)/\mathbf{d}\mathcal{F}(D) \mapsto \mathbf{d}\alpha \in \mathbf{d}\Omega^1(D)$ is well defined and has kernel the first de Rham cohomology group of D. Assuming that this first cohomology group is zero, the above map becomes an isomorphism. Thus, under this topological assumption on D, we can formally identify $\mathfrak{X}(D)^*$ with $\mathbf{d}\Omega^1(D)$. Summarizing, the weak pairing between $\mathfrak{X}_{\mathrm{div}}(D)$ and $\mathbf{d}\Omega^1(D)$ is given by

$$(u, \omega) \in \mathfrak{X}_{\mathrm{div}}(D) \times \mathbf{d}\Omega^1(D) \mapsto \int_D \alpha(u)\mu \in \mathbb{R}, \quad \text{for} \quad \omega := \mathbf{d}\alpha.$$

Therefore, the functional derivative $\delta F/\delta\omega \in \mathfrak{X}_{\mathrm{div}}(D)$ of $F : \mathfrak{X}(D)^* = \mathbf{d}\Omega^1(D) \to \mathbb{R}$ is defined by the identity

$$\mathbf{D}F(\omega) \cdot \delta\omega = \int_D \delta\alpha\left(\frac{\delta F}{\delta\omega}\right)\mu, \quad \text{for} \quad \delta\omega := \mathbf{d}(\delta\alpha).$$

Thus, the plus Lie-Poisson bracket of $F, H : \mathbf{d}\Omega^1(D) \to \mathbb{R}$ has the expression

$$\{F, H\}(\omega) = -\int_D \alpha\left(\left[\frac{\delta F}{\delta\omega}, \frac{\delta H}{\delta\omega}\right]\right)\mu, \quad \text{where} \quad \omega = \mathbf{d}\alpha.$$

However, since $\mathbf{d}\alpha(u, v) = u[\alpha(v)] - v[\alpha(u)] - \alpha([u, v])$, this formula becomes

$$\begin{aligned}\{F, H\}(\omega) = \int_D \omega\left(\frac{\delta F}{\delta\omega}, \frac{\delta H}{\delta\omega}\right)\mu - \int_D \frac{\delta F}{\delta\omega}\left[\alpha\left(\frac{\delta H}{\delta\omega}\right)\right]\mu \\ + \int_D \frac{\delta H}{\delta\omega}\left[\alpha\left(\frac{\delta F}{\delta\omega}\right)\right]\mu.\end{aligned}$$

The following argument shows that the last two terms vanish. For $u \in \mathfrak{X}_{\mathrm{div}}(D)$ and $f \in \mathcal{F}(D)$ we have by the Stokes theorem

$$\int_D u[f]\mu = \int_D \pounds_u(f\mu) = \int_D \mathbf{d}\mathbf{i}_u(f\mu) = \int_{\partial D} \mathbf{i}_u(f\mu) = \int_{\partial D} f\mathbf{i}_u\mu.$$

The definition of the boundary volume form $\mu_{\partial D}$ induced by the volume form μ implies that $\mathbf{i}_u\mu = g(u, n)\mu_{\partial D}$, where n is the outward unit normal to the boundary ∂D (see e.g. [AMR88] §7.2). Since u is tangent to the boundary, this term is zero. Thus the plus Lie-Poisson bracket has the expression

$$\{F, H\}(\omega) = \int_D \omega\left(\frac{\delta F}{\delta\omega}, \frac{\delta H}{\delta\omega}\right)\mu.$$

The term "vorticity representation" appears because of the following interpretation of the variable ω. Define the ***vorticity*** of a vector field u by $\omega_u := \mathbf{d}u^\flat$, where $u^\flat := g(u, \cdot) \in \Omega^1(D)$. Regard now $u \mapsto \omega_u$ as a change of variables to be implemented in the velocity representation of the Lie-Poisson bracket. To this end, if $F : \mathbf{d}\Omega^1(D) \to \mathbb{R}$, define $\widetilde{F} : \mathfrak{X}_{\mathrm{div}}(D) \to \mathbb{R}$

by $\widetilde{F}(u) := F(\omega_u)$. Then, if $\delta\omega := \mathbf{d}(\delta u^\flat)$, we get by the chain rule

$$\int_D g\left(\frac{\delta \widetilde{F}}{\delta u}, \delta u\right)\mu = \mathbf{D}\widetilde{F}(u)\cdot\delta u = \mathbf{D}F(\omega)\cdot\delta\omega$$
$$= \int_D \delta u^\flat \left(\frac{\delta F}{\delta\omega}\right)\mu = \int_D g\left(\delta u, \frac{\delta F}{\delta \omega}\right)\mu$$

and thus $\delta\widetilde{F}/\delta u = \delta F/\delta\omega$ which shows that the vorticity representation of the Lie-Poisson bracket is obtained from the velocity representation of the Lie-Poisson bracket by the linear change of variables $u \mapsto \omega_u$.

7. **Poisson-Vlasov Bracket.** For a Poisson manifold $(P, \{\,,\,\}_P)$ endowed with a volume form μ, the algebra $\mathcal{F}(P)$ is also a Lie algebra with Lie bracket the Poisson bracket. Consider the weak pairing between $\mathcal{F}(P)$ and the smooth densities $\mathcal{F}(P)^*$ on P given by

$$\langle \varphi, f\rangle = \int_P \varphi \bar{f}\mu, \quad \text{where} \quad \bar{f}\mu := f \in \mathcal{F}(P)^*, \quad \bar{f} \in \mathcal{F}(P).$$

The plus Lie-Poisson bracket in $\mathcal{F}(P)^*$ is thus given by

$$\{F, G\}(f) = \int_P f\left\{\frac{\delta F}{\delta f}, \frac{\delta G}{\delta f}\right\}_P.$$

This formal Poisson bracket is known as the Poisson-Vlasov bracket since the Poisson-Vlasov equations form a Hamiltonian system for it if $P = T^*\mathbb{R}^3$ and the Hamiltonian is given by

$$H(f) = \frac{1}{2}\iint \|v\|^2 f(x, v, t) d^3x\, d^3v + \frac{1}{2m}\int \|\nabla\phi_f\|^2 d^3x.$$

This system describes the motion of a collisionless plasma consisting of a single species of particles with mass m and charge e in the electrostatic limit (that is, one lets the speed of light $c \to \infty$). The physical significance of $f \geq 0$ is the plasma density that depends on the position $x \in \mathbb{R}^3$, the velocity $v \in \mathbb{R}^3$, and evolves in time $t \in \mathbb{R}$. The charge density is given by $\rho_f(x) := e\int f(x, v) d^3v$ and the electric potential $\phi_f(x)$ by the Poisson equation $-\Delta\phi_f = \rho_f$.

Let us carry out a formal computation to determine $\delta H/\delta f$ assuming that correct decay at infinity conditions are put on the relevant functions so that all integration by parts below are justified. We have

$$\begin{aligned}\left\langle \frac{\delta H}{\delta f}, \delta f \right\rangle &= \iint \|v\|^2 \delta f(x,v) d^3x\, d^3v + \frac{1}{m} \int \nabla \phi_f(x) \cdot \nabla \phi_{\delta f}(x) d^3x \\ &= \iint \|v\|^2 \delta f(x,v) d^3x\, d^3v - \frac{1}{m} \int (\Delta \phi_{\delta f}(x)) \phi_f(x) d^3x \\ &= \iint \|v\|^2 \delta f(x,v) d^3x\, d^3v + \frac{1}{m} \int \rho_{\delta f}(x) \phi_f(x) d^3x \\ &= \iint \|v\|^2 \delta f(x,v) d^3x\, d^3v + \frac{e}{m} \iint \delta f(x,v) \phi_f(x) d^3x\, d^3v,\end{aligned}$$

which shows that

$$\frac{\delta H}{\delta f} = \|v\|^2 + \frac{e}{m}\phi_f.$$

Thus Hamilton's equations $\dot{F} = \{F, H\}$ for an arbitrary functional F of f become $\dot{f} + \{f, \frac{\delta H}{\delta f}\} = 0$ (where the Poisson bracket is now the one on $T^*\mathbb{R}^3$). Replacing here the formula for $\delta H/\delta f$ just found yields the Poisson-Vlasov equations

$$\frac{\partial f}{\partial t} + v \cdot \nabla_x f - \frac{e}{m} \nabla_x \phi_f \cdot \nabla_v f = 0.$$

8. **Korteweg-de Vries Bracket.** Let $\mathcal{F}(\mathbb{R})$ be a space of smooth functions on $\mathbb{R}$ that satisfy together with their derivatives all necessary decay conditions at infinity guaranteeing that all integrals as well as the integrations by parts that will be carried out below make sense and the boundary terms appearing in these computations vanish.

   The KdV bracket on $\mathcal{F}(\mathbb{R})$ is given by

$$\{F, G\}(u) = \int_{-\infty}^{+\infty} \frac{\delta F}{\delta u} \frac{d}{dx} \left(\frac{\delta G}{\delta u} \right) dx,$$

   where the functional derivatives are taken relative to the L^2 product on real valued functions. Let us work out Hamilton's equations $\dot{F} = \{F, H\}$ for this bracket. We have

$$\begin{aligned}\int_{-\infty}^{+\infty} \frac{\delta F}{\delta u} \dot{u}\, dx = \mathbf{D}F(u(t)) \cdot \dot{u}(t) &= \frac{d}{dt} F(u(t)) = \{F, H\}(u(t)) \\ &= \int_{-\infty}^{+\infty} \frac{\delta F}{\delta u} \frac{d}{dx} \frac{\delta H}{\delta u} dx,\end{aligned}$$

   so Hamilton's equation are

$$u_t = \frac{d}{dx} \left(\frac{\delta H}{\delta u} \right) = \left(\frac{\delta H}{\delta u} \right)_x,$$

where $u_t := \frac{\partial u}{\partial t}$ and $u_x := \frac{\partial u}{\partial x}$.

In particular, if we take the Hamiltonian function

$$H_1 = -\frac{1}{6}\int_{-\infty}^{\infty} u^3 dx,$$

Hamilton's equations for the KdV bracket become the one dimensional transport equation $u_t + uu_x = 0$. Indeed, since

$$\int_{-\infty}^{+\infty} \frac{\delta H_1}{\delta u}\,\delta u\,dx = \mathbf{D}H_1(u)\cdot\delta u = -\frac{1}{2}\int_{-\infty}^{+\infty} u^2\delta u dx,$$

it follows that $\frac{\delta H_1}{\delta u} = -\frac{1}{2}\,u^2$ and so the equation of motion is

$$u_t = \frac{\partial}{\partial x}\left(-\frac{1}{2}\,u^2\right) = -u\,u_x.$$

If one takes the Hamiltonian equal to

$$H_2(u) = \int_{-\infty}^{\infty}\left(\frac{1}{2}u_x^2 - u^3\right)dx$$

then $\frac{\delta H_2}{\delta u} = -u_{xx} - 3u^2$ and Hamilton's equation is the Korteweg-de Vries (KdV) equation

$$u_t + 6u\,u_x + u_{xxx} = 0. \tag{4.10}$$

The KdV equation has infinitely many independent integrals F_i in involution (that is, $\{F_i\,F_j\} = 0$) and is a completely integrable system in infinite dimensions. Here are the first integrals:

$F_0(u) = \int_{-\infty}^{+\infty} u(x)\,dx$

$F_1(u) = \frac{1}{2}\int_{-\infty}^{+\infty} u^2(x)\,dx$

$F_2(u) = H_2(x) = \int_{-\infty}^{+\infty}\left(-u^3(x) + \frac{1}{2}\,u_x^2(x)\right)dx$

$F_3(u) = \int_{-\infty}^{+\infty}\left(\frac{1}{2}u_{xxx}^2(x) - 5u(x)u_x^2(x) + \frac{5}{2}\,u_x^4(x)\right)dx.$

The existence of such integrals is believed to be closely related to the presence of solitons, that is, "solitary waves which interact pairwise by passing through each other without changing shape" (see e.g. [AbMa78] and references therein).

Let us look for traveling wave solutions of the KdV equation (4.10), that is, solutions of the form $u(t,x) = \phi(x - ct)$, for $c > 0$ and $\phi \geq 0$. Substituting into (4.10) we get

$$c\phi' - 6\phi\phi' - \phi''' = 0.$$

Integrating once this equation gives

$$c\phi - 3\phi^2 - \phi'' = C, \tag{4.11}$$

where $C \in \mathbb{R}$ is a constant. This equation is equivalent to

$$\begin{cases} \frac{d\phi}{dx} = \phi' = \frac{\partial h}{\partial \phi'} \\ \frac{d\phi'}{dx} = \phi'' = c\phi - 3\phi^2 - C = -\frac{\partial h}{\partial \phi} \end{cases} \tag{4.12}$$

where

$$h(\phi, \phi') = \frac{1}{2}(\phi')^2 - \frac{c}{2}\phi^2 + \phi^3 + C\phi. \tag{4.13}$$

Thus (4.12) is Hamiltonian in the variables $(\phi, \phi') \in \mathbb{R}^2$ with Hamiltonian function (4.13). In particular, h can be viewed to be of the form kinetic energy $K(\phi, \phi') = \frac{1}{2}(\phi')^2$ plus potential energy $V(\phi) = \frac{-c}{2}\phi^2 + \phi^3 + C\phi$. Since energy is conserved, we have $h(\phi, \phi') = D$, for some constant $D \in \mathbb{R}$, which implies

$$\phi' = \pm\sqrt{c\phi^2 - 2\phi^3 + 2C\phi + 2D}.$$

Integrating we have

$$s = \pm \int \frac{d\phi}{\sqrt{c\phi^2 - 2\phi^3 + 2C\phi + 2D}}$$

where $s = x - ct$.

We seek solutions which together with their derivatives vanish at $\pm\infty$. Then $D = 0$ by (4.13) and $C = 0$ by (4.11). Thus we get

$$s = \pm \int \frac{d\phi}{\sqrt{c\phi^2 - \phi^3}} = \pm \frac{1}{\sqrt{c}} \log \left| \frac{\sqrt{c - 2\phi} - \sqrt{c}}{\sqrt{c - 2\phi} + \sqrt{c}} \right| + K. \tag{4.14}$$

The Hamiltonian system (4.12) has two equilibria when $C = D = 0$, namely $(\phi_e, \phi'_e) = (0, 0)$ and $(\phi_e, \phi'_e) = (c/3, 0)$. The matrix of the linearized system at $(\phi_e, \phi'_e) = (0, 0)$ and $(\phi_e, \phi'_e) = (c/3, 0)$ is

$$\begin{bmatrix} 0 & 1 \\ c & 0 \end{bmatrix} \quad \text{and} \quad \begin{bmatrix} 0 & 1 \\ -c & 0 \end{bmatrix},$$

respectively. So $(0, 0)$ is a saddle point while $(c/3, 0)$ is spectrally stable (the eigenvalues are on the imaginary axis). To see if $(c/3, 0)$ is stable or not we can use the second variation of the potential energy criterion. Since

$$\frac{\delta^2 V}{\delta \phi^2}(c/3, 0) = c > 0$$

it follows that $(c/3, 0)$ is a Lyapunov stable point.

Consider $(\phi(s), \phi'(s))$ to be a homoclinic orbit emanating and ending at $(0, 0)$ to which $(c/3, 0)$ belongs. Both equilibria belong to the zero level set of the energy and the homoclinic orbit is given by (4.14). Furthermore when $C = 0$ we have $h(c/2, 0) = 0$ that is $(c/2, 0)$ belongs also to this homoclinic orbit. Let us take $(c/2, 0)$ as initial condition $(\phi(0), \phi'(0))$. Then by (4.14) we get $K = 0$ and the homoclinic orbit is given

$$\pm\sqrt{c}s = \log\left|\frac{\sqrt{c - 2\phi(s)} - \sqrt{c}}{\sqrt{c - 2\phi(s)} + \sqrt{c}}\right|.$$

As $\phi > 0$ the value inside the modulus is negative and the homoclinic orbit has the expression

$$\begin{aligned} e^{\pm\sqrt{c}s} = -\frac{\sqrt{c-2\phi(s)}-\sqrt{c}}{\sqrt{c-2\phi(s)}+\sqrt{c}} \iff \phi(s) &= \frac{2ce^{\pm\sqrt{c}s}}{(1+e^{\pm\sqrt{c}s})^2} \\ &= \frac{c}{2}\operatorname{sech}^2(\sqrt{c}s/2), \end{aligned}$$

which gives the soliton: $u(x, t) = \phi(x - ct) = \frac{c}{2}\operatorname{sech}^2(\frac{\sqrt{c}}{2}(x - ct))$.

9. **Operator Algebra Brackets.** This example is taken from the papers [Bo00] and [OdzRa03]. Let $\mathcal{H}$ be a complex Hilbert space. Denote by $\mathfrak{S}(\mathcal{H})$, $\mathfrak{HS}(\mathcal{H})$, and $\mathfrak{B}(\mathcal{H})$ the involutive Banach algebras of the trace class operators, the Hilbert-Schmidt operators, and the bounded operators on $\mathcal{H}$, respectively. Recall that $\mathfrak{S}(\mathcal{H})$ and $\mathfrak{HS}(\mathcal{H})$ are self adjoint ideals in $\mathfrak{B}(\mathcal{H})$. Let $\mathfrak{K}(\mathcal{H}) \subset \mathfrak{B}(\mathcal{H})$ denote the ideal of all compact operators on $\mathcal{H}$. Then

$$\mathfrak{S}(\mathcal{H}) \subset \mathfrak{HS}(\mathcal{H}) \subset \mathfrak{K}(\mathcal{H}) \subset \mathfrak{B}(\mathcal{H})$$

and the following remarkable dualities hold:

$$\mathfrak{K}(\mathcal{H})^* \cong \mathfrak{S}(\mathcal{H}), \quad \mathfrak{HS}(\mathcal{H})^* \cong \mathfrak{HS}(\mathcal{H}), \quad \text{and} \quad \mathfrak{S}(\mathcal{H})^* \cong \mathfrak{B}(\mathcal{H});$$

the right hand sides are all Banach Lie algebras. These dualities are implemented by the strongly nondegenerate pairing

$$\langle x, \rho\rangle = \operatorname{trace}(x\rho)$$

where $x \in \mathfrak{S}(\mathcal{H})$, $\rho \in \mathfrak{K}(\mathcal{H})$ for the first isomorphism, $\rho, x \in \mathfrak{HS}(\mathcal{H})$

for the second isomorphism, and $x \in \mathfrak{B}(\mathcal{H})$, $\rho \in \mathfrak{S}(\mathcal{H})$ for the third isomorphism. Thus condition (4.5) holds and hence the Banach spaces $\mathfrak{S}(\mathcal{H})$, $\mathfrak{H}\mathfrak{S}(\mathcal{H})$, and $\mathfrak{K}(\mathcal{H})$ are Banach Lie-Poisson spaces in a rigorous functional analytic sense (see the discussion at the end of §4.2). The Lie-Poisson bracket (4.6) becomes in this case

$$\{F, H\}(\rho) = \pm \operatorname{trace}\left([\mathbf{D}F(\rho), \mathbf{D}H(\rho)]\rho\right) \tag{4.15}$$

where ρ is an element of $\mathfrak{S}(\mathcal{H})$, $\mathfrak{H}\mathfrak{S}(\mathcal{H})$, or $\mathfrak{K}(\mathcal{H})$, respectively. The bracket $[\mathbf{D}F(\rho), \mathbf{D}H(\rho)]$ denotes the commutator bracket of operators. The Hamiltonian vector field associated to H is given by

$$X_H(\rho) = \pm[\mathbf{D}H(\rho), \rho]. \tag{4.16}$$

4.4 Generalities on Lie-Poisson Structures

We shall collect here some of the most important properties of Lie-Poisson structures. Let $\mathfrak{g}^*$ be the dual of the finite dimensional Lie algebra $\mathfrak{g}$ of a Lie group G, $\operatorname{ad}_\xi \eta := [\xi, \eta]$ for $\xi, \eta \in \mathfrak{g}$, and $[\,,]$ the Lie bracket on $\mathfrak{g}$.

Proposition 4.9 *The equations of motion for the Hamiltonian $H : \mathfrak{g}^* \to \mathbb{R}$ with respect to the $(\pm)$ Lie-Poisson bracket on $\mathfrak{g}^*$ are*

$$\frac{d\mu}{dt} = X_H(\mu) = \mp \operatorname{ad}^*_{\delta H/\delta\mu}\mu. \tag{4.17}$$

Proof For an arbitrary function $F \in \mathcal{F}(\mathfrak{g}^*)$ and $\mu \in \mathfrak{g}^*$ we have:

$$\frac{dF}{dt} = \mathbf{D}F(\mu) \cdot \dot{\mu} = \left\langle \frac{\delta F}{\delta\mu}, \dot{\mu} \right\rangle. \tag{4.18}$$

On the other hand,

$$\begin{aligned} \{F, H\}_\pm(\mu) &= \pm\left\langle \mu, \left[\frac{\delta F}{\delta\mu}, \frac{\delta H}{\delta\mu}\right]\right\rangle = \pm\left\langle \mu, -\operatorname{ad}_{\delta H/\delta\mu}\frac{\delta F}{\delta\mu}\right\rangle \\ &= \mp\left\langle \operatorname{ad}^*_{\delta H/\delta\mu}\mu, \frac{\delta F}{\delta\mu}\right\rangle. \end{aligned}$$

This last equality and (4.18) gives the result. ∎

Let $\{\xi_a\}$, $a = 1, 2, \ldots, n$, be a basis of $\mathfrak{g}$ and $\{\xi^a\}$ its dual basis. The structure constants C^d_{ab} of $\mathfrak{g}$ are defined by $[\xi_a, \xi_b] = \sum_d C^d_{ab}\xi_d$.

For $\mu = \sum_a \mu_a \xi^a$ the Lie-Poisson brackets become

$$\{F, G\}_\pm = \pm\mu_d \frac{\delta F}{\delta\mu_a}\frac{\delta G}{\delta\mu_b} C^d_{ab},$$

where summation on repeated indices is understood. In particular,

$$\{\mu_a, \mu_b\}_{\pm} = \pm \mu_d C^d_{ab}.$$

So the equations of motion for H are

$$\dot{\mu}_a = \mp \mu_d C^d_{ab} \frac{\delta H}{\delta \mu_b}.$$

Next we study linear Poisson maps.

Proposition 4.10 *Let $\mathfrak{g}$ and $\mathfrak{h}$ be Lie algebras and $\alpha : \mathfrak{g} \to \mathfrak{h}$ a linear map. Then α is a homomorphism of Lie algebras if and only if its dual $\alpha^* : \mathfrak{h}^*_{\pm} \to \mathfrak{g}^*_{\pm}$ is a Poisson map.*

Proof By definition of the Lie-Poisson bracket on $\mathfrak{h}^*$ we have

$$\{F \circ \alpha^*, H \circ \alpha^*\}_{\pm}(\mu) = \pm \left\langle \mu, \left[\frac{\delta(F \circ \alpha^*)}{\delta \mu}, \frac{\delta(H \circ \alpha^*)}{\delta \mu} \right] \right\rangle \tag{4.19}$$

for any $F, H \in \mathcal{F}(\mathfrak{g}^*)$, $\mu \in \mathfrak{h}$.

By definition of the functional derivative, the chain rule, and the definition of the dual map one has

$$\begin{aligned} \left\langle \frac{\delta(F \circ \alpha^*)}{\delta \mu}, \delta\mu \right\rangle &= \mathbf{D}(F \circ \alpha^*) \cdot \delta\mu = \mathbf{D}F(\alpha^*(\mu)) \cdot \alpha^*(\delta\mu) \\ &= \left\langle \frac{\delta F}{\delta \nu}, \alpha^*(\delta\mu) \right\rangle = \left\langle \alpha\left(\frac{\delta F}{\delta \nu}\right), \delta\mu \right\rangle, \end{aligned}$$

where $\nu := \alpha^*(\mu) \in \mathfrak{g}^*$. Thus equation (4.19) becomes

$$\begin{aligned} \{F \circ \alpha^*, H \circ \alpha^*\}_{\pm}(\mu) &= \pm \left\langle \mu, \left[\frac{\delta(F \circ \alpha^*)}{\delta \mu}, \frac{\delta(H \circ \alpha^*)}{\delta \mu} \right] \right\rangle \\ &= \pm \left\langle \mu, \left[\alpha\left(\frac{\delta F}{\delta \nu}\right), \alpha\left(\frac{\delta H}{\delta \nu}\right) \right] \right\rangle. \end{aligned} \tag{4.20}$$

If α is a Lie algebra homomorphism then (4.20) is equal to

$$\pm \left\langle \mu, \alpha\left(\left[\frac{\delta F}{\delta \nu}, \frac{\delta H}{\delta \nu}\right]\right) \right\rangle = \pm \left\langle \alpha^*(\mu), \left[\frac{\delta F}{\delta \nu}, \frac{\delta H}{\delta \nu}\right] \right\rangle = \{F, H\}_{\pm}(\alpha^*(\mu)),$$

which shows that α^* is a Poisson map.

Conversely, if α^* is a Poisson map, by (4.20) we get

$$\begin{aligned} \pm \left\langle \mu, \left[\alpha\left(\frac{\delta F}{\delta \nu}\right), \alpha\left(\frac{\delta H}{\delta \nu}\right) \right] \right\rangle &= \{F, H\}_{\pm}(\alpha^*(\mu)) \\ &= \pm \left\langle \alpha^*(\mu), \left[\frac{\delta F}{\delta \nu}, \frac{\delta H}{\delta \nu}\right] \right\rangle = \pm \left\langle \mu, \alpha\left(\left[\frac{\delta F}{\delta \nu}, \frac{\delta H}{\delta \nu}\right]\right) \right\rangle \end{aligned}$$

for any $F, H \in \mathcal{F}(\mathfrak{g}^*)$. In particular, taking $F(\rho) := \langle \rho, \xi \rangle$ and $H(\rho) := \langle \rho, \eta \rangle$ for arbitrary $\xi, \eta \in \mathfrak{g}$, we get $\langle \mu, [\alpha(\xi), \alpha(\eta)] \rangle = \langle \mu, \alpha([\xi, \eta]) \rangle$ for any $\mu \in \mathfrak{h}$, which implies $[\alpha(\xi), \alpha(\eta)] = \alpha([\xi, \eta])$, that is, α is a Lie algebra homomorphism. ■

The last key property of Lie-Poisson brackets is that they are the only linear ones in the following sense. Consider a finite dimensional vector space V, V^* its dual, and let $\langle \, , \rangle$ be the duality pairing between V^* and V. One can then think of elements of V to be the linear functionals on V^*. A Poisson bracket in V^* is said to be ***linear*** if the bracket of two any linear functionals on V^* is a linear functional. Thus, if X', Y' are functionals on V^* then, as the pairing $\langle \, , \rangle$ is nondegenerate, there exist unique elements $X, Y \in V$ such that

$$X'(\mu) = \langle \mu, X \rangle \quad \text{and} \quad Y'(\mu) = \langle \mu, Y \rangle.$$

The linearity assumption on the Poisson bracket in V^* implies the existence of a unique element of V, say $[X, Y]$, such that

$$\{X' \, , \, Y'\} (\mu) = [X \, , \, Y]' (\mu) = \langle \mu \, , \, [X \, , \, Y] \rangle \, .$$

It is easy to prove that $[\, ,]$ so defined on V is a Lie algebra bracket and so the given linear Poisson bracket is the Lie-Poisson bracket for the Lie algebra V.

In the infinite dimensional case the same proof works if $\langle \, , \rangle : V^* \times V \to \mathbb{R}$ is a weak pairing between Banach spaces and one makes the extra hypothesis that the Poisson bracket of any two linear functionals on V^* belongs to the range of the Poisson tensor $\Lambda(\mu) : V \to V^*$ for all $\mu \in V^*$.

Proposition 4.11 *Let V and V^* be two Banach spaces, $\langle \, , \rangle$ a weak nondegenerate pairing of V^* with V, and assume that V^* has a linear Poisson bracket. If the Poisson bracket of any two linear functionals in V^* belongs to the range of $\langle \mu, \cdot \rangle$ for all $\mu \in V^*$, then V is a Banach Lie algebra and the Poisson bracket on V^* is the corresponding Lie-Poisson bracket.*

5 Momentum Maps

In this lecture we shall introduce the concept of momentum map and study its properties. We shall address the issue of existence and equivariance of momentum maps, give an explicit formula for the case of a cotangent bundle, and present several basic examples. The full power of the momentum map will appear only in the next chapter when dealing with reduction.

5.1 Actions and Infinitesimal Generators

Let $\Phi : G \times P \to P$ be a smooth left action of the Lie group G on the Poisson manifold P. The action is ***canonical*** if the map $\Phi_g : z \mapsto \Phi(g, z) = \Phi_g(z)$ is a Poisson map for all $g \in P$, that is

$$\{F \circ \Phi_g, G \circ \Phi_g\} = \{F, G\} \circ \Phi_g \quad \text{for all} \quad F, G \in \mathcal{F}(P). \tag{5.1}$$

Denote by $\mathfrak{g}$ the Lie algebra of G. Let us recall the following standard facts.

(i) $\mathrm{AD}_g : G \to G$ given for each $g \in G$ by

$$\mathrm{AD}_g(h) = ghg^{-1} = (L_g \circ R_{g^{-1}})(h) \tag{5.2}$$

is an inner automorphism. L_g and R_g denote, respectively, the left and right translations of G on itself.

(ii) Differentiating AD_g with respect to h at $h = e$ we get the ***adjoint representation*** $\mathrm{Ad}_g = T_{g^{-1}} L_g \circ T_e R_{g^{-1}} : \mathfrak{g} \to \mathfrak{g}$ of G on $\mathfrak{g}$. The inverse of the dual map defines the ***coadjoint representation*** $\mathrm{Ad}^*_{g^{-1}}$ of G on $\mathfrak{g}^*$.

(iii) Differentiating Ad_g with respect to g at e in the direction ξ yields

$$T_e \mathrm{Ad}(\xi) = \left.\frac{d}{dt}\right|_{t=0} \mathrm{Ad}_{\exp t\xi} = [\xi, \cdot] =: \mathrm{ad}_\xi : \mathfrak{g} \to \mathfrak{g}, \tag{5.3}$$

the ***adjoint representation*** of $\mathfrak{g}$ on $\mathfrak{g}$.

(iv) In these lectures, all Lie algebras are *left* Lie algebras. This means that $\mathfrak{g} = T_e G$ as a vector space and the Lie bracket on $\mathfrak{g}$ is given by the identity $[\xi, \eta] = [X_\xi, X_\eta](e)$, where $X_\xi, X_\eta \in \mathfrak{X}(G)$ are the left invariant vector fields whose values at the identity are ξ and η, respectively, that is, $X_\xi(g) = T_e L_g(\xi)$ and $X_\eta(g) = T_e L_g(\eta)$ for any $g \in G$. The vector fields X_ξ are complete, that is, each integral curve exists for all time.

(v) If $\xi \in \mathfrak{g}$, there is a unique integral curve $\gamma_\xi(t)$ of X_ξ with initial condition $\gamma_\xi(0) = e$. Then $\gamma_\xi(s + t) = \gamma_\xi(s)\gamma_\xi(t)$ for any $s, t \in \mathbb{R}$, that is, $\gamma_\xi : \mathbb{R} \to G$ is a ***one-parameter subgroup*** of G. Conversely, any Lie group homomorphism $\gamma : \mathbb{R} \to G$ is of the form γ_ξ, where $\xi = \gamma'(0)$. The ***exponential map*** is defined by $\exp(\xi) := \gamma_\xi(1)$. Then $\exp(t\xi) = \gamma_\xi(t)$ and the flow of X_ξ is given by $(t, g) \mapsto g \exp(t\xi)$.

(vi) If $F : G \to H$ is a homomorphism of Lie groups then $T_e F : \mathfrak{g} \to \mathfrak{h}$ is a Lie algebra homomorphism, that is,

$$T_e F\left([\xi\, , \eta]\right) = \left[T_e F(\xi)\, , T_e F(\eta)\right]. \tag{5.4}$$

In particular, $\mathrm{Ad}_g : \mathfrak{g} \to \mathfrak{g}$ is a Lie algebra isomorphism for every $g \in G$. In addition,

$$\exp_H \circ T_e F = F \circ \exp_G, \tag{5.5}$$

where $\exp_G : \mathfrak{g} \to G$ and $\exp_H : \mathfrak{h} \to H$ are the exponential maps. If one takes $G = H$ and $F = \mathrm{AD}_g$, this identity becomes

$$\exp(\mathrm{Ad}_g \xi) = \mathrm{AD}_g(\exp \xi) = g \exp(\xi) g^{-1}. \tag{5.6}$$

If one takes $H = \mathrm{Iso}(\mathfrak{g})$, the Lie group of Lie algebra isomorphisms of $\mathfrak{g}$, then its Lie algebra is the Lie algebra $\mathrm{Der}(\mathfrak{g})$ of derivations relative to the bracket $[\cdot, \cdot]$. Choosing in (5.5) $F = \mathrm{Ad} : G \to H = \mathrm{Iso}(\mathfrak{g})$, we get for any $\xi \in \mathfrak{g}$

$$e^{\mathrm{ad}_\xi} = \mathrm{Ad}_{\exp \xi}\,. \tag{5.7}$$

(vii) Given a Lie algebra element $\xi \in \mathfrak{g}$, $\exp(t\xi)$ defines a one-parameter subgroup in G and hence $t \mapsto \Phi_{\exp t\xi}$ is a flow on the manifold P. The vector field defined by this flow is denoted by ξ_P and is called the ***infinitesimal generator*** of the action determined by ξ. Thus we have

$$\xi_P(z) = \left.\frac{d}{dt}\right|_{t=0} \Phi_{\exp(t\xi)}(z).$$

The infinitesimal generator has the following properties

$$(\mathrm{Ad}_g \xi)_P = \Phi^*_{g^{-1}} \xi_P \quad \text{and} \quad [\xi_P\,, \eta_P] = -\,[\xi\,, \eta]_P \tag{5.8}$$

for any $g \in G$ and $\xi, \eta \in \mathfrak{g}$.

Examples 1. We deduce the formulas stated in §1.2. Let $G = SO(3)$ acting on $\mathbb{R}^3$ by matrix multiplication. The Lie algebra $\mathfrak{so}(3)$ is the set of 3×3 skew symmetric matrices with Lie bracket the commutator. It is isomorphic to $(\mathbb{R}^3\,, \times)$ via the map $\mathbf{u} \in \mathbb{R} \mapsto \hat{\mathbf{u}} \in \mathfrak{so}(3)$ given by (1.13). The adjoint action is hence

$$\mathrm{Ad}_A \hat{\mathbf{u}} = A \hat{\mathbf{u}} A^{-1} = (A\mathbf{u})\hat{}\,.$$

Therefore,

$$[\hat{\mathbf{u}}, \hat{\mathbf{v}}] = \mathrm{ad}_{\hat{\mathbf{u}}}\, \hat{\mathbf{v}} = \left.\frac{d}{dt}\right|_{t=0} \left(e^{t\hat{\mathbf{u}}}\mathbf{v}\right)\hat{} = \left(\hat{\mathbf{u}}\mathbf{v}\right)\hat{} = \left(\mathbf{u} \times \mathbf{v}\right)\hat{}\,.$$

The dual $\mathfrak{so}(3)^*$ is identified with $\mathbb{R}^3$ by the isomorphism $\mathbf{\Pi} \in \mathbb{R}^3 \mapsto \tilde{\mathbf{\Pi}} \in \mathfrak{so}(3)^*$ given by $\tilde{\mathbf{\Pi}}(\hat{\mathbf{u}}) := \mathbf{\Pi} \cdot \mathbf{u}$ for any $\mathbf{u} \in \mathbb{R}^3$. Then the coadjoint action of

$SO(3)$ on $\mathfrak{so}(3)^*$ is given by

$$\left(\mathrm{Ad}^*_{A^{-1}} \tilde{\mathbf{\Pi}}\right)(\hat{\mathbf{u}}) = \tilde{\mathbf{\Pi}} \cdot \mathrm{Ad}_{A^{-1}} \hat{\mathbf{u}} = \tilde{\mathbf{\Pi}} \cdot (A^{-1}\mathbf{u})\hat{} = \mathbf{\Pi} \cdot A^T \mathbf{u}$$
$$= A\mathbf{\Pi} \cdot \mathbf{u} = (A\mathbf{\Pi})\tilde{}\,(\hat{\mathbf{u}}),$$

that is, $\mathrm{Ad}^*_{A^{-1}} \tilde{\mathbf{\Pi}} = (A\mathbf{\Pi})\tilde{}$, thereby recovering formula (1.20) in §1.2.

The infinitesimal generator corresponding to $\mathbf{u} \in \mathbb{R}^3$ has the expression

$$\mathbf{u}_{\mathbb{R}^3}(\mathbf{x}) = \left.\frac{d}{dt}\right|_{t=0} e^{t\hat{\mathbf{u}}}\mathbf{x} = \hat{\mathbf{u}}\,\mathbf{x} = \mathbf{u} \times \mathbf{x}. \tag{5.9}$$

2. Let $G = \mathrm{Diff}_{\mathrm{vol}}(D)$ be the group of volume preserving diffeomorphisms of the oriented Riemannian manifold (D, g). On D there is a unique volume form μ which equals 1 on all positively oriented g-orthonormal bases of tangent vectors at all points of D; this volume form μ is called the ***Riemannian volume*** of (D, g) and we shall assume from now on that the orientation of D is given by μ.

Let us show formally that the Lie algebra of $\mathrm{Diff}_{\mathrm{vol}}(D)$ is $\mathfrak{X}_{\mathrm{div}}(D)$ endowed with the negative of the bracket of vector fields. First, as a vector space, the Lie algebra of the group $\mathrm{Diff}(D)$ equals the space $\mathfrak{X}(D)$ of vector fields on D. Indeed, the flow of an arbitrary vector field is a smooth path in $\mathrm{Diff}(D)$ whose tangent vector at time equal to zero is the given vector field.

Second, if $\eta_t \in \mathrm{Diff}_{\mathrm{vol}}(D)$ is the flow of the vector field v, then $\eta_t^*\mu = \mu$, so taking the derivative of this identity at $t = 0$ yields $(\mathrm{div}\, v)\mu = \boldsymbol{\pounds}_v \mu = 0$, that is, $v \in \mathfrak{X}_{\mathrm{div}}(D)$.

Third, since $\mathrm{AD}_\eta\, \varphi = \eta \circ \varphi \circ \eta^{-1}$, letting φ_t be the flow of v, we get

$$\mathrm{Ad}_\eta\, v = T_e\, \mathrm{AD}_\eta(v) = \left.\frac{d}{dt}\right|_{t=0} \mathrm{AD}_\eta\, \varphi_t = \left.\frac{d}{dt}\right|_{t=0} (\eta \circ \varphi_t \circ \eta^{-1})$$
$$= T\eta \circ v \circ \eta^{-1} = \eta_* v.$$

Fourth, if $u, v \in \mathrm{Diff}_{\mathrm{vol}}(D)$ and φ_t is the flow of u, the Lie algebra bracket of u and v in $\mathfrak{X}_{\mathrm{div}}(D)$ is given by

$$\left.\frac{d}{dt}\right|_{t=0} \mathrm{Ad}_{\varphi_t}\, v = \left.\frac{d}{dt}\right|_{t=0} (\varphi_t)_* v = \left.\frac{d}{dt}\right|_{t=0} (\varphi_{-t})^* v = -\boldsymbol{\pounds}_u v = -[u, v].$$

Thus *the left Lie algebra bracket on the space of vector fields equals the negative of the usual Jacobi-Lie bracket of vector fields.*

Identify the dual $\mathfrak{X}_{\mathrm{div}}(D)^*$ with $\mathbf{d}\Omega^1(D)$ (assuming that the first cohomology group of D is zero). The coadjoint action of $\mathrm{Diff}_{\mathrm{vol}}(D)$ on $\mathbf{d}\Omega^1(D)$ is computed in the following way. Let $\omega = \mathbf{d}\alpha \in \mathbf{d}\Omega^1(D)$ and $u \in \mathfrak{X}_{\mathrm{div}}(D)$.

Then

$$\left\langle u, \operatorname{Ad}^*_{\eta^{-1}} \omega \right\rangle = \left\langle \operatorname{Ad}_{\eta^{-1}} u, \omega \right\rangle = \int_D \alpha(\eta^* u)\mu = \int_D (\eta_* \alpha)(u)\mu = \langle u, \eta_* \omega \rangle$$

by the change of variables formula, taking into account that the Jacobian of η is one, and noting that $\mathbf{d}\eta_* \alpha = \eta_* \omega$. Therefore $\operatorname{Ad}^*_{\eta^{-1}} \omega = \eta_* \omega$.

5.2 Momentum Maps

Let the Lie group G with Lie algebra $\mathfrak{g}$ act on the Poisson manifold P in a canonical way, that is, (5.1) holds. Differentiating (5.1) with respect to g at the identity in the direction of $\xi \in \mathfrak{g}$ shows that the infinitesimal generator ξ_P is an infinitesimal Poisson automorphism, i.e.,

$$\xi_P \left[\{F_1 \, , \, F_2\} \right] = \{\xi_P[F_1] \, , \, F_2\} + \{F_1 \, , \, \xi_P[F_2]\}$$

for any $F_1, F_2 \in \mathcal{F}(P)$. Denote by $\mathcal{P}(P)$ the set of all vector fields satisfying this relation and call this Lie subalgebra of $\mathfrak{X}(P)$ the Lie algebra of ***Poisson bracket derivations*** or of ***infinitesimal Poisson automorphisms***.

If $\xi \in \mathfrak{g}$ we ask if the infinitesimal generator ξ_P is globally Hamiltonian. That is, we seek a Hamiltonian function $J^\xi \in \mathcal{F}(P)$ such that $X_{J^\xi} = \xi_P$ for every $\xi \in \mathfrak{g}$. Since the right hand side of this equation is linear in ξ, we shall require that the map $\xi \in \mathfrak{g} \mapsto J^\xi \in \mathcal{F}(P)$ be also linear.

Definition 5.1 Let G be a Lie group acting canonically on the Poisson manifold P. Suppose that there is a linear map $J : \mathfrak{g} \longrightarrow \mathcal{F}(P)$ such that

$$X_{J^\xi} = \xi_P, \tag{5.10}$$

for all $\xi \in \mathfrak{g}$, where ξ_P is the infinitesimal generator corresponding to ξ for then G-action on P. Then the map $\mathbf{J} : P \longrightarrow \mathfrak{g}^*$ defined by

$$\langle \mathbf{J}(z), \xi \rangle = J^\xi(z),$$

for all $\xi \in \mathfrak{g}$ and $z \in P$, is called a ***momentum map*** of the G-action.

One of the first questions that arise is whether or not equation (5.10) determines $\mathbf{J}$. Note that if $\mathbf{J}_1$ and $\mathbf{J}_2$ are functions verifying (5.10) then $X_{J_1^\xi - J_2^\xi} = 0$, which is equivalent with the statement that $J_1^\xi - J_2^\xi$ is a Casimir function. If P is symplectic and connected then the Casimirs are the constants and so equation (5.10) determines $\mathbf{J}$ only up to an element of $\mathfrak{g}^*$.

From the definition of $\mathbf{J}$ it follows that there is an isomorphism between the set of maps $P \longrightarrow \mathfrak{g}^*$ and the set of maps $\mathfrak{g} \longrightarrow \mathcal{F}(P)$. The collection of functions J^ξ as ξ varies on $\mathfrak{g}$ are the ***components of the momentum map***.

To give a momentum map is therefore equivalent to specifying a linear map $J : \mathfrak{g} \to \mathcal{F}(P)$ making the following diagram

$$\begin{array}{ccc} \mathcal{F}(P) & \xrightarrow{F \mapsto X_F} & \mathcal{P}(P) \\ & \nwarrow^{J} & \uparrow{\scriptstyle \xi \mapsto \xi_P} \\ & & \mathfrak{g} \end{array}$$

commutative. Two natural questions arise:

(A) What are the obstructions to the existence of a momentum map?

(B) If the G-action admits a momentum map, under what conditions is it a Lie algebra homomorphism?

Let us give some answers to these questions.

(A) The map $\mathcal{H} : \mathcal{F}(P) \to \mathcal{P}(P)$ given by $F \mapsto X_F$ is a Lie algebra anti-homomorphism. Denote by $\mathcal{H}(P)$ the Lie algebra of globally Hamiltonian vector fields. The existence of a momentum map is equivalent to be able to lift the anti-homomorphism of Lie algebras $\rho : \xi \in \mathfrak{g} \mapsto \xi_P \in \mathcal{P}(P)$ through $\mathcal{H}$ to a linear map $J : \mathfrak{g} \to \mathcal{F}(P)$. So consider the following diagram where i is the inclusion and π the projection

$$\begin{array}{ccccccccc} 0 & \longrightarrow & \mathcal{C}(P) & \xrightarrow{i} & \mathcal{F}(P) & \xrightarrow{\mathcal{H}} & \mathcal{P}(P) & \xrightarrow{\pi} & \mathcal{P}(P)/\mathcal{H}(P) \longrightarrow 0 \\ & & & & & \nwarrow^{J} & \uparrow{\scriptstyle \rho} & & \\ & & & & & & \mathfrak{g} & & \end{array}$$

If the linear map $J : \mathfrak{g} \to \mathcal{F}(P)$ is such that $\mathcal{H} \circ J = \rho$, then $\pi \circ \rho = \pi \circ \mathcal{H} \circ J = 0$ by the exactness of the sequence. Conversely, if $\pi \circ \rho = 0$, then $\rho(\mathfrak{g}) \subset \mathcal{H}(P)$, that is, each ξ_P is globally Hamiltonian, so there exists a function $J^\xi \in \mathcal{F}(P)$ such that $\xi_P = X_{J^\xi}$. Requiring that $\xi \mapsto J^\xi$ be linear, yields the existence of a momentum map. So under what conditions do we have that $\pi \circ \rho = 0$?

(i) If P is symplectic then $\mathcal{P}(P)$ coincides with the Lie algebra of locally Hamiltonian vector fields and $\mathcal{P}(P)/\mathcal{H}(P)$ is isomorphic to the first cohomology group $H^1(P)$, which is an Abelian Lie algebra. Thus, in the symplectic case, $\pi \circ \rho = 0$ if and only if the induced map $\mathfrak{g}/[\mathfrak{g}, \mathfrak{g}] \to H^1(P)$ vanishes. This happens, for instance, if $\mathfrak{g}$ is a semisimple Lie algebra because in that case, $[\mathfrak{g}, \mathfrak{g}] = \mathfrak{g}$.

(ii) If $\mathcal{P}(P)/\mathcal{H}(P) = 0$ then clearly $\pi \circ \rho = 0$. If P is symplectic this is equivalent to the vanishing of the first cohomology group $H^1(P)$.

(iii) If P is exact symplectic, i.e., the symplectic form is $\Omega = -\mathbf{d}\Theta$, and Θ is a $\mathfrak{g}$-invariant one-form, which means that $\boldsymbol{\pounds}_{\xi_P}\Theta = 0$ for all $\xi \in \mathfrak{g}$. Indeed, $\mathbf{d}\mathbf{i}_{\xi_P}\Theta + \mathbf{i}_{\xi_P}\mathbf{d}\Theta = 0$, implies that $\mathbf{i}_{\xi_P}\Omega = \mathbf{d}\left(\mathbf{i}_{\xi_P}\Theta\right)$, that is, the momentum map is given by $J^\xi = \mathbf{i}_{\xi_P}\Theta$.

(iv) An important special case of the previous situation is $P = T^*Q$ and the G-action on P is lifted from a G-action on Q, that is,

$$g \cdot \alpha_q = T^*_{g\cdot q}\Phi_{g^{-1}}\left(\alpha_q\right),$$

for $\alpha_q \in T^*_qQ$, $g \in G$, and $\Phi : G \times Q \to Q$ an action. By theorem 2.14 a cotangent lift preserves the canonical one-form Θ on T^*Q. Therefore, by the previous case, this action admits a momentum map which is given by $\langle \mathbf{J}, \xi \rangle = \mathbf{i}_{\xi_P}\Theta$. This expression can be further simplified using (2.8) and the equivariance of the projection $\pi : T^*Q \to Q$, that is, $\pi \circ T^*\Phi_{g^{-1}} = \Phi_g \circ \pi$ for all $g \in G$. The derivative of this relation relative to g at the identity in the direction $\xi \in \mathfrak{g}$ yields $T\pi \circ \xi_P = \xi_Q \circ \pi$. Therefore

$$\begin{aligned} \langle \mathbf{J}(\alpha_q), \xi \rangle &= \mathbf{i}_{\xi_P}\Theta(\alpha_q) = \langle \Theta(\alpha_q), \xi_P(\alpha_q) \rangle = \langle \alpha_q, (T\pi \circ \xi_P)(\alpha_q) \rangle \\ &= \langle \alpha_q, (\xi_Q \circ \pi)(\alpha_q) \rangle = \langle \alpha_q, \xi_Q(q) \rangle \,. \end{aligned} \tag{5.11}$$

(B) To say that $J : \mathfrak{g} \to \mathcal{F}(P)$ is a Lie algebra homomorphism is equivalent to the identity

$$J^{[\xi\,,\,\eta]} = \left\{ J^\xi, J^\eta \right\} \tag{5.12}$$

for all $\xi, \eta \in \mathfrak{g}$. How far are we from such a relation? To see this, note that

$$\begin{aligned} X_{J^{[\xi,\eta]}} &= [\xi, \eta]_P && \text{(by definition of } \mathbf{J}) \\ &= -\ [\xi_P, \eta_P] && \text{(by (5.8))} \\ &= -\ [X_{J^\xi}, X_{J^\eta}] && \text{(by definition of } \mathbf{J}) \\ &= X_{\{J^\xi, J^\eta\}} && (\mathcal{H} \text{ is an antihomorphism }). \end{aligned} \tag{5.13}$$

Equation (5.13) shows that $J^{[\xi,\eta]} - \left\{ J^\xi, J^\eta \right\}$ is a Casimir function, which we shall denote by $\Sigma(\xi, \eta)$. Thus $J : \mathfrak{g} \to \mathcal{F}(P)$ is a Lie algebra homomorphism if and only if $\Sigma(\xi, \eta) = 0$ for all $\xi, \eta \in \mathfrak{g}$.

The map $\Sigma : \mathfrak{g} \times \mathfrak{g} \to \mathcal{C}(P)$ has remarkable properties, easily deduced from the definition: it is bilinear, antisymmetric, and satisfies the ***cocycle identity***,

$$\Sigma(\xi\,, [\eta\,,\,\zeta]) + \Sigma(\eta\,, [\zeta\,,\,\xi]) + \Sigma(\zeta\,, [\xi\,,\,\eta]) = 0$$

for all $\xi, \eta, \zeta \in \mathfrak{g}$, that is, Σ is a *$\mathcal{C}(P)$-valued 2-cocycle* of $\mathfrak{g}$. So J is a Lie algebra homomorphism if and only if $[\Sigma] = 0$ in $H^2(\mathfrak{g}, \mathcal{C}(P))$, the second $\mathcal{C}(P)$-valued Lie algebra cohomology group of $\mathfrak{g}$.

When $\mathbf{J}$ verifies (5.12) we say that it is ***infinitesimally equivariant***. This terminology is justified in the following way. The momentum map $\mathbf{J} : P \to \mathfrak{g}^*$ is said to be ***equivariant***, if

$$\mathrm{Ad}^*_{g^{-1}} \circ \mathbf{J} = \mathbf{J} \circ \Phi_g$$

for all $g \in G$. Pairing this relation with $\eta \in \mathfrak{g}$, putting $g = \exp t\xi$, and taking the derivative of the resulting relation at $t = 0$, yields (5.12). Thus equivariance implies infinitesimal equivariance. The converse is also true if G is connected (see [MaRa94], Theorem 12.3.2).

Here are two classes of equivariant momentum maps that appear often in applications.

(i) The momentum map in point **(A)**(iii) is equivariant. Thus momentum maps of cotangent lifted actions (see (5.11)) are always equivariant. To see this, use (5.8) and G-invariance of Θ to get

$$\begin{aligned}\langle \mathbf{J}(g\cdot z), \xi\rangle &= \mathbf{i}_{\xi_P}\Theta(g\cdot z) = \Theta(g\cdot z)\left(\xi_P(g\cdot z)\right)\\ &= \Theta(g\cdot z)\left(T_z\Phi_g(\Phi_g^*\xi_P)(z)\right)\\ &= (\Phi_g^*\Theta)(z)\left((\mathrm{Ad}_{g^{-1}}\xi)_P(z)\right)\\ &= \Theta(z)\left((\mathrm{Ad}_{g^{-1}}\xi)_P(z)\right) = \langle \mathbf{J}(z), \mathrm{Ad}_{g^{-1}}\xi\rangle\\ &= \langle \mathrm{Ad}^*_{g^{-1}}\mathbf{J}(z), \xi\rangle,\end{aligned}$$

which shows that $\mathbf{J}(g\cdot z) = \mathrm{Ad}^*_{g^{-1}}\mathbf{J}(z)$.

(ii) For compact groups one can always choose the momentum map to be equivariant. More precisely, if the canonical G-action on the Poisson manifold P admits a momentum map $\mathbf{J} : P \to \mathfrak{g}^*$ and G is a compact Lie group, then $\mathbf{J}$ can be changed by the addition of an element in $L(\mathfrak{g}, \mathcal{C}(P))$ such that the resulting map is an equivariant momentum map for the same action. In particular, if P is symplectic, $\mathbf{J}$ can be changed by the addition of an element of $\mathfrak{g}^*$ on each connected component of P so that the resulting map is an equivariant momentum map.

To prove this statement, define for each $g \in G$

$$\mathbf{J}^g(z) := \mathrm{Ad}^*_{g^{-1}}\mathbf{J}(g^{-1}\cdot z)$$

or equivalently,

$$(J^g)^\xi := J^{\mathrm{Ad}_{g^{-1}}\xi} \circ \Phi_{g^{-1}}.$$

Then $\mathbf{J}^g$ is also a momentum map for the G-action on P. Indeed, if

$z \in P$, $\xi \in \mathfrak{g}$, and $F : P \to \mathbb{R}$, we have by (5.8)

$$\begin{aligned}
\{F, (J^g)^\xi\}(z) &= -\mathbf{d}(J^g)^\xi(z) \cdot X_F(z) \\
&= -\mathbf{d}J^{\mathrm{Ad}_{g^{-1}}\xi}(g^{-1} \cdot z) \cdot T_z\Phi_{g^{-1}} \cdot X_F(z) \\
&= -\mathbf{d}J^{\mathrm{Ad}_{g^{-1}}\xi}(g^{-1} \cdot z) \cdot (\Phi_g^* X_F)(g^{-1} \cdot z) \\
&= -\mathbf{d}J^{\mathrm{Ad}_{g^{-1}}\xi}(g^{-1} \cdot z) \cdot X_{\Phi_g^* F}(g^{-1} \cdot z) \\
&= \{\Phi_g^* F, J^{\mathrm{Ad}_{g^{-1}}\xi}\}(g^{-1} \cdot z) \\
&= X_{J^{\mathrm{Ad}_{g^{-1}}\xi}}[\Phi_g^* F](g^{-1} \cdot z) \\
&= (\mathrm{Ad}_{g^{-1}}\xi)_P[\Phi_g^* F](g^{-1} \cdot z) \\
&= (\Phi_g^* \xi_P)[\Phi_g^* F](g^{-1} \cdot z) \\
&= \mathbf{d}F(z) \cdot \xi_P(z) \\
&= \{F, J^\xi\}(z).
\end{aligned}$$

Therefore, $\{F, (J^g)^\xi - J^\xi\} = 0$ for every $F : P \to \mathbb{R}$, that is, $(J^g)^\xi - J^\xi$ is a Casimir function on P for every $g \in G$ and every $\xi \in \mathfrak{g}$. Therefore, since $\mathbf{J}$ is a momentum map, so is $\mathbf{J}^g$ for every $g \in G$. Now define

$$\langle \mathbf{J} \rangle := \int_G \mathbf{J}^g dg$$

where dg denotes the normalized Haar measure on G, that is, the volume of G is one. Equivalently, this definition states that

$$\langle J \rangle^\xi := \int_G (J^g)^\xi dg$$

for every $\xi \in \mathfrak{g}$. By linearity of the Poisson bracket in each factor, it follows that

$$\{F, \langle J \rangle^\xi\} = \int_G \{F, (J^g)^\xi\} dg = \int_G \{F, J^\xi\} dg = \{F, J^\xi\}$$

for every $F \in \mathcal{F}(P)$. Thus $\langle J \rangle^\xi - J^\xi$ is a Casimir on P for every $\xi \in \mathfrak{g}$ which shows that $\langle \mathbf{J} \rangle - \mathbf{J} \in L(\mathfrak{g}, \mathcal{C}(P)$ and that $\langle \mathbf{J} \rangle : P \to \mathfrak{g}^*$ is also a momentum map for the G-action.

Finally we show that the momentum map $\langle \mathbf{J} \rangle$ is equivariant. Indeed, begin by noting that

$$\mathbf{J}^g(h \cdot z) = \mathrm{Ad}_{h^{-1}}^* \mathbf{J}^{h^{-1}g}(z)$$

for every $g, h \in G$. Using invariance of the Haar measure on G under translations and inversion, we have for any $h \in G$

$$\begin{aligned} \langle \mathbf{J} \rangle (h \cdot z) &= \int_G \mathrm{Ad}^*_{h^{-1}} \mathbf{J}^{h^{-1}g}(z) dg = \mathrm{Ad}^*_{h^{-1}} \int_G \mathbf{J}^{h^{-1}g}(z) dg \\ &= \mathrm{Ad}^*_{h^{-1}} \int_G \mathbf{J}^k(z) dk = \mathrm{Ad}^*_{h^{-1}} \langle \mathbf{J} \rangle (z), \end{aligned}$$

where in the third equality we made the change of variables $g = hk$.

A crucial property of infinitesimally equivariant momentum maps is given in the following statement.

Theorem 5.2 *If* $\mathbf{J}$ *is an infinitesimally equivariant momentum map for the canonical* G*-action on the Poisson manifold* P *then* $\mathbf{J}$ *is a Poisson map, that is,*

$$\mathbf{J}^* \{F_1 \, , \, F_2\}_+ = \{\mathbf{J}^* F_1 \, , \, \mathbf{J}^* F_2\}$$

for all $F_1, F_2 \in \mathcal{F}(\mathfrak{g}^*)$, *where* $\{\, , \}_+$ *denotes the* $+$ *Lie-Poisson bracket on* $\mathfrak{g}^*$.

Proof For $F_1, F_2 : \mathfrak{g}^* \to \mathbb{R}$, $z \in P$, and $\mu = \mathbf{J}(z) \in \mathfrak{g}^*$, let $\xi := \frac{\delta F_1}{\delta \mu}$ and $\eta := \frac{\delta F_2}{\delta \mu}$. Then

$$\begin{aligned} \mathbf{J}^* \{F_1, F_2\}_+ (z) &= \{F_1, F_2\}_+ (\mathbf{J}(z)) = \left\langle \mu \, , \left[\frac{\delta F_1}{\delta \mu} \, , \, \frac{\delta F_2}{\delta \mu} \right] \right\rangle \\ &= \langle \mathbf{J}(z) \, , \, [\xi, \eta] \rangle = J^{[\xi, \eta]}(z) \\ &= \left\{ J^\xi, J^\eta \right\}, \end{aligned}$$

where the last equality follows by infinitesimal equivariance.

But, for $z \in P$ and $v_z \in T_z P$, we have

$$\begin{aligned} \mathbf{d} \left(F_1 \circ \mathbf{J} \right) (z)(v_z) &= \mathbf{d} F_1(\mu) \left(T_z \mathbf{J}(v_z) \right) = \left\langle T_z \mathbf{J}(v_z) \, , \, \frac{\delta F_1}{\delta \mu} \right\rangle \\ &= \langle T_z \mathbf{J}(v_z) \, , \xi \rangle = \mathbf{d} J^\xi(z)(v_z). \end{aligned}$$

Thus $\mathbf{d} \left(F_1 \circ \mathbf{J} \right) (z) = \mathbf{d} J^\xi(z)$. So, as the Poisson bracket on P depends only on the point values of the first derivatives, we have

$$\{F_1 \circ \mathbf{J} \, , \, F_2 \circ \mathbf{J}\} (z) = \left\{ J^\xi, J^\eta \right\} (z)$$

which proves the theorem. ■

Remark 5.3 The same result holds if G acts on the right, provided that we consider on $\mathfrak{g}^*$ the minus Lie-Poisson structure.

Theorem 5.4 (Noether's Theorem) *Let P be a Poisson manifold, G a Lie group acting canonically on P admitting a momentum map $\mathbf{J}$ and $H : P \to \mathbb{R}$ a G-invariant function. Then $\mathbf{J}$ is a constant of motion for H. That is, if ϕ_t is the flow of X_H then $\mathbf{J} \circ \phi_t = \mathbf{J}$.*

Proof If H is G-invariant then $\xi_P[H] = 0$ which implies

$$0 = \xi_P[H] = X_{J^\xi}[H] = \{H, J^\xi\} = -X_H[J^\xi].$$

So J^ξ is constant on the flow of X_H for every $\xi \in \mathfrak{g}$. ∎

5.3 Examples of Momentum Maps

1. The Hamiltonian

The flow ϕ_t of a complete vector field on a manifold P defines an $\mathbb{R}$-action on P given by $\phi(t, z) := \phi_t(z)$.

Consider the $\mathbb{R}$-action on a Poisson manifold P given by the flow of a complete Hamiltonian vector field X_H. Since the flow of X_H is canonical, this action preserves the Poisson bracket. Let us show that $H : P \to \mathbb{R}$ is an equivariant momentum map for this action. Indeed, if $s \in \mathbb{R}$, its infinitesimal generator is

$$s_P(z) = \left.\frac{d}{d\epsilon}\right|_{\epsilon=0} \phi_{\epsilon s}(z) = sX_H(z) = X_{sH}(z)$$

which shows that $J^s = sH$. Identifying $\mathbb{R}$ with $\mathbb{R}^*$ using the product of elements in $\mathbb{R}$, we get hence $\mathbf{J} = H$. Invariance of H is equivalent to the conservation of energy.

2. Linear momentum

Let $N \in \mathbb{N}$ and consider the N-particle system, with configuration space $Q = \mathbb{R}^{3N}$. Let $\mathbb{R}^3$ act on Q by translations, i.e., $\Phi : \mathbb{R}^3 \times Q \to Q$ is given by

$$(x, (\mathbf{q}_1, \ldots, \mathbf{q}_N)) \mapsto (\mathbf{q}_1 + x, \ldots, \mathbf{q}_N + x).$$

The infinitesimal generator corresponding to $\xi \in \mathbb{R}^3$ is:

$$\begin{aligned}
\xi_Q(\mathbf{q}_1, \ldots, \mathbf{q}_N) &= \left.\frac{d}{dt}\right|_{t=0} \Phi(t\xi, (\mathbf{q}_1, \ldots, \mathbf{q}_N)) \\
&= \left.\frac{d}{dt}\right|_{t=0} (\mathbf{q}_1 + t\xi, \ldots, \mathbf{q}_N + t\xi) \\
&= (\xi, \ldots, \xi) \in T_{(\mathbf{q}_1, \ldots, \mathbf{q}_N)}\mathbb{R}^{3N}.
\end{aligned}$$

Thus, by (5.11), the lifted $\mathbb{R}^{3N}$-action to $T^*\mathbb{R}^{3N}$ admits an invariant momentum map given by

$$\begin{aligned}\left\langle \mathbf{J}(\mathbf{q}_1,\ldots \mathbf{q}_N,\mathbf{p}^1,\ldots,\mathbf{p}^N),\xi\right\rangle &= \left\langle(\mathbf{p}^1,\ldots \mathbf{p}^N),\xi_Q(\mathbf{q}_1,\ldots,\mathbf{q}_N)\right\rangle \\ &= \mathbf{p}^1\cdot\xi+\cdots+\mathbf{p}^N\cdot\xi = (\mathbf{p}^1+\ldots \mathbf{p}^N)\cdot\xi,\end{aligned}$$

that is, $\mathbf{J}(\mathbf{q}_1,\ldots \mathbf{q}_N,\mathbf{p}^1,\ldots,\mathbf{p}^N) = \mathbf{p}^1+\ldots \mathbf{p}^N$, which is the classical linear momentum.

3. Angular momentum

Let $Q=\mathbb{R}^3$ and $\Phi : SO(3)\times\mathbb{R}^3\to\mathbb{R}^3$ be the standard action $\Phi(A,\mathbf{q}) := A\mathbf{q}$. Using the isomorphism of the Lie algebras $(\mathbb{R}^3,\times)$ and $(\mathfrak{so}(3)\,,\,[\,,\,])$ given by (1.13) and the expression (5.9) of the infinitesimal generator, the equivariant momentum map (5.11) for the lifted action of $SO(3)$ to T^*Q is given by

$$\langle \mathbf{J}(\mathbf{q},\mathbf{p}),\xi\rangle = \langle(\mathbf{q},\mathbf{p}),\xi_Q(\mathbf{q})\rangle = \mathbf{p}\cdot(\xi\times\mathbf{q}) = (\mathbf{q}\times\mathbf{p})\cdot\xi,$$

where $\xi\in\mathbb{R}^3$. Thus $\mathbf{J}(\mathbf{q},\mathbf{p}) = \mathbf{q}\times\mathbf{p}$ which is the classical angular momentum.

4. Momentum map for matrix groups

Denote by $GL(n,\mathbb{R})$ the group of linear isomorphisms of $\mathbb{R}^n$ to $\mathbb{R}^n$, that is the general linear group. Let G be a Lie subgroup of $GL(n,\mathbb{R})$, with Lie algebra $\mathfrak{g}\subset\mathfrak{gl}(n,\mathbb{R})$. Consider the action of G on $Q:=\mathbb{R}^n$ to be by matrix multiplication on the left, that is, $\Phi : (A,\mathbf{q})\in G\times\mathbb{R}^n\mapsto A\mathbf{q}\in\mathbb{R}^n$. For $\xi\in\mathfrak{g}$ the corresponding infinitesimal generator is given by:

$$\xi_Q(\mathbf{q}) = \left.\frac{d}{dt}\right|_{t=0}\exp(t\xi)\mathbf{q} = \xi\mathbf{q}.$$

Identify $\mathfrak{gl}(n,\mathbb{R})^*$ with $\mathfrak{gl}(n,\mathbb{R})$ via the positive definite inner product on $\mathfrak{gl}(n,\mathbb{R})$ given by

$$\langle a,b\rangle = \operatorname{trace}(a^Tb), \tag{5.14}$$

where a^T is the transpose of a. By (5.11), the momentum map for the cotangent lifted action is given by:

$$\begin{aligned}\langle \mathbf{J}(\mathbf{q},\mathbf{p}),\xi\rangle &= \mathbf{p}\cdot\xi_Q(\mathbf{q}) = \mathbf{p}\cdot\xi\mathbf{q} = \operatorname{trace}(\mathbf{p}^T\xi\mathbf{q}) \\ &= \operatorname{trace}(\mathbf{q}\mathbf{p}^T\xi) = \left\langle \mathbf{p}\mathbf{q}^T,\xi\right\rangle.\end{aligned}$$

for any $\xi\in\mathfrak{g}$. Now write $\mathfrak{gl}(n,\mathbb{R}) = \mathfrak{g}\oplus\mathfrak{g}^\perp$, where the perpendicular is taken relative to the inner product (5.14) and let $\Pi_{\mathfrak{g}} : \mathfrak{gl}(n,\mathbb{R})\to\mathfrak{g}$ be the corresponding orthogonal projection. Then $\mathfrak{g}$ is identified with $\mathfrak{g}^*$ and we get, $\mathbf{J}(\mathbf{q},\mathbf{p}) = \Pi_{\mathfrak{g}}(\mathbf{p}\mathbf{q}^T)$.

5. Canonical momentum map on $\mathfrak{g}^*$

The Lie group G acts on the dual $\mathfrak{g}^*$ of its Lie algebra $\mathfrak{g}$ by the coadjoint action. Since $\mathrm{Ad}_g : \mathfrak{g} \to \mathfrak{g}$ is a Lie algebra isomorphism for every $g \in G$, Proposition 4.10 insures that the coadjoint action is canonical relative to the Lie-Poisson bracket on $\mathfrak{g}^*$. The infinitesimal generator corresponding to $\xi \in \mathfrak{g}$ for the coadjoint action is, for $\mu \in \mathfrak{g}^*$ and $\eta \in \mathfrak{g}$, given by:

$$\begin{aligned} \langle \xi_{\mathfrak{g}^*}(\mu), \eta \rangle &= \frac{d}{dt}\bigg|_{t=0} \left\langle \mathrm{Ad}^*_{\exp(-t\xi)} \mu, \eta \right\rangle = \frac{d}{dt}\bigg|_{t=0} \langle \mu, \mathrm{Ad}_{\exp(-t\xi)} \eta \rangle \\ &= \langle \mu, [-\xi, \eta] \rangle = \langle \mu, -\mathrm{ad}_\xi \eta \rangle = \left\langle -\mathrm{ad}^*_\xi \mu, \eta \right\rangle, \end{aligned} \tag{5.15}$$

so $\xi_{\mathfrak{g}^*} = -\mathrm{ad}^*_\xi$ for every $\xi \in \mathfrak{g}$.

By Proposition 4.9, the Hamiltonian vector field for $H \in \mathcal{F}(\mathfrak{g}^*)$ has the expression

$$X_H(\mu) = \mp \mathrm{ad}^*_{\delta H/\delta \mu} \mu.$$

Therefore, the momentum map for the coadjoint action, if it exists, must satisfy

$$\mp \mathrm{ad}^*_{\delta J^\xi/\delta \mu} \mu = -\mathrm{ad}^*_\xi \mu \quad \text{for all} \quad \xi \in \mathfrak{g}, \quad \text{and} \quad \mu \in \mathfrak{g}^*,$$

which shows that the momentum map for the coadjoint action exists and is given by $\langle \mathbf{J}(\mu), \xi \rangle = \pm \langle \mu, \xi \rangle$. Therefore $\mathbf{J} = \pm\, \mathrm{id}_{\mathfrak{g}^*}$.

6. Momentum map for products

Let P_1 and P_2 be Poisson manifolds and $P_1 \times P_2$ be their product endowed with the product Poisson structure, that is, if $F, H : P_1 \times P_2 \to \mathbb{R}$, then

$$\{F, H\}_{P_1 \times P_2}(z_1, z_2) = \{F_{z_2}, H_{z_2}\}_{P_1}(z_1) + \{F_{z_1}, H_{z_1}\}_{P_2}(z_2),$$

where $F_{z_1} := F(z_1, \cdot) : P_2 \to \mathbb{R}$ and similarly for $F_{z_2} := F(\cdot, z_2) : P_1 \to \mathbb{R}$.

Let $\Phi : G \times P_1 \to P_1$ and $\Psi : G \times P_2 \to P_2$ be canonical G-actions admitting (equivariant) momentum maps $\mathbf{J}_1 : P_1 \to \mathfrak{g}_1^*$ and $\mathbf{J}_2 : P_2 \to \mathfrak{g}_2^*$ respectively. Then the product action $\Pi : G \times P_1 \times P_2 \to P_1 \times P_2$ given by $\Pi(g, (z_1, z_2)) := (\Phi(g, z_1), \Psi(g, z_2))$ admits an (equivariant) momentum map $\mathbf{J} : P_1 \times P_2 \to \mathfrak{g}^*$ given by $\mathbf{J}(z_1, z_2) = \mathbf{J}_1(z_1) + \mathbf{J}_2(z_2)$.

To prove this statement, we begin by showing that the action Π is canonical. Indeed, for every $g \in G$ we get

$$\begin{aligned} \{F, H\}_{P_1 \times P_2} (\Pi_g(z_1, z_2)) &= \{F, H\}_{P_1 \times P_2}(\Phi(g, z_1), \Psi(g, z_2)) \\ &= \{F_{g \cdot z_2}, H_{g \cdot z_2}\}_{P_1}(\Phi(g, z_1)) + \{F_{g \cdot z_1}, H_{g \cdot z_1}\}_{P_2}(\Psi(g, z_2)) \\ &= \{F_{g \cdot z_2} \circ \Phi_g, H_{g \cdot z_2} \circ \Phi_g\}_{P_1}(z_1) + \{F_{g \cdot z_1} \circ \Psi_g, H_{g \cdot z_1} \circ \psi_g\}_{P_2}(z_2) \\ &= \{(F \circ \Pi_g)_{z_2}, (H \circ \Pi_g)_{z_2}\}_{P_1}(z_1) + \{(F \circ \Pi_g)_{z_1}, (H \circ \Pi_g)_{z_1}\}_{P_2}(z_2) \\ &= \{F \circ \Pi_g, H \circ \Pi_g\}_{P_1 \times P_2}(z_1, z_2). \end{aligned}$$

For $\xi \in \mathfrak{g}$, the infinitesimal generator of Π corresponding to ξ is given by

$$\begin{aligned}
\xi_{P_1 \times P_2}(z_1, z_2) &= \left.\frac{d}{dt}\right|_{t=0} \Pi\left(\exp(t\xi), (z_1, z_2)\right) \\
&= \left.\frac{d}{dt}\right|_{t=0} \left(\Phi(\exp(t\xi), z_1), \Psi(\exp(t\xi), z_2)\right) \\
&= (\xi_{P_1}(z_1), \xi_{P_2}(z_2)) = \left(X_{J_1^\xi}(z_1), X_{J_2^\xi}(z_2)\right) \\
&= X_{(J_1^\xi, J_2^\xi)}(z_1, z_2),
\end{aligned}$$

where the Hamiltonian vector field in the last line is on $P_1 \times P_2$ for the function $(J_1^\xi, J_2^\xi)(z_1, z_2) := (J_1^\xi(z_1), J_2^\xi(z_2)) = \langle \mathbf{J}_1(z_1) + \mathbf{J}_2(z_2), \xi\rangle$. This shows that a momentum map for the product action is given indeed by $\mathbf{J}(z_1, z_2) = \mathbf{J}_1(z_1) + \mathbf{J}_2(z_2)$, as stated.

If $\mathbf{J}_1$ and $\mathbf{J}_2$ are equivariant so is $\mathbf{J}$, as an easy computation shows.

7. Momentum maps for the cotangent lift of the left and right translations of G to T^*G

Let G be a Lie group and denote by $L_g(h) := gh$ and $R_g(h) := hg$ the left and right translations of G on itself. Denote by $\mathbf{J}_L$ and $\mathbf{J}_R$ the corresponding equivariant momentum maps of the lifts of these actions to T^*G. To compute these momentum maps we use (5.11) to get for any $\alpha_g \in T_g^*G$ and $\xi \in \mathfrak{g}$

$$\begin{aligned}
\langle \mathbf{J}_L(\alpha_g), \xi\rangle &= \left\langle \alpha_g, \left.\frac{d}{dt}\right|_{t=0} L_{\exp(t\xi)}g \right\rangle = \langle \alpha_g, T_e R_g \xi\rangle = \langle T_e^* R_g \alpha_g, \xi\rangle \\
\langle \mathbf{J}_R(\alpha_g), \xi\rangle &= \left\langle \alpha_g, \left.\frac{d}{dt}\right|_{t=0} R_{\exp(t\xi)}g \right\rangle = \langle \alpha_g, T_e L_g \cdot \xi\rangle = \langle T_e^* L_g \alpha_g, \xi\rangle,
\end{aligned}$$

which shows that

$$\mathbf{J}_L(\alpha_g) = T_e^* R_g \alpha_g \quad \text{and} \quad \mathbf{J}_R(\alpha_g) = T_e^* L_g \alpha_g. \tag{5.16}$$

8. Momentum map in Maxwell's equations

Let $\mathcal{A}$ be the space of vector potentials $\mathbf{A}$ on $\mathbb{R}^3$, that is, smooth functions $\mathbf{A} : \mathbb{R}^3 \to \mathbb{R}^3$. Let $P := T^*\mathcal{A}$, whose elements are denoted $(\mathbf{A}, -\mathbf{E})$ with $\mathbf{A}$ and $\mathbf{E}$ vector fields on $\mathbb{R}^3$. Let $G = \mathcal{F}(\mathbb{R}^3)$ act on $\mathcal{A}$ by

$$\phi \cdot \mathbf{A} = \mathbf{A} + \operatorname{grad} \phi.$$

The Lie algebra $\mathfrak{g}$ of G coincides with $\mathcal{F}(\mathbb{R}^3)$ and we formally think of $\mathfrak{g}^*$ as $\mathfrak{g}$ via the weakly nondegenerate L^2 pairing. Thus given $\xi \in \mathfrak{g}$, the corresponding infinitesimal generator is:

$$\xi_{\mathcal{A}}(\mathbf{A}) = \operatorname{grad} \xi.$$

Assuming that all computations below are justified by imposing the relevant decay conditions at infinity, the momentum map (5.11) becomes in this case

$$\langle \mathbf{J}(\mathbf{A}, -\mathbf{E}), \xi \rangle = \int -\mathbf{E} \cdot \operatorname{grad} \xi \, d^3x = \int (\operatorname{div} \mathbf{E}) \, \xi \, d^3x.$$

Thus the invariant momentum map $\mathbf{J} : T^*\mathcal{A} \to \mathcal{F}(\mathbb{R}^3)$ is $\mathbf{J}(\mathbf{A}, -\mathbf{E}) = \operatorname{div} \mathbf{E}$.

9. Clairaut's Theorem

Let Q be the surface of revolution obtained by rotating the graph of the smooth function $r = f(z)$ about the z-axis. Pull back the usual Riemannian metric given by the Euclidean inner product on $\mathbb{R}^3$ to Q and identify T^*Q with TQ using this induced metric. The circle S^1 acts on Q and the Riemannian metric on Q is obviously invariant under this action. Consider the geodesic flow on Q, so the Hamiltonian of this vector field on TQ is given by the kinetic energy of the metric; thus it is also S^1 invariant. The infinitesimal generator of $\xi \in \mathbb{R}$, the Lie algebra of S^1, is given by

$$\xi_Q(\mathbf{q}) = \left.\frac{d}{dt}\right|_{t=0} \begin{bmatrix} \cos t\xi & -\sin t\xi & 0 \\ \sin t\xi & \cos t\xi & 0 \\ 0 & 0 & 1 \end{bmatrix} \mathbf{q} = \xi \hat{\mathbf{k}} \mathbf{q} = \xi \mathbf{k} \times \mathbf{q}.$$

Therefore, the momentum map $\mathbf{J} : TQ \to \mathbb{R}$, given by (5.11), has the expression

$$\mathbf{J}(\mathbf{q}, \mathbf{v}) = \mathbf{v} \cdot \xi \mathbf{k} \times \mathbf{q} = \xi r \|\mathbf{v}\| \cos\theta$$

since r is the distance of $\mathbf{q}$ to the z-axis and where θ is the angle between $\mathbf{v}$ and the horizontal plane. Recall that $\|\mathbf{v}\|$ is conserved since the kinetic energy is constant on the geodesic flow. By Noether's theorem it follows that $\mathbf{J}$ is conserved which then implies that $r\cos\theta$ *is conserved along any geodesic on* Q. This is the statement of the classical Clairaut's theorem.

10. Momentum map for symplectic representations

Let (V, Ω) be a symplectic vector space and let G be a Lie group acting linearly and symplectically on V. This action admits an equivariant momentum map $\mathbf{J} : V \to \mathfrak{g}$ given by

$$J^\xi(v) = \langle \mathbf{J}(v), \xi \rangle = \frac{1}{2}\Omega(\xi \cdot v, v),$$

where $\xi \cdot v$ denotes the Lie algebra representation of the element $\xi \in \mathfrak{g}$ on the vector $v \in V$. To verify this, note that the infinitesimal generator $\xi_V(v) = \xi \cdot v$, by the definition of the Lie algebra representation induced by the given Lie group representation, and that $\Omega(\xi \cdot u, v) = -\Omega(u, \xi \cdot v)$ for all $u, v \in V$. Therefore

$$\mathbf{d}J^{\xi}(u)(v) = \frac{1}{2}\Omega(\xi \cdot u, v) + \frac{1}{2}\Omega(\xi \cdot v, u) = \Omega(\xi \cdot u, v).$$

Equivariance of $\mathbf{J}$ follows from the obvious relation $g^{-1} \cdot \xi \cdot g \cdot v = (\mathrm{Ad}_{g^{-1}} \xi) \cdot v$ for any $g \in G$, $\xi \in \mathfrak{g}$, and $v \in V$.

11. Cayley-Klein parameters and the Hopf fibration

Consider the natural action of $SU(2)$ on $\mathbb{C}^2$. Since this action is by isometries of the Hermitian metric, it is automatically symplectic and therefore has a momentum map $\mathbf{J} : \mathbb{C}^2 \to \mathfrak{su}(2)^*$ given in example 10, that is,

$$\langle \mathbf{J}(z, w), \xi \rangle = \frac{1}{2}\Omega(\xi \cdot (z, w), (z, w)),$$

where $z, w \in \mathbb{C}$ and $\xi \in \mathfrak{su}(2)$. Now recall from §2.1 that the symplectic form on $\mathbb{C}^2$ is given by minus the imaginary part of the Hermitian inner product. The Lie algebra $\mathfrak{su}(2)$ of $SU(2)$ consists of 2×2 skew Hermitian matrices of trace zero. This Lie algebra is isomorphic to $\mathfrak{so}(3)$ and therefore to $(\mathbb{R}^3, \times)$ by the isomorphism given by

$$\mathbf{x} = (x^1, x^2, x^3) \in \mathbb{R}^3 \mapsto \widetilde{\mathbf{x}} := \frac{1}{2}\begin{bmatrix} -ix^3 & -ix^1 - x^2 \\ -ix^1 + x^2 & ix^3 \end{bmatrix} \in \mathfrak{su}(2).$$

Thus we have $[\widetilde{\mathbf{x}}, \widetilde{\mathbf{y}}] = (\mathbf{x} \times \mathbf{y})^{\sim}$ for any $\mathbf{x}, \mathbf{y} \in \mathbb{R}^3$. Other useful relations are $\det(2\widetilde{\mathbf{x}}) = \|\mathbf{x}\|^2$ and $\mathrm{trace}(\widetilde{\mathbf{x}}\widetilde{\mathbf{y}}) = -\frac{1}{2}\mathbf{x} \cdot \mathbf{y}$. Identify $\mathfrak{su}(2)^*$ with $\mathbb{R}^3$ by the map $\mu \in \mathfrak{su}(2)^* \mapsto \check{\mu} \in \mathbb{R}^3$ defined by

$$\check{\mu} \cdot \mathbf{x} := -2\langle \mu, \widetilde{\mathbf{x}} \rangle$$

for any $\mathbf{x} \in \mathbb{R}^3$. With these notations, the momentum map $\check{\mathbf{J}} : \mathbb{C}^2 \to \mathbb{R}^3$ can be explicitly computed in coordinates: for any $\mathbf{x} \in \mathbb{R}^3$ we have

$$\begin{aligned}
\check{\mathbf{J}}(z, w) \cdot \mathbf{x} &= -2\langle \mathbf{J}(z, w), \widetilde{\mathbf{x}} \rangle \\
&= \frac{1}{2}\operatorname{Im}\left(\begin{bmatrix} -ix^3 & -ix^1 - x^2 \\ -ix^1 + x^2 & ix^3 \end{bmatrix} \begin{bmatrix} z \\ w \end{bmatrix} \cdot \begin{bmatrix} z \\ w \end{bmatrix} \right) \\
&= -\frac{1}{2}(2\operatorname{Re}(w\overline{z}), 2\operatorname{Im}(w\overline{z}), |z|^2 - |w|^2) \cdot \mathbf{x}.
\end{aligned}$$

Therefore

$$\check{\mathbf{J}}(z, w) = -\frac{1}{2}(2w\overline{z}, |z|^2 - |w|^2) \in \mathbb{R}^3.$$

By Theorem 5.2, $\check{\mathbf{J}}$ is a Poisson map from $\mathbb{C}^2$, endowed with the canonical symplectic structure, to $\mathbb{R}^3$, endowed with the $+$ Lie Poisson structure. Therefore, $-\check{\mathbf{J}} : \mathbb{C}^2 \to \mathbb{R}^3$ is a canonical map, if $\mathbb{R}^3$ has the $-$ Lie-Poisson bracket relative to which the free rigid body equations are Hamiltonian. Pulling back the Hamiltonian $H(\mathbf{\Pi}) = \mathbf{\Pi} \cdot \mathbb{I}^{-1}\mathbf{\Pi}/2$ to $\mathbb{C}^2$ gives a Hamiltonian function (called collective) on $\mathbb{C}^2$. The classical Hamilton equations for this function are therefore projected by $-\check{\mathbf{J}}$ to the rigid body equations $\dot{\mathbf{\Pi}} = \mathbf{\Pi} \times \mathbb{I}^{-1}\mathbf{\Pi}$. In this context, the variables (z, w) are called the ***Cayley-Klein parameters***. They represent a first attempt to understand the rigid body equations as a Hamiltonian system, before the introduction of Poisson manifolds. In quantum mechanics, the same variables are called the ***Kustaanheimo-Stiefel coordinates***. A similar construction was carried out in fluid dynamics making the Euler equations a Hamiltonian system relative to the so-called ***Clebsch variables***.

Now notice that if $(z, w) \in S^3 := \{(z, w) \in \mathbb{C}^2 \mid |z|^2 + |w|^2 = 1\}$, then $\| - \check{\mathbf{J}}(z, w)\| = 1/2$, so that $-\check{\mathbf{J}}|_{S^3} : S^3 \to S^2_{1/2}$, where $S^2_{1/2}$ is the sphere in $\mathbb{R}^3$ of radius $1/2$. It is also easy to see that $-\check{\mathbf{J}}|_{S^3}$ is surjective and that its fibers are circles. Indeed, given $(x^1, x^2, x^3) = (x^1 + ix^2, x^3) = (re^{i\psi}, x^3) \in S^2_{1/2}$, the inverse image of this point is

$$
\begin{aligned}
&-\check{\mathbf{J}}^{-1}(re^{i\psi}, x^3) = \\
&\qquad \left\{ \left(e^{i\theta}\sqrt{\frac{1}{2} + x^3}, e^{i\varphi}\sqrt{\frac{1}{2} - x^3} \right) \in S^3 \mid e^{i(\theta - \varphi + \psi)} = 1 \right\}.
\end{aligned}
$$

One recognizes now that $-\check{\mathbf{J}}|_{S^3} : S^3 \to S^2$ is the ***Hopf fibration***. In other words, *the momentum map of the $SU(2)$-action on $\mathbb{C}^2$, the Cayley-Klein parameters, the Kustaanheimo-Stiefel coordinates, and the family of Hopf fibrations on concentric three-spheres in $\mathbb{C}^2$ are the same map.*

6 Lie-Poisson and Euler-Poincaré Reduction

In this lecture we shall present the simplest case of reduction, namely the Lie-Poisson reduction theorem. It states that the quotient of a cotangent bundle of a Lie group by the lift of the left or right translation is Poisson isomorphic to the dual of the Lie algebra endowed with the $\pm$ Lie-Poisson bracket. The symplectic leaves of this Poisson structure are the connected components of the coadjoint orbits. The Lagrangian version of this result is a reduced constrained variational principle that is equivalent to first order equations on the dual of the Lie algebra, called Euler-Poincaré equations. We shall carry out in detail several examples both in finite and in infinite dimensions.

6.1 Lie-Poisson Reduction

One way to construct new Poisson manifolds out of known ones is by symmetry reduction.

Let G be a Lie group acting canonically on a Poisson manifold P. Assume that the orbit space P/G is a smooth manifold and the quotient projection $\pi : P \to P/G$ a surjective submersion. This is the case, for example, if the G-action is proper and free, or proper with all isotropy groups conjugate. Then there exists a unique Poisson bracket $\{\cdot, \cdot\}_{P/G}$ on P/G relative to which π is a Poisson map. The Poisson bracket on P/G is given in the following way. If $\hat{F}, \hat{H} \in \mathcal{F}(P/G)$, then $\hat{F} \circ \pi, \hat{H} \circ \pi \in \mathcal{F}(P)$ are G-invariant functions and, due to the fact that the action is canonical, their Poisson bracket $\{\hat{F} \circ \pi, \hat{H} \circ \pi\}$ is also G-invariant. Therefore, this function descends to a smooth function on the quotient P/G; this is, by definition, $\{\hat{F}, \hat{H}\}_{P/G}$ and we have, by construction, $\{\hat{F} \circ \pi, \hat{H} \circ \pi\} = \{\hat{F}, \hat{H}\}_{P/G} \circ \pi$. It is easy to see that $\{\cdot, \cdot\}_{P/G}$ so defined satisfies all the axioms of a Poisson bracket. This proves in a constructive way the existence of the Poisson bracket on the quotient. In addition, because $\pi : P \to P/G$ is a surjective Poisson submersion, the bracket on the quotient is necessarily unique with the requirement that π is a Poisson map.

When the manifold P is the cotangent bundle T^*G of a Lie group G and the action of G on T^*G is by cotangent lift of the left (or right) translation of G on itself, the reduced space $(T^*G)/G$ is naturally diffeomorphic to the dual $\mathfrak{g}^*$ of the Lie algebra $\mathfrak{g}$ of G. The goal of this section is to show that *the quotient Poisson bracket is the minus (or plus) Lie-Poisson bracket*. To do this, we follow the presentation in [MaRa94], §13, and we will give two proofs.

First Proof. The left and right translations by $g \in G$ are denoted by $L_g(h) := gh$ and $R_g(h) = hg$. Let $\mathcal{F}_L(T^*G)$ be the space of smooth left-invariant functions on T^*G, that is, $F_L \in \mathcal{F}_L(T^*G)$ if and only if $F_L \circ T^*L_g = F_L$ for all $g \in G$, where T^*L_g is the cotangent lift of L_g. Similarly, a right-invariant function F_R verifies $F_R \circ T^*R_g = F_R$ and the space of all smooth right invariant functions on T^*G is denoted by $\mathcal{F}_R(T^*G)$. Note that $\mathcal{F}_L(T^*G)$ and $\mathcal{F}_R(T^*G)$ are closed under Poisson bracket.

Any $F \in \mathcal{F}(\mathfrak{g}^*)$ can be uniquely extended to a left (respectively right) invariant function F_L (respectively F_R) on T^*G by setting

$$F_L(\alpha_g) := F(T_e^*L_g\alpha_g) = (F \circ \mathbf{J}_R)(\alpha_g)$$

(respectively $F_R(\alpha_g) := F(T_e^*R_g\alpha_g) = (F \circ \mathbf{J}_L)(\alpha_g)$). Here $\mathbf{J}_L$ and $\mathbf{J}_R$ are the momentum maps for the left and right translations given by (5.16).

So, composition with $\mathbf{J}_R$ (respectively with $\mathbf{J}_L$) defines, by Theorem 5.2, an isomorphism of Poisson algebras $\mathcal{F}(\mathfrak{g}^*_-) \to \mathcal{F}_L(T^*G)$ (respectively, $\mathcal{F}(\mathfrak{g}^*_+) \to$

$\mathcal{F}_R(T^*G))$ whose inverse is the restriction to the fiber $T_e^*G = \mathfrak{g}^*$:

$$\{F, H\}_- \circ \mathbf{J}_R = \{F \circ \mathbf{J}_R, H \circ \mathbf{J}_R\} = \{F_L, H_L\},$$
$$\{F, H\}_+ \circ \mathbf{J}_L = \{F \circ \mathbf{J}_L, H \circ \mathbf{J}_L\} = \{F_R, H_R\},$$

$$\{F, H\}_- = \{F_L, H_L\}|_{\mathfrak{g}^*}, \quad \text{and} \quad \{F, H\}_+ = \{F_R, H_R\}|_{\mathfrak{g}^*},$$

where $\{\cdot, \cdot\}_\pm$ are the Lie-Poisson brackets on $\mathfrak{g}^*$ and $\{\cdot, \cdot\}$ is the Poisson bracket on T^*G. ■

While mathematically correct, this proof is unsatisfactory for it requires to know a priori that $\mathfrak{g}^*$ is a Poisson manifold. This is why we shall give below a second proof in which the Lie-Poisson bracket is *discovered* by carrying out the identification of $(T^*G)/G$ with $\mathfrak{g}^*$ explicitly.

Second Proof. This is done in several steps. We begin by noting that the map $\mathcal{P} : X \in \mathfrak{X}(Q) \mapsto \langle \cdot, X \rangle \in \mathcal{L}(T^*Q)$, where

$$\mathcal{L}(T^*Q) := \{f \in \mathcal{F}(T^*Q) \mid f \text{ linear on the fibers}\}$$

is a Poisson subalgebra of $\mathcal{F}(T^*Q)$, is a Lie algebra anti-isomorphism. To see this, work in coordinates and note that $F, H \in \mathcal{L}(T^*Q)$ if and only if $F(q, p) = X^i(q)p_i, H(q, p) = Y^i(q)p_i$ and hence

$$\{F, H\}(q, p) = \frac{\partial F}{\partial q^j}\frac{\partial H}{\partial p_j} - \frac{\partial H}{\partial q^j}\frac{\partial F}{\partial p_j} = \left(\frac{\partial X^i}{\partial q^j}Y^i - \frac{\partial Y^i}{\partial q^j}X^i\right) p_i.$$

Thus, $\{\mathcal{P}(X), \mathcal{P}(Y)\} = -\mathcal{P}([X, Y])$. This immediately implies that the linear isomorphism $Y \in \mathfrak{X}(Q) \mapsto X_{\mathcal{P}(Y)} \in \{X_F \mid F \in \mathcal{L}(T^*Q)\}$ preserves the Lie brackets. Indeed,

$$[X, Y] \mapsto X_{\mathcal{P}([X,Y])} = -X_{\{\mathcal{P}(X),\mathcal{P}(Y)\}} = [X_{\mathcal{P}(X)}, X_{\mathcal{P}(Y)}]$$

for any $X, Y \in \mathfrak{X}(Q)$. Thus $Y \in \mathfrak{X}(Q) \mapsto X_{\mathcal{P}(Y)} \in \{X_F \mid F \in \mathcal{L}(T^*Q)\}$ is a Lie algebra isomorphism.

Next we prove that if the flow of $X \in \mathfrak{X}(Q)$ is ϕ_t then its cotangent lift $T^*\phi_{-t}$ is the flow of $X_{\mathcal{P}(X)}$.

To see this, let $\pi : T^*Q \to Q$ be the canonical projection. Differentiating at $t = 0$ the equation $\pi \circ T^*\phi_{-t} = \phi_t \circ \tau_Q$, we get

$$T\pi \circ Y = X \circ \pi \quad \text{where} \quad Y(\alpha_q) = \left.\frac{d}{dt}\right|_{t=0} T^*\phi_{-t}(\alpha_q).$$

So $T^*\phi_{-t}$ is the flow of Y. As $T^*\phi_{-t}$ preserves the canonical one-form, it follows that $\boldsymbol{\pounds}_Y\Theta = 0$ and hence $\mathbf{i}_Y\Omega = \mathbf{d}(\mathbf{i}_Y\Theta)$. This shows that Y is

Hamiltonian with energy $\mathbf{i}_Y\Theta(\alpha_q) = \langle \alpha_q, (T\pi \circ Y)(\alpha_q)\rangle = \langle \alpha_q, X(q)\rangle = \mathcal{P}(X)(\alpha_q)$, that is, $Y = X_{\mathcal{P}(X)}$, which proves the statement.

Finally we shall implement the diffeomorphism between $(T^*G)/G$ and $\mathfrak{g}^*$ given by dropping to the quotient the left invariant map $\mathbf{J}_R : T^*G \to \mathfrak{g}^*$. Concretely, we shall prove that the push-forward by this diffeomorphism of the quotient Poisson bracket on $(T^*G)/G$ gives the known formula

$$\{F, H\}_-(\mu) = -\left\langle \mu, \left[\frac{\delta F}{\delta \mu}, \frac{\delta H}{\delta \mu}\right]\right\rangle.$$

To achieve this, we shall show that if $F, H \in \mathcal{F}(\mathfrak{g}^*)$, we get the identity $\{F_L, H_L\}|_{\mathfrak{g}^*} = \{F, H\}_-$.

This is done in the following way. Since the Poisson bracket of any $F, H \in \mathcal{F}(\mathfrak{g}^*)$ depends only on the differentials of F and H, it is enough to prove the statement for linear functions on $\mathfrak{g}^*$. So we can replace the general smooth function $F : \mathfrak{g}^* \to \mathbb{R}$ with its linear part

$$F^\ell(\mu) := \left\langle \mu, \frac{\delta F}{\delta \mu}\right\rangle.$$

Then, denoting by ξ_L the left invariant vector field on G whose value at the identity is $\xi \in \mathfrak{g}$, that is, $\xi_L(g) := T_eL_g\xi$, we get

$$\begin{aligned} F^\ell_L(\alpha_g) &= F^\ell\left(T_eL_g^*(\alpha_g)\right) = \left\langle T_eL_g^*(\alpha_g), \frac{\delta F}{\delta \mu}\right\rangle \\ &= \left\langle \alpha_g, T_eL_g\left(\frac{\delta F}{\delta \mu}\right)\right\rangle = \left\langle \alpha_g, \left(\frac{\delta F}{\delta \mu}\right)_L(g)\right\rangle \\ &= \mathcal{P}\left(\left(\frac{\delta F}{\delta \mu}\right)_L\right)(\alpha_g). \end{aligned}$$

Therefore if $\mu \in \mathfrak{g}^*$ we have

$$\begin{aligned} \{F^\ell_L, H^\ell_L\}|_{\mathfrak{g}^*}(\mu) &= \{F^\ell_L, H^\ell_L\}(\mu) = \left\{\mathcal{P}\left(\left(\frac{\delta F}{\delta \mu}\right)_L\right), \mathcal{P}\left(\left(\frac{\delta H}{\delta \mu}\right)_L\right)\right\}(\mu) \\ &= -\mathcal{P}\left(\left[\left(\frac{\delta F}{\delta \mu}\right)_L, \left(\frac{\delta H}{\delta \mu}\right)_L\right]\right)(\mu) \\ &= -\mathcal{P}\left(\left[\frac{\delta F}{\delta \mu}, \frac{\delta H}{\delta \mu}\right]_L\right)(\mu) = -\left\langle \mu, \left[\frac{\delta F}{\delta \mu}, \frac{\delta H}{\delta \mu}\right]_L(e)\right\rangle \\ &= -\left\langle \mu, \left[\frac{\delta F}{\delta \mu}, \frac{\delta H}{\delta \mu}\right]\right\rangle = \{F, H\}_-(\mu) = \{F^\ell, H^\ell\}(\mu), \end{aligned}$$

which ends the proof. ■

Thus, *the identification of the set of real-valued functions on $\mathfrak{g}^*$ with the* *left (respectively, right) invariant functions on T^*G endows $\mathfrak{g}^*$ with the minus (respectively the plus) Lie-Poisson bracket.*

6.2 Lie-Poisson Reduction of Dynamics

In this section we shall discuss the Lie-Poisson reduction of dynamics. Since the momentum maps $\mathbf{J}_R : T^*G \to \mathfrak{g}^*_-$ and $\mathbf{J}_L : T^*G \to \mathfrak{g}^*_+$ are Poisson maps, they will map integral curves of left and right invariant Hamiltonian vector fields to integral curves of Lie-Poisson Hamiltonian systems. This immediately yields the following theorem.

Theorem 6.1 (Lie-Poisson reduction of dynamics) *If $H : T^*G \to \mathbb{R}$ is a left (respectively, right) G-invariant function its restriction $H^- := H|_{\mathfrak{g}^*}$ (respectively, $H^+ := H|_{\mathfrak{g}^*}$) to $\mathfrak{g}^*$ satisfies*

$$H = H^- \circ \mathbf{J}_R \quad (respectively \quad H^+ = H \circ \mathbf{J}_L),$$

where $\mathbf{J}_R = T^*L_g\alpha_g$ and $\mathbf{J}_L = T^*R_g\alpha_g$ for all $\alpha_g \in T^*_gG$.

The flow F_t of X_H on T^*G and the flow, F_t^- of X_{H^-} on $\mathfrak{g}^*_-$ (respectively, F_t^+ of X_{H^+} on $\mathfrak{g}^*_+$) are related by $\mathbf{J}_R \circ F_t = F_t^- \circ \mathbf{J}_R$ (respectively, $\mathbf{J}_L \circ F_t = F_t^+ \circ \mathbf{J}_L$.

As the original Hamiltonian and the reduced Hamiltonian are related by a momentum map we can get some additional information using the fact that the momentum map is a conserved quantity.

Proposition 6.2 *Let $H : T^*G \to \mathbb{R}$ be left-invariant, $H^- = H|_{\mathfrak{g}^*}$, $\alpha(t) \in T^*_{g(t)}G$ an integral curve of X_H, $\mu(t) = \mathbf{J}_R(\alpha(t))$, and $\nu = \mathbf{J}_L(\alpha(t))$. Then*

$$\nu = \mathrm{Ad}^*_{g(t)^{-1}}\,\mu(t).$$

Proof The curve $\nu(t) := \mathbf{J}_L(\alpha(t)) = T^*_eR_{g(t)}\alpha(t)$ is constant by the Noether theorem, say equal to ν. As $\mu(t) = \mathbf{J}_R(\alpha(t)) = T^*_eL_{g(t)}\alpha(t)$, we get

$$\nu = T^*_eR_{g(t)}\alpha(t) = \left(T^*_eR_{g(t)} \circ T^*_{g(t)}L_{g^{-1}(t)}\right)\mu(t) = \mathrm{Ad}^*_{g^{-1}(t)}\,\mu(t)$$

which proves the statement. ■

It is interesting to relate the reduced dynamics to its left (right) trivialization. Explicitly, T^*G is diffeomorphic to $G \times \mathfrak{g}^*$ via the ***left trivialization*** diffeomorphism

$$\lambda : T^*G \to G \times \mathfrak{g}^*, \qquad \lambda(\alpha_g) := (g, T^*_eL_g(\alpha_g)) = (g, \mathbf{J}_R(\alpha_g)).$$

Since $\mathbf{J}_R$ is equivariant, λ is an equivariant diffeomorphism for the cotangent lift of the left translation and the following action of G on $G \times \mathfrak{g}^*$

$$g \cdot (h\,,\,\mu) := (gh, \mu).$$

Thus $(T^*G)/G$ is diffeomorphic to $(G \times \mathfrak{g}^*)/G$. As G does not act on $\mathfrak{g}^*$, it follows that $(G \times \mathfrak{g}^*)/G$ is equal to $\mathfrak{g}^*$ and we see again that $(T^*G)/G$ is diffeomorphic to $\mathfrak{g}^*$.

If X_H is the Hamiltonian vector field on T^*G for a left invariant Hamiltonian H, a lengthy but elementary computation (see, e.g. [MaRa94], Proposition 13.4.3) shows that the left trivialization $\lambda_* X_H$ equals

$$(\lambda_* X_H) = \left(T_e L_g \frac{\delta H^-}{\delta \mu}, \mu, \operatorname{ad}^*_{\delta H^-/\delta\mu} \mu \right) \in T_g G \times T_\mu \mathfrak{g}^*, \tag{6.1}$$

which says that Hamilton's equations on $G \times \mathfrak{g}^*$ for the push forward Hamiltonian function $\lambda_* H$ and the push forward symplectic form $\lambda_* \Omega$ are

$$\dot{\mu} = \operatorname{ad}^*_{\delta H^-/\delta\mu} \mu, \qquad \dot{g} = T_e L_g \frac{\delta H^-}{\delta \mu}. \tag{6.2}$$

Note that the first equation is just the Lie-Poisson reduced Hamiltonian vector field and hence does not depend on $g \in G$. Once the first equation is solved, the second one yields a *linear* equation with time dependent coefficients, that is, the second equation is what one usually calls a "quadrature". We summarize these remarks in the following *Reconstruction Theorem.*

Theorem 6.3 *Let $H : T^*G \to \mathbb{R}$ be a left-invariant Hamiltonian, $H^- := H|_{\mathfrak{g}^*}$, and $\mu(t)$ the integral curve of the Lie-Poisson equations*

$$\frac{d\mu}{dt} = \operatorname{ad}^*_{\delta H^-/\delta\mu} \mu$$

with initial condition $\mu(0) = T_e^* L_{g_0}(\alpha_{g_0})$. Then the integral curve $\alpha(t) \in T^*_{g(t)}G$ of X_H with initial condition $\alpha(0) := \alpha_{g_0}$ is given by

$$\alpha(t) = T^*_{g(t)} L_{g(t)^{-1}} \mu(t),$$

where $g(t)$ is the solution of the equation

$$\frac{dg(t)}{dt} = T_e L_{g(t)} \frac{\delta H^-}{\delta \mu},$$

with initial condition $g(0) = g_0$.

Proof A curve $\alpha(t)$ is the unique integral curve of X_H with initial condition $\alpha(0) = \alpha_{g_0}$ if and only if

$$\lambda(\alpha(t)) = (g(t)\,,\,T_e^* L_{g(t)} \alpha(t)) = (g(t), J_R(\alpha(t))) = (g(t)\,,\,\mu(t))$$

is the unique integral curve of $\lambda_* X_H$ with initial condition

$$\lambda(\alpha(0)) = (g_0\,,\, T_e^* L_{g_0} \alpha_{g_0}).$$

So, the result follows from equation (6.1). ■

A similar statement holds for right invariant Hamiltonians by replacing everywhere "left" by "right" and $-$ by $+$ in the Lie-Poisson equations.

6.3 Coadjoint Orbits

In §4.2 we studied the internal structure of a Poisson manifold, namely, its stratification into a disjoint union of symplectic leaves. In this section we shall see that the symplectic leaves of the Poisson manifold $\mathfrak{g}^*$ endowed with the Lie-Poisson bracket are the connected components of the coadjoint orbits.

The ***coadjoint orbit*** $\mathcal{O}(\mu)$ through $\mu \in \mathfrak{g}^*$ is the subset of $\mathfrak{g}^*$ defined by

$$\mathcal{O}(\mu) := G \cdot \mu := \left\{ \mathrm{Ad}^*_{g^{-1}}(\mu) : g \in G \right\}.$$

Like the orbit of any Lie group, $\mathcal{O}(\mu)$ is an immersed submanifold of $\mathfrak{g}^*$ but is not, in general, a submanifold of $\mathfrak{g}^*$. If G is compact then $\mathcal{O}(\mu)$ is a closed embedded submanifold of $\mathfrak{g}^*$. This is, in general, not true for an arbitrary Lie group. Coadjoint orbits of algebraic groups are also embedded submanifolds.

For any smooth Lie group action $\Phi : G \times M \to M$ on a manifold M, the ***orbit*** through a point $m \in M$ is the set

$$\mathcal{O}_m = \{\Phi(g, m) \mid g \in G\} \subset M.$$

For $m \in M$, the ***isotropy subgroup*** of Φ at m is

$$G_m = \{g \in G \mid \Phi(g, m) = m \in G\} \subset G.$$

Since the map $\Phi^m : G \to M$, given by $\Phi^m(g) := \Phi(g, m)$, is smooth, $G_m = (\Phi^m)^{-1}(m)$ is a closed subgroup and hence a Lie subgroup of G. The bijective map $[g] \in G/G_m \mapsto \Phi(g, m) \in \mathcal{O}_m$ induces a manifold structure on $\mathcal{O}_m$ that makes it diffeomorphic to the smooth homogeneous manifold G/G_m.

Recall that for $\xi \in \mathfrak{g}$ the family of diffeomorphisms $t \mapsto \Phi_{\exp(t\xi)}$ on M defines a flow and the corresponding vector field $\xi_M \in \mathfrak{X}(M)$ is the ***infinitesimal generator*** of the action. Thus

$$\xi_M(m) := \left.\frac{d}{dt}\right|_{t=0} \Phi_{\exp(t\xi)}(m).$$

This definition shows that the tangent space to the orbit $\mathcal{O}_m$ is given by

$$T_m \mathcal{O}_m = \{\xi_M(m) \mid \xi \in \mathfrak{g}\}.$$

We apply these general considerations to $M = \mathfrak{g}^*$ and the G-action the coadjoint action. Then the orbit through $\mu \in \mathfrak{g}^*$ is diffeomorphic to G/G_μ, where G_μ is the isotropy subgroup of μ

$$G_\mu = \{g \in G \mid \mathrm{Ad}_g^* \mu = \mu\}.$$

We recall that for $\xi \in \mathfrak{g}$, the infinitesimal generator for the coadjoint action corresponding to ξ is given by (5.15), that is,

$$\xi_{\mathfrak{g}^*}(\mu) = -\operatorname{ad}_\xi^* \mu.$$

Therefore

$$T_\mu \mathcal{O}_\mu = \{-\operatorname{ad}_\xi^* \mu \mid \xi \in \mathfrak{g}\} = \mathfrak{g}_\mu^\circ,$$

where $\mathfrak{g}_\mu^\circ := \{\nu \in \mathfrak{g}^* \mid \langle \nu, \eta \rangle = 0 \text{ for all } \eta \in \mathfrak{g}_\mu\}$, $\mathfrak{g}_\mu = \{\xi \in \mathfrak{g} \mid \operatorname{ad}_\xi^* \mu = 0\}$, and $\langle \cdot, \cdot \rangle : \mathfrak{g}^* \times \mathfrak{g} \to \mathbb{R}$ is a strongly nondegenerate pairing (see [MaRa94], Proposition 14.2.1).

The next theorems show that the symplectic leaves of the Poisson manifold $\mathfrak{g}^*$ are the connected components of the coadjoint orbits and give explicitly the symplectic form. The proofs can be found in [MaRa94].

Theorem 6.4 *Let G be a Lie group and let $\mathcal{O} \subset \mathfrak{g}^*$ be a coadjoint orbit. Then $\mathcal{O}$ is a symplectic manifold relative to the* ***orbit symplectic form***

$$\omega^{\pm}(\mu)\big(\xi_{\mathfrak{g}^*}(\mu), \eta_{\mathfrak{g}^*}(\mu)\big) := \pm\langle \mu, [\xi, \eta]\rangle \tag{6.3}$$

for all $\mu \in \mathcal{O}$ and $\xi, \eta \in \mathfrak{g}$.

The symplectic form (6.3) is also known as the *Kostant-Kirillov-Souriau symplectic form.*

Theorem 6.5 *The Lie-Poisson bracket and the coadjoint orbit symplectic structure are consistent in the following sense: for $F, H : \mathfrak{g}^* \to \mathbb{R}$ and $\mathcal{O}$ a coadjoint orbit in $\mathfrak{g}^*$, we have*

$$\{F, H\}_+|_\mathcal{O} = \{F|_\mathcal{O}, H|_\mathcal{O}\}^+,$$

where, $\{\cdot, \cdot\}_+$ is the $+$ Lie-Poisson bracket, while $\{\cdot, \cdot\}^+$ is the Poisson bracket defined by the $+$ coadjoint symplectic orbit structure ω^+ on $\mathcal{O}$. Similarly,

$$\{F, H\}_-|_\mathcal{O} = \{F|_\mathcal{O}, H|_\mathcal{O}\}^-.$$

We summarize below some results for coadjoint orbits:

- For $\mu, \nu \in \mathfrak{g}^*_-$ and $H : \mathfrak{g}^* \to \mathbb{R}$, the Hamiltonian vector field for H is $X_H(\nu) = \mathrm{ad}^*_{\delta H/\delta \nu}(\nu)$. Therefore, if $\nu \in \mathcal{O}$ then $X_H(\nu)$ is tangent to $\mathcal{O}$. So the trajectory of X_H starting at $\mu \in \mathcal{O} \subset \mathfrak{g}^*$ stays in $\mathcal{O}$.
- Recall that a function $C \in \mathcal{F}(\mathfrak{g}^*)$ is a Casimir if and only if $0 = X_C(\mu) = \mathrm{ad}^*_{\delta C/\delta \mu}\mu$. Thus if C is a Casimir of $\mathfrak{g}^*$ then $\delta C/\delta \mu \in \mathfrak{g}_\mu$ for all $\mu \in \mathfrak{g}^*$.
- **(Duflo-Vergne Theorem)** Let $\mathfrak{g}$ be a finite-dimensional Lie algebra with dual $\mathfrak{g}^*$ and let $r := \min\{\dim \mathfrak{g}_\mu \mid \mu \in \mathfrak{g}^*\}$. The set $\{\mu \in \mathfrak{g}^* \mid \dim \mathfrak{g}_\mu = r\}$ is Zariski open and thus open and dense in the usual topology of $\mathfrak{g}^*$. If $\dim \mathfrak{g}_\mu = r$, then $\mathfrak{g}_\mu$ is Abelian.
- If $C \in \mathcal{F}(\mathfrak{g}^*)$ is Ad^*-invariant, i.e $C(\mathrm{Ad}^*_{g^{-1}}\mu) = C(\mu)$, then the differentiation of this equality with respect to g at $g = e$ shows that C is a Casimir function. Thus, a function that is constant on coadjoint orbits is necessarily a Casimir function.

  In general Ad^*-invariance of C is a stronger condition than C being a Casimir. A theorem of Kostant gives a characterization of which Ad^*-invariant functions are Casimirs. Namely, an Ad^*-invariant function C is a Casimir if and only if $\delta C/\delta \mu$ lies in the center of $\mathfrak{g}_\mu$ for all $\mu \in \mathfrak{g}^*$ (see [MaRa94], Proposition 14.4.4).

The rest of this section is dedicated to working out a few examples.

(1) Rotation group. Recall from §5.1 that the coadjoint action of $SO(3)$ on $\mathfrak{so}(3)^*$ has the expression $\mathrm{Ad}^*_{A^{-1}}\tilde{\mathbf{\Pi}} = (A\mathbf{\Pi})\tilde{}$, where the isomorphism $\tilde{}\; : \mathbb{R}^3 \to \mathfrak{so}(3)^*$ is given by $\tilde{\mathbf{\Pi}}(\hat{\mathbf{u}}) := \mathbf{\Pi} \cdot \mathbf{u}$ for any $\mathbf{u} \in \mathbb{R}^3$ and $\hat{}\; : (\mathbb{R}^3, \times) \to (\mathfrak{so}(3), [\cdot,\cdot])$ is the Lie algebra isomorphism (1.13). Therefore, the coadjoint orbit $\mathcal{O} = \{A\mathbf{\Pi} \mid A \in SO(3)\} \subset \mathbb{R}^3$ of $SO(3)$ through $\mathbf{\Pi} \in \mathbb{R}^3$ is a 2-sphere of radius $\|\mathbf{\Pi}\|$.

To compute the coadjoint action of $\mathfrak{so}(3)$ on its dual, let $\mathbf{u}, \mathbf{v} \in \mathbb{R}^3$ and note that

$$\begin{aligned}\left\langle \mathrm{ad}^*_{\hat{\mathbf{u}}}\tilde{\mathbf{\Pi}}, \hat{\mathbf{v}} \right\rangle &= \left\langle \tilde{\mathbf{\Pi}}, [\hat{\mathbf{u}}, \hat{\mathbf{v}}] \right\rangle = \left\langle \tilde{\mathbf{\Pi}}, (\mathbf{u} \times \mathbf{v})\hat{} \right\rangle = \mathbf{\Pi} \cdot (\mathbf{u} \times \mathbf{v}) \\ &= (\mathbf{\Pi} \times \mathbf{u}) \cdot \mathbf{v} = \langle (\mathbf{\Pi} \times \mathbf{u})\tilde{}, \hat{\mathbf{v}} \rangle,\end{aligned}$$

which shows that $\mathrm{ad}^*_{\hat{\mathbf{u}}}\tilde{\mathbf{\Pi}} = (\mathbf{\Pi} \times \mathbf{u})\tilde{}$, proving (1.21). Therefore, $T_{\mathbf{\Pi}}\mathcal{O} = \{\mathbf{\Pi} \times \mathbf{u} \mid \mathbf{u} \in \mathbb{R}^3\}$ as expected, since the plane perpendicular to $\mathbf{\Pi}$, that is, the tangent space to the sphere centered at the origin of radius $\|\mathbf{\Pi}\|$, is indeed given by $\{\mathbf{\Pi} \times \mathbf{u} \mid \mathbf{u} \in \mathbb{R}^3\}$.

The minus orbit symplectic structure on $\mathcal{O}$ is given hence by

$$\omega^-(\mathbf{\Pi})(\mathbf{\Pi} \times \mathbf{u}, \mathbf{\Pi} \times \mathbf{v}) = -\mathbf{\Pi} \cdot (\mathbf{u} \times \mathbf{v}).$$

How does this exactly relate to the area form on the sphere $\mathcal{O}$? To see this, recall that the oriented area of a planar parallelogram spanned by two vectors $\mathbf{a}, \mathbf{b}$ (in this order) is given by $\mathbf{a} \times \mathbf{b}$. Thus, the oriented area spanned by $\mathbf{\Pi} \times \mathbf{u}$ and $\mathbf{\Pi} \times \mathbf{v}$ is

$$(\mathbf{\Pi} \times \mathbf{u}) \times (\mathbf{\Pi} \times \mathbf{v}) = (\mathbf{\Pi} \cdot (\mathbf{u} \times \mathbf{v}))\mathbf{\Pi}.$$

The area element dS on the sphere assigns to each ordered pair of tangent vectors $\mathbf{a}, \mathbf{b}$ the number $dS(\mathbf{a}, \mathbf{b}) = \mathbf{n} \cdot (\mathbf{a} \times \mathbf{b})$, where $\mathbf{n}$ is the outward unit normal. Therefore

$$dS(\mathbf{\Pi} \times \mathbf{u}, \mathbf{\Pi} \times \mathbf{v}) = \frac{\mathbf{\Pi}}{\|\mathbf{\Pi}\|} \cdot ((\mathbf{\Pi} \times \mathbf{u}) \times (\mathbf{\Pi} \times \mathbf{v})) = \|\mathbf{\Pi}\| \mathbf{\Pi} \cdot (\mathbf{u} \times \mathbf{v}).$$

This shows that

$$\omega^{-}(\mathbf{\Pi}) = -\frac{1}{\|\mathbf{\Pi}\|} dS.$$

We have computed in §1.2 the kinetic energy of a heavy top. If the center of mass is the point of suspension of the top, that is $\ell = 0$, then one obtains the free rigid body and the total energy is the kinetic energy given by (1.38)

$$K(A, \dot{A}) = -\frac{1}{4} \operatorname{trace}((JA^{-1}\dot{A} + A^{-1}\dot{A}J)A^{-1}\dot{A}).$$

This expression is on $TSO(3)$ but using the Riemannian metric on $SO(3)$ obtained by left translating the inner product (1.34) on $\mathfrak{so}(3)$ one obtains a bundle metric on $T^*SO(3)$ whose kinetic energy is the Hamiltonian of the free rigid body. Its restriction to $\mathfrak{so}(3)^* \cong \mathbb{R}^3$ is given by

$$H(\mathbf{\Pi}) = \frac{1}{2} \mathbf{\Pi} \cdot \mathbb{I}^{-1} \mathbf{\Pi}.$$

By (6.2) the geodesic equations in left trivialization are

$$\dot{\mathbf{\Pi}} = \mathbf{\Pi} \times \mathbb{I}^{-1}\mathbf{\Pi} \quad \text{and} \quad \dot{A} = A\mathbb{I}^{-1}\mathbf{\Pi}$$

since $\nabla H(\mathbf{\Pi}) = \mathbb{I}^{-1}\mathbf{\Pi}$. The first equation is the Lie-Poisson equation on $\mathbb{R}^3$ and the second one is a linear equation with time dependent coefficients and gives the attitude matrix of the body. Since the concentric spheres centered at the origin are the coadjoint orbits, the integral curves of the first equation necessarily lie on them. In addition, the first equation is Hamiltonian on these spheres relative to the orbit symplectic form and the Hamiltonian H. Thus the solutions of the Lie-Poisson equation $\dot{\mathbf{\Pi}} = \mathbf{\Pi} \times \mathbb{I}^{-1}\mathbf{\Pi}$ are obtained by intersecting each sphere with the ellipsoids $H(\mathbf{\Pi}) = constant$. This shows that there are six equilibria four of which are stable (rotation about the long and short axes) and two of which are saddles (rotation about the middle axis).

(2) Affine group on $\mathbb{R}$. Consider the Lie group G of the transformations T :

$\mathbb{R} \to \mathbb{R}$, $T(x) = ax + b$ with $a \neq 0$. We can identify G with the set of pairs (a, b). As

$$(T_2 \circ T_1)(x) = T_2(a_1 x + b_1) = (a_2 a_1 x + a_2 b_1 + b_2),$$

then the group multiplication on $G = \{(a, b) \in \mathbb{R}^2 \mid a \neq 0\}$ is given by

$$(a_2, b_2)(a_1, b_1) = (a_2 a_1 \, , \, a_2 b_1 + b_2).$$

The identity element of G is $e = (1, 0)$ and the inverse of (a, b) is $(a, b)^{-1} = (1/a, -b/a)$. The inner conjugation automorphism $\mathrm{AD}_{(a,b)}$ is given by

$$\begin{aligned} \mathrm{AD}_{(a,b)}(c, d) &= (a, b)(c, d)(a, b)^{-1} = (a, b)(c, d) \left(\frac{1}{a}, \frac{-b}{a} \right) \\ &= (c, -cb + ad + b). \end{aligned}$$

The adjoint action is obtained differentiating $\mathrm{AD}_{(a,b)}(c, d)$ with respect to (c, d) at the identity $e = (1, 0)$ in the direction to (u, v) which gives

$$\begin{aligned} \mathrm{Ad}_{(a,b)}(u, v) &= \left. \frac{d}{dt} \right|_{t=0} \mathrm{AD}_{(a,b)}(c + tu, d + tv) \\ &= \left. \frac{d}{dt} \right|_{t=0} (c + tu, -(c + tu)b + a(d + tv) + b) \\ &= (u, av - bu). \end{aligned}$$

Thus, *the adjoint orbit through* (u, v) *is* $\{u\} \times \mathbb{R}$ *if* $(u, v) \neq (0, 0)$ *and is the origin if* $(u, v) = (0, 0)$. We shall see below that the coadjoint orbits are very different.

The underlying vector space of the Lie algebra $\mathfrak{g}$ of G is $\mathbb{R}^2$ since G is obviously open in $\mathbb{R}^2$. The Lie bracket is obtained by differentiating $\mathrm{Ad}_{(a,b)}(u, v)$ with respect to (a, b) at the identity in the direction of (r, s) which gives

$$\left. \frac{d}{dt} \right|_{t=0} (u, (a + tr)v - (b + ts)u) = (0, rv - su) = [(r, s), (u, v)] \, , \tag{6.4}$$

for $(r, s), (u, v) \in \mathfrak{g} = \mathbb{R}^2$.

Consider the pairing between $\mathfrak{g}^*$ and $\mathfrak{g}$ to be the standard inner product in $\mathbb{R}^2$, that is, $\mathfrak{g}^* = \mathbb{R}^2$. So, for $(\alpha, \beta) \in \mathfrak{g}^*$ and $(u, v) \in \mathfrak{g}$, we have

$$\begin{aligned} \left\langle \mathrm{Ad}^*_{(a,b)}(\alpha, \beta), (u, v) \right\rangle &= \langle (\alpha, \beta), \mathrm{Ad}_{(a,b)}(u, v) \rangle \\ &= \langle (\alpha, \beta), (u, av - bu) \rangle = (\alpha - \beta b)u + \beta a v, \end{aligned}$$

which shows that

$$\mathrm{Ad}^*_{(a,b)}(\alpha, \beta) = (\alpha - \beta b, \beta a).$$

So, *if* $\beta = 0$, *the coadjoint orbit through* (α, β) *is the single point* $(\alpha, 0)$, *while if* $\beta \neq 0$, *the coadjoint orbit through* (α, β) *is* $\mathbb{R}^2$ *minus the* α*-axis; this latter orbit is open in* $\mathfrak{g}^*$. This example shows that the dimensions of the adjoint and coadjoint orbits can be different.

To compute the coadjoint action of $\mathfrak{g}$ on $\mathfrak{g}^*$ we use (6.4) and the Euclidean inner product for the duality pairing to get

$$\left\langle \mathrm{ad}^*_{(u,v)}(\alpha, \beta), (r, s) \right\rangle = \langle (\alpha, \beta), [(u, v), (r, s)] \rangle = \langle (\alpha, \beta), (0, su - rv) \rangle$$
$$= su\beta - rv\beta = \langle (-v\beta, u\beta), (r, s) \rangle ,$$

that is,

$$\mathrm{ad}^*_{(u,v)}(\alpha, \beta) = (-v\beta, u\beta).$$

This shows that $T_{(\alpha,\beta)}\mathcal{O}_{(\alpha,\beta)}$ equals is $\{(0{,}0)\}$ if $\beta = 0$ or $\mathbb{R}^2$ if $\beta \neq 0$, as expected.

The minus Lie-Poisson bracket of $F, H : \mathfrak{g}^* \to \mathbb{R}$ is hence given by

$$\{F, H\}_-(\alpha, \beta) = -\beta \left(\frac{\partial F}{\partial \alpha}\frac{\partial H}{\partial \beta} - \frac{\partial F}{\partial \beta}\frac{\partial H}{\partial \alpha} \right).$$

The orbit symplectic structure (6.3) for the open orbit $\mathcal{O}$, that is, the orbit passing through (α, β) with $\beta \neq 0$, equals

$$\omega^-(\alpha, \beta)\left((\mathrm{ad}^*_{(r,s)}(\alpha, \beta), \mathrm{ad}^*_{(u,v)}(\alpha, \beta) \right) = -\langle (\alpha, \beta), [(r, s), (u, v)] \rangle$$
$$= -\beta(rv - su),$$

or, in canonical coordinates (α, β) on $\mathcal{O}_{(\alpha,\beta)}$,

$$\omega^- = -\frac{1}{\beta}\mathbf{d}\alpha \wedge \mathbf{d}\beta.$$

Given a smooth function $H : \mathfrak{g}^* = \mathbb{R}^2 \to \mathbb{R}$, the Hamiltonian vector field relative to the minus Lie-Poisson bracket is given by

$$X_{H}(\alpha, \beta) = \beta(-\partial H/\partial \beta, \partial H/\partial \alpha).$$

As is obvious from this expression, an integral curve whose initial condition (α_0, β_0) satisfies $\beta_0 > 0$ (respectively $\beta_0 < 0$) will satisfy the same condition for all time. This verifies the standard fact that the open symplectic leaves are invariant under the flow. In addition, all points on the line $\beta = 0$ are equilibria, that is, the zero dimensional orbits are also invariant under the flow, as expected.

(3) The group $\mathrm{Diff}_{\mathrm{vol}}(D)$**, vorticity representation.** Let $G = \mathrm{Diff}_{\mathrm{vol}}(D)$ be the group of volume preserving diffeomorphisms of a k-dimensional oriented

Riemannian manifold (D, g) with smooth boundary ∂D. The Riemannian volume form μ on D is the unique volume form on D which is equal to 1 on all positively oriented g-orthonormal bases of tangent vectors at all points of D.

The Riemannian volume μ naturally induces a Riemannian volume form on the boundary ∂D (relative to the induced metric) given in the following way. Let $i : \partial D \to D$ be the inclusion. If $v_1 \ldots, v_{k-1} \in T_x(\partial D)$ is a basis such that $\mu(x)(n, v_1, \ldots, v_{k-1}) > 0$ for n the outward pointing unit normal, define $\mu_{\partial D}(x)(v_1, \ldots, v_{k-1}) := \mu(x)(n, v_1, \ldots, v_{k-1})$ and extend it by skew symmetry and multilinearity to any other k-tuple of tangent vectors in $T_x \partial D$. Recall that the normal n is ***pointing outward*** if, in a (and hence any) chart on D intersecting ∂D whose image lies in the upper half space $\{(x^1, \ldots, x^k) \in \mathbb{R}^k \mid x^k \geq 0\}$, the vector n is collinear with $-\partial/\partial x^k$. The key relation that relates g, μ, and $\mu_{\partial D}$ is

$$i^*(\mathbf{i}_v \mu) = g(v, n)\mu_{\partial D} \tag{6.5}$$

for any $v \in \mathfrak{X}(D)$.

As we have seen in §5.1, formally, the Lie algebra of G is the Lie algebra $\mathfrak{X}_{\mathrm{div}}(D)$ of divergence free vector fields tangent to the boundary ∂D, endowed with minus the usual bracket of vector fields. We have identified the dual $\mathfrak{X}_{\mathrm{div}}(D)^*$ with $\mathbf{d}\Omega^1(D)$, assuming that the first cohomology group of D is zero. The weak pairing (see §4.3) between $\mathfrak{X}_{\mathrm{div}}(D)$ and $\mathbf{d}\Omega^1(D)$ was given by

$$(u, \omega) \in \mathfrak{X}_{\mathrm{div}}(D) \times \mathbf{d}\Omega^1(D) \mapsto \int_D \alpha(u)\mu \in \mathbb{R}, \quad \text{for} \quad \omega := \mathbf{d}\alpha \tag{6.6}$$

and the plus Lie-Poisson bracket by

$$\{F, H\}(\omega) = \int_D \omega\left(\frac{\delta F}{\delta \omega}, \frac{\delta H}{\delta \omega}\right)\mu, \tag{6.7}$$

for any $F, H \in \mathcal{F}(\mathfrak{X}_{\mathrm{div}}(D))$.

The coadjoint action of the diffeomorphism $\eta \in \mathrm{Diff}_{\mathrm{vol}}(D)$ on $\omega \in \mathbf{d}\Omega^1(D)$ is given by $\mathrm{Ad}^*_{\eta^{-1}} \omega = \eta_* \omega$ (see §5.1). Therefore, the coadjoint orbit passing through ω equals $\mathcal{O} = \{\eta_* \omega \mid \eta \in \mathrm{Diff}_{\mathrm{vol}}(D)\}$. The coadjoint action of $\mathfrak{X}_{\mathrm{div}}(D)$ on $\mathbf{d}\Omega^1(D) \cong \mathfrak{X}_{\mathrm{div}}(D)^*$ is hence given by

$$-\mathrm{ad}^*_v \omega = -\boldsymbol{\pounds}_v \omega. \tag{6.8}$$

Note the $-$ sign on the right hand side. Normally, one should expect a $+$ sign since

$$\left.\frac{d}{dt}\right|_{t=0} (\eta_t)_* \omega = -\boldsymbol{\pounds}_v \omega,$$

where η_t is the flow of v. However, all formulas derived abstractly use the *left* Lie algebra and, as we have seen in Section 5.1, Example 2, the left Lie algebra bracket on vector fields is *minus* the usual Lie bracket. This is why one needs to change the sign in (6.8). One can easily derive (6.8) directly: for any $u, v \in \mathfrak{X}_{\mathrm{div}}(D)$ and $\omega = \mathbf{d}\alpha \in \mathbf{d}\Omega^1(D)$, the identities $\mathrm{ad}_v u = -[v, u]$ (note the minus sign), $\boldsymbol{\pounds}_v(\alpha(u)) = (\boldsymbol{\pounds}_v\alpha)(u) + \alpha([v, u])$, (6.5), and the Stokes theorem give

$$\begin{aligned}
\langle \mathrm{ad}_v^* \omega, u \rangle = \langle \omega, \mathrm{ad}_v u \rangle = \langle \omega, -[v, u] \rangle &= -\int_D \alpha([v, u])\mu \\
&= \int_D (\boldsymbol{\pounds}_v\alpha)(u)\mu - \int_D \boldsymbol{\pounds}_v(\alpha(u))\mu \\
&= \int_D (\boldsymbol{\pounds}_v\alpha)(u)\mu - \int_D \boldsymbol{\pounds}_v(\alpha(u)\mu) \\
&= \int_D (\boldsymbol{\pounds}_v\alpha)(u)\mu - \int_{\partial D} \alpha(u)\mathbf{i}_v\mu \\
&= \langle \mathbf{d}\boldsymbol{\pounds}_v\alpha, u \rangle = \langle \boldsymbol{\pounds}_v\omega, u \rangle ,
\end{aligned}$$

which, by weak non-degeneracy of the pairing, proves (6.8).

Thus, the tangent space to the orbit $\mathcal{O}$ is

$$T_\omega \mathcal{O} = \{ \boldsymbol{\pounds}_v\omega = \mathbf{d}\mathbf{i}_v\omega \mid v \in \mathfrak{X}_{\mathrm{div}}(D) \}.$$

The orbit symplectic structure (6.3) has therefore the expression

$$\omega^+(\omega)(\boldsymbol{\pounds}_u\omega, \boldsymbol{\pounds}_v\omega) = -\int_D \langle \alpha, [u, v] \rangle \mu \quad \text{for} \quad \omega = \mathbf{d}\alpha.$$

However, if $\omega = \mathbf{d}\alpha$, for any $u, v \in \mathfrak{X}_{\mathrm{div}}(D)$ we have

$$\langle \alpha, [u, v] \rangle + \omega(u, v) = u[\langle \alpha, v \rangle] - v[\langle \alpha, u \rangle]$$

so that by the Stokes theorem and (6.5)

$$\begin{aligned}
\int_D (\langle \alpha, [u,v]\rangle + \omega(u,v))\mu &= \int_D u[\langle \alpha, v\rangle]\mu - \int_D v[\langle \alpha, u\rangle]\mu \\
&= \int_D \boldsymbol{\pounds}_u(\langle \alpha, v\rangle \mu) - \int_D \boldsymbol{\pounds}_v(\langle \alpha, u\rangle \mu) \\
&= \int_D \mathbf{d}\mathbf{i}_u(\langle \alpha, v\rangle \mu) - \int_D \mathbf{d}\mathbf{i}_v(\langle \alpha, u\rangle \mu) \\
&= \int_{\partial D} i^*(\mathbf{i}_u\langle \alpha, v\rangle \mu) - \int_{\partial D} i^*(\mathbf{i}_v\langle \alpha, u\rangle \mu) \\
&= \int_{\partial D} (i^*\langle \alpha, v\rangle) i^*(\mathbf{i}_u \mu) - \int_{\partial D} (i^*\langle \alpha, u\rangle) i^*(\mathbf{i}_v \mu) \\
&= \int_{\partial D} (i^*\langle \alpha, v\rangle) g(u,n)\mu_{\partial D} - \int_{\partial D} (i^*\langle \alpha, u\rangle) g(v,n)\mu_{\partial D} = 0
\end{aligned}$$

since, by hypothesis, $g(u,n) = g(v,n) = 0$ on ∂D. Therefore, the orbit symplectic structure is given by

$$\omega^+(\omega)(\boldsymbol{\pounds}_u\omega, \boldsymbol{\pounds}_v\omega) = \int_D \omega(u,v)\mu. \tag{6.9}$$

Let us compute the plus Lie-Poisson equations for the geodesic flow, that is, the equations $\dot{F} = \{F, H\}$ for any $F \in \mathcal{F}(\mathrm{Diff}_{\mathrm{vol}}(D))$, where

$$H(\omega) = \frac{1}{2}\int_D \|v\|^2 \mu = \frac{1}{2}\int_D v^\flat(v)\mu = \frac{1}{2}\left\langle \mathbf{d}v^\flat, v\right\rangle \tag{6.10}$$

by (6.6). Define the ***vorticity*** associated to the spatial velocity vector field v of the incompressible perfect fluid by $\omega := \mathbf{d}v^\flat$. Note that the pairing (6.6) satisfies $\langle \mathbf{d}u^\flat, v\rangle = \langle \mathbf{d}v^\flat, u\rangle$ for any $u, v \in \mathfrak{X}_{\mathrm{div}}(D)$, so that letting $\delta\omega := \mathbf{d}(\delta v)^\flat$ we get

$$\left\langle \delta\omega, \frac{\delta H}{\delta \omega}\right\rangle = \mathbf{D}H(\omega)\cdot \delta\omega = \langle \delta\omega, v\rangle,$$

that is, $\delta H/\delta\omega = v$. Thus, the plus Lie-Poisson equations (4.17) become

$$\frac{\partial \omega}{\partial t} + \boldsymbol{\pounds}_v \omega = 0, \quad \text{where} \quad \omega := \mathbf{d}v^\flat, \tag{6.11}$$

which are the *Euler equations for an incompressible homogeneous perfect fluid in vorticity formulation*. Therefore the geodesic $\eta_t \in \mathrm{Diff}_{\mathrm{vol}}(D)$ is given by solving the equation $\partial\eta_t/\partial t = v_t \circ \eta_t$ with the velocity v_t found after solving for ω equation (6.11) and then inverting the relation $\omega = \mathbf{d}v^\flat$ with boundary condition $v \cdot n = 0$, where n is the outward unit normal to the boundary ∂D.

Equation (6.11) is equivalent to any of the following statements:

(i) *The vorticity ω_t is transported by the flow.* Indeed, if ω_0 is an initial condition at $t = 0$ of (6.11), then

$$\left.\frac{d}{dt}\right|_{t=0} (\eta_t)_* \omega_0 = -(\eta_t)_* \boldsymbol{\pounds}_v \omega_0 = -\boldsymbol{\pounds}_v (\eta_t)_* \omega_0,$$

which shows that $\omega_t = (\eta_t)_* \omega_0$ solves (6.11). By uniqueness, this is the only solution of (6.11) with ω_0 as initial condition.

(ii) *Solution curves of the vorticity equation* (6.11) *remain on coadjoint orbits in $\mathfrak{X}^*_{\mathrm{div}}(D)$.* Indeed, since the solution of (6.11) is $\omega_t = (\eta_t)_* \omega_0$, where η_t is the flow of v, it follows that ω_t necessarily lies on the coadjoint orbit containing ω_0.

(iii) *Kelvin's circulation theorem: For any loop C in D bounding a surface S, the circulation*

$$\int_{C_t} v_t^\flat = \text{constant},$$

where $C_t := \eta_t(C)$ and η_t is the flow of v. Indeed, by change of variables and Stokes' theorem, for $S_t := \eta_t(S)$, we have

$$\int_{C_t} v_t^\flat = \int_{S_t} \mathbf{d} v_t^\flat = \int_{S_t} \omega_t = \int_{S_t} (\eta_t)_* \omega_0 = \int_S \omega_0 = \text{constant}.$$

(4) The group $\mathrm{Diff}_{\mathrm{vol}}(D)$, velocity representation. In §4.3 we have also identified $\mathfrak{X}_{\mathrm{div}}(D)$ with itself by the weak L^2 pairing $\langle\, , \rangle : \mathfrak{X}_{\mathrm{div}}(D) \times \mathfrak{X}_{\mathrm{div}}(D) \to \mathbb{R}$ given by

$$\langle u, v \rangle = \int_D g(u, v)\, \mu.$$

The plus Lie-Poisson bracket is given by (4.9), namely

$$\{F, H\}(v) = -\int_D g\left(v, \left[\frac{\delta F}{\delta v}, \frac{\delta H}{\delta v}\right]\right) \mu,$$

for $F, H \in \mathcal{F}(\mathrm{Diff}_{\mathrm{vol}}(D))$. Using the change of variables formula and the fact that η is volume preserving, the coadjoint action is computed to be

$$\begin{aligned}
\langle \mathrm{Ad}^*_{\eta^{-1}} u, v \rangle &= \langle u, \mathrm{Ad}_{\eta^{-1}} v \rangle = \langle u, \eta^* v \rangle = \int_D g(u, \eta^* v) \mu \\
&= \int_D g(u, T\eta^{-1} \circ v \circ \eta) \mu \\
&= \int_D g((T\eta^{-1})^\dagger \circ u \circ \eta^{-1}, v) \mu = \langle \mathbb{P}\left((T\eta^{-1})^\dagger \circ u \circ \eta^{-1}\right), v \rangle,
\end{aligned}$$

where $(T\eta)^\dagger$ is the pointwise adjoint relative to the metric g of the fiberwise linear map $T\eta : TD \to TD$ and $\mathbb{P} : \mathfrak{X}(D) \to \mathfrak{X}_{\mathrm{div}}(D)$ is the Helmholtz projector. In defining $\mathbb{P}$ we used the *Helmholtz decomposition* which is the Hodge decomposition on forms for the special case of one-forms and formulated in terms of vector fields: *every vector field on D can be uniquely decomposed as an L^2 orthogonal sum of a divergence free vector field tangent to the boundary and the gradient of a function.* Thus, since the L^2 pairing is weakly nondegenerate on $\mathfrak{X}_{\mathrm{div}}(D)$, we conclude

$$\mathrm{Ad}^*_{\eta^{-1}} u = \mathbb{P}\left((T\eta^{-1})^\dagger \circ u \circ \eta^{-1}\right)$$

for any $u \in \mathfrak{X}_{\mathrm{div}}(D) \cong \mathfrak{X}_{\mathrm{div}}(D)^*$. Therefore, the coadjoint orbit $\mathcal{O}$ passing through $w \in \mathfrak{X}_{\mathrm{div}}(D)$ equals $\mathcal{O} = \{\mathbb{P}\left((T\eta)^\dagger \circ w \circ \eta\right) \mid \eta \in \mathrm{Diff}_{\mathrm{vol}}(D)\}$. Compared to the vorticity representation, the expression of the coadjoint action and of the coadjoint orbit are more complicated. This also shows that different identifications of the dual can give rise to different expressions for the coadjoint orbits. We shall remark below why it is important to work with both representations when considering the Euler equations.

The orbit symplectic structure (6.3) has hence the expression

$$\omega^+(w)(u_{\mathcal{O}}(w), v_{\mathcal{O}}(w)) = -\int_D g(w, [u, v])\mu \tag{6.12}$$

for any $w \in \mathcal{O}$ and any $u, v \in \mathfrak{X}_{\mathrm{div}}(D)$.

It is interesting to give the expression of the value of the infinitesimal generator $v_{\mathcal{O}}(w)$ at $w \in \mathcal{O}$. If η_t is the flow of w, we have (see [MaRa95])

$$\left.\frac{d}{dt}\right|_{t=0} \left[(T\eta_t)^\dagger \circ v \circ \eta_t\right] = \nabla_w v + (v \cdot \nabla w)^\sharp,$$

where $v \cdot \nabla w$ is the contraction of v with the upper index of ∇w, that is, if

$$(\nabla w)^j_k = \frac{\partial w^j}{\partial x^k} + \Gamma^j_{k\ell} w^\ell$$

then

$$(v \cdot (\nabla w))_k = g_{mj} v^m (\nabla w)^j_k.$$

Thus $v \cdot \nabla w \in \Omega^1(D)$ and its associated vector field using the metric is $(v \cdot \nabla w)^\sharp \in \mathfrak{X}(D)$. This shows that

$$v_{\mathcal{O}}(w) = -\operatorname{ad}^*_v w = -\mathbb{P}\left(\nabla_w v + (v \cdot \nabla w)^\sharp\right). \tag{6.13}$$

We shall not compute here the plus Lie-Poisson equations for the Hamiltonian (6.10) because we shall carry out an identical computation in Section 6.4, Example 4, when dealing with the Euler-Poincaré equations. We mention only that they are the classical Euler equations for an incompressible homogeneous perfect fluid

$$\begin{cases} \frac{\partial v}{\partial t} + \nabla_v v = -\nabla p \\ \\ \operatorname{div} v = 0, \quad v \cdot n = 0, \end{cases} \tag{6.14}$$

where p is the ***pressure*** and n is the outward unit normal to the boundary ∂D. The pressure p exists and is determined up to a constant since it is a solution of the following Neumann problem

$$\Delta p = \operatorname{div} \nabla_v v \quad \text{in} \quad D, \qquad \text{with} \qquad \frac{\partial p}{\partial n} = (\nabla_v v) \cdot n \quad \text{on} \quad \partial D$$

obtained by taking the divergence and the inner product with n on the boundary of the Euler equation; here $\Delta := -\operatorname{div} \circ \operatorname{grad}$ is the Laplacian on functions. Thus p is a nonlinear functional of the Eulerian velocity v.

Since the solutions of Lie-Poisson equations always lie on coadjoint orbits, we can conclude that the solution of the Euler equations with initial condition $v_0 \in \mathfrak{X}_{\text{div}}(D)$ necessarily lies on the coadjoint orbit $\{\mathbb{P}\left((T\eta)^\dagger \circ v_0 \circ \eta\right) \mid \eta \in \text{Diff}_{\text{vol}}(D)\}$.

6.4 Euler-Poincaré Reduction

A Hamiltonian H on T^*Q often arises from a Lagrangian L on TQ. Namely, as seen in section 3.3, the two formalisms are equivalent when the Legendre transform $\mathbb{F}L : TQ \to T^*Q$ is a diffeomorphism. In the previous sections we saw how to reduce a Poisson bracket on T^*G to one on $\mathfrak{g}^*$ via Lie-Poisson reduction and how Hamiltonian dynamics on T^*G induces Lie-Poisson dynamics on $\mathfrak{g}^*$. In this section we study the passage from TG to $\mathfrak{g}$ in a context appropriate for the Lagrangian formalism. As Lagrangian mechanics is based on variational principles, it is natural that the basic objects to be reduced here are the variational principles rather than the Poisson bracket or symplectic form as was the case in Lie-Poisson reduction.

Euler-Poincaré Reduction starts with a left (respectively right) invariant Lagrangian $L : TG \to \mathbb{R}$ on the tangent bundle of a Lie group G. Recall that this means that $L(T_g L_h(v)) = L(v)$, respectively $L(T_g R_h(v)) = L(v)$, for all $g, h \in G$ and all $v \in T_g G$.

Theorem 6.6 *Let G be a Lie group, $L : TG \to \mathbb{R}$ a left-invariant Lagrangian, and $l := L|_{\mathfrak{g}} : \mathfrak{g} \to \mathbb{R}$ be its restriction to $\mathfrak{g}$. For a curve $g(t) \in G$, let $\xi(t) = g(t)^{-1} \cdot \dot{g}(t) := T_{g(t)} L_{g(t)^{-1}} \dot{g}(t) \in \mathfrak{g}$. Then the following are equivalent:*

(i) *$g(t)$ satisfies the Euler-Lagrange equations for L on G.*

(ii) *The variational principle*

$$\delta \int_a^b L(g(t), \dot{g}(t))dt = 0$$

holds, for variations with fixed endpoints.

(iii) *The* ***Euler-Poincaré equations*** *hold:*

$$\frac{d}{dt}\frac{\delta l}{\delta \xi} = \mathrm{ad}_\xi^* \frac{\delta l}{\delta \xi}.$$

(iv) *The variational principle*

$$\delta \int_a^b l(\xi(t))dt = 0$$

holds on $\mathfrak{g}$, using variations of the form $\delta\xi = \dot{\eta} + [\xi, \eta]$, where $\eta(t)$ is an arbitrary path in $\mathfrak{g}$ that vanishes at the endpoints, i.e $\eta(a) = \eta(b) = 0$.

Proof The equivalence **(i)** $\Longleftrightarrow$ **(ii)** is the variational principle of Hamilton (see Theorem 3.5). To show that **(ii)** $\Longleftrightarrow$ **(iv)** we need to compute variations $\delta\xi$ induced by δg. We will do it for matrix groups to simplify the exposition.

Let $\xi = g^{-1}\dot{g}$ and g_ϵ a family of curves in G such that $g_0(t) = g(t)$ and denote $\delta g := (dg_\epsilon(t)/d\epsilon)|_{\epsilon=0}$. Then we have

$$\delta\xi = \left.\frac{d}{d\epsilon}\right|_{\epsilon=0} (g_\epsilon^{-1}\dot{g}_\epsilon) = -g^{-1}(\delta g)g^{-1}\dot{g} + g^{-1} \left.\frac{d^2 g}{dt d\epsilon}\right|_{\epsilon=0}. \tag{6.15}$$

Let $\eta := g^{-1}\delta g$, that is, $\eta(t)$ is an arbitrary curve in $\mathfrak{g}$ with the only restriction that it vanishes at the endpoints. Then we get

$$\frac{d\eta}{dt} = \frac{d}{dt}\left(g^{-1} \left.\frac{d}{d\epsilon}\right|_{\epsilon=0} g_\epsilon\right) = -g^{-1}\dot{g}g^{-1}(\delta g) + g^{-1}\left.\frac{d^2 g}{dt d\epsilon}\right|_{\epsilon=0}. \tag{6.16}$$

Taking the difference of (6.15) and (6.16) we get

$$\delta\xi - \dot{\eta} = -g^{-1}(\delta g)g^{-1}\dot{g} + g^{-1}\dot{g}g^{-1}(\delta g) = \xi\eta - \eta\xi = [\xi, \eta],$$

that is, $\delta\xi = \dot{\eta} + [\xi, \eta]$.

Left invariance of L together with the formula just deduced prove the equivalence of **(ii)** and **(iv)**.

To avoid the assumption that G is a matrix group and to do the general case for any Lie group, the same proof works using the following lemma.

Lemma 6.7 *Let $g : U \subset \mathbb{R}^2 \to G$ be a smooth map and denote its partial derivatives by*

$$\xi(t,\varepsilon) := T_{g(t,\varepsilon)} L_{g(t,\varepsilon)^{-1}} \frac{\partial g(t,\varepsilon)}{\partial t}, \; \eta(t,\varepsilon) := T_{g(t,\varepsilon)} L_{g(t,\varepsilon)^{-1}} \frac{\partial g(t,\varepsilon)}{\partial \varepsilon}. \tag{6.17}$$

Then

$$\frac{\partial \xi}{\partial \varepsilon} - \frac{\partial \eta}{\partial t} = [\xi, \eta]. \tag{6.18}$$

Conversely, if $U \subset \mathbb{R}^2$ is simply connected and $\xi, \eta : U \to \mathfrak{g}$ are smooth functions satisfying (6.18)*, then there exists a smooth function $g : U \to G$ such that* (6.17) *holds.*

Let us show that **(iii)** $\Longleftrightarrow$ **(iv)**. We have

$$\begin{aligned} \delta \int_a^b l(\xi(t))dt &= \int_a^b \left\langle \frac{\delta l}{\delta \xi}, \delta \xi \right\rangle dt = \int_a^b \left\langle \frac{\delta l}{\delta \xi}, \dot{\eta} + \mathrm{ad}_\xi \eta \right\rangle dt \\ &= \int_a^b \left\langle \frac{\delta l}{\delta \xi}, \dot{\eta} \right\rangle dt + \int_a^b \left\langle \frac{\delta l}{\delta \xi}, \mathrm{ad}_\xi \eta \right\rangle dt \\ &= \int_a^b \left\langle -\frac{d}{dt}\frac{\delta l}{\delta \xi} + \mathrm{ad}^*_\xi \frac{\delta l}{\delta \xi}, \eta \right\rangle dt, \end{aligned}$$

where the last equality follows from integration by parts and because the curve $\eta(t)$ vanishes at the endpoints. Thus, $\delta \int_a^b l(\xi(t))dt = 0$ if and only if the right hand side of the previous equality vanishes for any $\eta(t)$ that vanishes at the endpoints, which proves that this is equivalent to

$$\frac{d}{dt}\frac{\delta l}{\delta \xi} = \mathrm{ad}^*_\xi \frac{\delta l}{\delta \xi}.$$

which are the Euler-Poincaré equations. ∎

In case of right invariant Lagrangians on TG the same theorem holds with the changes that the Euler-Poincaré equations are

$$\frac{d}{dt}\frac{\delta l}{\delta \xi} = -\mathrm{ad}^*_\xi \frac{\delta l}{\delta \xi}$$

and the variations are taken of the form $\delta \xi = \dot{\eta} - [\xi, \; \eta]$.

As was the case for Lie-Poisson dynamics, there is a reconstruction procedure in this case. The goal is to find the solution $v(t) \in T_{g(t)}G$ of the Euler-Lagrange equations with initial conditions $g(0) = g_0$ and $\dot{g}(0) = v_0$ knowing the solution of the Euler-Poincaré equations. To do this, first solve the initial value problem for the Euler-Poincaré equations:

$$\begin{cases} \frac{d}{dt}\frac{\delta l}{\delta \xi} = \operatorname{ad}_{\xi}^{*}\frac{\delta l}{\delta \xi} \\ \\ \xi(0) = \xi_0 := g_0^{-1}v_0 \end{cases}$$

and then solve the "linear differential equation with time-dependent coefficients"

$$\begin{cases} \dot{g}(t) = g(t)\xi(t) \\ \\ g(0) = g_0. \end{cases}$$

The Euler-Poincaré reduction theorem guarantees then that $v(t) = \dot{g}(t) = g(t)\cdot\xi(t)$ is a solution of the Euler-Lagrange equations with initial condition $v_0 = g_0\xi_0$.

A similar statement holds, with obvious changes for right invariant Lagrangian systems on TG.

The relationship between Lie-Poisson and Euler-Poincaré reduction, similar to the link between the Hamiltonian and Lagrangian formulations discussed in Section 3.3, is the following. Define the Legendre transformation $\mathbb{F}l : \mathfrak{g} \to \mathfrak{g}^*$ by

$$\mathbb{F}l(\xi) = \frac{\delta l}{\delta \xi} = \mu,$$

and let $h(\mu) := \langle \mu, \xi\rangle - l(\xi)$. Assuming that $\mathbb{F}l$ is a diffeomorphism, we get

$$\frac{\delta h}{\delta \mu} = \xi + \left\langle \mu, \frac{\delta \xi}{\delta \mu}\right\rangle - \left\langle \frac{\delta l}{\delta \xi}, \frac{\delta \xi}{\delta \mu}\right\rangle = \xi.$$

So the Euler-Poincaré equations for l are equivalent to the Lie-Poisson equations for h:

$$\frac{d}{dt}\left(\frac{\delta l}{\delta \xi}\right) = \operatorname{ad}_{\xi}^{*}\frac{\delta l}{\delta \xi} \Longleftrightarrow \dot{\mu} = \operatorname{ad}_{\frac{\delta h}{\delta \mu}}^{*}\mu.$$

There is one more element to be discussed: the reduction of Hamilton's phase space principle due to [Cendra et al.], called *Hamilton-Poincaré reduction*.

Theorem 6.8 *Let G be a Lie group, $H : T^*G \to \mathbb{R}$ a left-invariant Hamiltonian, and $h := H|_{\mathfrak{g}^*} : \mathfrak{g}^* \to \mathbb{R}$ be its restriction to $\mathfrak{g}^*$. For a curve $\alpha(t) \in T^*G$, let $g(t) = \pi(\alpha(t)) \in G$, where $\pi : T^*G \to G$ is the cotangent bundle projection, and define $\mu(t) = g(t)^{-1}\cdot\alpha(t) := T_eL_{g(t)}\alpha(t) \in \mathfrak{g}^*$, and $\xi(t) = g(t)^{-1}\cdot\dot{g}(t) := T_{g(t)}L_{g(t)^{-1}}\dot{g}(t) \in \mathfrak{g}$. The following are equivalent:*

(i) Hamilton's Phase Space Principle. *The curve $\alpha(t) \in T^*G$ is a critical point of the action*

$$\int_{t_0}^{t_1} \left(\langle \Theta, \dot{\alpha}(t) \rangle - H(\alpha(t)) \right) dt,$$

where the variations $\delta\alpha$ satisfy $T\pi_G(\delta\alpha(t_i)) = 0$, for $i = 0, 1$.

(ii) *Hamilton's equations hold on T^*G.*

(iii) The Hamilton-Poincaré Variational Principle. *The curve $(\mu(t), \xi(t)) \in \mathfrak{g}^* \times \mathfrak{g}$ is a critical point of the action*

$$\int_{t_0}^{t_1} \left(\langle \mu(t), \xi(t) \rangle - h(\mu(t)) \right) dt,$$

with variations $\delta\xi(t) = \dot{\eta}(t) + [\xi(t), \eta(t)] \in \mathfrak{g}$, where $\eta(t)$ is an arbitrary curve satisfying $\eta(t_i) = 0$, for $i = 0, 1$, and $\delta\mu(t) \in \mathfrak{g}^*$ is arbitrary.

(iv) *The Lie-Poisson equations hold:*

$$\dot{\mu} = -\operatorname{ad}^*_{\frac{\delta h}{\delta \mu}} \mu.$$

There is, of course, a similar statement for right invariant systems where one has the change the sign in front of the Lie-Poisson equation and in front of the bracket defining $\delta\xi$.

Proof The equivalence **(i)** $\Longleftrightarrow$ **(ii)** is Hamilton's phase space variational principle which holds on any cotangent bundle (see Theorem 3.7). The equivalence **(ii)** $\Longleftrightarrow$ **(iv)** is the Lie-Poisson reduction and reconstruction of dynamics (see Theorems 6.1 and 6.3). We now show that **(iii)** $\Longleftrightarrow$ **(iv)**. The variation

$$\begin{aligned}
\delta \int_{t_0}^{t_1} & \left(\langle \mu(t), \xi(t) \rangle - h(\mu(t)) \right) dt \\
&= \int_{t_0}^{t_1} \left(\langle \delta\mu, \xi \rangle + \langle \mu, \delta\xi \rangle - \left\langle \delta\mu, \frac{\delta h}{\delta \mu} \right\rangle \right) dt \\
&= \int_{t_0}^{t_1} \left(\left\langle \delta\mu, \xi - \frac{\delta h}{\delta \mu} \right\rangle + \langle \mu, \dot{\eta} + \operatorname{ad}_\xi \eta \rangle \right) dt \\
&= \int_{t_0}^{t_1} \left(\left\langle \delta\mu, \xi - \frac{\delta h}{\delta \mu} \right\rangle + \left\langle -\dot{\mu} + \operatorname{ad}^*_\xi \mu, \eta \right\rangle \right) dt
\end{aligned}$$

vanishes for any functions $\delta\mu(t) \in \mathfrak{g}^*$ and $\eta(t) \in \mathfrak{g}$ (vanishing at the endpoints t_0 and t_1) if and only if $\xi = \delta h / \delta\mu$ and $\dot{\mu} = -\operatorname{ad}^*_\xi \mu$, that is, when the Lie Poisson equations in **(iv)** hold. ■

If the Lagrangian L is hyperregular then any of the statements in Theorem

6.8 are equivalent to any of the statements in Theorem 6.6. The link between the reduced Lagrangian l and the reduced Hamiltonian h is given by $h(\mu) = \langle \mu, \xi \rangle - l(\xi)$, where $\mu = \delta l / \delta \xi$, as was already remarked earlier.

Let us work out a few examples in detail.

(1) Free rigid body. As we saw in Example 1 of Section 6.3, the restriction of the Lagrangian of the free rigid body to $\mathbb{R}^3 \cong \mathfrak{so}(3)$ is $L(\mathbf{\Omega}) = \mathbf{\Omega} \cdot \mathbb{I}\mathbf{\Omega}/2$ (see (1.36) with $\mathbf{\Pi} = \mathbb{I}\mathbf{\Omega}$) and hence, using the identity $\mathrm{ad}^*_{\hat{\mathbf{u}}} \tilde{\mathbf{\Pi}} = (\mathbf{\Pi} \times \mathbf{u})\tilde{}$ and Theorem 6.6, we get the Euler-Poincaré equations

$$\mathbb{I}\dot{\mathbf{\Omega}} = \mathbb{I}\mathbf{\Omega} \times \mathbf{\Omega}$$

and we recognize the Euler equations, this time formulated in terms of the body angular velocity $\mathbf{\Omega}$ as opposed to the body angular momentum $\mathbf{\Pi}$. The Legendre transformation is $\mathbf{\Pi} = \mathbb{I}\mathbf{\Omega}$. Looking back at the computations involving the variational principle carried out at the end of Section 1.2 and setting in those computations $\ell = 0$, one recognizes the variational principle on $\mathfrak{so}(3)$ for the free rigid body Lagrangian.

(2) Lagrangian systems on the affine algebra. Consider the affine Lie algebra introduced in example 2 of Section 6.3 and let $L : TG \to \mathbb{R}$ be a left invariant Lagrangian. The Euler-Poincaré equations for $l : \mathbb{R}^2 \to \mathbb{R}$ are

$$\frac{d}{dt}\frac{\partial l}{\partial u} = -v\frac{\partial l}{\partial v}, \qquad \frac{d}{dt}\frac{\partial l}{\partial v} = u\frac{\partial l}{\partial v}.$$

(3) Incompressible homogeneous fluids, vorticity representation. In Section 6.3, Example 3, we have studied the motion of an incompressible homogeneous fluid in vorticity representation as a Lie-Poisson system. Now we shall derive the same equations (6.11) as Euler-Poincaré equations. We recall that $\mathrm{ad}^*_v \omega = \boldsymbol{\pounds}_v \omega$ (see (6.8)) and

$$l(v) = \frac{1}{2}\int_D \|v\|^2 \mu = \frac{1}{2}\langle \omega, v \rangle,$$

where $\mathbf{d}v^\flat = \omega$ (see (6.10)). Therefore, $\delta l/\delta v = \omega$ and hence the Euler-Poincaré equations (for a right invariant system on the tangent bundle of a Lie group) are given by

$$\frac{\partial \omega}{\partial t} + \boldsymbol{\pounds}_v \omega = 0.$$

(4) Incompressible homogeneous fluids, velocity representation. In Section 6.3, Example 4, we have presented in Lie-Poisson setting the equations of motion of an ideal incompressible homogeneous fluid in velocity representation.

These are the Euler equations (6.14) which were not explicitly derived there because this will be done now. To do this, one could proceed by using the general expression for the Euler-Poincaré equations and formula (6.13) for the coadjoint action. Even though (6.13) was not proved in these notes, let us do it first this way. Since

$$l(v) = \frac{1}{2} \int_D \|v\|^2 \mu$$

is the kinetic energy of the fluid and we use the L^2 weak pairing to identify $\mathfrak{X}_{\mathrm{div}}(D)$ with itself, we have $\delta l / \delta v = v$ and so the Euler-Poincaré equations are

$$\frac{\partial v}{\partial t} = -\operatorname{ad}_v^* v = -\mathbb{P}\left(\nabla_v v + (v \cdot \nabla v)^\sharp\right).$$

Write, according to the Helmholtz decomposition,

$$\nabla_v v + (v \cdot \nabla v)^\sharp = \mathbb{P}\left(\nabla_v v + (v \cdot \nabla v)^\sharp\right) - \nabla q$$

for some smooth function $q : D \to \mathbb{R}$ and note that $(v \cdot \nabla v)^\sharp = \nabla(\|v\|^2/2)$. The Euler-Poincaré equations become hence

$$\frac{\partial v}{\partial t} + \nabla_v v = -\nabla\left(\frac{1}{2}\|v\|^2 + q\right)$$

which are the Euler equations for the pressure $q + \|v\|^2/2$. Therefore, *the Euler equations for an ideal incompressible homogeneous fluid are the spatial representation of the geodesic spray on* $\mathrm{Diff}_{\mathrm{vol}}(D)$ *for the right invariant metric whose value on* $\mathfrak{X}_{\mathrm{div}}(D)$ *is the* L^2 *inner product. This proves that the solutions of the Euler equations are geodesics on* $\mathrm{Diff}_{\mathrm{vol}}(D)$.

Even though the computation above proves the claim that the Euler-Poincaré equations are the Euler equations for an ideal incompressible homogeneous fluid, it is unsatisfactory since it relies on the unproven formula (6.13). Due to its importance we shall derive it once more directly.

We begin by recalling that $\delta l / \delta v = v$ and hence for any $w \in \mathfrak{X}_{\mathrm{div}}(D)$ we have

$$\left\langle -\operatorname{ad}_v^* \frac{\delta l}{\delta v}, w \right\rangle = \left\langle \frac{\delta l}{\delta v}, -\operatorname{ad}_v w \right\rangle = \langle v, [v, w] \rangle . \tag{6.19}$$

In order to isolate w in $\langle v, [v, w] \rangle$, let us compute the Lie derivative of $v^\flat(w)\mu$ along the vector field v in two different ways. We have

$$\begin{aligned} \boldsymbol{\pounds}_v \left(v^\flat(w)\mu\right) &= \mathbf{i}_v \mathbf{d}\left(v^\flat(w)\mu\right) + \mathbf{d}\mathbf{i}_v\left(v^\flat(w)\mu\right) \\ &= \mathbf{d}\left(v^\flat(w)\mathbf{i}_v\mu\right). \end{aligned} \tag{6.20}$$

On the other hand, since $\boldsymbol{\pounds}_v\mu = (\operatorname{div} v)\mu = 0$, we get

$$\begin{aligned}\boldsymbol{\pounds}_v\left(v^\flat(w)\mu\right) &= \left(\boldsymbol{\pounds}_v v^\flat\right)(w)\mu + v^\flat\left(\boldsymbol{\pounds}_v w\right)\mu + v^\flat(w)\boldsymbol{\pounds}_v\mu \\ &= \left(\boldsymbol{\pounds}_v v^\flat\right)(w)\mu + v^\flat([v,w])\mu. \end{aligned} \tag{6.21}$$

Then, from equations (6.20) and (6.21), we get

$$\mathbf{d}\left(v^\flat(w)\mathbf{i}_v\mu\right) = \left(\boldsymbol{\pounds}_v v^\flat\right)(w)\mu + v^\flat([v,w])\mu$$

and hence by integration

$$\int_D \mathbf{d}\left(v^\flat(w)\mathbf{i}_v\mu\right) = \int_D \left(\boldsymbol{\pounds}_v v^\flat\right)(w)\mu + \int_D v^\flat([v,w])\mu.$$

The left hand side is zero by Stokes' theorem and (6.5) and so

$$\begin{aligned}\langle v, [v,w]\rangle &= \int_D g(v,[v,w])\mu = \int_D v^\flat([v,w])\mu \\ &= -\int_D \left(\boldsymbol{\pounds}_v v^\flat\right)(w)\mu = -\int_D g\left(\left(\boldsymbol{\pounds}_v v^\flat\right)^\sharp, w\right)\mu \\ &= -\int_D g\left(\mathbb{P}\left(\left(\boldsymbol{\pounds}_v v^\flat\right)^\sharp\right), w\right)\mu = -\left\langle \mathbb{P}\left(\left(\boldsymbol{\pounds}_v v^\flat\right)^\sharp\right), w\right\rangle\end{aligned}$$

which, using (6.19), shows that

$$\operatorname{ad}_v^*\frac{\delta l}{\delta v} = \mathbb{P}\left(\left(\boldsymbol{\pounds}_v v^\flat\right)^\sharp\right)$$

by weak non-degeneracy of the L^2 pairing on divergence free vector fields tangent to the boundary. Thus the Euler-Poincaré equations are

$$\frac{\partial v}{\partial t} + \mathbb{P}\left(\left(\boldsymbol{\pounds}_v v^\flat\right)^\sharp\right) = 0.$$

Since $(\boldsymbol{\pounds}_v v^\flat)^\sharp = \nabla_v v + \frac{1}{2}\nabla\|v\|^2$, writing $\nabla_v v = \mathbb{P}(\nabla_v v) - \nabla p$, for some smooth function $p : D \to \mathbb{R}$, the above equation becomes

$$\begin{aligned} 0 &= \frac{\partial v}{\partial t} + \mathbb{P}\left(\nabla_v v + \frac{1}{2}\nabla\|v\|^2\right) = \frac{\partial v}{\partial t} + \mathbb{P}\left(\mathbb{P}(\nabla_v v) - \nabla p + \frac{1}{2}\nabla\|v\|^2\right) \\ &= \frac{\partial v}{\partial t} + \mathbb{P}(\nabla_v v) = \frac{\partial v}{\partial t} + \nabla_v v + \nabla p \end{aligned}$$

which are the Euler equations (6.14).

(5) KdV equation and the Virasoro algebra. Here we show, following [OKh87], that the periodic KdV equation is the Euler-Poincaré equation on the Virasoro algebra $\mathfrak{v}$ corresponding to the geodesic flow of the L^2 right invariant metric on the Virasoro group.

The Lie algebra $\mathfrak{X}(S^1)$ of vector fields on the circle, identified with the periodic functions of period 1, with Lie bracket given by $[u, v] = uv' - u'v$, has a unique central extension by $\mathbb{R}$. This unique central extension is the ***Virasoro algebra***

$$\mathfrak{v} := \left\{(u, a) \in \mathfrak{X}(S^1) \times \mathbb{R}\right\},$$

with Lie bracket

$$[(u, a), (v, b)] := \left(-uv' + u'v, \gamma \int_0^1 u'(x)v''(x)\, dx\right),$$

where the first argument is the (left) Lie bracket on $\mathfrak{X}(S^1)$, the second is the Gelfand-Fuchs cocycle, and $\gamma \in \mathbb{R}$ is a constant.

We identify $\mathfrak{v}^*$ with $\mathfrak{v}$ using the weak L^2 pairing

$$\langle (u, a)\, ,\, (v, b)\rangle = ab + \int_0^1 u(x)v(x)\, dx.$$

We also consider the right invariant weak Riemannian metric whose value on $\mathfrak{v}$ is the expression above. We are interested in the geodesic equations for this metric. To see who they are, we compute the Euler-Poincaré equations for the kinetic energy of this metric.

We begin by computing the coadjoint action of $\mathfrak{v}$ on $\mathfrak{v}^*$:

$$\begin{aligned}
\left\langle \mathrm{ad}^*_{(u,a)}(v, b), (w, c)\right\rangle &= \langle (v, b), [(u, a)\, ,\, (w, c)]\rangle \\
&= \left\langle (v, b), \left(-uw' + u'w, \gamma \int_0^1 u'(x)w''(x)\, dx\right)\right\rangle \\
&= b\gamma \int_0^1 u'w''dx - \int_0^1 vuw'dx + \int_0^1 vu'wdx \\
&= \int_0^1 (b\gamma u''' + 2u'v + uv')\, wdx \\
&= \langle (b\gamma u''' + 2u'v + uv', 0), (w, c)\rangle\,,
\end{aligned}$$

where from the third to the fourth equality we have used integration by parts, twice for the first term and once for the second term, as well as the null conditions on the boundary. So,

$$\mathrm{ad}^*_{(u,a)}(v, b) = (b\gamma u''' + 2u'v + uv', 0) \tag{6.22}$$

and the Euler-Poincaré equations (for a right invariant system) determined by

$l : \mathfrak{v} \to \mathbb{R}$ are

$$\frac{d}{dt}\left(\frac{\delta l}{\delta u}, \frac{\delta l}{\delta a}\right) = -\mathrm{ad}^*_{(u,a)}\left(\frac{\delta l}{\delta u}, \frac{\delta l}{\delta a}\right)$$
$$= -\left(\gamma\frac{\delta l}{\delta a}u''' + 2u'\frac{\delta l}{\delta u} + u\left(\frac{\delta l}{\delta u}\right)', 0\right),$$

that is,

$$\frac{d}{dt}\frac{\delta l}{\delta a} = 0, \qquad \frac{d}{dt}\frac{\delta l}{\delta u} = -\left(\gamma\frac{\delta l}{\delta a}u''' + 2u'\frac{\delta l}{\delta u} + u\left(\frac{\delta l}{\delta u}\right)'\right). \tag{6.23}$$

If we are interested in the L^2 geodesic flow, then we take

$$l(u,a) = \frac{1}{2}\left(a^2 + \int_0^1 u^2(x)dx\right)$$

so that $\frac{\delta l}{\delta a} = a$ and $\frac{\delta l}{\delta u} = u$. Thus the corresponding Euler-Poincaré equations are:

$$\frac{da}{dt} = 0, \qquad \frac{du}{dt} = -\gamma a u''' - 3u'u.$$

Hence a is constant and we get

$$u_t + 3u_x u + \gamma a u_{xxx} = 0,$$

which is one of the forms of the KdV equation (for example by choosing $a = \gamma = 1$). To get the expression (4.10), that is,

$$u_t + 6u_x u + u_{xxx} = 0$$

one needs to rescale time ($\tau(t) = t/2$) and make an appropriate choice of the constants ($a = 1/(2\gamma)$). This shows that *the solutions of the KdV equation are geodesics of the L^2 right invariant metric on the Virasoro group.*

(6) Camassa-Holm equation and the Virasoro algebra. Let us compute the Euler-Poincaré equations for the kinetic energy of the H^1 metric

$$\langle (u,a),(v,b)\rangle := ab + \int_0^1 (u(x)v(x) + u'(x)v'(x))dx$$

on the Virasoro algebra. As before, we identify $\mathfrak{v}^*$ with $\mathfrak{v}$ using the L^2 inner product. Thus $\delta l/\delta a = a$ and $\delta l/\delta u = u - u''$. The Euler-Poincaré equations (6.23) are given by $a =$ constant and

$$\frac{d}{dt}(u - u'') = -\left(\gamma a u''' + 2u'(u - u'') + u\,(u - u'')'\right)$$
$$= -3uu' + 2u'u'' + uu''' - \gamma a u'''.$$

This is the Camassa-Holm equation:

$$u_t - u_{txx} = -3uu_x + 2u_x u_{xx} + uu_{xxx} - \gamma a u_{xxx}.$$

Shifting $u \mapsto u + \gamma a$ brings the Camassa-Holm equation into yet another form, often used in the literature

$$u_t - u_{txx} = -3uu_x + 2u_x u_{xx} + uu_{xxx} - 3\gamma a u_x.$$

Thus, *the solutions of the Camassa-Holm equation are geodesics of the H^1 right invariant metric on the Virasoro group*, a result due to [Mis02].

An identical computation shows that *the solutions of Hunter-Saxon equation*

$$u_{txx} = -2u_x u_{xx} - uu_{xxx}$$

are geodesics of the right invariant degenerate metric on the Virasoro group whose value on $\mathfrak{v}$ is

$$\langle (u, a), (v, b) \rangle = ab + \int_0^1 u'(x)v'(x)dx,$$

a result due to [KhMis03]. Eliminating the degeneracy of the metric means looking at the Hunter-Saxon equation on the homogeneous space which is the quotient of the Virasoro group by rotations of the circle.

To see this, note first that if l is the kinetic energy of this degenerate metric then $\delta l/\delta u = -u''$, so again by (6.23) we get

$$-u_{txx} = -2u_x u_{xx} - uu_{xxx} - \gamma a u_{xxx}.$$

Reversing time and shifting $u \mapsto u - \gamma a$ yields the Hunter-Saxon equation.

7 Symplectic Reduction

In previous lectures we focussed our attention to the particular cases of reduction of Hamiltonian and Lagrangian systems where the phase space is either the cotangent space or the tangent space of a given Lie group. In this lecture we shall present the general case of symplectic reduction as formulated by [MaWei74]. All the hypotheses will require regularity assumptions. The singular case is considerably more involved and its complete treatment can be found in [OR04]. We shall present with detailed proofs the Marsden-Weinstein theorem, nowadays often called "point reduction". This will be done at geometric and dynamic level. Then we shall present three important examples of reduced manifolds with all computations done in detail. The orbit reduction method and the so-called "shifting trick" will then be presented without proofs. This chapter ends with a discussion of the semidirect product reduction theorem and its application to the motion of the heavy top presented in the Introduction.

7.1 Point Reduction

In this section we shall review the symplectic point reduction theorem and give its formulation and classical proof due to [MaWei74]. This procedure is of paramount importance in symplectic geometry and geometric mechanics. It underlies all the other reduction methods that one can find now in the literature as well as all the various generalizations that have proved their usefulness in areas as varied as algebraic geometry and topology, differential and symplectic topology, classical, continuum, and quantum mechanics, field theory, dynamical systems, bifurcation theory, and control theory.

The setup of the problem is the following. Let (P, Ω) be a symplectic manifold on which a Lie group G with Lie algebra $\mathfrak{g}$ acts in a Hamiltonian fashion with associated equivariant momentum map $\mathbf{J} : P \to \mathfrak{g}^*$. If $\mu \in \mathbf{J}(P) \subset \mathfrak{g}^*$ denote by $G_\mu := \{g \in G \mid \mathrm{Ad}_g^* \mu = \mu\}$ the coadjoint isotropy subgroup of μ. The Marsden-Weinstein reduction theorem is the following.

Theorem 7.1 (Symplectic point reduction) *Assume that μ is a regular value of $\mathbf{J}$ and that the coadjoint isotropy subgroup G_μ acts freely and properly on $\mathbf{J}^{-1}(\mu)$. Then the quotient manifold $P_\mu := \mathbf{J}^{-1}(\mu)/G_\mu$ has a unique symplectic form Ω_μ characterized by the identity $\iota_\mu^* \Omega = \pi_\mu^* \Omega_\mu$, where $\iota_\mu : \mathbf{J}^{-1}(\mu) \hookrightarrow P$ is the inclusion and $\pi_\mu : \mathbf{J}^{-1}(\mu) \to P_\mu$ the projection. The symplectic manifold (P_μ, Ω_μ) is called the* ***symplectic point reduced space*** *at μ.*

In what follows we shall need the following notations. If (V, Ω) is a finite dimensional symplectic vector space and $W \subset V$ a subspace, define its ***symplectic orthogonal*** by $W^\Omega = \{v \in V \mid \Omega(v, w) = 0 \text{ for all } w \in W\}$. An elementary linear algebra argument, using the identity $\dim W + \dim W^\Omega = \dim V$, shows that $(W^\Omega)^\Omega = W$. If $z \in P$, denote by $G \cdot z$ and $G_\mu \cdot z$ the G- and G_μ-orbits through z respectively. It is important to note that the set $\mathbf{J}^{-1}(\mu)$ is G- invariant if and only if $G_\mu = G$. In general, $\mathbf{J}^{-1}(\mu)$ is only G_μ-invariant. The key ingredient in the proof of the reduction theorem is the following result that we shall state in the setting of Poisson manifolds because of its usefulness in that general situation (not covered in these lectures).

Lemma 7.2 (Reduction Lemma) *Let P be a Poisson manifold and let $\mathbf{J} : P \to \mathfrak{g}^*$ an equivariant momentum map of the canonical G-action on P. Let $G \cdot \mu$ denote the coadjoint orbit through a regular value $\mu \in \mathfrak{g}^*$ of $\mathbf{J}$. Then*

(i) $\mathbf{J}^{-1}(G \cdot \mu) = G \cdot \mathbf{J}^{-1}(\mu) := \{g \cdot z \mid g \in G \text{ and } \mathbf{J}(z) = \mu\}$;

(ii) $G_\mu \cdot z = (G \cdot z) \cap \mathbf{J}^{-1}(\mu)$;

(iii) $\mathbf{J}^{-1}(\mu)$ *and* $G \cdot z$ ***intersect cleanly****, i.e.,*

$$T_z(G_\mu \cdot z) = T_z(G \cdot z) \cap T_z(\mathbf{J}^{-1}(\mu));$$

(iv) *if* (P, Ω) *is symplectic, then* $T_z(\mathbf{J}^{-1}(\mu)) = (T_z(G \cdot z))^\Omega$.

Proof **(i)** Since $\mathbf{J}^{-1}(G \cdot \mu)$ is a G-invariant set by equivariance of $\mathbf{J}$ and $\mathbf{J}^{-1}(\mu) \subset \mathbf{J}^{-1}(G \cdot \mu)$, it follows that $G \cdot \mathbf{J}^{-1}(\mu) \subset \mathbf{J}^{-1}(G \cdot \mu)$. Conversely, $z \in \mathbf{J}^{-1}(G \cdot \mu)$ if and only if $\mathbf{J}(z) = \mathrm{Ad}^*_{g^{-1}}\mu$ for some $g \in G$, which is equivalent to $\mu = \mathrm{Ad}^*_{g^{-1}}\mathbf{J}(z) = \mathbf{J}(g^{-1} \cdot z)$, *i.e.*, $g^{-1} \cdot z \in \mathbf{J}^{-1}(\mu)$ and hence $z = g \cdot (g^{-1} \cdot z) \in G \cdot \mathbf{J}^{-1}(\mu)$.

(ii) $g \cdot z \in \mathbf{J}^{-1}(\mu) \Leftrightarrow \mu = \mathbf{J}(g \cdot z) = \mathrm{Ad}^*_{g^{-1}}\mathbf{J}(z) = \mathrm{Ad}^*_{g^{-1}}\mu \Leftrightarrow g \in G_\mu$.

(iii) First suppose that $v_z \in T_z(G \cdot z) \cap T_z(\mathbf{J}^{-1}(\mu))$. Then $v_z = \xi_P(z)$ for some $\xi \in \mathfrak{g}$ and $T_z\mathbf{J}(v_z) = 0$ which, by infinitesimal equivariance (written in the form $T_z\mathbf{J}(\xi_P(z)) = -\mathrm{ad}^*_\xi \mathbf{J}(z))$ gives $\mathrm{ad}^*_\xi\mu = 0$; *i.e.*, $\xi \in \mathfrak{g}_\mu$. If $v_z = \xi_P(z)$ for $\xi \in \mathfrak{g}_\mu$ then $v_z \in T_z(G_\mu \cdot z)$. The reverse inclusion is immediate since, by **(ii)**, $G_\mu \cdot z$ is included in both $G \cdot z$ and $\mathbf{J}^{-1}(\mu)$.

(iv) The condition $v_z \in (T_z(G \cdot z))^\Omega$ means that $\Omega(z)(\xi_P(z), v_z) = 0$ for all $\xi \in \mathfrak{g}$. This is equivalent to $\langle T_z\mathbf{J}(v_z), \xi\rangle = \mathbf{d}J^\xi(z)(v_z) = 0$ for all $\xi \in \mathfrak{g}$ by definition of the momentum map. Thus, $v_z \in (T_z(G \cdot z))^\Omega$ if and only if $v_z \in \ker T_z\mathbf{J} = T_z(\mathbf{J}^{-1}(\mu))$. ∎

We are now ready to prove the Symplectic Point Reduction Theorem.

Proof Since π_μ is a surjective submersion, if Ω_μ exists, it is uniquely determined by the condition $\pi_\mu^*\Omega_\mu = \iota_\mu^*\Omega$. This relation also defines Ω_μ in the following way. For $v \in T_z\mathbf{J}^{-1}(\mu)$, let $[v] = T_z\pi_\mu(v) \in T_{[z]}P_\mu$, where $[z] = \pi_\mu(z)$. Then $\pi_\mu^*\Omega_\mu = \iota_\mu^*\Omega$ is equivalent to

$$\Omega_\mu([z])([v], [w]) = \Omega(z)(v, w)$$

for all $v, w \in T_z\mathbf{J}^{-1}(\mu)$. To see that this relation defines Ω_μ, that is, it is independent of the choices made to define it, let $y = \Phi_g(z)$, $v' = T_z\Phi_g(v)$, and $w' = T_z\Phi_g(w)$, where $g \in G_\mu$. If, in addition $[v''] = [v']$ and $[w''] = [w']$, then $v'' - v', w'' - w' \in \ker T_{g \cdot z}\pi_\mu = T_{g \cdot z}(G_\mu \cdot z)$ and thus

$$\begin{aligned}\Omega(y)(v'', w'') &= \Omega(y)((v'' - v') + v', (w'' - w') + w') \\ &= \Omega(y)(v'' - v', w'' - w') + \Omega(y)(v'' - v', w') \\ &\quad + \Omega(y)(v', w'' - w') + \Omega(y)(v', w').\end{aligned}$$

The second and third terms vanish by Lemma 7.2 **(iv)**. The first term vanishes by Lemma 7.2 **(iii)** and **(iv)**. Thus we have

$$\Omega(y)(v'', w'') = \Omega(y)(v', w')$$

and we conclude

$$\begin{aligned}\Omega(y)(v'', w'') &= \Omega(y)(v', w') = \Omega(\Phi_g(z))(T_z\Phi_g(v), T_z\Phi_g(w)) \\ &= (\Phi_g^*\Omega)(z)(v, w) = \Omega(z)(v, w)\end{aligned}$$

since the action is symplectic. This proves that $\Omega_\mu([z])([v],[w])$ is well defined and satisfies the relation in the statement of the theorem. It is smooth since $\pi_\mu^*\Omega_\mu = \iota_\mu^*\Omega$ is smooth. Thus we have a well defined smooth two-form Ω_μ on P_μ.

Since $\mathbf{d}\Omega = 0$, we get

$$\pi_\mu^*\mathbf{d}\Omega_\mu = \mathbf{d}\pi_\mu^*\Omega_\mu = \mathbf{d}\iota_\mu^*\Omega = \iota_\mu^*\mathbf{d}\Omega = 0.$$

Since π_μ is a surjective submersion, the pull-back map π_μ^* on forms is injective, so we can conclude that $\mathbf{d}\Omega_\mu = 0$.

Finally, we prove that Ω_μ is non-degenerate. Suppose that $\Omega_\mu([z])([v],[w]) = 0$ for all $w \in T_z(\mathbf{J}^{-1}(\mu))$. This means that $\Omega(z)(v,w) = 0$ for all $w \in T_z(\mathbf{J}^{-1}(\mu))$, which is equivalent to $v \in (T_z(\mathbf{J}^{-1}(\mu)))^\Omega = T_z(G\cdot z)$ by Lemma 7.2 **(iv)**. Hence $v \in T_z(\mathbf{J}^{-1}(\mu)) \cap T_z(G\cdot z) = T_z(G_\mu \cdot z)$ by Lemma 7.2 **(iii)** so that $[v] = 0$, thus proving the weak non-degeneracy of Ω_μ. ∎

There are several important comments related to the Symplectic Point Reduction Theorem which will be addressed below.

(1) Define the ***symmetry algebra*** at $z \in P$ by $\mathfrak{g}_z := \{\xi \in \mathfrak{g} \mid \xi_P(z) = 0\}$. An element $\mu \in \mathfrak{g}^*$ is a regular value of $\mathbf{J}$ if and only if $\mathfrak{g}_z = 0$ for all $z \in \mathbf{J}^{-1}(\mu)$.

To prove this, recall that z is a regular point if and only if $T_z\mathbf{J}$ is surjective which is equivalent to $\{0\} = \{\xi \in \mathfrak{g} \mid \langle \xi, T_z\mathbf{J}(v)\rangle = 0, \text{ for all } v \in T_zP\}$. Since $\langle \xi, T_z\mathbf{J}(v)\rangle = \Omega(z)(\xi_P(z), v)$ by the definition of the momentum map, it thus follows that z is a regular point of $\mathbf{J}$ if and only if $\{0\} = \{\xi \in \mathfrak{g} \mid \Omega(z)(\xi_P(z), v) = 0 \text{ for all } v \in T_zP\}$. As $\Omega(z)$ is nondegenerate, this is in turn equivalent to $\mathfrak{g}_z = \{0\}$.

(2) The previous statement affirms that only points with trivial symmetry algebra are regular points of $\mathbf{J}$. This is important in concrete examples because it isolates the singular points easily. Another way to look at this statement is to interpret it as saying that points with symmetry are bifurcation points of $\mathbf{J}$. This simple observation turns out to have many important consequences, for example in the convexity theorems for momentum maps. Another consequence of statement **(1)** is that if $\mu \in \mathbf{J}(P) \subset \mathfrak{g}^*$ is a regular value of $\mathbf{J}$ then the action is ***locally free***, which means, by definition, that the symmetry algebras of all the points in $\mathbf{J}^{-1}(\mu)$ vanish. In this case the reduction construction can be carried out locally.

(3) Even if $\Omega = -\mathbf{d}\Theta$ and the action of G leaves Θ invariant, Ω_μ *need not be exact.* We shall prove in the next section that coadjoint orbits with their orbit symplectic form are reduced spaces. This immediately gives an example of a situation where the original symplectic form is exact but the reduced one is not. The sphere is the reduced space of $T^*SO(3)$. The canonical symplectic form on $T^*SO(3)$ is clearly exact whereas the area form on S^2 is not.

(4) If one looks at the proof of the theorem carefully, one notices that the hypothesis that μ is a regular value of $\mathbf{J}$ was not really used. What was necessary is that μ is a ***clean value*** of $\mathbf{J}$ which means, by definition, that $\mathbf{J}^{-1}(\mu)$ is a manifold and $T_z(\mathbf{J}^{-1}(\mu)) = \ker T_z\mathbf{J}$.

(5) The freeness and properness of the G_μ action on $\mathbf{J}^{-1}(\mu)$ are used only to guarantee that P_μ is a manifold. So these hypotheses can be replaced by the requirement that P_μ is a manifold and that $\pi_\mu : \mathbf{J}^{-1}(\mu) \to P_\mu$ a submersion.

(6) A point in $\mathfrak{g}^*$ is said to be ***generic*** if its coadjoint orbit is of maximal dimension. Duflo and Vergne have shown that the set of generic points is Zariski open in $\mathfrak{g}^*$ and that the coadjoint isotropy algebra of a generic point is necessarily Abelian. Regarding the momentum map, one should be warned that if μ is a regular value of $\mathbf{J}$, it need not be a generic point in $\mathfrak{g}^*$. As we shall see in the next section, the cotangent lift of the left (or right) translation of a Lie group G has all its values regular. However, among those, there are points that are not generic, such as the origin, in $\mathfrak{g}^*$.

(7) The connected components of the point reduced spaces P_μ can be viewed in a natural way as symplectic leaves of the Poisson manifold $(P/G, \{\cdot,\cdot\}_{P/G})$, provided that G acts freely and properly on P. Indeed, the smooth map $\kappa_\mu : P_\mu \to P/G$ naturally defined by the commutative diagram

$$\begin{array}{ccc} \mathbf{J}^{-1}(\mu) & \xrightarrow{\iota_\mu} & P \\ \downarrow{\scriptstyle \pi_\mu} & & \downarrow{\scriptstyle \pi} \\ P_\mu & \xrightarrow{\kappa_\mu} & P/G \end{array}$$

is a Poisson injective immersion. Moreover, the κ_μ-images in P/G of the connected components of the symplectic manifolds (P_μ, Ω_μ) are its symplectic leaves (see e.g. [OR04]). We will return to this in §7.4, where we will summarize the orbit reduction process and link it with these observations.

Note that, in general, κ_μ is only an injective immersion. So the topology of the image of κ_μ, homeomorphic to the topology of P_μ, is stronger than the subspace topology induced by the ambient space P/G. For example, we can have a subset of $\kappa_\mu(P_\mu)$ which is compact in the induced topology from P/G and not compact in the intrinsic topology of $\kappa_\mu(P_\mu)$ (relative to which it is homeomorphic to P_μ endowed with the quotient topology).

(8) We describe how reduction can be carried out for a non-equivariant momentum map $\mathbf{J} : P \to \mathfrak{g}^*$. If P is connected, the expression $\mathbf{J}(g \cdot z) - \mathrm{Ad}^*_{g^{-1}} \mathbf{J}(z)$ turns out to be independent of $z \in P$. Setting $\sigma(g) := \mathbf{J}(g \cdot z) - \mathrm{Ad}^*_{g^{-1}} \mathbf{J}(z)$ one obtains a group one-cocycle with values in $\mathfrak{g}^*$, that is, σ satisfies the cocycle identity $\sigma(gh) = \sigma(g) + \mathrm{Ad}_{g^{-1}} \sigma(h)$ (see, e.g. [MaRa94]). If this cocycle is a coboundary, that is, there is some $\lambda \in \mathfrak{g}^*$ such that $\sigma(g) = \lambda - \mathrm{Ad}^*_{g^{-1}} \lambda$, then the momentum map can be modified by the addition of $-\lambda$ to become equivariant. If σ is not a coboundary then there is no way one can modify $\mathbf{J}$ to make it equivariant. This $\mathfrak{g}^*$-valued group one-cocycle $\sigma : G \to \mathfrak{g}^*$ is called the ***nonequivariance group one-cocycle*** defined by $\mathbf{J}$.

To carry out reduction, modify the coadjoint action of G on $\mathfrak{g}^*$ in the following way: $g \cdot \mu := \mathrm{Ad}^*_{g^{-1}} \mu + \sigma(g)$. Relative to this affine action the momentum map $\mathbf{J}$ is equivariant and the reduction procedure works by dividing the level set $\mathbf{J}^{-1}(\mu)$ by the μ-isotropy subgroup for this affine action.

(9) If the regularity assumptions in the Symplectic Point Reduction Theorem do not hold, then P_μ is a stratified space, all of whose strata are symplectic manifolds. This result is considerably more difficult to prove and we refer to [OR04] and references therein for an exposition of this theory. One could even further relax the requirements, namely, do not even assume that there is a momentum map for a given canonical Lie group action on (P, Ω). In this case there is a generalization of the momentum map due to Condevaux, Dazord, and Molino, the so-called cylinder valued momentum map, for which the reduction procedure can be implemented. We refer again to [OR04] and references therein for a treatment of this subject.

7.2 Reduction and Reconstruction of Dynamics

The geometric theorem presented in the previous section has a dynamic counterpart that will be discussed now. We keep the same conventions and notations as in §7.1.

Theorem 7.3 (Point reduction of dynamics) *Let $\Phi : G \times P \to P$ be a free proper canonical action of the Lie group G on the connected symplectic manifold (P, Ω). Assume that this action has an associated momentum map $\mathbf{J} : P \to \mathfrak{g}^*$, with nonequivariance one-cocycle $\sigma : G \to \mathfrak{g}^*$. Let $\mu \in \mathfrak{g}^*$ be a value of $\mathbf{J}$ and denote by G_μ the isotropy subgroup of μ under the affine action of G on $\mathfrak{g}^*$.*

(i) *Let $H : P \to \mathbb{R}$ be a smooth G-invariant function. The flow F_t of the Hamiltonian vector field X_H leaves the connected components of $\mathbf{J}^{-1}(\mu)$ invariant and commutes with the G-action, so it induces a flow F_t^μ on the reduced space $P_\mu = \mathbf{J}^{-1}(\mu)/G_\mu$ defined by*

$$\pi_\mu \circ F_t \circ \iota_\mu = F_t^\mu \circ \pi_\mu.$$

The vector field generated by the flow F_t^μ on (P_μ, Ω_μ) is Hamiltonian with associated ***reduced Hamiltonian function*** *$H_\mu : P_\mu \to \mathbb{R}$ defined by*

$$H_\mu \circ \pi_\mu = H \circ \iota_\mu.$$

The vector fields X_H and X_{H_μ} are π_μ-related.

(ii) *Let $F : P \to \mathbb{R}$ be another smooth G-invariant function. Then $\{F, H\}$ is also G-invariant and $\{F, H\}_\mu = \{F_\mu, H_\mu\}_{P_\mu}$, where $\{\cdot, \cdot\}_{P_\mu}$ denotes the Poisson bracket associated to the reduced symplectic form Ω_μ on P_μ.*

Proof **(i)** By Noether's Theorem 5.4, the flow F_t leaves the connected components of $\mathbf{J}^{-1}(\mu)$ invariant. Since H is G-invariant and the G-action is canonical, it follows by Proposition 2.7 that F_t commutes with the G-action. Thus F_t induces a flow F_t^μ on P_μ that makes the following diagram commutative:

$$\begin{array}{ccc} \mathbf{J}^{-1}(\mu) & \xrightarrow{F_t \circ \iota_\mu} & \mathbf{J}^{-1}(\mu) \\ {\scriptstyle \pi_\mu}\downarrow & & \downarrow{\scriptstyle \pi_\mu} \\ P_\mu & \xrightarrow{F_t^\mu} & P_\mu . \end{array}$$

The G-invariance of H implies the existence of a smooth function $H_\mu : P_\mu \to \mathbb{R}$ uniquely determined by the identity $H_\mu \circ \pi_\mu = H \circ \iota_\mu$. Let $Y \in$

$\mathfrak{X}(P_\mu)$ be the vector field on P_μ whose flow is F_t^μ. By construction, Y is π_μ-related to X_H. Indeed, differentiating the relation given by the diagram above relative to t at $t = 0$, we obtain

$$T\pi_\mu \circ X_H \circ \iota_\mu = Y \circ \pi_\mu .$$

Let us check that $Y = X_{H_\mu}$. For $z \in \mathbf{J}^{-1}(\mu)$ and $v \in T_z\mathbf{J}^{-1}(\mu)$ we have

$$\begin{aligned}\Omega_\mu(\pi_\mu(z))\left(Y(\pi_\mu(z)), T_z\pi_\mu(v)\right) &= \Omega_\mu(\pi_\mu(z))\left(T_z\pi_\mu(X_H(z)), T_z\pi_\mu(v)\right)\\ &= \Omega(z)(X_H(z), v) = \mathbf{d}H(z)(v) = \mathbf{d}(H_\mu \circ \pi_\mu)(z)(v)\\ &= \mathbf{d}H_\mu(\pi_\mu(z))\left(T_z\pi_\mu(v)\right) = \Omega_\mu(\pi_\mu(z))\left(X_{H_\mu}(\pi_\mu(z)), T_z\pi_\mu(v)\right),\end{aligned}$$

which, by nondegeneracy of Ω_μ, shows that $Y = X_{H_\mu}$.

(ii) The G-invariance of $\{F, H\}$ is a straightforward corollary of Proposition 2.8. Recall that the function $\{F, H\}_\mu$ is uniquely characterized by the identity $\{F, H\}_\mu \circ \pi_\mu = \{F, H\} \circ \iota_\mu$. By the definition of the Poisson bracket on (P_μ, Ω_μ), π_μ-relatedness of the relevant Hamiltonian vector fields, and the identity $\iota_\mu^*\Omega = \pi_\mu^*\Omega_\mu$, we have for any $z \in \mathbf{J}^{-1}(\mu)$ denoting $[z]_\mu := \pi_\mu(z)$,

$$\begin{aligned}\{F_\mu, H_\mu\}_{P_\mu}([z]_\mu) &= \Omega_\mu([z]_\mu)\left(X_{F_\mu}([z]_\mu), X_{H_\mu}([z]_\mu)\right)\\ &= \Omega_\mu([z]_\mu)\left(T_z\pi_\mu(X_F(z)), T_z\pi_\mu(X_H(z))\right)\\ &= (\pi_\mu^*\Omega_\mu)(z)\left(X_F(z), X_H(z)\right) = (\iota_\mu^*\Omega)(z)\left(X_F(z), X_H(z)\right)\\ &= \Omega(z)\left(X_F(z), X_H(z)\right) = \{F, H\}(z),\end{aligned}$$

that is, the function $\{F_\mu, H_\mu\}_{P_\mu}$ also satisfies the relation $\{F_\mu, H_\mu\}_{P_\mu} \circ \pi_\mu = \{F, H\} \circ \iota_\mu$, which proves the desired equality $\{F_\mu, H_\mu\}_{P_\mu} = \{F, H\}_\mu$. ■

This theorem shows how dynamics on P descends to dynamics on all reduced manifolds P_μ for any $\mu \in \mathfrak{g}^*$, if the group action is free and proper. For the singular case see [OR04].

Let us now pose the converse question. Assume that an integral curve $c_\mu(t)$ of the reduced Hamiltonian system X_{H_μ} on (P_μ, Ω_μ) is known. Let $z_0 \in \mathbf{J}^{-1}(\mu)$ be given. Can one determine from this data the integral curve of the Hamiltonian system X_H with initial condition z_0? The answer to this question is affirmative as we shall see below.

The general method of ***reconstruction of dynamics*** is the following ([AbMa78] §4.3, [MaMoRa90], [MaRa03]). Pick a smooth curve $d(t)$ in $\mathbf{J}^{-1}(\mu)$ such that $d(0) = z_0$ and $\pi_\mu(d(t)) = c_\mu(t)$. We shall give later concrete choices for such curves in terms of connections. Then, if $c(t)$ denotes the integral curve of X_H with $c(0) = z_0$, we can write $c(t) = g(t) \cdot d(t)$ for some smooth curve $g(t)$ in G_μ. We shall determine now $g(t)$ and therefore $c(t)$. Below, $\Phi : G \times P \to P$

denotes the left action of G on P and $\Phi_g : P \to P$ is the diffeomorphism of P given by the group element $g \in G$. We have

$$\begin{aligned} X_H(c(t)) &= \dot{c}(t) = T_{d(t)}\Phi_{g(t)}\dot{d}(t) + T_{d(t)}\Phi_{g(t)}\left(T_{g(t)}L_{g(t)^{-1}}\dot{g}(t)\right)_P(d(t)) \\ &= T_{d(t)}\Phi_{g(t)}\left(\dot{d}(t) + (T_{g(t)}L_{g(t)^{-1}}\dot{g}(t))_P(d(t))\right) \end{aligned}$$

which implies

$$\begin{aligned} \dot{d}(t) + (T_{g(t)}L_{g(t)^{-1}}\dot{g}(t))_P(d(t)) &= T_{g(t)\cdot d(t)}\Phi_{g(t)^{-1}}X_H(c(t)) \\ &= T_{g(t)\cdot d(t)}\Phi_{g(t)^{-1}}X_H(g(t)\cdot d(t)) \\ &= \left(\Phi^*_{g(t)}X_H\right)(d(t)) = X_H(d(t)) \end{aligned}$$

since, by hypothesis, $H = \Phi_g^* H$ and thus, by Proposition 2.7, $X_H = X_{\Phi_g^* H} = \Phi_g^* X_H$ for any $g \in G$. This equation is solved in two steps as follows:

- **Step 1:** *Find a smooth curve $\xi(t)$ in $\mathfrak{g}_\mu$ such that*

$$\xi(t)_P(d(t)) = X_H(d(t)) - \dot{d}(t). \tag{7.1}$$

- **Step 2:** *With $\xi(t) \in \mathfrak{g}_\mu$ determined above, solve the nonautonomous differential equation on G_μ*

$$\dot{g}(t) = T_e L_{g(t)}\xi(t), \qquad \textit{with} \qquad g(0) = e. \tag{7.2}$$

Here are some useful remarks regarding the solution of each step.

(1) The first step is of algebraic nature. For example, if G is a matrix Lie group, (7.1) is just a matrix equation. If one is willing to work with more geometric structure, this equation can be solved explicitly. Typically, one endows the left principal G_μ-bundle $\pi_\mu : \mathbf{J}^{-1}(\mu) \to P_\mu$ with a connection. Recall that a (left) connection on this bundle is given by a $\mathfrak{g}_\mu$-valued one-form $A \in \Omega^1(\mathbf{J}^{-1}(\mu); \mathfrak{g}_\mu)$ satisfying for all $z \in \mathbf{J}^{-1}(\mu)$ the relations

$$A(z)\left(\xi_{\mathbf{J}^{-1}(\mu)}(z)\right) = \xi, \quad \text{for all} \quad \xi \in \mathfrak{g}_\mu$$

and

$$A(g \cdot z)(T_z\Phi_g(v_z)) = \mathrm{Ad}_g(A(z)(v_z)), \quad \text{for all} \quad g \in G_\mu, v_z \in T_z\mathbf{J}^{-1}(\mu).$$

Let G_μ act on P by restriction so that $\xi_P = \xi_{\mathbf{J}^{-1}(\mu)}$ for any $\xi \in \mathfrak{g}_\mu$. Choose in Step 1 of the reconstruction method the curve $d(t)$ to be the horizontal lift of $c_\mu(t)$ through z_0, that is, $d(t)$ is uniquely characterized by the conditions $A(d(t))(\dot{d}(t)) = 0$, $\pi_\mu(d(t)) = c_\mu(t)$, for all t, and $d(0) = z_0$. Then the solution of (7.1) is given by

$$\xi(t) = A(d(t))\big(X_H(d(t))\big).$$

(2) The second step is the main difficulty in finding a complete answer to the reconstruction problem; equation (7.2) cannot be solved explicitly, in general. For matrix groups this is a linear system with time dependent coefficients. However, if G is Abelian, this equation can be solved *by quadratures*. To see how this works, we need the formula of the derivative of the exponential map at any point in the Lie algebra.

If G is a Lie group and $\exp : \mathfrak{g} \to G$ the exponential map then

$$T_\xi \exp = T_e L_{\exp \xi} \circ \sum_{n=0}^{\infty} \frac{(-1)^n}{(n+1!} \mathrm{ad}_\xi^n$$

for any $\xi \in \mathfrak{g}$. If G is Abelian, then $\mathrm{ad}_\xi = 0$ and so $T_\xi \exp = T_e L_{\exp \xi}$, a formula that we shall use below.

So let us return to the second step in the reconstruction method for an Abelian group G. Since the connected component of the p-dimensional Lie group G is a cylinder $\mathbb{R}^k \times \mathbb{T}^{p-k}$, the exponential map $\exp(\xi_1, \ldots, \xi_p) = (\xi_1, \ldots, \xi_k, \xi_{k+1}(\mathrm{mod}\, 2\pi), \ldots, \xi_p(\mathrm{mod}\, 2\pi))$ is onto, so we can write $g(t) = \exp \eta(t)$ for some smooth curve $\eta(t) \in \mathfrak{g}$ satisfying $\eta(0) = 0$. Equation (7.2) gives then $\xi(t) = T_{g(t)} L_{g(t)^{-1}} \dot{g}(t) = \dot{\eta}(t)$ since $\dot{g}(t) = T_{\eta(t)} \exp \dot{\eta}(t) = T_e L_{\exp \eta(t)} \dot{\eta}(t)$ by the comments above. Therefore, in this case, the solution of (7.2) is given by

$$g(t) = \exp\left(\int_0^t \xi(s) ds\right).$$

This reconstruction method is crucial in the determination of various geometric phases in mechanical problems; see [MaMoRa90] for details.

7.3 Examples of Reduced Manifolds

The projective space. Consider $\mathbb{C}^{2n} = \mathbb{R}^{4n} = T^*\mathbb{R}^{2n}$ endowed with the canonical symplectic structure $\mathbf{dq} \wedge \mathbf{dp}$ for $(\mathbf{q}, \mathbf{p}) \in T^*\mathbb{R}^{2n}$. The flow of the Hamiltonian vector field given by the harmonic oscillator Hamiltonian $H(\mathbf{q}, \mathbf{p}) := (\|\mathbf{q}\|^2 + \|\mathbf{p}\|^2)/2$ is 2π-periodic and hence defines the circle action $\theta \cdot (\mathbf{q}, \mathbf{p}) \mapsto (\mathbf{q}\cos\theta + \mathbf{p}\sin\theta, -\mathbf{q}\sin\theta + \mathbf{p}\cos\theta)$ on $\mathbb{R}^{2n}$ or $\theta \cdot (\mathbf{q} + i\mathbf{p}) := e^{-i\theta}(\mathbf{q} + i\mathbf{p})$ on $\mathbb{C}^n$. Since the circle is compact this action is necessarily proper and it is obvious that it is free away from the origin. The infinitesimal generator of this action is X_H, which shows that $H : T^*\mathbb{R}^n \to \mathbb{R}$ is an invariant momentum map for this circle action. All hypotheses of the Symplectic Point Reduction Theorem hold and therefore $H^{-1}(1/2)$ is a symplectic manifold. However, since $H^{-1}(1/2)$ is diffeomorphic to the unit sphere S^{2n-1}, we can immediately conclude that $H^{-1}(1/2)/S^1 \cong S^{2n-1}/S^1$ This suggests that this reduced manifold is in fact symplectically diffeomorphic to complex projective space $\mathbb{CP}^{n-1}$. We shall prove this below.

Recall that $\mathbb{CP}^{n-1}$ is the space of complex lines through the origin in $\mathbb{C}^n$. Let $\pi : \mathbb{C}^n \setminus \{\mathbf{0}\} \to \mathbb{CP}^{n-1}$ be the tautological projection that sends the vector $\mathbf{z} \neq \mathbf{0}$ to the complex line it spans, denoted by $[\mathbf{z}]$ when thought of as an element of $\mathbb{CP}^{n-1}$. Consider the inclusion $\iota_{1/2} : H^{-1}(1/2) = S^{2n-1} \hookrightarrow \mathbb{C}^n \setminus \{\mathbf{0}\}$ and note that it preserves the equivalence relations, that is, the span of $e^{-i\theta}\mathbf{z}$ equals the span of $\mathbf{z} \neq \mathbf{0}$. Therefore, $\iota_{1/2}$ induces a smooth map $\hat{\iota}_{1/2} : S^{2n-1}/S^1 \to \mathbb{CP}^{n-1}$, uniquely characterized by the relation $\pi \circ \iota_{1/2} = \hat{\iota}_{1/2} \circ \pi_{1/2}$, which is easily seen to be bijective. Since the inverse of this map is the quotient of the smooth map $\mathbf{z} \in \mathbb{C}^n \setminus \{\mathbf{0}\} \mapsto S^1 \cdot (\mathbf{z}/\|\mathbf{z}\|) \in S^{2n-1}/S^1$ by the equivalence relation defining projective space, it follows that this inverse is also smooth and therefore $\hat{\iota}_{1/2} : S^{2n-1}/S^1 \to \mathbb{CP}^{n-1}$ is a diffeomorphism. In what follows it is convenient to define the map $\varphi : S^{2n-1} \to \mathbb{CP}^{n-1}$ by $\varphi := \pi \circ \iota_{1/2} = \hat{\iota}_{1/2} \circ \pi_{1/2}$. Let us record all of these maps in the following commutative diagram:

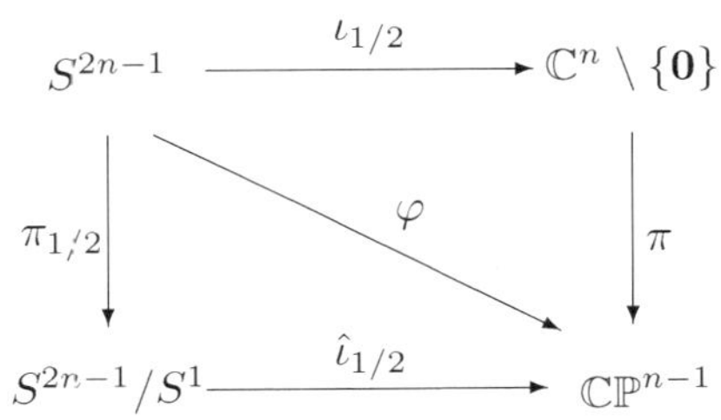

Finally, we recall the symplectic form on projective space (see e.g. [MaRa94], §5.3). Let $[\mathbf{z}] \in \mathbb{CP}^{n-1}$ be such that $\|\mathbf{z}\| = 1$ and let $\mathbf{w}_1, \mathbf{w}_2 \in (\mathbb{C}\mathbf{z})^\perp$. Then the symplectic form Ω_{FS} on $\mathbb{CP}^{n-1}$ is given by

$$\Omega_{FS}([\mathbf{z}])\,(T_{\mathbf{z}}\pi(\mathbf{w}_1), T_{\mathbf{z}}\pi(\mathbf{w}_2)) = -\operatorname{Im}(\mathbf{w}_1 \cdot \mathbf{w}_2).$$

This symplectic form is associated to the ***Fubini-Study*** Kähler metric on $\mathbb{CP}^{n-1}$, a subject not discussed in these lectures; it is the negative of its imaginary part.

We need to prove that $\hat{\iota}_{1/2}^*\Omega_{FS} = \Omega_{1/2}$. By the characterization of the reduced symplectic form $\Omega_{1/2}$, this is equivalent to $\iota_{1/2}^*(\mathbf{dq}\wedge\mathbf{dp}) = \pi_{1/2}^*\Omega_{1/2} = \pi_{1/2}^*\hat{\iota}_{1/2}^*\Omega_{FS} = (\hat{\iota}_{1/2} \circ \pi_{1/2})^*\Omega_{FS} = \varphi^*\Omega_{FS}$, which is the identity that shall be verified below. Since $\varphi = \pi \circ \iota_{1/2}$, we have for any $\mathbf{z} \in S^{2n-1}$ and any $\mathbf{w}_1, \mathbf{w}_2 \in T_{\mathbf{z}}S^{2n-1} = (\mathbb{C}\mathbf{z})^\perp$ the identity $T_{\mathbf{z}}\varphi(\mathbf{w}_j) = T_{\mathbf{z}}\pi(\mathbf{w}_j)$ for $j = 1, 2$, and hence

$$\begin{aligned}(\varphi^*\Omega_{FS})(\mathbf{z})\,(\mathbf{w}_1,\mathbf{w}_2) &= \Omega_{FS}([\mathbf{z}])\,(T_{\mathbf{z}}\varphi(\mathbf{w}_1),T_{\mathbf{z}}\varphi(\mathbf{w}_2))\\ &= \Omega_{FS}([\mathbf{z}])\,(T_{\mathbf{z}}\pi(\mathbf{w}_1),T_{\mathbf{z}}\pi(\mathbf{w}_2))\\ &= -\operatorname{Im}(\mathbf{w}_1\cdot\mathbf{w}_2) = \iota_{1/2}^*(\mathbf{dq}\wedge\mathbf{dp})\,(\mathbf{w}_1,\mathbf{w}_2)\end{aligned}$$

as was remarked at the beginning of §2.1. This proves that *the symplectic reduced space $(H^{-1}(1/2)/S^1,\Omega_{1/2})$ is symplectically diffeomorphic to the complex projective space $(\mathbb{CP}^{n-1},\Omega_{FS})$.*

Kaluza-Klein construction in electromagnetism. Let us revisit the motion of a particle with charge e and mass m moving in a given time independent divergence free magnetic field $\mathbf{B} := B_x\mathbf{i} + B_y\mathbf{j} + B_z\mathbf{k}$, where $\mathbf{i},\mathbf{j},\mathbf{k}$ is the usual orthonormal basis of $\mathbb{R}^3$. In §2.5 we have shown that Newton's equations (2.13) for the Lorentz force law

$$m\frac{d\mathbf{v}}{dt} = \frac{e}{c}\mathbf{v}\times\mathbf{B},$$

where $\mathbf{v} := \dot{\mathbf{q}}$ is the velocity of the particle, are equivalent to Hamilton's equations in $T^*\mathbb{R}^3 := \{(\mathbf{q},\mathbf{p}) \mid \mathbf{x} = (x,y,z), \mathbf{p} := m\mathbf{v} = (m\dot{x},m\dot{y},m\dot{z}) = (p_x,p_y,p_z)\in\mathbb{R}^3\}$ endowed with the magnetic symplectic form

$$\Omega_B = \mathbf{d}x\wedge\mathbf{d}p_x + \mathbf{d}y\wedge\mathbf{d}p_y + \mathbf{d}z\wedge\mathbf{d}p_z - \frac{e}{c}B$$

and the Hamiltonian given by the kinetic energy of the particle

$$H = \frac{1}{2m}\|\mathbf{p}\|^2 = \frac{m}{2}(\dot{x}^2+\dot{z}^2+\dot{y}^2);$$

B denotes the closed two-form on $\mathbb{R}^3$ associated to the divergence free vector field $\mathbf{B}$, that is,

$$B := \mathbf{i}_{\mathbf{B}}(\mathbf{d}x\wedge\mathbf{d}y\wedge\mathbf{d}z) = B_x\mathbf{d}y\wedge\mathbf{d}z + B_y\mathbf{d}z\wedge\mathbf{d}x + B_z\mathbf{d}x\wedge\mathbf{d}y.$$

In addition, we have shown that writing $B = \mathbf{d}A$, or equivalently, $\mathbf{B} = \nabla\times\mathbf{A}$, where $\mathbf{A}^\flat = A$, that is, $\mathbf{A} = A_x\mathbf{i}+A_y\mathbf{j}+A_z\mathbf{k}$ and $A = A_x\mathbf{d}x+A_y\mathbf{d}y+A_z\mathbf{d}z$, the same Lorentz force law equations are Hamiltonian on $T^*\mathbb{R}^3$ endowed with the canonical symplectic structure but the momentum shifted Hamiltonian

$$H_A(\mathbf{q},\mathbf{p}) = \frac{1}{2m}\left\|\mathbf{p}-\frac{e}{c}A\right\|^2.$$

In §3.3, writing these equations in Lagrangian form, we remarked that they are not geodesic, essentially because of the magnetic symplectic form. Then we used the Kaluza-Klein construction to find a new Lagrangian L_{KK} (see

(3.6)) on the enlarged configuration space $Q_{KK} := \mathbb{R}^3 \times S^1 = \{(\mathbf{q}, \theta) \mid \mathbf{q} \in \mathbb{R}^3, \theta \in S^1\}$ which turned out to be the quadratic form of an $\mathbf{A}$-dependent Riemannian metric on Q_{KK}, called the Kaluza-Klein metric. Thus the Euler-Lagrange equations for L_{KK} are the geodesic equations of this metric on $\mathbb{R}^3 \times S^1$. Furthermore, we Legendre transformed L_{KK} to get a Hamiltonian (see (3.8)) on $T^*\mathbb{R}^3 = \{\mathbf{q}, \theta, \mathbf{p}, \pi) \mid \mathbf{q}, \mathbf{p} \in \mathbb{R}^3, \theta \in S^1, \pi \in \mathbb{R}\}$ given by

$$H_{KK}(\mathbf{q}, \mathbf{p}, \theta, \pi) = \frac{1}{2m}\|\mathbf{p} - \pi\mathbf{A}\|^2 + \frac{1}{2}\pi^2.$$

Relative to the canonical symplectic form, Hamilton's equations for H_{KK} are the geodesic equations for the Kaluza-Klein metric (expressed in Hamiltonian form). Since H_{KK} does not depend on θ, π is conserved and setting $\pi = e/c$, we noted that H_{KK}, regarded as a function of only the variables $(\mathbf{q}, \mathbf{p})$, is up to a constant (namely $\pi^2/2$), equal to the momentum shifted Lorentz force Hamiltonian H_A (see (3.4)). These were just observations obtained by direct calculations.

Now we shall show how all of this is obtained from reduction theory. We start from the Hamiltonian system on $T^*Q_{KK} = T^*(\mathbb{R}^3 \times S^1)$ and note that S^1 acts on Q_{KK} by $\psi \cdot (\mathbf{q}, \theta) := (\mathbf{q}, \theta + \psi)$ where $\theta + \psi$ is taken modulo 1 (so we normalize the length of the circle to be 1). The infinitesimal generator defined by $\xi \in \mathbb{R}$ for this action is $\xi_{Q_{KK}}(\mathbf{q}, \theta) = (\mathbf{q}, \theta; \mathbf{0}, \xi)$. The Hamiltonian H_{KK} is obviously invariant under this action. The momentum map for this action, given by (5.11), is in this case

$$\langle \mathbf{J}(\mathbf{q}, \theta, \mathbf{p}, \pi), \xi \rangle = (\mathbf{p}\mathbf{d}\mathbf{q} + \pi\mathbf{d}\theta)\xi\frac{\partial}{\partial\theta} = \xi\pi,$$

that is, $\mathbf{J}(\mathbf{q}, \theta, \mathbf{p}, \pi) = \pi$, thus recovering the direct observation that π is a conserved quantity. Moreover, any value of $\mathbf{J}$ in $\mathbb{R}$ is a regular value and we have $\mathbf{J}^{-1}(e/c) = T^*\mathbb{R}^3 \times S^1 \times \{e/c\}$ on which S^1 acts on the S^1-factor only. Thus the reduced space is $(T^*Q_{KK})_{e/c} = T^*\mathbb{R}^3$. The reduced Hamiltonian is obviously $H_A + e^2/2c^2$, so the reduced dynamics is given by H_A. Finally we need to compute the reduced symplectic form Ω_{red}. We have

$$\begin{aligned}
&(\pi_{e/c}^*\Omega_{red})(\mathbf{q}, \theta, \mathbf{p}, \pi)\left(\mathbf{X}\frac{\partial}{\partial\mathbf{q}} + a\frac{\partial}{\partial\theta} + \mathbf{Y}\frac{\partial}{\partial\mathbf{p}}, \mathbf{X}'\frac{\partial}{\partial\mathbf{q}} + a'\frac{\partial}{\partial\theta} + \mathbf{Y}'\frac{\partial}{\partial\mathbf{p}}\right) \\
&\quad = \iota_{e/c}^*(\mathbf{d}\mathbf{q} \wedge \mathbf{d}\mathbf{p} - \mathbf{d}\theta \wedge \mathbf{d}\pi)\left(\mathbf{X}\frac{\partial}{\partial\mathbf{q}} + a\frac{\partial}{\partial\theta} + \mathbf{Y}\frac{\partial}{\partial\mathbf{p}},\right. \\
&\qquad\qquad\qquad\qquad\qquad\qquad \left.\mathbf{X}'\frac{\partial}{\partial\mathbf{q}} + a'\frac{\partial}{\partial\theta} + \mathbf{Y}'\frac{\partial}{\partial\mathbf{p}}\right) \\
&\quad = \mathbf{X} \cdot \mathbf{Y}' - \mathbf{X}' \cdot \mathbf{Y}.
\end{aligned}$$

Since

$$T_{(\mathbf{q},\theta,\mathbf{p},e/c)}\pi_{e/c}\left(\mathbf{X}\frac{\partial}{\partial\mathbf{q}}+a\frac{\partial}{\partial\theta}+\mathbf{Y}\frac{\partial}{\partial\mathbf{p}}\right)=\mathbf{X}\frac{\partial}{\partial\mathbf{q}}+\mathbf{Y}\frac{\partial}{\partial\mathbf{p}}$$

this identity shows that Ω_{red} is the canonical symplectic form on the reduced space $T^*\mathbb{R}^3$, thus recovering the second result in §2.5: *the Lorentz force law equations are Hamiltonian on $T^*\mathbb{R}^3$ relative to the canonical symplectic form and the Hamiltonian function H_A.*

Let us carry out the reduction in a different manner, by insisting that we get on the reduced space the kinetic energy Hamiltonian $\|\mathbf{p}\|^2/2\dot{m}$. To do this, requires that we project $\mathbf{J}^{-1}(e/c)\to T^*\mathbb{R}^3$ by the S^1-invariant smooth map

$$\varphi\left(\mathbf{q},\theta,\mathbf{p},\frac{e}{c}\right):=\left(\mathbf{q},\mathbf{p}-\frac{e}{c}\mathbf{A}\right).$$

It is clear from the computation above that the reduced symplectic form will not be the canonical one anymore, since we are using a different map to identify $(T^*Q_{KK})_{e/c}$ with $T^*\mathbb{R}^3$. So what is the symplectic form now?

We compute it by using φ as the projection from the reduction theorem and get as before

$$\begin{aligned}
&(\varphi^*\Omega_{e/c})(\mathbf{q},\theta,\mathbf{p},\pi)\left(\mathbf{X}\frac{\partial}{\partial\mathbf{q}}+a\frac{\partial}{\partial\theta}+\mathbf{Y}\frac{\partial}{\partial\mathbf{p}},\mathbf{X}'\frac{\partial}{\partial\mathbf{q}}+a'\frac{\partial}{\partial\theta}+\mathbf{Y}'\frac{\partial}{\partial\mathbf{p}}\right)\\
&\quad=\iota_{e/c}^*(\mathbf{dq}\wedge\mathbf{dp}+\mathbf{d}\theta\wedge\mathbf{d}\pi)\left(\mathbf{X}\frac{\partial}{\partial\mathbf{q}}+a\frac{\partial}{\partial\theta}+\mathbf{Y}\frac{\partial}{\partial\mathbf{p}},\right.\\
&\qquad\qquad\left.\mathbf{X}'\frac{\partial}{\partial\mathbf{q}}+a'\frac{\partial}{\partial\theta}+\mathbf{Y}'\frac{\partial}{\partial\mathbf{p}}\right)\\
&\quad=\mathbf{X}\cdot\mathbf{Y}'-\mathbf{X}'\cdot\mathbf{Y}.
\end{aligned}$$

However,

$$T_{(\mathbf{q},\theta,\mathbf{p},e/c)}\varphi\left(\mathbf{X}\frac{\partial}{\partial\mathbf{q}}+a\frac{\partial}{\partial\theta}+\mathbf{Y}\frac{\partial}{\partial\mathbf{p}}\right)=\mathbf{X}\frac{\partial}{\partial\mathbf{q}}+\left(\mathbf{Y}-\frac{e}{c}(\mathbf{X}\cdot\nabla)\mathbf{A}\right)\frac{\partial}{\partial\mathbf{p}},$$

so the previous equality yields

$$\begin{aligned}
&\Omega_{e/c}\left(\mathbf{q},\mathbf{p}-\frac{e}{c}\mathbf{A}\right)\left(\mathbf{X}\frac{\partial}{\partial\mathbf{q}}+\left(\mathbf{Y}-\frac{e}{c}(\mathbf{X}\cdot\nabla)\mathbf{A}\right)\frac{\partial}{\partial\mathbf{p}},\right.\\
&\qquad\qquad\left.\mathbf{X}'\frac{\partial}{\partial\mathbf{q}}+\left(\mathbf{Y}'-\frac{e}{c}(\mathbf{X}'\cdot\nabla)\mathbf{A}\right)\frac{\partial}{\partial\mathbf{p}}\right)\\
&\quad=\mathbf{X}\cdot\mathbf{Y}'-\mathbf{X}'\cdot\mathbf{Y}.
\end{aligned}$$

Replacing now $\mathbf{p}$ by $\mathbf{p}+\frac{e}{c}\mathbf{A}$, noting that the right hand side of this equality does not depend on $\mathbf{p}$, replacing $\mathbf{Y}$ by $\mathbf{Y} + \frac{e}{c}(\mathbf{X} \cdot \nabla)\mathbf{A}$, and $\mathbf{Y}'$ by $\mathbf{Y}' + \frac{e}{c}(\mathbf{X}' \cdot \nabla)\mathbf{A}$, yields

$$\begin{aligned}
\Omega_{e/c}(\mathbf{q}, \mathbf{p}) &\left(\mathbf{X}\frac{\partial}{\partial \mathbf{q}} + \mathbf{Y}\frac{\partial}{\partial \mathbf{p}}, \mathbf{X}'\frac{\partial}{\partial \mathbf{q}} + \mathbf{Y}'\frac{\partial}{\partial \mathbf{p}} \right) \\
&= \mathbf{X} \cdot \left(\mathbf{Y}' + \frac{e}{c}(\mathbf{X}' \cdot \nabla)\mathbf{A} \right) - \mathbf{X}' \cdot \left(\mathbf{Y} + \frac{e}{c}(\mathbf{X} \cdot \nabla)\mathbf{A} \right) \\
&= \mathbf{X} \cdot \mathbf{Y}' - \mathbf{X}' \cdot \mathbf{Y} - \frac{e}{c}\left(\mathbf{X}' \cdot ((\mathbf{X} \cdot \nabla)\mathbf{A}) - \mathbf{X} \cdot ((\mathbf{X}' \cdot \nabla)\mathbf{A}) \right) \\
&= \left(\mathbf{dq} \wedge \mathbf{dp} - \frac{e}{c}B \right) \left(\mathbf{X}\frac{\partial}{\partial \mathbf{q}} + \mathbf{Y}\frac{\partial}{\partial \mathbf{p}}, \mathbf{X}'\frac{\partial}{\partial \mathbf{q}} + \mathbf{Y}'\frac{\partial}{\partial \mathbf{p}} \right)
\end{aligned}$$

since

$$B\left(\mathbf{X}\frac{\partial}{\partial \mathbf{q}}, \mathbf{X}'\frac{\partial}{\partial \mathbf{q}} \right) = \mathbf{X}' \cdot ((\mathbf{X} \cdot \nabla)\mathbf{A}) - \mathbf{X} \cdot ((\mathbf{X}' \cdot \nabla)\mathbf{A})$$

as a straightforward computation in coordinates shows. Thus

$$\Omega_{e/c} = \mathbf{dq} \wedge \mathbf{dp} - \frac{e}{c}B = \Omega_B.$$

which proves the first assertion in the example of §2.5 and explains the appearance of the magnetic term B in the symplectic form: *the Lorentz force law equations are Hamiltonian on $T^*\mathbb{R}^3$ relative to the symplectic form Ω_B and the kinetic energy $\|\mathbf{p}\|^2/2m$ as Hamiltonian function.*

The phenomenon occurring here is very general and has to do with cotangent bundle reduction, a topic not covered in these lectures. What is happening here is the following: one has a principal bundle with a connection and searches for explicit realizations of the reduced spaces of the cotangent bundle of the total space by the structure group. It turns out that there are two natural ways to carry out this reduction, both at Poisson and at symplectic level. For details of this theory see [MaRa03] and [PR04]; for a quick summary without proofs see [OR04].

Coadjoint Orbits. For the Lie group G denote by $L_g, R_g : G \to G$ the left and right translations on G by $g \in G$. The lifts of these actions to T^*G, denoted by $\bar{L}, \bar{R} : G \times T^*G \to T^*G$ respectively, are proper, free, and have equivariant momentum maps $\mathbf{J}_L, \mathbf{J}_R : T^*G \to \mathfrak{g}^*$ given by (see (5.16))

$$\mathbf{J}_L(\alpha_g) = T_e^*R_g(\alpha_g), \qquad \mathbf{J}_R(\alpha_g) = T_e^*L_g(\alpha_g).$$

Recall that

$$\bar{L}_h(\alpha_g) = T_{hg}^*L_{h^{-1}}(\alpha_g), \qquad \bar{R}_h(\alpha_g) = T_{gh}^*R_{h^{-1}}(\alpha_g)$$

for any $h \in G$ and $\alpha_g \in T^*G$.

Let us compute the point reduced space $(T^*G)_\mu$ for any $\mu \in \mathfrak{g}^*$ relative to the left action $\bar{L}$. Notice that we do *not* require μ to be a generic point in $\mathfrak{g}^*$; that is, arbitrarily nearby coadjoint orbits may have a different dimensions. Since the action $\bar{L}$ is free, the symmetry algebra of every point in T^*G is zero and thus every $\mu \in \mathfrak{g}^*$ is a regular value of $\mathbf{J}_L$. Thus, the hypotheses of the Symplectic Point Reduction Theorem hold.

The submanifold $\mathbf{J}_L^{-1}(\mu) = \{\alpha_g \in T^*G \mid T_e^*R_g(\alpha_g) = \mu\} = \{T_g^*R_{g^{-1}}\mu \mid g \in G\}$ is the graph of the right invariant one-form α_μ on G whose value at the identity is μ. Thus $\alpha_\mu : G \to \mathbf{J}_L^{-1}(\mu)$ is a diffeomorphism. Let us show that it is G_μ-equivariant, that is,

$$\alpha_\mu \circ L_h = \bar{L}_h \circ \alpha_\mu \quad \text{for all} \quad h \in G_\mu.$$

Indeed, for any $g \in G$ we have

$$\begin{aligned}(\bar{L}_h \circ \alpha_\mu)(g) &= \lambda_h(T_g^*R_{g^{-1}}\mu) = T_{hg}^*L_{h^{-1}}T_g^*R_{g^{-1}}\mu = T_{hg}^*R_{g^{-1}}T_h^*L_{h^{-1}}\mu \\ &= T_{hg}^*R_{g^{-1}}T_h^*R_{h^{-1}}T_e^*R_hT_h^*L_{h^{-1}}\mu = T_{hg}^*R_{(hg)^{-1}}\operatorname{Ad}_{h^{-1}}^*\mu \\ &= T_{hg}^*R_{(hg)^{-1}}\mu = \alpha_\mu(hg).\end{aligned}$$

Therefore $\alpha_\mu : G \to \mathbf{J}_L^{-1}(\mu)$ induces a diffeomorphism $\overline{\alpha}_\mu : G/G_\mu \to (T^*G)_\mu$, where $G/G_\mu := \{G_\mu g \mid g \in G\}$. Recall now that the map $\varepsilon_\mu : G/G_\mu \to \mathcal{O}_\mu$ given by $\varepsilon_\mu(G_\mu g) = \operatorname{Ad}_g^*\mu$ is the diffeomorphism that defines the smooth manifold structure of the orbit $\mathcal{O}_\mu$. Define the diffeomorphism $\overline{\varphi} : (T^*G)_\mu = \mathbf{J}_L^{-1}(\mu)/G_\mu \to \mathcal{O}_\mu$ by $\overline{\varphi} := \varepsilon_\mu \circ (\overline{\alpha}_\mu)^{-1}$ and note that $\varphi := \overline{\varphi} \circ \pi_\mu = \mathbf{J}_R|_{\mathbf{J}_L^{-1}(\mu)} : \mathbf{J}_L^{-1}(\mu) \to \mathcal{O}_\mu$ has the expression $\varphi(T_g^*R_{g^{-1}}\mu) = \operatorname{Ad}_g^*\mu$.

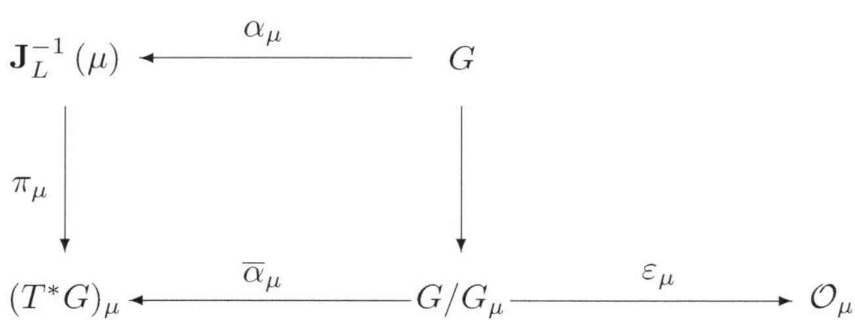

We claim that $\overline{\varphi}_*\Omega_\mu = \omega^-$, where Ω_μ is the reduced symplectic form and ω^- is the minus orbit symplectic form on $\mathcal{O}_\mu$. Since Ω_μ is characterized by the identity $\pi_\mu\Omega_\mu = \iota_\mu^*\Omega = -\iota_\mu^*\mathbf{d}\Theta$, where Θ is the canonical one-form on T^*G, this relation is equivalent to $\varphi^*\omega^- = -\iota_\mu^*\mathbf{d}\Theta$. We shall now prove this identity. In what follows we shall denote by $\mathrm{AD}_g : G \to G$ the conjugation automorphism $\mathrm{AD}_g(h) := ghg^{-1}$, for any $h \in G$. Thus $T_e \mathrm{AD}_g = \mathrm{Ad}_g : \mathfrak{g} \to \mathfrak{g}$ is the adjoint representation of G on $\mathfrak{g}$.

Let

$$\alpha_\xi(t) = T^*_{g\exp(t\xi)} R_{\exp(-t\xi)g^{-1}} \mu$$

be an arbitrary smooth curve in $\mathbf{J}_L^{-1}(\mu)$ passing through $\alpha_\xi(0) = T_g^* R_{g^{-1}} \mu$, where $\xi \in \mathfrak{g}$. Since $R_{\exp(-t\xi)g^{-1}} = \mathrm{AD}_g \circ R_{\exp(-t\xi)} \circ L_{g^{-1}}$, we get

$$T_{g\exp(t\xi)} R_{\exp(-t\xi)g^{-1}} = \mathrm{Ad}_g \circ T_{\exp(t\xi)} R_{\exp(-t\xi)} \circ T_{g\exp(t\xi)} L_{g^{-1}},$$

so letting $\nu = \mathrm{Ad}_g^* \mu$, we have

$$\alpha_\xi(t) = T^*_{g\exp(t\xi)} L_{g^{-1}} T^*_{\exp(t\xi)} R_{\exp(-t\xi)} \nu = \bar{L}_g \bar{R}_{\exp(t\xi)} \nu.$$

Therefore, an arbitrary tangent vector at $T_g^* R_{g^{-1}} \mu \in \mathbf{J}_L^{-1}(\mu)$ to the submanifold $\mathbf{J}_L^{-1}(\mu)$ has the expression

$$\alpha'_\xi(0) = T_\nu \bar{L}_g (\xi^R_{T^*G}(\nu)), \tag{7.3}$$

where $\xi^R_{T^*G}$ is the infinitesimal generator of the right action $\bar{R}$ on T^*G.

Now note that $\varphi(\alpha_\xi(t)) = \mathrm{Ad}^*_{g\exp(t\xi)} \mu = \mathrm{Ad}^*_{\exp(t\xi)} \mathrm{Ad}^*_g \mu = \mathrm{Ad}^*_{\exp(t\xi)} \nu$ so that

$$T_{\alpha_\xi(0)} \varphi(\alpha'_\xi(0)) = \left.\frac{d}{dt}\right|_{t=0} \varphi(\alpha_\xi(t)) = \left.\frac{d}{dt}\right|_{t=0} \mathrm{Ad}^*_{\exp(t\xi)} \nu = \mathrm{ad}^*_\xi \nu. \tag{7.4}$$

By (7.4), we have for any $\xi, \eta \in \mathfrak{g}$ and $\nu = \mathrm{Ad}_g^* \mu$,

$$\begin{aligned}
&(\varphi^* \omega^-)(T_g^* R_{g^{-1}} \mu)(\alpha'_\xi(0), \alpha'_\eta(0)) \\
&= \omega^-(\nu) \left(T_{\alpha_\xi(0)} \varphi(\alpha'_\xi(0)), T_{\alpha_\eta(0)} \varphi(\alpha'_\eta(0)) \right) \\
&= \omega^-(\nu) \left(\mathrm{ad}^*_\xi \nu, \mathrm{ad}^*_\eta \nu \right) = -\langle \nu, [\xi, \eta] \rangle \\
&= -J_R^{[\xi,\eta]}(\nu) = \{J_R^\xi, J_R^\eta\}(\nu),
\end{aligned} \tag{7.5}$$

since for *right* actions we have $J_R^{[\xi,\eta]} = -\{J_R^\xi, J_R^\eta\}$ (see (5.12) for *left* actions).

On the other hand, since $T_g^* R_{g^{-1}} \mu = \bar{L}_g \nu$, the expression (7.3) and left invariance of Θ give

$$\begin{aligned}
&(\iota^* \mathbf{d}\Theta)(T_g^* R_{g^{-1}} \mu)(\alpha'_\xi(0), \alpha'_\eta(0)) \\
&= \mathbf{d}\Theta(\bar{L}_g \nu) \left(T_\nu \bar{L}_g(\xi^R_{T^*G}(\nu)), T_\nu \bar{L}_g(\eta^R_{T^*G}(\nu)) \right) \\
&= \mathbf{d}\Theta(\nu) \left(\xi^R_{T^*G}(\nu), \eta^R_{T^*G}(\nu) \right) \\
&= \xi^R_{T^*G}[\Theta(\eta^R_{T^*G})](\nu) - \eta^R_{T^*G}[\Theta(\xi^R_{T^*G})](\nu) - \Theta([\xi^R_{T^*G}, \eta^R_{T^*G}])(\nu).
\end{aligned}$$

If $\pi : T^*G \to G$ is the cotangent bundle projection, the definition (2.8) of Θ gives for any $\alpha_g \in T^*G$,

$$\Theta(\eta^R_{T^*G})(\alpha_g) = \left\langle \alpha_g, T_{\alpha_g}\pi(\eta^R_{T^*G})\right\rangle = \left\langle \alpha_g, \eta^R_G(g))\right\rangle = J^\eta_R(\alpha_g)$$

since $\eta^R_{T^*G}$ and η^R_G are π-related. Thus, using the identity $[\xi^R_{T^*G}, \eta^R_{T^*G}] = [\xi, \eta]^R_{T^*G}$ valid for *right* actions (see (5.8) for *left* actions), as well as the definition of the momentum map, we can continue the computation above and write

$$\begin{aligned}
&\xi^R_{T^*G}[\Theta(\eta^R_{T^*G})](\nu) - \eta^R_{T^*G}[\Theta(\xi^R_{T^*G})](\nu) - \Theta([\xi,\eta]^R_{T^*G})(\nu)\\
&\quad = X_{J^\xi_R}[J^\eta_R](\nu) - X_{J^\eta_R}[J^\xi_R](\nu) - J^{[\xi,\eta]}_R(\nu)\\
&\quad = \{J^\eta_R, J^\xi_R\}(\nu) - \{J^\xi_R, J^\eta_R\}(\nu) + \{J^\xi_R, J^\eta_R\}(\nu) = -\{J^\xi_R, J^\eta_R\}(\nu).
\end{aligned}$$

Thus we have proved the identity

$$-(\iota^*_\mu \mathbf{d}\Theta)(T^*_g R_{g^{-1}}\mu)(\alpha'_\xi(0), \alpha'_\eta(0)) = \{J^\xi_R, J^\eta_R\}(\nu). \tag{7.6}$$

Equations (7.5) and (7.6) prove that $\varphi^*\omega^- = -\iota^*_\mu \mathbf{d}\Theta$.

7.4 Orbit Reduction

Let us return to Remark **(7)** in §7.1 where we have commented on the fact that the inclusion $\iota_\mu : \mathbf{J}^{-1}(\mu) \hookrightarrow P$ induces a Poisson injective immersion $\kappa_\mu : P_\mu \to P/G$. So, the κ_μ-images in P/G of the connected components of the point reduced symplectic manifolds (P_μ, Ω_μ) are the symplectic leaves of P/G. In this section we shall present, without proofs, how this is actually carried out concretely.

We begin with the easy observation that, as sets, $\kappa_\mu(P_\mu) = \mathbf{J}^{-1}(\mathcal{O}_\mu)/G$, where $\mathcal{O}_\mu$ is the coadjoint orbit through $\mu \in \mathfrak{g}^*$. If the momentum map $\mathbf{J} : P \to \mathfrak{g}^*$ is not equivariant then instead of the coadjoint orbit through μ one considers the orbit of the affine action $g \cdot \mu := \mathrm{Ad}^*_{g^{-1}}\mu + \sigma(g)$, where σ is the $\mathfrak{g}^*$-valued nonequivariance group one-cocycle defined by $\mathbf{J}$; we need to assume here that P is connected (see Remark **(8)** in §7.1). Recall that $\sigma(g) := \mathbf{J}(g \cdot z) - \mathrm{Ad}^*_{g^{-1}}\mathbf{J}(z)$ and that if P is connected the right hand side of this equation is independent of $z \in P$. The group one-cocycle σ induces by derivation a real valued Lie algebra two-cocycle $\Sigma : \mathfrak{g} \times \mathfrak{g} \to \mathbb{R}$ which can be shown to equal $\Sigma(\xi, \eta) = J^{[\xi,\eta]}(z) - \{J^\xi, J^\eta\}(z)$ for every $z \in P$ and $\xi, \eta \in \mathfrak{g}$. Denote by $\xi_{\mathfrak{g}^*}(\nu) := -\mathrm{ad}^*_\xi \nu + \Sigma(\xi, \cdot)$ the infinitesimal generator of the affine action of G on $\mathfrak{g}^*$, where $\nu \in \mathfrak{g}^*$. The affine action orbit $\mathcal{O}_\mu$ carries two symplectic forms given by

$$\omega^\pm_{\mathcal{O}_\mu}(\nu)(\xi_{\mathfrak{g}^*}(\nu), \eta_{\mathfrak{g}^*}(\nu)) = \pm\langle \nu, [\xi, \eta]\rangle \mp \Sigma(\xi, \eta), \tag{7.7}$$

for any $\xi, \eta \in \mathfrak{g}$. They are the natural modifications of the usual orbit symplectic forms on coadjoint orbits. For the proofs of the statements above see [AbMa78], [LiMa87], or [OR04]. From now on we shall not make any equivariance hypotheses on $\mathbf{J}$ and shall work with the affine orbit $\mathcal{O}_\mu \subset \mathfrak{g}^*$ through μ. The set $P_{\mathcal{O}_\mu} := \mathbf{J}^{-1}(\mathcal{O}_\mu)/G$ is called the ***orbit reduced space*** associated to the orbit $\mathcal{O}_\mu$. The smooth manifold structure (and hence the topology) on $P_{\mathcal{O}_\mu}$ is the one that makes the bijective map $\kappa_\mu : P_\mu \to P_{\mathcal{O}_\mu}$ into a diffeomorphism.

The next theorem characterizes the symplectic form and the Hamiltonian dynamics on $P_{\mathcal{O}_\mu}$.

Theorem 7.4 (Symplectic orbit reduction) *Assume that the free proper symplectic action of the Lie group G on the symplectic manifold (P, Ω) admits an associated momentum map $\mathbf{J} : P \to \mathfrak{g}^*$.*

(i) *On $\mathbf{J}^{-1}(\mathcal{O}_\mu)$ there is a unique immersed smooth manifold structure such that the projection $\pi_{\mathcal{O}_\mu} : \mathbf{J}^{-1}(\mathcal{O}_\mu) \to P_{\mathcal{O}_\mu}$ is a surjective submersion, where $P_{\mathcal{O}_\mu}$ is endowed with the manifold structure making κ_μ into a diffeomorphism. This smooth manifold structure does not depend on the choice of μ in the orbit $\mathcal{O}_\mu$. If $\mathbf{J}^{-1}(\mathcal{O}_\mu)$ is a submanifold of P in its own right, then the immersed topology by κ_μ and the induced topology on $P_{\mathcal{O}_\mu}$ coincide.*

(ii) *$P_{\mathcal{O}_\mu}$ is a symplectic manifold with the symplectic form $\Omega_{\mathcal{O}_\mu}^{\maltese}$ uniquely characterized by the relation*

$$\iota_{\mathcal{O}_\mu}^* \Omega = \pi_{\mathcal{O}_\mu}^* \Omega_{\mathcal{O}_\mu}^{\maltese} + \mathbf{J}_{\mathcal{O}_\mu}^* \omega_{\mathcal{O}_\mu}^+,$$

where $\mathbf{J}_{\mathcal{O}_\mu}$ is the restriction of $\mathbf{J}$ to $\mathbf{J}^{-1}(\mathcal{O}_\mu)$, $\iota_{\mathcal{O}_\mu} : \mathbf{J}^{-1}(\mathcal{O}_\mu) \hookrightarrow P$ is the inclusion, and $\omega_{\mathcal{O}_\mu}^+$ is the $+$orbit symplectic form on $\mathcal{O}_\mu$ given by (7.7). The symplectic manifolds P_μ and $P_{\mathcal{O}_\mu}$ are symplectically diffeomorphic by κ_μ.

(iii) *Let H be a G-invariant function on P and define $\widetilde{H} : P/G \to \mathbb{R}$ by $H = \widetilde{H} \circ \pi$. Then the Hamiltonian vector field X_H is also G-invariant and hence induces a vector field on P/G which coincides with the Hamiltonian vector field $X_{\widetilde{H}}$. Moreover, the flow of $X_{\widetilde{H}}$ leaves the symplectic leaves $P_{\mathcal{O}_\mu}$ of P/G invariant. This flow restricted to the symplectic leaves is again Hamiltonian relative to the symplectic form $\Omega_{\mathcal{O}_\mu}^{\maltese}$ and the Hamiltonian function $H_{\mathcal{O}_\mu}$ given by*

$$H_{\mathcal{O}_\mu} \circ \pi_{\mathcal{O}_\mu} = H \circ i_{\mathcal{O}_\mu} \qquad or \qquad H_{\mathcal{O}_\mu} = \widetilde{H}|_{\mathcal{O}_\mu}.$$

Moreover, if $F : P \to \mathbb{R}$ is another smooth G-invariant function, then

$\{F, H\}$ is also G-invariant and $\{F, H\}_{\mathcal{O}_\mu} = \{F_{\mathcal{O}_\mu}, H_{\mathcal{O}_\mu}\}_{P_{\mathcal{O}_\mu}}$, where $\{\cdot, \cdot\}_{\mathcal{O}_\mu}$ is the Poisson bracket on the symplectic manifold $P_{\mathcal{O}_\mu} = \mathbf{J}^{-1}(\mathcal{O}_\mu)/G$.

The proof of this theorem in the regular case and when $\mathcal{O}_\mu$ is an embedded submanifold of $\mathfrak{g}^*$ can be found in [Marle76], [KaKoSt78], and [Marsden81]. For the general case, when $\mathcal{O}_\mu$ is not a submanifold of $\mathfrak{g}^*$ see [OR04]. Here is the main idea of the proof. Consider for each value $\mu \in \mathfrak{g}^*$ of $\mathbf{J}$ the G-equivariant bijection

$$s : [g, z] \in G \times_{G_\mu} \mathbf{J}^{-1}(\mu) \mapsto g \cdot z \in \mathbf{J}^{-1}(\mathcal{O}_\mu),$$

where $G \times_{G_\mu} \mathbf{J}^{-1}(\mu) := (G \times \mathbf{J}^{-1}(\mu))/G_\mu$, the G_μ-action being the diagonal action. Endow $\mathbf{J}^{-1}(\mathcal{O}_\mu)$ with the smooth manifold structure that makes the bijection s into a diffeomorphism. Then $\mathbf{J}^{-1}(\mathcal{O}_\mu)$ with this smooth structure is an immersed submanifold of P. This is the manifold structure on $\mathbf{J}^{-1}(\mathcal{O}_\mu)$ used in the statement of Theorem 7.4.

In the particular case when $\mathbf{J}^{-1}(\mathcal{O}_\mu)$ is a smooth submanifold of P in its own right, this manifold structure coincides with the one induced by the mapping s described above since in this situation the bijection s becomes a diffeomorphism relative to the a priori given smooth manifold structure on $\mathbf{J}^{-1}(\mathcal{O}_\mu)$.

If μ is a regular value of $\mathbf{J}$ and $\mathcal{O}_\mu$ is an embedded submanifold of $\mathfrak{g}^*$, then $\mathbf{J}$ is transverse to $\mathcal{O}_\mu$ and hence $\mathbf{J}^{-1}(\mathcal{O}_\mu)$ is automatically an embedded submanifold of P.

The orbit $\mathcal{O}_\mu$ can be used to transform point reduction at an arbitrary $\mu \in \mathfrak{g}^*$ to point reduction at zero for a larger manifold. Suppose that we are in the hypotheses of the Symplectic Point Reduction Theorem 7.1. Form the point reduced space P_μ and consider the G-orbit through μ in $\mathfrak{g}^*$ (in general the orbit under the affine action) endowed with the $+$ orbit symplectic form. The group G acts canonically on the left on $\mathcal{O}_\mu$ with momentum map given by the inclusion $i : \mathcal{O}_\mu \hookrightarrow \mathfrak{g}^*$. Let $P \ominus \mathcal{O}_\mu$ denote the symplectic manifold $P \times \mathcal{O}_\mu$ endowed with the symplectic structure $\Omega - \omega^+_{\mathcal{O}_\mu} := \pi_1^*\Omega - \pi_2^*\omega^+_{\mathcal{O}_\mu}$, where $\pi_1 : P \times \mathcal{O}_\mu \to P$ and $\pi_2 : P \times \mathcal{O}_\mu \to \mathcal{O}_\mu$ are the projections on the first and second factors respectively. The Lie group G acts canonically on $P \ominus \mathcal{O}_\mu$ by $g \cdot (z, \nu) := (g \cdot z, \mathrm{Ad}^*_{g^{-1}} \nu)$. As discussed in §5.3, example **(6)**, this action has the momentum map $\mathbf{J} - i : P \ominus \mathcal{O}_\mu \to \mathfrak{g}^*$. This momentum map is equivariant if $\mathbf{J}$ is, in which case, $\mathcal{O}_\mu$ is taken to be the coadjoint orbit. With these notations we have the following result.

Theorem 7.5 (Shifting theorem) *The reduced symplectic manifolds P_μ and $(P \ominus \mathcal{O}_\mu)_0$ are symplectically diffeomorphic.*

One should not read into this theorem more than it states. It is tempting to quote it in order to dismiss the reduction procedure at all points $\mu \neq 0$. This would be an error, for the price one pays to reduce only at zero is heavy: the original phase space is enlarged by multiplication with the orbit $\mathcal{O}_\mu$ whose topology can be quite involved and who is, in general, not an embedded submanifold of $\mathfrak{g}^*$. Specifically, when dealing with singular reduction it is important to study reduction at non-zero values of the momentum map carefully. The Shifting Theorem 7.5 only hides the differential topological difficulties by burying them into $\mathcal{O}_\mu$.

7.5 Semidirect Product Reduction

In this section we present the general Semidirect Product Reduction Theorem as found in [MaRaWe84a, MaRaWe84b]. We do not attempt to give a history of the subject here since it can be found in many other papers and books. To avoid any technical complications, all of this section deals only with finite dimensional objects, even though the range of applicability of the theorems presented here goes far beyond that to many continuum and quantum mechanical systems.

Let V be a vector space and assume that $\sigma : G \to \mathrm{Aut}(V)$ is a representation of the Lie group G on V; $\mathrm{Aut}(V)$ denotes the Lie group of linear isomorphisms of V onto itself whose Lie algebra is $\mathrm{End}(V)$, the space of all linear maps of V to itself. Denote by $\sigma' : \mathfrak{g} \to \mathrm{End}(V)$ the induced Lie algebra representation, that is,

$$\xi \cdot v := \xi_V(v) := \sigma'(\xi)v := \left.\frac{d}{dt}\right|_{t=0} \sigma(\exp t\xi)v$$

Given G, V, and ρ define the semidirect product $S := G \circledS V$ as the Lie group whose underlying manifold is $G \times V$ and multiplication

$$(g_1, v_1)(g_2, v_2) := (g_1 g_2, v_1 + \sigma(g_1)v_2)$$

for $g_1, g_2 \in G$ and $v_1, v_2 \in V$. The identity element is $(e, 0)$ and $(g, v)^{-1} = (g^{-1}, -\sigma(g^{-1})v)$. Note that V is a normal subgroup of S and that $S/V = G$.

Let $\mathfrak{g}$ be the Lie algebra of G and let $\mathfrak{s} := \mathfrak{g} \circledS V$ be the Lie algebra of S; it is the semidirect product of $\mathfrak{g}$ with V using the representation σ' and its underlying vector space is $\mathfrak{g} \times V$. The Lie bracket on $\mathfrak{s}$ is given by

$$[(\xi_1, v_1), (\xi_2, v_2)] = ([\xi_1, \xi_2], \sigma'(\xi_1)v_2 - \sigma'(\xi_2)v_1)$$

for $\xi_1, \xi_2 \in \mathfrak{g}$ and $v_1, v_2 \in V$. Identify $\mathfrak{s}^*$ with $\mathfrak{g}^* \times V^*$ by using the duality pairing on each factor. The following formulas are useful for our next considerations. They are obtained by straightforward (and sometimes lengthy) computations:

- the adjoint action of S on $\mathfrak{s}$:

$$\mathrm{Ad}_{(g,u)}(\xi, v) = (\mathrm{Ad}_g \xi, \sigma(g)v - \sigma'(\mathrm{Ad}_g \xi)u), \text{ for } (g,u) \in S, (\xi, v) \in \mathfrak{s};$$

- the coadjoint action of S on $\mathfrak{s}^*$:

$$\mathrm{Ad}^*_{(g,u)^{-1}}(\nu, a) = \left(\mathrm{Ad}^*_{g^{-1}} \nu + (\sigma'_u)^* \sigma_*(g)a, \sigma_*(g)a\right),$$

  for $(g,u) \in S$, $(\nu, a) \in \mathfrak{s}^*$, where $\sigma_*(g) := \sigma(g^{-1})^* \in \mathrm{Aut}(V^*)$, $\sigma'_u : \mathfrak{g} \to V$ is the linear map given by $\sigma'_u(\xi) := \sigma'(\xi)u$ and $(\sigma'_u)^* : V^* \to \mathfrak{g}^*$ is its dual;
- the lift λ of left translation of S on T^*S:

$$\lambda((g,u), (\alpha_h, v, a)) = \left(T^*_{gh} L_{g^{-1}} \alpha_h, u + \sigma(g)v, \sigma_*(g)a\right)$$

  for $(g,u) \in S$, and $(\alpha_h, v, a) \in T^*_{(h,v)} S = T^*_h G \times \{v\} \times V^*$; λ induces a canonical S-action on the product Poisson manifold $T^*G \times V^*$ by ignoring the third factor (V^* has the trivial Poisson bracket);
- the lift ρ of right translation of S on T^*S:

$$\rho((g,u), (\alpha_h, v, a)) = \left(T^*_{gh} R_{g^{-1}} \alpha_h - \mathbf{d} f^a_{\sigma(g^{-1})u}(hg), v + \sigma(h)u, a\right)$$

  for $(g,u) \in S$, and $(\alpha_h, v, a) \in T^*_{(h,v)} S = T^*_h G \times \{v\} \times V^*$, and where $f^a_u : G \to \mathbb{R}$ is the "matrix element function" $f^a_u(g) := \langle a, \sigma(g)u \rangle$; ρ induces a canonical S-action on $T^*G \times V^*$ by ignoring the third factor (V^* has the trivial Poisson bracket);
- the momentum map $\mathbf{J}_L : T^*S \to \mathfrak{s}^*_+$ for the left translation λ:

$$\mathbf{J}_L(\alpha_g, v, a) = T^*_{(e,0)} R_{(g,v)}(\alpha_g, v, a) = (T^*_e R_g \alpha_g + (\sigma'_v)^* a, a); \quad (7.8)$$

- the momentum map $\mathbf{J}_R : T^*S \to \mathfrak{s}^*_-$ for the right translation ρ:

$$\mathbf{J}_R(\alpha_g, v, a) = T^*_{(e,0)} R_{(g,v)}(\alpha_g, v, a) = (T^*_e L_g \alpha_g, \sigma(g)^* a); \quad (7.9)$$

- the $\pm$ Lie-Poisson bracket of $F, H : \mathfrak{s}^* \to \mathbb{R}$:

$$\begin{aligned}\{F, H\}_\pm(\mu, a) = \pm \left\langle \mu, \left[\frac{\delta F}{\delta \mu}, \frac{\delta H}{\delta \mu}\right]\right\rangle \pm \left\langle a, \sigma'\left(\frac{\delta F}{\delta \mu}\right)\frac{\delta H}{\delta a}\right\rangle \\ \mp \left\langle a, \sigma'\left(\frac{\delta H}{\delta \mu}\right)\frac{\delta F}{\delta a}\right\rangle \text{ for } \mu \in \mathfrak{g}^*, \ a \in V^*; \quad (7.10)\end{aligned}$$

- the Hamiltonian vector field determined by $H : \mathfrak{s}^* \to \mathbb{R}$:

$$X_H(\mu, a) = \mp \left(\mathrm{ad}^*_{\frac{\delta H}{\delta \mu}} \mu - \left(\sigma'_{\frac{\delta H}{\delta a}}\right)^* a, \sigma'\left(\frac{\delta H}{\delta \mu}\right)^* a\right). \quad (7.11)$$

Now we shall reduce in two steps. We start with the left action of S on T^*S. As we already know from general theory, the momentum map $\mathbf{J}_R$ is invariant under λ. The normal subgroup V of S acts on S by left translations and the lift of this action admits an equivariant momentum map (in this case invariant since V is Abelian), given by the second component $(\alpha_g, v, a) \mapsto a$ of $\mathbf{J}_L$. In addition, the projection $T^*S \to T^*G$ is clearly canonical so that the map

$$P_L : (\alpha_g, v, a) \in T^*S \mapsto (\alpha_g, a) \in T^*G \times V^*$$

is also canonical; $T^*G \times V^*$ has the product Poisson structure (see §5.3, example **(6)**). It is easy to see that $\mathbf{J}_R$ factors through P_L, that is, there is a smooth map

$$\widetilde{\mathbf{J}}_R : (\alpha_g, a) \in T^*G \times V^* \mapsto (T_e^* L_g \alpha_g, \sigma(g)^* a) \in \mathfrak{s}^*_-$$

such that the following diagram is commutative:

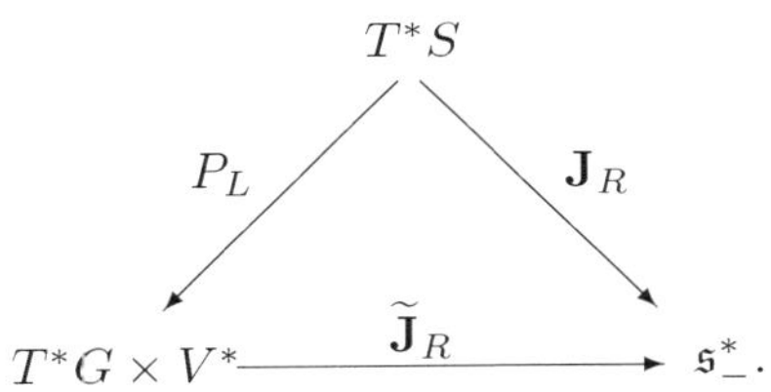

Since $\mathbf{J}_R = \widetilde{\mathbf{J}}_R \circ P_L$, all three maps are canonical, and P_L is onto, it follows that $\widetilde{\mathbf{J}}_R$ is also canonical.

The same phenomenon occurs when working with the right action ρ. Since there is a lot a asymmetry in the expression of all the maps involved, we shall repeat the argument. The momentum map $\mathbf{J}_L$ is right invariant. The normal subgroup V of S acts on the right on T^*S with momentum map $(\alpha_g, v, a) \mapsto \sigma(g)^* a$ given by the second component of $\mathbf{J}_R$. This map is therefore canonical. Moreover, the map

$$(\alpha_g, u, a) \mapsto \alpha_g + \mathbf{d} f^a_{\sigma(g^{-1})u}(g) = \alpha_g + T_g^* R_{g^{-1}} (\sigma'_u)^* a$$

is a projection followed by a translation with an exact differential on the fibers and is hence a canonical map from T^*S to T^*G (see Proposition 2.15). Therefore

$$P_R : (\alpha_g, u, a) \in T^*S \mapsto (\alpha_g + T_g^* R_{g^{-1}} (\sigma'_u)^* a, \sigma(g)^* a) \in T^*G \times V^*$$

is a canonical map. Now notice that $\mathbf{J}_L$ factors through P_R, that is, there is a smooth map

$$\widetilde{\mathbf{J}}_L : (\alpha_g, a) \in T^*G \times V^* \mapsto (T_e^* R_g \alpha_g, \sigma(g^{-1})^* a) \in \mathfrak{s}_+^*$$

such that the following diagram is commutative:

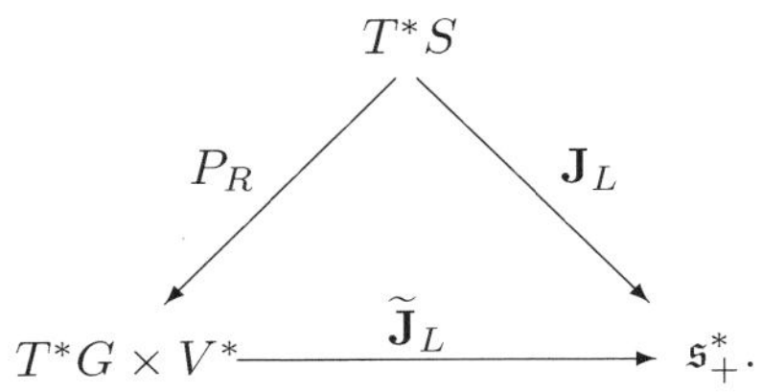

As before, this implies that $\widetilde{\mathbf{J}}_L$ is a canonical map.

The origin of the maps P_L and P_R is also transparent. The space $T^*G \times V^*$ is diffeomorphic to the orbit space of T^*S by the left or right V-action. The diffeomorphisms that implement this identification are easily seen to be

$$[\alpha_g, u, a] \mapsto (\alpha_g, a)$$

for the left V-action and

$$[\alpha_g, u, a] \mapsto (\alpha_g + \mathbf{d} f^a_{\sigma(g^{-1})u}(g), \sigma(g)^* a)$$

for the right V-action, where $[\alpha_g, u, a]$ denotes the left or right V-orbit through (α_g, u, a). Using these diffeomorphisms, the projections onto the orbit spaces become P_L and P_R respectively.

We summarize these considerations in the following theorem.

Theorem 7.6 *The maps $\widetilde{\mathbf{J}}_L, \widetilde{\mathbf{J}}_R : T^*G \times V^* \to \mathfrak{s}_\pm^*$ given by*

$$\widetilde{\mathbf{J}}_L(\alpha_g, a) = \left(T_e^* R_g \alpha_g, \sigma(g^{-1})^* a\right)$$
$$\widetilde{\mathbf{J}}_R(\alpha_g, a) = ((T_e^* L_g \alpha_g, \sigma(g)^* a)$$

are canonical. These maps are reductions of momentum maps by the action of the normal subgroup V and are themselves momentum maps for the left, respectively right, actions of S on the product Poisson manifold $T^*G \times V^*$, where V^* carries the trivial Poisson bracket.

The procedure used here to reduce in two steps is very general and can be applied to many other situations, such as central extensions of groups, for example. We refer to [Marsden et. al.] and references therein for the general theory of reduction by stages and many other examples.

Let us study the reduction of dynamics implied by this theorem. So, consider a Hamiltonian $H : T^*G \times V^* \to \mathbb{R}$ and assume that it is invariant under the left action of S on $T^*G \times V^*$. In particular, for each $a \in V^*$ the function $H_a : T^*G \to \mathbb{R}$ given by $H_a(\alpha_g) := H(\alpha_g, a)$ is invariant under the lift to T^*G of the left action of the stabilizer $G_a := \{g \in G \mid \sigma(g)^* a = a\}$ on G. Then it follows that H induces a smooth function $H_L : \mathfrak{s}^*_- \to \mathbb{R}$ defined by $H_L \circ \widetilde{\mathbf{J}}_R = H$, that is, $H_L(T_e^* L_g \alpha_g, \sigma(g)^* a) = H(\alpha_g, a)$. For right invariant systems, one interchanges, as usual, "left" by "right" and "$-$" by "$+$". However, in this case, because the maps involved are different we record H_R separately: $H_R \circ \widetilde{\mathbf{J}}_L = H$, that is, $H_R(T_e^* R_g \alpha_g, \sigma(g^{-1})^* a) = H(\alpha_g, a)$.

It turns out that the evolution of $a \in V^*$ is particularly simple. We begin with the left action and work on $\mathfrak{s}^*_-$. Let $c_a(t) \in T^*G$ denote an integral curve of the Hamiltonian system for H_a and let $g_a(t)$ be its projection on G. Then $t \mapsto (c_a(t), a)$ is an integral curve of H on $T^*G \times V^*$ so that the curve $t \mapsto \widetilde{\mathbf{J}}_R(c_a(t), a)$ is an integral curve of H_L on $\mathfrak{s}^*_-$. Thus, $t \mapsto \sigma(g_a(t))^* a$ is the evolution of the initial condition $a \in V^*$ in $\mathfrak{s}^*_-$.

For right actions the situation is identical, but we shall find another formula. If $c_a(t)$ and $g_a(t)$ are as before, the curve $t \mapsto \widetilde{\mathbf{J}}_L(c_a(t), a)$ is an integral curve of H_R on $\mathfrak{s}^*_+$. Hence $t \mapsto \sigma(g_a(t)^{-1})^* a$ is the evolution of $a \in V^*$ in $\mathfrak{s}^*_+$. This proves the following theorem.

Theorem 7.7 *Let $H : T^*G \times V^* \to \mathbb{R}$ be a left invariant function relative to the S-action on $T^*G \times V^*$. Then H induces a Hamiltonian $H_L : \mathfrak{s}^*_- \to \mathbb{R}$ defined by $H_L(T_e^* L_g \alpha_g, \sigma(g)^* a) = H(\alpha_g, a)$ which then yields Lie-Poisson equations on $\mathfrak{s}^*_-$. The curve $(c_a(t), a) \in T^*G \times V^*$ is a solution of Hamilton's equations defined by H on the product Poisson manifold $T^*G \times V^*$, where V^* is endowed with the trivial Poisson bracket, if and only if $\widetilde{\mathbf{J}}_R(c_a(t), a)$ is a solution of the Lie-Poisson system on $\mathfrak{s}^*_-$ defined by H_L. In particular, the evolution of $a \in V^*$ is given by $\sigma(g_a(t))^* a$, where $g_a(t)$ is the projection of $c_a(t)$ on G. For right invariant systems one interchanges "left" by "right", "$-$" by "$+$", and defines $H_R : \mathfrak{s}^*_+ \to \mathbb{R}$ by $H_R(T_e^* R_g \alpha_g, \sigma(g^{-1})^* a) = H(\alpha_g, a)$. In this case, the evolution of $a \in V^*$ is given by $\sigma(g_a(t)^{-1})^* a$.*

The combination of these two theorems is quite powerful in examples. Often, a physical system is given by a Hamiltonian on $T^*G \times V^*$, where V^* is usually a space of parameters of the system. This Hamiltonian is left or right invariant under the $G \circledS V$-action on $T^*G \times V^*$. Then, the theorems just proved, guarantee that one can reduce the given system to Lie-Poisson equations on $(\mathfrak{g} \circledS V)^*$ and one knows already that the second equation has as solution the "dragging along by the action" of the initial condition. For systems in continuum mechanics, this appears usually as a "Lie transport" equation, such as the conservation of mass, of entropy, or the frozen magnetic lines in the fluid in the magneto-hydrodynamics approximation.

We close these considerations by presenting a symplectic counterpart of Theorem 7.6. We shall use the Symplectic Orbit Reduction Theorem 7.4 to determine, up to connected components, the symplectic leaves of $(T^*G)/G_a$ for any $a \in V^*$. Fix in all that follows an $a \in V^*$ and let $\mathfrak{g}_a := \{\xi \in \mathfrak{g} \mid \sigma'(\xi)^* a = 0\}$ be the Lie algebra of G_a. The lift to T^*G of left translation of G_a on G has the equivariant momentum map $\mathbf{J}_L^a : T^*G \to \mathfrak{g}_a^*$ given by restriction $\mathbf{J}_L^a(\alpha_g) = (T_e^* R_g \alpha_g)|_{\mathfrak{g}_a}$. The map $i_L^a : T^*G \to T^*S$ given by $i_L^a(\alpha_g) := (\alpha_g, 0, a)$ is a Poisson embedding which is equivariant relative to the left action of G_a on T^*G and the lifted left action λ of S on its cotangent bundle T^*S. Therefore i_L^a induces a Poisson embedding on the quotients $\bar{i}_L^a : (T^*G)/G_a \to T^*S/S \cong \mathfrak{s}_-^*$. From the Symplectic Orbit Reduction Theorem we know that

$$
\begin{aligned}
(\mathbf{J}_L^a)^{-1}(\mu|_{\mathfrak{g}_a})/(G_a)_{\mu|_{\mathfrak{g}_a}} &\cong (\mathbf{J}_L^a)^{-1}(\mathcal{O}_{\mu|_{\mathfrak{g}_a}})/G_a \hookrightarrow (T^*G)/G_a \\
&\xrightarrow{\bar{i}_L^a} (T^*S)/S \xrightarrow{\bar{\mathbf{J}}_R} \mathfrak{s}_-^*
\end{aligned}
$$

where the first diffeomorphism is symplectic and given by the Orbit Symplectic Reduction Theorem and $\bar{\mathbf{J}}_R$ is the quotient of $\mathbf{J}_R : T^*S \to \mathfrak{s}_-^*$ implementing the Lie-Poisson reduction theorem (see §6.1). Where does the reduced space $(\mathbf{J}_L^a)^{-1}(\mathcal{O}_{\mu|_{\mathfrak{g}_a}})/G_a$ land by this sequence of symplectic and Poisson diffeomorphisms and embeddings? To see this, we compute

$$
\begin{aligned}
(\bar{\mathbf{J}}_R \circ \bar{i}_L^a)\left((\mathbf{J}_L^a)^{-1}(\mathcal{O}_{\mu|_{\mathfrak{g}_a}})/G_a\right) &= (\mathbf{J}_R \circ i_L^a)\left((\mathbf{J}_L^a)^{-1}(\mathcal{O}_{\mu|_{\mathfrak{g}_a}})\right) \\
&= \{(\nu, b) \in \mathfrak{s}^* \mid \text{there exists } g \in G \text{ such that } \sigma_*(g)a = b,\ \mathrm{Ad}_g^* \nu \in \mathcal{O}_{\mu|_{\mathfrak{g}_a}}\} \\
&= \bigcup_{\chi|_{\mathfrak{g}_a} = \mu|_{\mathfrak{g}_a}} S \cdot (\chi, a),
\end{aligned}
$$

where $S \cdot (\chi, a)$ denotes the S-coadjoint orbit through (χ, a) in $\mathfrak{s}_-^*$. However, the identity

$$
\{(\sigma_u')^* a \mid u \in V\} = \{\nu \in \mathfrak{g}^* \mid \nu|_{\mathfrak{g}_a} = 0\}
$$

shows that $S \cdot (\chi, a) = S \cdot (\mu, a)$ for all $\chi \in \mathfrak{g}^*$ satisfying $\chi|_{\mathfrak{g}_a} = \mu|_{\mathfrak{g}_a}$. Therefore the union above is actually one single orbit, namely $S \cdot (\mu, a)$, and we have shown that the reduced space $(\mathbf{J}_L^a)^{-1}(\mathcal{O}_{\mu|_{\mathfrak{g}_a}})/G_a$ lands in $S \cdot (\mu, a)$.

For right actions the same thing happens but we need the map $i_R^a : T^*G \to T^*S$ given by $i_R^a(\alpha_g) := (\alpha_g, 0, \sigma_*(g)a)$ to embed right (G_a, S)-equivariantly

T^*G into T^*S. We shall also need another notation for the quotients relative to right actions and we shall adopt here $S\backslash T^*S$ and $G_a\backslash T^*G$. Similarly, a sign on a coadjoint orbit signifies the sign in front of the orbit symplectic structure. We have proved the following theorem.

Theorem 7.8 *The map* $\overline{\mathbf{J}}_R \circ \overline{i}^a_L : (\mathbf{J}^a_L)^{-1}(\mathcal{O}_{\mu|_{\mathfrak{g}_a}})/G_a \to S \cdot (\mu, a)_-$ *is a symplectic diffeomorphism thereby realizing this reduced space as a coadjoint orbit in* $\mathfrak{s}^*_-$. *The map* $\overline{\mathbf{J}}_L \circ \overline{i}^a_R : G_a\backslash(\mathbf{J}^a_R)^{-1}(\mathcal{O}_{\mu|_{\mathfrak{g}_a}}) \to S \cdot (\mu, a)_+$ *is a symplectic diffeomorphism thereby realizing this reduced space as a coadjoint orbit in* $\mathfrak{s}^*_+$.

In other words, forgetting about the precise maps involved and the orbit reduction formulation of this result, this theorem states that *there is a symplectic diffeomorphism between the coadjoint orbit* $S \cdot (\mu, a) \subset (\mathfrak{g} \circledS V)^*$ *and the reduced space obtained by reducing* T^*G *by the subgroup* G_a *at the point* $\mu|_{\mathfrak{g}_a} \in \mathfrak{g}^*_a$.

There is a Lagrangian version of this theorem, that is, a formulation in terms of Euler-Poincaré type equations. It is *not true* that the Euler-Poincaré equations that we shall deduce below for $\mathfrak{g} \circledS V$ are simply the general Euler-Poincaré equations explicitly written out for a semidirect product. The reduced Lagrangian formulation in the case of semidirect products is more subtle and was done in [HMR98]. We present only the situation of left representations and left invariant Lagrangians. There are clearly three other versions and, unfortunately, they are important because of various relative sign differences in the equations and the constrained variations. Since in these lectures we shall only deal with the heavy top, we refer to the above mentioned paper for additional details and examples.

The set-up of the problem is the following. Given are:

- a left representation $\sigma : G \to \mathrm{Aut}(V)$ of a Lie group G on a vector space V which induces the left action of G on $TG \times V^*$ given by $h \cdot (v_g, a) := (T_gL_h(v_g), \sigma_*(g)a)$, for $v_g \in T_gG$ and $a \in V^*$;
- a smooth left invariant function $L : TG \times V^* \to \mathbb{R}$ relative to this action;
- in particular, if $a_0 \in V^*$ the function $L_{a_0} : TG \to \mathbb{R}$ given by $L_{a_0}(v_g) := L(v_g, a_0)$ is invariant under the lift to TG of left translation of G_{a_0} on G;
- by left G-invariance of L the formula

$$l(T_gL_{g^{-1}}v_g, \sigma(g)^*a) = L(v_g, a)$$

defines a smooth function $l : \mathfrak{g} \times V^* \to \mathbb{R}$ and conversely any such function $l : \mathfrak{g} \times V^* \to \mathbb{R}$ determines a left invariant function $L : TG \times V^* \to \mathbb{R}$;

- for a curve $g(t) \in G$ with $g(0) = e$, let $\xi(t) := T_{g(t)} L_{g(t)^{-1}} \dot{g}(t) \in \mathfrak{g}$ and define the curve $a(t) \in V^*$ by

$$a(t) := \sigma(g(t))^* a_0 \tag{7.12}$$

for some given $a_0 \in V^*$; the unique solution of the linear differential equation with time dependent coefficients

$$\dot{a}(t) = \sigma'(\xi(t))^* a(t) \tag{7.13}$$

with initial condition a_0 is this curve $a(t)$.

With these notations we have the following.

Theorem 7.9 *The following statements are equivalent:*

(i) *With $a_0 \in V^*$ fixed, Hamilton's variational principle*

$$\delta \int_{t_1}^{t_2} L_{a_0}(g(t), \dot{g}(t)) dt = 0$$

holds, for variations $\delta g(t)$ of $g(t)$ vanishing at the endpoints.

(ii) *The curve $g(t)$ satisfies the Euler-Lagrange equations for L_{a_0} on G.*

(iii) *The constrained variational principle*

$$\delta \int_{t_1}^{t_2} l(\xi(t), a(t)) dt = 0$$

holds on $\mathfrak{g} \times V^*$, using variations of the form

$$\delta\xi = \dot{\eta} + [\xi, \eta], \qquad \delta a = \sigma'(\eta)^* a,$$

where $\eta(t) \in \mathfrak{g}$ is any curve vanishing at the endpoints.

(iv) *The **semidirect Euler-Poincaré equations***

$$\frac{d}{dt}\frac{\delta l}{\delta \xi} = \mathrm{ad}^*_\xi \frac{\delta l}{\delta \xi} + \left(\sigma'_{\frac{\delta l}{\delta a}}\right)^* a \tag{7.14}$$

hold on $\mathfrak{g} \times V^*$.

Proof The proof follows the same pattern as that of Theorem 6.6. The equivalence of **(i)** and **(ii)** is Hamilton's classical variational principle that holds for any manifold (see Theorem 3.5). To prove that **(iii)** and **(iv)** are equivalent, we compute the variation of l, integrate by parts, and use the conditions

$\eta(t_1) = \eta(t_2) = 0$ to get

$$\begin{aligned}
\delta \int_{t_1}^{t_2} l(\xi(t), a(t))dt &= \int_{t_1}^{t_2} \left(\left\langle \frac{\delta l}{\delta \xi}, \delta \xi \right\rangle + \left\langle \delta a, \frac{\delta l}{\delta a} \right\rangle \right) dt \\
&= \int_{t_1}^{t_2} \left(\left\langle \frac{\delta l}{\delta \xi}, \dot{\eta} + [\xi, \eta] \right\rangle + \left\langle \sigma'(\eta)^* a, \frac{\delta l}{\delta a} \right\rangle \right) dt \\
&= \int_{t_1}^{t_2} \left(\left\langle -\frac{d}{dt}\frac{\delta l}{\delta \xi} + \mathrm{ad}_\xi^* \frac{\delta l}{\delta \xi}, \eta \right\rangle + \left\langle \left(\sigma'_{\frac{\delta l}{\delta a}} \right)^* a, \eta \right\rangle \right) dt \\
&= \int_{t_1}^{t_2} \left\langle -\frac{d}{dt}\frac{\delta l}{\delta \xi} + \mathrm{ad}_\xi^* \frac{\delta l}{\delta \xi} + \left(\sigma'_{\frac{\delta l}{\delta a}} \right)^* a, \eta \right\rangle dt.
\end{aligned}$$

Since this is valid for any smooth path $\eta(t)$ vanishing at the endpoints, the variation of the integral of l vanishes subject to the constrained variations of ξ and a if and only if the semidirect Euler-Poincaré equations hold.

It remains to be shown that **(i)** and **(iii)** are equivalent. We begin by noticing that due to the G-invariance of L and the relation $a(t) = \sigma'(g(t))^* a_0$ the integrands in the two variational principles are equal. Now let, $\eta(t) := T_{g(t)} L_{g(t)^{-1}} \delta g(t) \in \mathfrak{g}$. At this point one could proceed with the proof exactly as was done in the one for Theorem 6.6 by assuming that we work only with matrix groups, or use Lemma 6.7 to do the general case. Let us work abstractly this time around. So, by Lemma 6.7, all variations $\delta g(t) \in TG$ of $g(t)$ with fixed endpoints induce and are induced by variations $\delta \xi(t) \in \mathfrak{g}$ of $\xi(t)$ of the form $\delta \xi = \dot{\eta} + [\xi, \eta]$ with $\eta(t)$ a smooth curve vanishing at the endpoints.

Thus if **(i)** holds, define $\eta(t) := T_{g(t)} L_{g(t)^{-1}} \delta g(t) \in \mathfrak{g}$ for a variation $\delta g(t)$ vanishing at the endpoints and set $\delta \xi(t) = T_{g(t)} \dot{g}(t)$. By Lemma 6.7 we have $\delta \xi = \dot{\eta} + [\xi, \eta]$ with $\eta(t)$ a smooth curve vanishing at the endpoints. In addition, the variation of $a(t) = \sigma'(g(t))^* a_0$ is $\delta a(t) = \sigma'(\eta(t)) a(t)$. Thus **(iii)** holds.

Conversely, assume that **(iii)** holds. So if $\delta \xi = \dot{\eta} + [\xi, \eta]$ with $\eta(t)$ a smooth curve vanishing at the endpoints, define $\delta g(t) = T_e L_{g(t)} \eta(t) \in TG$. Lemma 6.7 guarantees then that this $\delta g(t)$ is the general variation of $g(t)$ vanishing at the endpoints. Finally, the relation $\delta a(t) = \sigma'(\eta(t))^* a(t)$ shows that the variation of $\sigma_*(g(t)) a(t) = a_0$ vanishes, which is consistent with the fact that L_{a_0} depends only on $g(t)$ and $\dot{g}(t)$. Thus **(i)** holds. ∎

We close this section by showing how the heavy top equations fit into the semidirect Lie-Poisson and Euler-Poincaré framework. In the process, many of the remarkable statements in §1.2 that appeared as computational coincidences will be explained through the theory that was just presented in this section. To do this, we shall use all the explicit formulas deduced in §1.2. The configuration space is $G = SO(3)$ and it represents the attitude of the heavy top. In coordinates it is given by Euler angles, as explained in §1.2. The parameter of the problem is $Mg\ell\boldsymbol{\chi}$, where $M \in \mathbb{R}$ is the mass of the heavy top, $g \in \mathbb{R}$ is the value of the gravitational acceleration, $\ell \in \mathbb{R}$ is the distance from the fixed point (that is, the point of suspension of the rigid body) to the center of mass of the body, and $\boldsymbol{\chi} \in \mathbb{R}^3$ is the unit vector pointing from the fixed point to the center of mass. Therefore, the parameter space of this problem is $V^* := \mathbb{R}^3$. Identifying $\mathbb{R}^3$ with itself using the usual inner product, gives $V := \mathbb{R}^3$. The representation $\sigma : SO(3) \to \mathrm{Aut}(\mathbb{R}^3)$ is usual matrix multiplication on vectors, that is, $\sigma(A)\mathbf{v} := A\mathbf{v}$, for any $A \in SO(3)$ and $\mathbf{v} \in \mathbb{R}^3$. Dualizing we get $\sigma(A)^*\boldsymbol{\Gamma} = A^*\boldsymbol{\Gamma} = A^{-1}\boldsymbol{\Gamma}$, for any $\boldsymbol{\Gamma} \in V^* \cong \mathbb{R}^3$. The induced Lie algebra representation $\sigma' : \mathbb{R}^3 \cong \mathfrak{so}(3) \to \mathrm{End}(\mathbb{R}^3)$ is given by $\sigma'(\boldsymbol{\Omega})\mathbf{v} = \sigma'_{\mathbf{v}}\boldsymbol{\Omega} = \boldsymbol{\Omega}\times\mathbf{v}$, for any $\boldsymbol{\Omega}, \mathbf{v} \in \mathbb{R}^3$. Therefore, $(\sigma'_{\mathbf{v}})^*\,\boldsymbol{\Gamma} = \mathbf{v}\times\boldsymbol{\Gamma}$ and $\sigma'(\boldsymbol{\Omega})^*\boldsymbol{\Gamma} = \boldsymbol{\Gamma}\times\boldsymbol{\Omega}$, for any $\mathbf{v} \in V \cong \mathbb{R}^3$, $\boldsymbol{\Omega} \in \mathbb{R}^3 \cong \mathfrak{so}(3)$, and $\boldsymbol{\Gamma} \in V^* \cong \mathbb{R}^3$. Recall also that $\mathrm{ad}^*_{\boldsymbol{\Omega}}\,\boldsymbol{\Pi} = \boldsymbol{\Pi} \times \boldsymbol{\Omega}$ by using the isomorphism (1.14); see (1.21).

The expressions of the Hamiltonian and Lagrangian functions on $\mathfrak{se}(3)^* \cong \mathbb{R}^3 \times \mathbb{R}^3$ and $\mathfrak{se}(3) \cong \mathbb{R}^3 \times \mathbb{R}^3$ respectively (see (1.42) and (1.46)),

$$H(\boldsymbol{\Pi},\boldsymbol{\Gamma}) = \frac{1}{2}\boldsymbol{\Pi}\cdot\mathbb{I}^{-1}\boldsymbol{\Pi} + Mg\ell\boldsymbol{\Gamma}\cdot\boldsymbol{\chi}$$
$$L(\boldsymbol{\Omega},\boldsymbol{\Gamma}) = \frac{1}{2}\mathbb{I}\boldsymbol{\Omega}\cdot\boldsymbol{\Omega} - Mg\ell\boldsymbol{\Gamma}\cdot\boldsymbol{\chi}$$

yield $\delta H/\delta\boldsymbol{\Pi} = \mathbb{I}^{-1}\boldsymbol{\Pi} = \boldsymbol{\Omega}$, $\delta H/\delta\boldsymbol{\Gamma} = Mg\ell\boldsymbol{\chi}$, $\delta L/\delta\boldsymbol{\Omega} = \mathbb{I}\boldsymbol{\Omega} = \boldsymbol{\Pi}$, and $\delta L/\delta\boldsymbol{\Gamma} = -Mg\ell\boldsymbol{\chi}$, where, in this case, due to the dot product pairing, the partial functional derivatives are given by $\delta/\delta\boldsymbol{\Pi} = \nabla_{\boldsymbol{\Pi}}$ and $\delta/\delta\boldsymbol{\Gamma} = \nabla_{\boldsymbol{\Gamma}}$. So we can immediately write both the semidirect product Lie-Poisson (7.11) and Euler-Poincaré equations (7.14), (7.13) to get

$$\dot{\boldsymbol{\Pi}} = \boldsymbol{\Pi}\times\boldsymbol{\Omega} + Mg\ell\boldsymbol{\Gamma}\times\boldsymbol{\chi}, \qquad \dot{\boldsymbol{\Gamma}} = \boldsymbol{\Gamma}\times\boldsymbol{\Omega}$$

which are the Euler-Poisson equations (1.44). The solution (7.12) of the second Euler-Poisson equation with initial condition $\boldsymbol{\Gamma}(0) = \mathbf{k}$ is $\boldsymbol{\Gamma} = A^{-1}\mathbf{k}$ which was the definition of $\boldsymbol{\Gamma}$ used in the expression (1.39) of the potential energy. The minus Lie-Poisson bracket (7.10) becomes in this case

$$\{F,H\}(\boldsymbol{\Pi},\boldsymbol{\Gamma}) = -\boldsymbol{\Pi}\cdot(\nabla_{\boldsymbol{\Pi}}F\times\nabla_{\boldsymbol{\Pi}}H) - \boldsymbol{\Gamma}\cdot(\nabla_{\boldsymbol{\Pi}}F\times\nabla_{\boldsymbol{\Gamma}}H + \nabla_{\boldsymbol{\Gamma}}F\times\nabla_{\boldsymbol{\Pi}}H),$$

which is formula (1.45). One recognizes in the computations at the end of §1.2 part of the proof of Theorem 7.9. The remarkable map (1.43) that sends the Euler angles and their conjugate momenta to the variables $(\boldsymbol{\Pi},\boldsymbol{\Gamma})$ is none other than the momentum map $\mathbf{J}_R$ (see (7.9)) expressed in the chart given by the Euler angles. So, of course, it will map Hamilton's equations on $T^*SO(3)$ to minus Lie-Poisson equations on $\mathfrak{se}(3)^*$, as the general theory presented in these lectures stipulates.

Since the Euler-Poisson equations are of Lie-Poisson type, their solutions must lie on coadjoint orbits. The generic ones are given by the surfaces defined by $\|\mathbf{\Gamma}\| = constant$ and $\mathbf{\Pi} \cdot \mathbf{\Gamma} = constant$. Any function of $\|\mathbf{\Gamma}\|$ and $\mathbf{\Pi} \cdot \mathbf{\Gamma}$ is a Casimir function for the Lie-Poisson bracket, as an easy verification shows. In particular, the Euler-Poisson equations always have these two functions as conserved quantities. Restricted to such a generic coadjoint orbit, the Euler-Poisson equations are Hamiltonian relative to the orbit symplectic form and have the total energy H conserved. To be completely integrable, one needs therefore one more conserved quantity, independent of H and commuting with it. It is known that this is possible only in three cases: the ***Euler case*** characterized by $\ell = 0$, that is, the center of mass coincides with the point of suspension of the rigid body, or equivalently, no forces act on the body and one has fixed its center of mass, the ***Lagrange case*** characterized by $\boldsymbol{\chi} = (0, 0, 1)$ and $I_1 = I_2$, that is, the body has an additional S^1-symmetry around the line connecting the point of suspension of the body with its center of mass, and the ***Kowalewski case***, characterized by the conditions $I_1 = I_2 = 2I_3$ and the center of mass lies in the plane of the equal moments of inertia, so it can be assumed to be $\boldsymbol{\chi} = (1, 0, 0)$ by simply adjusting the frame of reference. The last two cases have an additional integral. In the Lagrange top case, this is the momentum map of the S^1-action. In the Kowalewski case, the origin of this additional integral remains to this day a mystery from the point of view of momentum maps, that is, it is not known a priori (that is, without solving explicitly the system) what one-dimensional Lie group action has as its momentum map this additional quartic integral.

To completely describe the kinematics of this system we shall record here the coadjoint orbits of the special Euclidean group $SE(3)$; for the proofs of all the formulas below see [MaRa94], §14.7. Let $\{\mathbf{e}_1, \mathbf{e}_2, \mathbf{e}_3, \mathbf{f}_1, \mathbf{f}_2, \mathbf{f}_3\}$ be an orthonormal basis of $\mathfrak{se}(3) \cong \mathbb{R}^3 \times \mathbb{R}^3$ such that $\mathbf{e}_i = \mathbf{f}_i$, $i = 1, 2, 3$. The dual basis of $\mathfrak{se}(3)^*$ via the dot product is again $\{\mathbf{e}_1, \mathbf{e}_2, \mathbf{e}_3, \mathbf{f}_1, \mathbf{f}_2, \mathbf{f}_3\}$. There is a single zero dimensional coadjoint orbit, namely the origin. The other orbits are two and four dimensional. There is no six dimensional coadjoint orbit since the Poisson bracket is degenerate having the two Casimir functions given above. There are three types of coadjoint orbits.

Type I: The orbit $\mathcal{O}$ through $(\mathbf{e}, \mathbf{0})$ equals

$$\mathrm{SE}(3) \cdot (\mathbf{e}, \mathbf{0}) = \{\, (\mathbf{A}\mathbf{e}, \mathbf{0}) \mid \mathbf{A} \in \mathrm{SO}(3) \,\} = S^2_{\|\mathbf{e}\|} \times \{\mathbf{0}\}, \tag{7.15}$$

the two-sphere of radius $\|\mathbf{e}\|$. The tangent space to $\mathcal{O}$ at $(\mathbf{e}, \mathbf{0})$ is the tangent space to the sphere of radius $\|\mathbf{e}\|$ at the point $\mathbf{e}$ in the first factor. The minus orbit symplectic form is

$$\omega^-(\mathbf{e}, \mathbf{0})(\mathrm{ad}^*_{(\mathbf{x},\mathbf{y})}(\mathbf{e}, \mathbf{0}), \mathrm{ad}^*_{(\mathbf{x}',\mathbf{y}')}(\mathbf{e}, \mathbf{0})) = -\mathbf{e} \cdot (\mathbf{x} \times \mathbf{x}')$$

which equals $-1/\|\mathbf{e}\|$ times the area element of the sphere of radius $\|\mathbf{e}\|$ (see §6.3, example **1**).

Type II: The orbit $\mathcal{O}$ through $(\mathbf{0}, \mathbf{f})$ is given by

$$\begin{aligned} \mathrm{SE}(3) \cdot (\mathbf{0}, \mathbf{f}) &= \{ (\mathbf{a} \times \mathbf{Af}, \mathbf{Af}) \mid \mathbf{A} \in \mathrm{SO}(3), \mathbf{a} \in \mathbb{R}^3 \} \\ &= \{ (\mathbf{u}, \mathbf{Af}) \mid \mathbf{A} \in \mathrm{SO}(3), \mathbf{u} \perp \mathbf{Af} \} = TS^2_{\|\mathbf{f}\|}, \end{aligned} \tag{7.16}$$

the tangent bundle of the two-sphere of radius $\|\mathbf{f}\|$; note that the vector part is the first component. The tangent space to $\mathcal{O}$ at $(\mathbf{0}, \mathbf{f})$ equals $\mathbf{f}^\perp \times \mathbf{f}^\perp$, where $\mathbf{f}^\perp$ denotes the plane perpendicular to $\mathbf{f}$. Let $(\mathbf{u}, \mathbf{v}) \in \mathcal{O}$, that is, $\|\mathbf{v}\| = \|\mathbf{f}\|$ and $\mathbf{u} \perp \mathbf{v}$. The symplectic form in this case is

$$\begin{aligned} &\omega^-(\mathbf{u}, \mathbf{v})(\mathrm{ad}^*_{(\mathbf{x},\mathbf{y})}(\mathbf{u}, \mathbf{v}), \mathrm{ad}^*_{(\mathbf{x}',\mathbf{y}')}(\mathbf{u}, \mathbf{v})) \\ &\qquad = -\mathbf{u} \cdot (\mathbf{x} \times \mathbf{x}') - \mathbf{v} \cdot (\mathbf{x} \times \mathbf{y}' - \mathbf{x}' \times \mathbf{y}). \end{aligned} \tag{7.17}$$

It can be shown that this form is exact, namely, $\omega^- = -\mathbf{d}\theta$, where

$$\theta(\mathbf{u}, \mathbf{v}) \left(\mathrm{ad}^*_{(\mathbf{x},\mathbf{y})}(\mathbf{u}, \mathbf{v}) \right) = \mathbf{u} \cdot \mathbf{x}.$$

Thus $\mathcal{O}$ is symplectically diffeomorphic to T^*S^2 endowed with the canonical cotangent bundle symplectic structure (we identify T^*S^2 with TS^2 using the natural Riemannian metric on the sphere S^2).

Type III: The orbit $\mathcal{O}$ through $(\mathbf{e}, \mathbf{f})$, where $\mathbf{e} \neq \mathbf{0}, \mathbf{f} \neq \mathbf{0}$, equals

$$\mathrm{SE}(3) \cdot (\mathbf{e}, \mathbf{f}) = \{ (\mathbf{Ae} + \mathbf{a} \times \mathbf{Af}, \mathbf{Af}) \mid \mathbf{A} \in \mathrm{SO}(3), \mathbf{a} \in \mathbb{R}^3 \}. \tag{7.18}$$

To get a better description of this orbit, consider the smooth map

$$\varphi : (\mathbf{A}, \mathbf{a}) \in \mathrm{SE}(3) \mapsto \left(\mathbf{Ae} + \mathbf{a} \times \mathbf{Af} - \frac{\mathbf{e} \cdot \mathbf{f}}{\|\mathbf{f}\|^2} \mathbf{Af}, \mathbf{Af} \right) \in TS^2_{\|\mathbf{f}\|},$$

which is right invariant under the isotropy group

$$\mathrm{SE}(3)_{(\mathbf{e},\mathbf{f})} = \{ (\mathbf{B}, \mathbf{b}) \mid \mathbf{Be} + \mathbf{b} \times \mathbf{f} = \mathbf{e},\ \mathbf{Bf} = \mathbf{f} \}$$

and induces hence a diffeomorphism $\bar{\varphi} : \mathrm{SE}(3)/\mathrm{SE}(3)_{(\mathbf{e},\mathbf{f})} \to TS^2_{\|\mathbf{f}\|}$. The orbit $\mathcal{O}$ through $(\mathbf{e}, \mathbf{f})$ is diffeomorphic to $\mathrm{SE}(3)/\mathrm{SE}(3)_{(\mathbf{e},\mathbf{f})}$ by the diffeomorphism

$$(\mathbf{A}, \mathbf{a}) \mapsto \mathrm{Ad}^*_{(\mathbf{A},\mathbf{a})^{-1}}(\mathbf{e}, \mathbf{f}).$$

Composing these two maps and identifying TS^2 and T^*S^2 by the natural Riemannian metric on S^2, we get the diffeomorphism $\Phi : \mathcal{O} \to T^*S^2_{\|\mathbf{f}\|}$ given by

$$\Phi(\mathrm{Ad}^*_{(\mathbf{A},\mathbf{a})^{-1}}(\mathbf{e},\mathbf{f})) = \left(\mathbf{Ae} + \mathbf{a} \times \mathbf{Af} - \frac{\mathbf{e}\cdot\mathbf{f}}{\|\mathbf{f}\|^2}\mathbf{Af}, \mathbf{Af}\right).$$

Thus this orbit is also diffeomorphic to $T^*S^2_{\|\mathbf{f}\|}$. The tangent space at $(\mathbf{e},\mathbf{f})$ to $\mathcal{O}$ is $\{(\mathbf{u},\mathbf{v}) \mid \mathbf{u}\cdot\mathbf{f} + \mathbf{v}\cdot\mathbf{e} = \mathbf{0} \text{ and } \mathbf{v}\cdot\mathbf{f} = \mathbf{0}\}$. If $(\overline{\mathbf{u}},\overline{\mathbf{v}}) \in \mathcal{O}$, the orbit symplectic structure is given by formula (7.17), where $\overline{\mathbf{u}} = \mathbf{Ae} + \mathbf{a}\times\mathbf{Af}$ and $\overline{\mathbf{v}} = \mathbf{Af}$, for some $\mathbf{A} \in \mathrm{SO}(3)$, $\mathbf{a} \in \mathbb{R}^3$. Let

$$\mathbf{u} = \mathbf{Ae} + \mathbf{a}\times\mathbf{Af} - \frac{\mathbf{e}\cdot\mathbf{f}}{\|\mathbf{f}\|^2}\mathbf{Af} = \overline{\mathbf{u}} - \frac{\mathbf{e}\cdot\mathbf{f}}{\|\mathbf{f}\|^2}\overline{\mathbf{v}}, \qquad \mathbf{v} = \mathbf{Af} = \overline{\mathbf{v}},$$

be a pair of vectors $(\mathbf{u},\mathbf{v})$ representing an element of $TS^2_{\|\mathbf{f}\|}$. Note that $\|\mathbf{v}\| = \|\mathbf{f}\|$ and $\mathbf{u}\cdot\mathbf{v} = 0$. Then a tangent vector to $TS^2_{\|\mathbf{f}\|}$ at $(\mathbf{u},\mathbf{v})$ can be represented as $\mathrm{ad}^*_{(\mathbf{x},\mathbf{y})}(\mathbf{u},\mathbf{v}) = (\mathbf{u}\times\mathbf{x} + \mathbf{v}\times\mathbf{y}, \mathbf{v}\times\mathbf{x})$. The push-forward of the orbit symplectic form ω^- to $TS^2_{\|\mathbf{f}\|}$ is computed then to be

$$\begin{aligned}&(\Phi_*\omega^-)(\mathbf{u},\mathbf{v})(\mathrm{ad}^*_{(\mathbf{x},\mathbf{y})}(\mathbf{u},\mathbf{v}), \mathrm{ad}^*_{(\mathbf{x}',\mathbf{y}')}(\mathbf{u},\mathbf{v}))\\ &\quad = -\mathbf{u}\cdot(\mathbf{x}\times\mathbf{x}') - \mathbf{v}\cdot(\mathbf{x}\times\mathbf{y}' - \mathbf{x}'\times\mathbf{y}) - \frac{\mathbf{e}\cdot\mathbf{f}}{\|\mathbf{f}\|^2}\mathbf{v}\cdot(\mathbf{x}\times\mathbf{x}').\end{aligned}$$

A comparison with (7.17) shows that the first two terms represent the canonical cotangent bundle symplectic form on $T^*S^2_{\|\mathbf{f}\|}$. The last term is the following closed two-form on $TS^2_{\|\mathbf{f}\|}$:

$$\beta(\mathbf{u},\mathbf{v})\left(\mathrm{ad}^*_{(\mathbf{x},\mathbf{y})}(\mathbf{u},\mathbf{v}), \mathrm{ad}^*_{(\mathbf{x}',\mathbf{y}')}(\mathbf{u},\mathbf{v})\right) = -\frac{\mathbf{e}\cdot\mathbf{f}}{\|\mathbf{f}\|^2}\mathbf{v}\cdot(\mathbf{x}\times\mathbf{x}').$$

This two-form β is a magnetic term as in §2.5. Therefore $\mathcal{O}$ is the cotangent bundle of the two-sphere of radius $\|\mathbf{f}\|$ endowed with a magnetic symplectic form. The type II and III coadjoint orbits are diffeomorphic but not symplectomorphic.

The motion of the heavy top always lies on these coadjoint orbits and is a Hamiltonian system relative to the total energy H and the orbit symplectic structures presented here.

Bibliography

[AbMa78] Abraham, R. and Marsden, J.E. [1978], *Foundations of Mechanics*. Addison-Wesley, second edition.

[AMR88] Abraham, R., Marsden, J.E. and Ratiu, T.S. [1988], *Manifolds, Tensor Analysis, and Applications*. Volume 75 of *Applied Mathematical Sciences*. Springer-Verlag, second edition.

[AmCZ96] Ambrosetti, A. and Coti Zelati, V. [1996], *Periodic Solutions of Singular Lagrangian Systems*. Birkhäuser

[Arnold79] Arnold V.I. [1979], *Mathematical Methods of Classical Mechanics*. Volume 60 of *Graduate Texts in Mathematics*. Springer-Verlag.

[BelRa04] Beltiţă, D. and Ratiu, T.S. [2004], Symplectic leaves in real Banach Lie-Poisson spaces, *Geom and Funct. Anal.*, to appear.

[Ben86] Benci, V. [1986], Periodic solutions of Lagrangian Systems on a compact manifold, *Journal of Differential Equations* **63**, 135–161.

[BlBr92] Blanchard, P. and Bruning, E. [1992], *Variational Methods in Mathematical Physics*. Springer Verlag.

[Bo00] Bona, P. [2000], Extended quantum mechanics. *Acta Physica Slovaca*, **50** (1), 1–198.

[Bou71] Bourbaki, N. [1971], *Variétés différentielles et analytiques. Fascicule de résultats*. Hermann.

[Bou89] Bourbaki, N. [1989], *Lie Groups and Lie Algebras. Chapters 1–3*. Springer Verlag.

[Cendra et al.] Cendra, H, Marsden, J.E., Pekarsky, S., and Ratiu, T.S. [2003], Variational principles for Lie-Poisson and Hamilton-Poincaré equations, *Moskow Math. Journ.*, **3**(3), 833–867.

[ChMa74] Chernoff, P. R. and Marsden, J.E. [1974], *Properties of Infinite Dimensional Hamiltonian Systems*, Lecture Notes in Mathematics, **425**, Springer-Verlag.

[CB97] Cushman, R.H. and Bates, L.M. [1997], *Global Aspects of Integrable Systems*. Birkhäuser.

[DeMe93] Dellnitz M. and Melbourne I. [1993], The equivariant Darboux Theorem, *Lectures in Appl. Math.*, **29**, 163–169.

[DrJo89] Drazin, P.G. and Johnson, R.S. [1989], *Solitons: An Introduction.* Cambridge University Press.

[DFN95] Dubrovin, B., Novikov, S.P., and Fomenko, A.T. [1995], *Modern Geometry I, II, III*, Volumes 93, 104, 124 of *Graduate Texts in Mathematics.* Springer-Verlag.

[EbMa70] Ebin, D.G. and Marsden, J.E. [1970], Groups of diffeomorphisms and the motion of an incompressible fluid, *Ann. of Math.* **92**, 102–163.

[GuSt84] Guillemin, V. and Sternberg, S. [1984], *Symplectic Techniques in Physics.* Cambridge University Press.

[HMR98] Holm, D.D., Marsden, J.E., and Ratiu, T.S. [1998], The Euler-Poincaré equations and semidirect products with applications to continuum theories, *Advances in Math.*, **137**, 1–81.

[JoSa98] José, J.V. and Saletan, E.J. [1998], *Classical Dynamics : A Contemporary Approach.* Cambridge University Press.

[Jost] Jost, J. [1998], *Riemannian Geometry and Geometric Analysis. University Text.* Springer-Verlag, second edition.

[KaKoSt78] Kazhdan, D. Kostant, B., and Sternberg, S. [1978], Hamiltonian group actions and dynamical systems of Calogero type, *Comm. Pure Appl. Math.*, **31**, 481–508.

[KhMis03] Khesin, B. and Misiolek, G. [2003], Euler equations on homogeneous spaces and Virasoro orbits, *Adv. in Math.*, **176**, 116–144.

[LiMa87] Libermann, P. and Marle, C.-M. [1987], *Symplectic Geometry and Analytical Mechanics.* Reidel.

[Lang] Lang, S. [1999], *Fundamentals of Differential Geometry.* Volume 191 of *Graduate Texts in Mathematics.* Springer-Verlag, New York.

[Marle76] Marle, C.-M. [1976], Symplectic manifolds, dynamical groups, and Hamiltonian mechanics, in *Differential Geometry and Relativity*, Cahen, M. and Flato, M. eds., D. Reidel, Boston, 249–269.

[Marsden81] Marsden, J.E. [1981], *Lectures on Geometric Methods in Mathematical Physics.* Volume 37, SIAM, Philadelphia, 1981.

[Marsden92] Marsden, J.E. [1992], *Lectures on Mechanics.* Volume 174 of *London Mathematical Society Lecture Note Series.* Cambridge University Press.

[Marsden et. al.] Marsden, J.E., Misiolek, G., Ortega, J.-P., Perlmutter, M., and Ratiu, T.S. [2004], *Hamiltonian Reduction by Stages.* Lecture Notes in Mathematics, to appear. Springer-Verlag.

[MaMoRa90] Marsden, J.E., Montgomery, R., and Ratiu, T.S. [1984], Reduction, symmetry, and phases in mechanics, *Memoirs Amer. Math. Soc.*, **88**(436), 1–110.

[MaRa94] Marsden, J.E. and Ratiu, T.S. [1994], *Introduction to Mechanics and Symmetry.* Volume 75 of *Texts in Applied Mathematics*, second printing of second edition 2003. Springer-Verlag.

[MaRa95] Marsden, J.E. and Ratiu, T.S. [2003], *Geometric Fluid Dynamics.* Unpublished notes.

[MaRa03] Marsden, J.E. and Ratiu, T.S. [2003], *Mechanics and Symmetry. Reduction Theory.* In preparation.

[MaRaWe84a] Marsden, J.E., Ratiu, T.S., and Weinstein, A. [1984a], Semidirect products and reduction in mechanics, *Trans. Amer. Math. Soc.*, **281**(1), 147–177.

[MaRaWe84b] Marsden, J.E., Ratiu, T.S., and Weinstein, A. [1984b], Reduction and

Hamiltonian structures on duals of semidirect product Lie algebras, *Contemporary Math.*, **28**, 55–100.

[MaWei74] Marsden, J.E. and Weinstein, A. [1983], Reduction of symplectic manifolds with symmetry, *Rep. Math. Phys.*, **5**, 121–130.

[MaWei83] Marsden, J.E. and Weinstein, A. [1983], Coadjoint orbits, vortices and Clebsch variables for incompressible fluids, *Physica D*, **7**, 305–323.

[MawWil989] Mawhin, J. and Willem, M. [1989], *Critical Point Theory and Hamiltonin Systems*. Volume 74 of *Applied Mathematical Sciences*. Springer-Verlag, second edition.

[McDSal95] McDuff, D. and Salamon, D. [1995], *Introduction to Symplectic Topology*. Clarendon Press.

[MD93] Melbourne, I. and Dellnitz, M. [1993], Normal forms for linear Hamiltonian vector fields commuting with the action of a compact Lie group, *Proc. Camb. Phil. Soc.*, **114**, 235–268.

[Milnor] Milnor, J. [1963], *Morse Theory*. Princeton University Press.

[Mis02] Misiolek, G. [2002], Classical solutions of the periodic Camassa-Holm equation, *Geom. Funct. Anal.*, **12**, 1080–1104.

[OdzRa03] Odzijewicz A. and Ratiu T.S. [2003], Banach Lie-Poisson spaces and reduction, *Comm. Math. Phys.*, **243**, 1–54.

[OdzRa004] Odzijewicz, A. and Ratiu, T.S. [2004], Extensions of Banach Lie-Poisson spaces, *J. Func. Anal.*, **217**(1), 103–125.

[OR04] Ortega, J.-P. and Ratiu, T.S. [2004], *Momentum Maps and Hamiltonian Reduction*. Volume 222 of *Progress in Mathematics*. Birkhäuser.

[OKh87] Ovsienko, V.Y. and Khesin, B.A. [1987], Korteweg-de Vries superequations as an Euler equation, *Funct. Anal. Appl.*, **21**, 329–331.

[Palais68] Palais, R. [1968], *Foundations of Global Non-Linear Analysis*, W. A. Benjamin, Inc., New York-Amsterdam.

[PR04] Perlmutter, M. and Ratiu, T.S. [2004], Gauged Poisson structures, preprint.

[Schm87] Schmid, R. [1987], *Infinite Dimensional Hamiltonian Systems*. Bibliopolis.

[Serre] Serre, J.-P. [1992], *Lie Algebras and Lie Groups*, Volume 1500 of *Lecture Notes in Mathematics*. Springer-Verlag.

[Sp79] Spivak, M. [1979], *Differential Geometry*, Volume I. New printing with corrections. Publish or Perish, Inc. Houston, Texas.

[V96] Vaisman, I. [1996], *Lectures on the Geometry of Poisson Manifolds*. Volume 118 of *Progress in Mathematics*, Birkhäuser.

[W83] Warner, F.W. [1983], *Foundation of Differentiable Manifolds and Lie Groups*. Volume 94 of *Graduate Texts in Mathematics*, Springer-Verlag.

[W] Weinstein, A. [1983], The local structure of Poisson manifolds, *Journ. Diff. Geom* **18**, 523–557.

III

The Euler-Poincaré variational framework for modeling fluid dynamics

Darryl Holm

Abstract

The global climate involves fluid motions that occur over a huge range of interacting length and time scales. The multiscale aspect of the challenge of modeling the global climate summons a unified approach that should have the capability to address a sequence of nested subproblems in fluid dynamics. The approach should be based on fundamental principles and it should have the capability to incorporate physical processes at many different scales. The Euler-Poincaré theorem provides the framework for such an approach. After introducing the global climate problem from the viewpoint of modeling global ocean circulation, we review the Euler-Poincaré theorem and apply it to address a sequence of modeling challenges that ranges from balance equations for geophysical fluid dynamics, to large eddy simulation models for three-dimensional turbulence, to Hamiltonian dynamics of solitons.

1 The problem of ocean circulation & global climate

Figure 1.1 will help us focus our minds on a serious problem — the problem of the role of ocean circulation in climate modeling and global warming. The ocean and the atmosphere transport heat from the Earth's equator to the poles at about equal rates. This coupled ocean-atmosphere interaction is the basis of the global climate problem. This problem is timely. Recently the "temperate zones" have been experiencing extreme weather (floods in Europe, droughts in America) and the issue of global warming is often in the news. We may all be studying this problem eventually, simply to decide where to live, if global warming becomes a reality. (The Sun also plays a large role in this problem. But we will never be able to affect the solar contribution; so we shall not consider the variability of the Sun here.) Of course, these lecture notes will not even begin discussing the intricacies of this complex problem. We shall only explain how geometry (in particular, the geometry of variational principles and geodesic motion) provides a framework in which we may think about some basic aspects of this problem and formulate equations to solve some of its subproblems. These subproblems are interesting in their own right and we shall pursue some of their fundamental features.

To study the climate, one must understand quantitatively how the circulation of the coupled ocean-atmosphere system transports heat from the equator to the poles by fluid motion. We shall explain the geometric approach for attacking this problem. This approach is based on extending the classical geodesic equation for motion (which is driven purely by kinetic energy) to include the effects of thermodynamics and potential energy. This extension yields the Euler-Poincaré (EP) framework for modeling and analyzing fluid dynamics introduced in [32].

The EP framework we shall describe is useful in formulating mathematical models for numerical simulations of weather and climate in Earth's coupled ocean-atmosphere system. The EP framework was already used to formulate a number of GFD (Geophysical Fluid Dynamics) models of ocean-atmosphere circulation in [3] and in [33]. This framework was also used to formulate the geometric elements of nonlinearity in Lagrangian-averaged turbulence models of the LES type (for Large Eddy Simulation) in [10]. These LES turbulence models may be useful tools in solving some of the subproblems of the global climate problem. A measure of the difficulty of the global climate problem is that it contains three dimensional turbulence as a subproblem.

Geostrophic balance Figure 1.1 shows a snapshot of the sea surface elevation in the South Atlantic Ocean around Antarctica. The sea surface elevation provides a wealth of readily observable information about certain aspects of ocean flows. The Earth's rotation is rapid compared to the flow speeds in the ocean and atmosphere. Consequently, both the

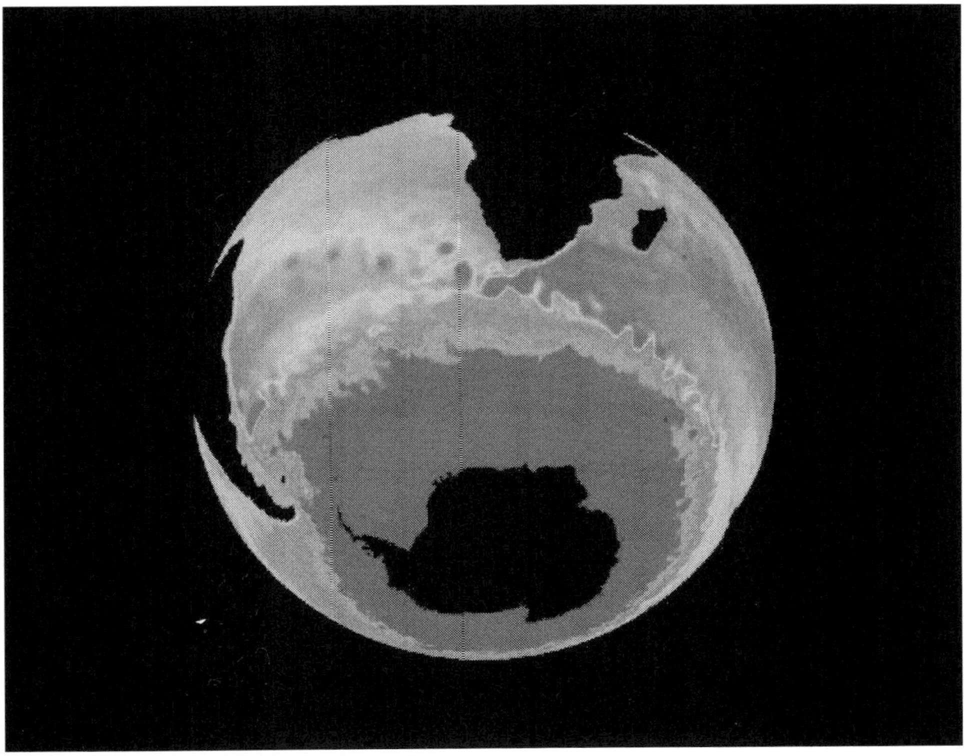

Fig. 1.1. The colors show sea surface height as a function of position — red is higher, while blue is lower than the equilibrium geopotential. In geostrophic balance, the flow is along the isoclines, whose deviations show the complex meanderings of the currents and their tendency to form vortices. Geostrophically balanced vortices about 300 *km* in diameter appear in this numerically generated figure as confined regions of local elevation and depression of the sea surface height. The dark regions are the continents of South America, Antarctica and Africa, as well as Madagascar and a few other islands. In the Southern Hemisphere, the Antarctic Circumpolar Current (ACC, appearing here in blue, green and purple) flows rapidly in the only globally open channel. The ACC interacts with the Agulhas Current off the tip of South Africa to produce a sequence of vortices (shown in red, being elevations that circulate anticlockwise in the Southern Hemisphere). These anticlockwise vortices retain their integrity as they drift northwestward across the South Atlantic. The recirculation at the western boundary (near South America) rejoins the ACC to form the eastward flowing Malvinas Current in the Southeast Atlantic. Several other anticlockwise vortices emerge due to the meanderings of the ACC and the Malvinas Current.

This figure is available in colour for download from www.cambridge.org/9780521539579

ocean and atmosphere satisfy the condition of "geostrophic balance," which determines the local velocity of the ocean from its surface elevation. Namely, in geostrophic balance, the hydrostatic pressure gradient *cancels* the Coriolis force, thereby allowing one to solve for the velocity of the ocean from the gradient of its sea surface elevation at each fixed (Eulerian) position in the Earth's co-rotating frame. In geostrophic balance, a fluid moves along isoclines of its surface elevation in the co-rotating frame. Hence, the gyres (or, large-scale ocean circulations) can be observed as sea surface elevations in Figure 1.1 and their local speeds can be inferred from geostrophic balance.

The sea surface height as a function of position is being measured by the Topex-Poseidon satellite experiments using radar altimetry as part of the World Ocean Circulation Experiment (WOCE). The satellite measurements imply the local speed of clockwise geostrophic flow around an elevation in sea surface height in the Northern Hemisphere, and anti-clockwise flow in the Southern Hemisphere. These geostrophic flows are slow, but persistent. The gyres out in the middle of the ocean move at horizontal speeds of about $U = 10\,cm/sec$. In the Southern Hemisphere, the Antarctic Circumpolar Current (ACC) shown in Figure 1.1 flows in the only globally open channel. This current is the fastest, at $2\,m/sec$, although the current in the Madagascar channel between the island and the eastern coast of Africa is almost as fast. The flux of the ACC is about one hundred million tons of water per second (through the Drake Passage, for example, below the tip of South America) in its circulation around Antarctica.

1.1 Eddy formation

Various types of phenomena may be recognized as aspects of the problem of global ocean circulation. One of these is the prolific formation of eddies in ocean flows. The two main mechanisms for forming eddies are both seen in Figure 1.1, which shows a snapshot of the sea surface elevation in the South Atlantic Ocean. First, one sees a westward current that comes around South Africa from the Indian Ocean. In the Agulas Straight, this westward current runs into the eastward flowing Antarctic Circumpolar Current. The collision of these two strong counter-flowing currents produces eddies about 200-300 *km* in diameter via the mechanism of Kelvin-Helmholtz instability. These mesoscale eddies slowly drift toward the northwest and move across the South Atlantic Ocean in about *three years*. The eddies last essentially as long as it takes until they run into the coastline someplace near Venezuela, where they start feeling the drag due to bottom topography. Hence, motion at this scale is very nondissipative: large gyres in the ocean basins are driven by the winds, but the only significant dissipation mechanism for the currents and eddies in the ocean is their interaction with the bottom. The gyres are thousands of kilometers in diameter. The eddies created by the currents at their edges are about 200-300 km across, so they are also large enough to be in geostrophic balance, although they are much smaller than the gyres. However, these *mesoscale* eddies still involve a great deal of mass and momentum — so they certainly would not be stopped by molecular viscosity any time soon. The Reynolds number for an eddy 100 km across turning at 10 cm/sec in water of viscosity 0.01 cm^2/sec is about 10^{10}. The only significant dissipation mechanism for this system of gyres, eddies and currents is their interaction with the bottom, or along the shallows near the lateral boundaries.

A second mechanism for creating mesoscale eddies is also seen in Figure 1.1, in the vicinity of the Malvinas Current, off the coast of Agentina. This snapshot illustrates the double gyre mechanism, in which recirculation in the ocean basin creates a region of low surface elevation, coupled with a region of high surface elevation that circulates in the opposite sense. This double gyre mechanism forms the great currents in the ocean, such as the Gulf Stream in the North Atlantic and the Kuroshio, or Japanese Current in the North Pacific. The currents that form in the middle regions between the ocean's great gyres spontaneously form unstable lateral displacements. This instability increases the complexity of these flows, because the meandering unstable displacements grow into loops that may pinch off to form vortices. These vortices, in turn, drift westward and may interact with the current, until they run into the western boundary, where they tend to interrupt the recirculation along this boundary that sustains the double gyre itself.

So, in approaching the problem of ocean circulation as a subproblem of the global climate, one must consider a thin, rapidly rotating domain of flow driven at its free surface and dissipated by bottom drag and drag at the lateral boundaries. In this thin rotating domain, unstable complex three-dimensional fluid motions occur that remain nearly in horizontal geostrophic balance. We shall use the Euler-Poincaré (EP) framework to formulate the fundamental principles for developing models and analyzing the equations for these complex geophysical flows. This is a field that oceanographers have already been investigating for many years, primarily by truncating asymptotic expansions of the solutions of the exact equations, in powers of the various small parameters that are available in this problem. For the physical motivations for these models, discussions of the recent progress of their investigations and reviews of the historical literature in this field, see [51].

Outline In seeking a method that will allow systematic investigations of the global climate problem introduced and described in section 1 as a sequence of nested subproblems in fluid dynamics, we shall begin in section 2 by reviewing the recently developed Euler-Poincaré (EP) framework. We shall show in section 3 how the EP framework provides a unified approach for deriving and comparing the historical GFD models, and for developing and analyzing the balance models needed in future GFD applications. In section 1, we explained that three dimensional turbulence is a subproblem of the ocean and atmosphere circulation problems. In section 4, we will describe the approach of Lagrangian reduction and EP approximations in deriving turbulence closure models that could augment and enhance the traditional GFD models. The key idea here is to perform Lagrangian averaging in the Euler-Poincaré framework. Specifically, we will review some of the recent models, called "alpha-models" for describing the mean effects of turbulence at lengthscales that are larger than the model's fundamental lengthscale, alpha. The lengthscale, alpha, is the variance of a Lagrangian trajectory in the flow from its mean. For the Navier-Stokes (NS) equations of incompressible fluid motion in three dimensions, these alpha models provide extensions of the classical ideas for regularizing Navier-Stokes equations that were first introduced by Leray in [44]. The Leray regularization approach for NS can be reframed as a mathematical basis for studying and developing the modern class of computational turbulence models called Large Eddy Simulations (LES) also discussed in section 4. Since large eddies in geostrophic balance lie at the heart of ocean circulation dynamics, one may expect that the LES method could develop into a fruitful approach for future numerical investigations of the climate modelling problem. A crucial step in the development of an LES model is the choice of filter that one uses in discriminating between resolved and subgrid-scale effects. Section 4 sets the stage for future LES models to benefit, by introducing these filtering choices in the EP framework. Given the choice of filter, the EP framework provides the motion equations for these LES models and endows them with the required properties (energetic balance, Kelvin circulation theorem, potential vorticity conservation law, etc.) for properly modeling ocean and atmosphere circulation in the Eulerian representation.

Finally, in a change of pace that allows one to develop insight by hands-on manipulation of the solutions, in sections 5 and 6 we shall employ the Euler-Poincaré approach to study the subproblem of geodesic motion of pressureless compressible fluids. The one dimensional considerations of this problem in section 5 illustrate the mechanism of nonlinear balance arising in the Euler-alpha models, in the context of soliton dynamics. This nonlinear balance is isolated in the dispersionless limit of the dynamics of unidirectional traveling wave pulses in shallow water. Section 5 discusses this one dimensional compressible dynamics, and its pulson and peakon solutions in the absence of linear dispersion. Pulsons are solutions of the EP equation for the horizontal fluid velocity for geodesic motion in one dimension. The corresponding momenta of the $N-$pulson solution is defined on a set of N Dirac measures that undergo *finite-dimensional* canonical Hamiltonian particle dynamics. Peakons arise as a special case of the pulson solutions that are genuine solitons, and whose velocity profiles have a discontinuity in derivative at its peak. That is, peakons propagate as confined pulses of peaked shape that interact elastically, and they correspond to spectral data of an isospectral eigenvalue problem. In section 6 we shall extend to higher dimensions the idea of pulsons as measure-valued momentum solutions that interact elastically in the geodesic EP dynamics of pressureless compressible fluids. For example, the collective solutions corresponding to N one-dimensional pulsons become N momentum filaments in two dimensions. The momenta for these solutions are defined on delta functions along a set of evolving curves in the plane. These solutions evolve by canonical Hamiltonian dynamics that produces a system of $2N$ integral-partial-differential equations (IPDE). The momentum filaments dominate the solution of the initial value problem and arise as emergent patterns in the geodesic motion of pressureless compressible fluids. This emergence is demonstrated for the limit of shallow water dynamics in two dimensions that corresponds to the peakons in one dimension.

1.2 Kelvin circulation theorem and "good" equations

Suppose we took an analytical approach to the problem of modeling global ocean circulation and other GFD (Geophysical Fluid Dynamics) flows. We would identify the various balances (such as the geostrophic balance) and identify the small dimensionless numbers that are associated with these flows. For example, we would take advantage of having a thin domain that is rotating rapidly (f_0, once per day). This rotation rate is very much more rapid than the rate (U/L, once per month) for even relatively small mesoscale eddies of, say, $L \approx 100\,km$ to turn around. That is, we would use the available separation of time scales to create the small dimensionless ratio $\epsilon_1 = U/(Lf_0) \approx 1/30 \ll 1$, called the Rossby number. The Rossby number also measures the ratio of the nonlinearity to the Coriolis force. We are interested in length scales for eddies that are hundreds of kilometers in diameter, but the ocean depth is only about $D \approx 4\,km$. So our rapidly rotating domain is also *thin* — its aspect ratio is of order $\epsilon_2 = D/L \approx 4/100 \ll 1$. We would expand the exact equations for a rotating, surface-driven, stratified fluid moving under gravity in these two small ratios of length scales and time scales, the Rossby number, ϵ_1, and the aspect ratio, ϵ_2. Because the acceleration of gravity g is acting on these flows, we might also introduce the dimensionless ratio, $F = f_0^2L^2/(gD) = O(1)$, the rotational Froude number. At leading order in this expansion, the geostrophic and hydrostatic balances would appear. At the next orders, a family of approximate equations would develop. Eventually, we would emerge from all our asymptotic expansion efforts and balance considerations at each power of ϵ_1 and ϵ_2 with approximate equations of motion containing these three dimensionless numbers. As mentioned earlier, the inverse of the Reynolds number is negligible, so viscosity would not enter in these asymptotic balance equations for the large scale motion and its effects would need to be diagnosed separately.

The family of GFD balance equations that people have derived via this asymptotic procedure is investigated by Allen, Barth and Newberger (ABN) in [1], who provide an impressive list of candidates. On consulting this "ABN list" of balance equations, one would want to know about the properties of these equations and the behavior of their solutions. In fact, when the Climate Change Prediction Program first started in the early 1990's, people who were developing the sort of numerical simulations shown in Figure 1.1, on meeting with a new set of GFD balance equations, would ask, "Ok, what about this one? Is it good?" And then someone would be required to learn enough about the differences of this new one from the various other balance models, to answer the question. This question was being asked perennially throughout the GFD modeling community, during the past few decades as the GFD balance models developed in concert with the development of the computer capabilities for their simulation. Eventually, two fundamental criteria emerged for selecting "good" approximate sets of balance equations. Namely, in dealing with the circulation of the ocean or atmosphere, whatever governing equations one shall choose will need two properties: a circulation theorem and proper energy balance. These two properties are closely linked in the Euler-Poincaré (EP) framework. Indeed, a theorem exists in the EP framework that governs the circulation and energy balance, both for the exact fluid equations, and for any of their approximations in this framework. We will talk about this EP theorem and describe its use in making approximations and developing turbulence models for ocean circulation dynamics in the remainder of these lectures.

The EP approach to mathematical modeling for fluids provides the unified structure required for all the balanced models for ocean circulation in the putative ABN list, and it also eliminates some of them. The remaining models on this list each possesses a Kelvin circulation theorem and satisfies energetics that are consistent with the leading order geostrophic balance. We shall see that the EP theorem for deriving fluid equations that possess proper circulation and energy laws follows directly from Hamilton's principle in the Eulerian fluid description.

In GFD, one is typically dealing with fluids in the Eulerian picture. For example, geostrophic balance is an Eulerian concept, even though it governs the velocity of the fluid's Lagrangian motion. All the GFD models on the ABN list that survive the selection procedure for proper energetics and circulation are EP equations with advected parameters. That is, these equations all follow from Hamilton's principle (HP) for a fluid action that depends parametrically on the advected quantities such as mass, salt and heat that are carried as material properties of the fluid's motion. The EP equations for fluid dynamics with advected parameters result from applying geometrical methods of reduction by symmetry to reduce this HP from the material (or Lagrangian) picture of fluid dynamics, to the spatial (or Eulerian) picture. This is the content of the Euler-Poincaré theorem, reviewed in the next section.

2 Euler-Poincaré fluid dynamics following Holm, Marsden & Ratiu [32]

Almost all of the GFD models on the ABN list admit the following general assumptions. These assumptions form the basis of the Euler-Poincaré theorem for Continua that we shall state later in this section, after introducing the notation necessary for dealing geometrically with the reduction of HP from the material (or Lagrangian) picture of fluid dynamics, to the spatial (or Eulerian) picture. This theorem was first stated and proved in [32], to which we refer for additional details, as well as for abstract definitions and proofs.

Basic assumptions underlying the Euler-Poincaré theorem

- There is a *right* representation of a Lie group G on the vector space V and G acts in the natural way on the *right* on $TG \times V^*$: $(U_g, a)h = (U_g h, ah)$.
- The Lagrangian function $L : TG \times V^* \to \mathbb{R}$ is right G-invariant.[1]
- In particular, if $a_0 \in V^*$, define the Lagrangian $L_{a_0} : TG \to \mathbb{R}$ by $L_{a_0}(U_g) = L(U_g, a_0)$. Then L_{a_0} is right invariant under the lift to TG of the right action of G_{a_0} on G, where G_{a_0} is the isotropy group of a_0.
- Right G-invariance of L permits one to define the Lagrangian on the Lie algebra $\mathfrak{g}$ of the group G. Namely, $\ell : \mathfrak{g} \times V^* \to \mathbb{R}$ is defined by,

  $$\ell(u, a) = L\big(U_g g^{-1}(t), a_0 g^{-1}(t)\big) = L(U_g, a_0)\,,$$

  where $u = U_g g^{-1}(t)$ and $a = a_0 g^{-1}(t)$. Conversely, this relation defines for any $\ell : \mathfrak{g} \times V^* \to \mathbb{R}$ a right G-invariant function $L : TG \times V^* \to \mathbb{R}$.
- For a curve $g(t) \in G$, let $u(t) := \dot{g}(t)g(t)^{-1}$ and define the curve $a(t)$ as the unique solution of the linear differential equation with time dependent coefficients $\dot{a}(t) = -a(t)u(t)$, where the action of an element of the Lie algebra $u \in \mathfrak{g}$ on an advected quantity $a \in V^*$ is denoted by concatenation from the right. The solution with initial condition $a(0) = a_0 \in V^*$ can be written as $a(t) = a_0 g(t)^{-1}$.

Notation for reduction of HP by symmetries

Let $\mathfrak{g}(\mathcal{D})$ denote the space of vector fields on $\mathcal{D}$ of some fixed differentiability class. These vector fields are endowed with the ***Lie bracket*** given in components by (summing on repeated indices)

$$[\mathbf{u}, \mathbf{v}]^i = u^j \frac{\partial v^i}{\partial x^j} - v^j \frac{\partial u^i}{\partial x^j}\,. \tag{2.1}$$

The notation $\mathrm{ad}_{\mathbf{u}}\mathbf{v} := [\mathbf{u}, \mathbf{v}]$ formally denotes the adjoint action of the *right* Lie algebra of $\mathrm{Diff}(\mathcal{D})$ on itself.

[1] For fluid dynamics, right G-invariance of the Lagrangian function L is traditionally called "particle relabeling symmetry."

We shall identify the Lie algebra of vector fields $\mathfrak{g}$ with its dual $\mathfrak{g}^*$ by using the L^2 pairing

$$\langle \mathbf{u}, \mathbf{v} \rangle = \int_{\mathcal{D}} \mathbf{u} \cdot \mathbf{v} \, d^3x. \tag{2.2}$$

We shall also let $\mathfrak{g}(\mathcal{D})^*$ denote the geometric dual space of $\mathfrak{g}(\mathcal{D})$, that is, $\mathfrak{g}(\mathcal{D})^* := \Lambda^1(\mathcal{D}) \otimes \mathrm{Den}(\mathcal{D})$. This is the space of one-form densities on $\mathcal{D}$. If $\mathbf{m} \otimes dV \in \Lambda^1(\mathcal{D}) \otimes \mathrm{Den}(\mathcal{D})$, then the pairing of $\mathbf{m} \otimes dV$ with $\mathbf{u} \in \mathfrak{g}(\mathcal{D})$ is given by the L^2 pairing,

$$\langle \mathbf{m} \otimes dV, \mathbf{u} \rangle = \int_{\mathcal{D}} \mathbf{m} \cdot \mathbf{u} \, dV \tag{2.3}$$

where $\mathbf{m} \cdot \mathbf{u}$ is the standard contraction of a one-form $\boldsymbol{m}$ with a vector field $\mathbf{u}$. For $\mathbf{u} \in \mathfrak{g}(\mathcal{D})$ and $\mathbf{m} \otimes dV \in \mathfrak{g}(\mathcal{D})^*$, the dual of the adjoint representation is defined by

$$\langle \mathrm{ad}^*_{\mathbf{u}}(\mathbf{m} \otimes dV), \mathbf{v} \rangle = -\int_{\mathcal{D}} \mathbf{m} \cdot \mathrm{ad}_{\mathbf{u}} \mathbf{v} \, dV = -\int_{\mathcal{D}} \mathbf{m} \cdot [\mathbf{u}, \mathbf{v}] \, dV \tag{2.4}$$

and its expression is

$$\mathrm{ad}^*_{\mathbf{u}}(\mathbf{m} \otimes dV) = (\pounds_{\mathbf{u}} \mathbf{m} + (\mathrm{div}_{dV}\, \mathbf{u}) \mathbf{m}) \otimes dV = \pounds_{\mathbf{u}}(\mathbf{m} \otimes dV)\,, \tag{2.5}$$

where $\mathrm{div}_{dV} \mathbf{u}$ is the divergence of $\mathbf{u}$ relative to the measure dV, that is, $\pounds_{\mathbf{u}} dV = (\mathrm{div}_{dV} \mathbf{u}) dV$. Hence, $\mathrm{ad}^*_{\mathbf{u}}$ coincides with the Lie-derivative $\pounds_{\mathbf{u}}$ for one-form densities. If $\mathbf{u} = u^j \partial / \partial x^j$, $\mathbf{m} = m_i dx^i$, then the one-form factor in the preceding formula for $\mathrm{ad}^*_{\mathbf{u}}(\mathbf{m} \otimes dV)$ has the coordinate expression

$$\left(u^j \frac{\partial m_i}{\partial x^j} + m_j \frac{\partial u^j}{\partial x^i} + (\mathrm{div}_{dV}\, \mathbf{u}) m_i \right) dx^i = \left(\frac{\partial}{\partial x^j}(u^j m_i) + m_j \frac{\partial u^j}{\partial x^i} \right) dx^i \,. \tag{2.6}$$

The last equality assumes that the divergence is taken relative to the standard measure $dV = d^n\mathbf{x}$ in $\mathbb{R}^n$. (On a Riemannian manifold the metric divergence needs to be used.)

Throughout the rest of the lecture notes, we shall follow [32] in using the conventions and terminology for the standard quantities in continuum mechanics. Elements of $\mathcal{D}$ representing the material particles of the system are denoted by X; their coordinates $X^A, A = 1, ..., n$ may thus be regarded as the particle labels. A ***configuration***, which we typically denote by η, or g, is an element of $\mathrm{Diff}(\mathcal{D})$. A ***motion***, denoted as η_t or alternatively as $g(t)$, is a time dependent curve in $\mathrm{Diff}(\mathcal{D})$. The ***Lagrangian***, or ***material velocity*** $\mathbf{U}(X, t)$ of the continuum along the motion η_t or $g(t)$ is defined by taking the time derivative of the motion keeping the particle labels X fixed:

$$\mathbf{U}(X,t) := \frac{d\eta_t(X)}{dt} := \frac{\partial}{\partial t}\bigg|_X \eta_t(X) := \dot{g}(t)\cdot X\,.$$

These are convenient shorthand notations for the time derivative at fixed X.

Consistent with this definition of velocity, the tangent space to $\mathrm{Diff}(\mathcal{D})$ at $\eta \in \mathrm{Diff}(\mathcal{D})$ is given by

$$T_\eta \,\mathrm{Diff}(\mathcal{D}) = \{\mathbf{U}_\eta : \mathcal{D} \to T\mathcal{D} \mid \mathbf{U}_\eta(X) \in T_{\eta(X)}\mathcal{D}\}.$$

Elements of $T_\eta \,\mathrm{Diff}(\mathcal{D})$ are usually thought of as vector fields on $\mathcal{D}$ covering η. The tangent lift of right translations on $T\,\mathrm{Diff}(\mathcal{D})$ by $\varphi \in \mathrm{Diff}(\mathcal{D})$ is given by

$$\mathbf{U}_\eta\varphi := T_\eta R_\varphi(\mathbf{U}_\eta) = \mathbf{U}_\eta \circ \varphi\,.$$

During a motion η_t or $g(t)$, the particle labeled by X describes a path in $\mathcal{D}$ whose points

$$x(X,t) := \eta_t(X) := g(t)\cdot X$$

are called the ***Eulerian*** or ***spatial points*** of this path, which is also called the ***Lagrangian trajectory***, because a Lagrangian fluid parcel follows this path in space. The derivative $\mathbf{u}(x,t)$ of this path, evaluated at fixed Eulerian point x, is called the ***Eulerian*** or ***spatial velocity*** of the system:

$$\mathbf{u}(x,t) := \mathbf{u}(\eta_t(X),t) := \mathbf{U}(X,t) := \frac{\partial}{\partial t}\bigg|_X \eta_t(X) := \dot{g}(t)\cdot X := \dot{g}(t)g^{-1}(t)\cdot x\,.$$

Thus the Eulerian velocity $\mathbf{u}$ is a time dependent vector field on $\mathcal{D}$: $\mathbf{u}_t \in \mathfrak{g}(\mathcal{D})$, where $\mathbf{u}_t(x) := \mathbf{u}(x,t)$. We also have the fundamental relationships

$$\mathbf{U}_t = \mathbf{u}_t \circ \eta_t \quad \text{and} \quad \mathbf{u}_t = \dot{g}(t)g^{-1}(t)\,,$$

where $\mathbf{U}_t(X) := \mathbf{U}(X,t)$.

The representation space V^* of $\mathrm{Diff}(\mathcal{D})$ in continuum mechanics is often some subspace of $\mathfrak{T}(\mathcal{D}) \otimes \mathrm{Den}(\mathcal{D})$, the tensor field densities on $\mathcal{D}$ and the representation is given by pull back. It is thus a *right* representation of $\mathrm{Diff}(\mathcal{D})$ on $\mathfrak{T}(\mathcal{D}) \otimes \mathrm{Den}(\mathcal{D})$. The right action of the Lie algebra $\mathfrak{g}(\mathcal{D})$ on V^* is given by concatenation from the right. Thus, $a\mathbf{u} := \pounds_{\mathbf{u}} a$ is the Lie derivative of the tensor field density a along the vector field $\mathbf{u}$.

The Lagrangian of a continuum mechanical system is a function $L : T\,\mathrm{Diff}(\mathcal{D}) \times V^* \to \mathbb{R}$ which is right invariant relative to the tangent lift of right translation of $\mathrm{Diff}(\mathcal{D})$ on itself and pull back on the tensor field densities. Invariance of the Lagrangian L induces a function $\ell : \mathfrak{g}(\mathcal{D}) \times V^* \to \mathbb{R}$ given by

$$\ell(\mathbf{u}, a) = L(\mathbf{u} \circ \eta, \eta^* a) = L(\mathbf{U}, a_0)\,,$$

where $\mathbf{u} \in \mathfrak{g}(\mathcal{D})$ and $a \in V^* \subset \mathfrak{T}(\mathcal{D}) \otimes \mathrm{Den}(\mathcal{D})$, and where $\eta^* a$ denotes the pull back of a by the diffeomorphism η and $\mathbf{u}$ is the Eulerian velocity. That is,

$$\mathbf{U} = \mathbf{u} \circ \eta \quad \text{and} \quad a_0 = \eta^* a\,. \tag{2.7}$$

The evolution of a is by right action, given by the equation

$$\dot{a} = -\pounds_{\mathbf{u}}\, a = -a\mathbf{u}. \tag{2.8}$$

The solution of this equation, for the initial condition a_0, is

$$a(t) = \eta_{t*} a_0 = a_0 g^{-1}(t)\,, \tag{2.9}$$

where the lower star denotes the push forward operation and η_t is the flow of $\mathbf{u} = \dot{g} g^{-1}(t)$.

Advected Eulerian quantities are defined in continuum mechanics to be those variables which are Lie transported by the flow of the Eulerian velocity field. Using this standard terminology, equation (2.8), or its solution (2.9) states that the tensor field density $a(t)$ (which may include mass density and other Eulerian quantities) is advected.

As remarked, typically $V^* \subset \mathfrak{T}(\mathcal{D}) \otimes \mathrm{Den}(\mathcal{D})$ for continuum mechanics. On a general manifold, tensors of a given type have natural duals. For example, symmetric covariant tensors are dual to symmetric contravariant tensor densities, the pairing being given by the integration of the natural contraction of these tensors. Likewise, k-forms are naturally dual to $(n-k)$-forms, the pairing being given by taking the integral of their wedge product.

The ***diamond operation*** $\diamond$ between elements of V and V^* produces an element of the dual Lie algebra $\mathfrak{g}(\mathcal{D})^*$ and is defined as

$$\langle b \diamond a, \mathbf{w} \rangle = -\int_{\mathcal{D}} b \cdot \pounds_{\mathbf{w}}\, a\;, \tag{2.10}$$

where $b \cdot \pounds_{\mathbf{w}}\, a$ denotes the contraction, as described above, of elements of V and elements of V^* and $\mathbf{w} \in \mathfrak{g}(\mathcal{D})$. (These operations do *not* depend on a Riemannian structure.)

For a path $\eta_t \in \mathrm{Diff}(\mathcal{D})$ let $\mathbf{u}(x,t)$ be its Eulerian velocity and consider the curve $a(t)$ with initial condition a_0 given by the equation

$$\dot{a} + \pounds_{\mathbf{u}} a = 0. \tag{2.11}$$

Let the Lagrangian $L_{a_0}(\mathbf{U}) := L(\mathbf{U}, a_0)$ be right-invariant under $\mathrm{Diff}(\mathcal{D})$. We can now state the Euler-Poincaré Theorem for Continua of [32].

Theorem 2.1 (Euler-Poincaré Theorem for Continua.) *Consider a path η_t in $\mathrm{Diff}(\mathcal{D})$ with Lagrangian velocity $\mathbf{U}$ and Eulerian velocity $\mathbf{u}$. The following are equivalent:*

i *Hamilton's variational principle*

$$\delta \int_{t_1}^{t_2} L\left(X, \mathbf{U}_t(X), a_0(X)\right)\, dt = 0 \tag{2.12}$$

holds, for variations $\delta\eta_t$ vanishing at the endpoints.

ii *η_t satisfies the Euler-Lagrange equations for L_{a_0} on $\mathrm{Diff}(\mathcal{D})$.*

iii *The constrained variational principle in Eulerian coordinates*

$$\delta \int_{t_1}^{t_2} \ell(\mathbf{u}, a)\, dt = 0 \tag{2.13}$$

holds on $\mathfrak{g}(\mathcal{D}) \times V^*$, using variations of the form

$$\delta \mathbf{u} = \frac{\partial \mathbf{w}}{\partial t} + [\mathbf{u}, \mathbf{w}] = \frac{\partial \mathbf{w}}{\partial t} + \mathrm{ad}_{\mathbf{u}} \mathbf{w}\,, \qquad \delta a = -\pounds_{\mathbf{w}}\, a, \tag{2.14}$$

where $\mathbf{w}_t = \delta\eta_t \circ \eta_t^{-1}$ vanishes at the endpoints.

iv *The Euler-Poincaré equations for continua*

$$\frac{\partial}{\partial t}\frac{\delta \ell}{\delta \mathbf{u}} = -\,\mathrm{ad}^*_{\mathbf{u}} \frac{\delta \ell}{\delta \mathbf{u}} + \frac{\delta \ell}{\delta a} \diamond a = -\pounds_{\mathbf{u}} \frac{\delta \ell}{\delta \mathbf{u}} + \frac{\delta \ell}{\delta a} \diamond a\,, \tag{2.15}$$

hold, with auxiliary equations $(\partial_t + \pounds_{\mathbf{u}})a = 0$ for each advected quantity $a(t)$. The $\diamond$ operation defined in (2.10) needs to be determined on a case by case basis, depending on the nature of the tensor $a(t)$. The variation $\mathbf{m} = \delta\ell/\delta\mathbf{u}$ is a one-form density and we have used relation (2.5) in the last step of equation (2.15).

Discussion of the Euler-Poincaré equations

In the absence of dissipation, most Eulerian fluid equations [1] can be written in the EP form in equation (2.15),

$$\frac{\partial}{\partial t}\frac{\delta \ell}{\delta \mathbf{u}} + \mathrm{ad}^*_{\mathbf{u}}\frac{\delta \ell}{\delta \mathbf{u}} = \frac{\delta \ell}{\delta a}\diamond a\,, \quad \text{with} \quad \big(\partial_t + \pounds_{\mathbf{u}}\big)a = 0\,. \tag{2.16}$$

Equation (2.16) is ***Newton's Law***: The Eulerian time derivative of the momentum density $\mathbf{m} = \delta\ell/\delta\mathbf{u}$ (a one-form density dual to the velocity $\mathbf{u}$) is equal to the force density $(\delta\ell/\delta a)\diamond a$, with the $\diamond$ operation defined in (2.10). Thus, Newton's Law is written in the Eulerian fluid representation as,[1]

$$\frac{d}{dt}\Big|_{Lag}\mathbf{m} := \big(\partial_t + \pounds_{\mathbf{u}}\big)\mathbf{m} = \frac{\delta \ell}{\delta a}\diamond a\,, \quad \text{with} \quad \frac{d}{dt}\Big|_{Lag} a := \big(\partial_t + \pounds_{\mathbf{u}}\big)a = 0\,. \tag{2.17}$$

The left side of the EP equation in (2.17) describes the fluid's dynamics due to its kinetic energy. A fluid's kinetic energy typically defines a norm for the Eulerian fluid velocity, $KE = \frac{1}{2}\|\mathbf{u}\|^2$. The left side of the EP equation is the ***geodesic*** part of its evolution, with respect to this norm. See [5] for discussions of this interpretation of ideal incompressible flow and references to the literature. However, in a gravitational field, for example, there will also be dynamics due to potential energy. And this dynamics will by governed by the right side of the EP equation.

The right side of the EP equation in (2.17) modifies the geodesic motion. Naturally, the right side of the EP equation is also a geometrical quantity. The diamond operation $\diamond$ represents the dual of the Lie algebra action of vectors fields on the tensor a. Here $\delta\ell/\delta a$ is the dual tensor, under the natural pairing (usually, L^2 pairing) $\langle\,\cdot\,,\cdot\,\rangle$ that is induced by the variational derivative of the Lagrangian $\ell(\mathbf{u}, a)$. The diamond operation $\diamond$ is defined in terms of this pairing in (2.10). For the L^2 pairing, this is integration by parts of (minus) the Lie derivative in (2.10).

[1] Exceptions to this statement are certain multiphase fluids, and complex fluids with active internal degrees of freedom such as liquid crystals. These require a further extension, as discussed in [28].

[1] In coordinates, a one-form density takes the form $\mathbf{m}\cdot d\mathbf{x}\otimes dV$ and the EP equation (2.15) is given neumonically by

$$\frac{d}{dt}\Big|_{Lag}\big(\mathbf{m}\cdot d\mathbf{x}\otimes dV\big) = \frac{d\mathbf{m}}{dt}\Big|_{Lag}\cdot d\mathbf{x}\otimes dV + \mathbf{m}\cdot d\mathbf{u}\otimes dV + \mathbf{m}\cdot d\mathbf{x}\otimes(\nabla\cdot\mathbf{u})dV = \frac{\delta\ell}{\delta a}\diamond a$$

with $\frac{d}{dt}\Big|_{Lag} d\mathbf{x} := (\partial_t + \pounds_{\mathbf{u}})d\mathbf{x} = d\mathbf{u} = \mathbf{u}_{,j}dx^j$, upon using commutation of Lie derivative and exterior derivative. Compare this formula with the definition of $\mathrm{ad}^*_{\mathbf{u}}(\mathbf{m}\otimes dV)$ in equation (2.6).

The quantity a is typically a tensor (e.g., a density, a scalar, or a differential form) and we shall sum over the various types of tensors a that are involved in the fluid description. The second equation in (2.17) states that each tensor a is carried along by the Eulerian fluid velocity $\mathbf{u}$. Thus, a is for fluid "attribute," and its Eulerian evolution is given by minus its Lie derivative, $-\pounds_{\mathbf{u}}a$. That is, a stands for the set of fluid attributes that each Lagrangian fluid parcel carries around (advects), such as its buoyancy, which is determined by its individual salt, or heat content, in ocean circulation.

Many examples of how equation (2.17) arises in the dynamics of continuous media are given in [32]. The EP form of the Eulerian fluid description in (2.17) is analogous to the classical dynamics of rigid bodies (and tops, under gravity) in body coordinates. Rigid bodies and tops are also governed by Euler-Poincaré equations, as Poincaré showed in a two-page paper with no references, over a century ago [52]. For modern discussions of the EP theory, see the notes of Tudor Ratiu's lectures in this volume, [45], or [32].

2.1 Corollary of the EP theorem: the Kelvin-Noether circulation theorem

Theorem 2.2 (Kelvin-Noether Circulation Theorem.) *Assume $\mathbf{u}(x,t)$ satisfies the Euler-Poincaré equations for continua:*

$$\frac{\partial}{\partial t}\left(\frac{\delta \ell}{\delta \mathbf{u}}\right) = -\pounds_{\mathbf{u}}\left(\frac{\delta \ell}{\delta \mathbf{u}}\right) + \frac{\delta \ell}{\delta a} \diamond a$$

and the quantity a satisfies the advection relation

$$\frac{\partial a}{\partial t} + \pounds_{\mathbf{u}} a = 0. \tag{2.18}$$

Let η_t be the flow of the Eulerian velocity field $\mathbf{u}$, that is, $\mathbf{u} = (d\eta_t/dt)\circ\eta_t^{-1}$. Define the advected fluid loop $\gamma_t := \eta_t \circ \gamma_0$ and the circulation map $I(t)$ by

$$I(t) = \oint_{\gamma_t} \frac{1}{D}\frac{\delta \ell}{\delta \mathbf{u}}\,. \tag{2.19}$$

In the circulation map $I(t)$ the advected mass density D_t satisfies the push forward relation $D_t = \eta_* D_0$. This implies the advection relation (2.18) with $a = D$. Then the map $I(t)$ satisfies the ***Kelvin circulation***

relation,

$$\frac{d}{dt}I(t) = \oint_{\gamma_t} \frac{1}{D}\frac{\delta \ell}{\delta a} \diamond a\,. \tag{2.20}$$

Both an abstract proof of the Kelvin-Noether Circulation Theorem and a proof tailored for the case of continuum mechanical systems are given in [32].

The Kelvin-Noether theorem is called so here because its derivation relies on the invariance of the Lagrangian L under the particle relabeling symmetry, and Noether's theorem is associated with this symmetry. However, the result (2.20) is the ***Kelvin circulation theorem***: the circulation integral $I(t)$ around any fluid loop (γ_t, moving with the velocity of the fluid parcels $\mathbf{u}$) is invariant under the fluid motion. These two statements are equivalent. We note that ***two velocities*** appear in the integrand $I(t)$: the fluid velocity $\mathbf{u}$ and $D^{-1}\delta\ell/\delta\mathbf{u}$. The latter velocity is the momentum density $\mathbf{m} = \delta\ell/\delta\mathbf{u}$ divided by the mass density D. These two velocities are the basic ingredients for performing modeling and analysis in the ocean circulation problem. We simply need to put these ingredients together in the Euler-Poincaré theorem and its corollary, the Kelvin-Noether theorem.

Corollary 2.3 (Kelvin-Noether form.) *Since the last expression holds for every loop* γ_t, *we may write it as*

$$\left(\frac{\partial}{\partial t} + \pounds_{\mathbf{u}}\right)\frac{1}{D}\frac{\delta \ell}{\delta \mathbf{u}} = \frac{1}{D}\frac{\delta \ell}{\delta a} \diamond a\,. \tag{2.21}$$

This is the ***Kelvin-Noether form*** of the Euler-Poincaré equations for ideal continuum dynamics. By defining the covariant vector velocity,

$$\mathbf{v} := \frac{1}{D}\mathbf{m} := \frac{1}{D}\frac{\delta \ell}{\delta \mathbf{u}}\,, \tag{2.22}$$

we may write equation (2.21) in vector notation as

$$\frac{\partial}{\partial t}\mathbf{v} + \mathbf{u}\cdot\nabla\mathbf{v} + \nabla\mathbf{u}^T\cdot\mathbf{v} = \frac{1}{D}\frac{\delta \ell}{\delta a} \diamond a\,. \tag{2.23}$$

In three-dimensional vector notation, this is also expressed as

$$\frac{\partial}{\partial t}\mathbf{v} - \mathbf{u}\times\operatorname{curl}\mathbf{v} + \nabla(\mathbf{u}\cdot\mathbf{v}) = \frac{1}{D}\frac{\delta \ell}{\delta a} \diamond a\,. \tag{2.24}$$

Most of the ideal GFD model equations on the ABN list [1] may be written in this form. For examples and more theoretical details, see [33].

3 Applications of the Euler-Poincaré theorem in GFD

Variational Formulae in Three Dimensions We compute explicit formulae for the variations δa in the cases that the set of tensors a is drawn from a set of scalar fields and densities on $\mathbb{R}^3$. We shall denote this symbolically by writing

$$a \in \{b, D\, d^3x\}\,. \tag{3.1}$$

We have seen that invariance of the set a in the Lagrangian picture under the dynamics of $\mathbf{u}$ implies in the Eulerian picture that

$$\left(\frac{\partial}{\partial t} + \pounds_{\mathbf{u}}\right) a = 0\,,$$

where $\pounds_{\mathbf{u}}$ denotes Lie derivative with respect to the velocity vector field $\mathbf{u}$. Hence, for a fluid dynamical Eulerian action $\mathfrak{S} = \int dt\ \ell(\mathbf{u}; b, D)$, the advected variables b and D satisfy the following Lie-derivative relations,

$$\left(\frac{\partial}{\partial t} + \pounds_{\mathbf{u}}\right) b = 0, \quad \text{or} \quad \frac{\partial b}{\partial t} = -\ \mathbf{u}\cdot\nabla\, b\,, \tag{3.2}$$

$$\left(\frac{\partial}{\partial t} + \pounds_{\mathbf{u}}\right) D\, d^3x = 0, \quad \text{or} \quad \frac{\partial D}{\partial t} = -\ \nabla\cdot(D\mathbf{u})\,. \tag{3.3}$$

In fluid dynamical applications, the advected Eulerian variables b and $D\, d^3x$ represent the buoyancy b (or specific entropy, for the compressible case) and volume element (or mass density) $D\, d^3x$, respectively. According to Theorem 2.1, equation (2.13), the variations of the tensor functions a at fixed $\mathbf{x}$ and t are also given by Lie derivatives, namely $\delta a = -\ \pounds_{\mathbf{w}}\, a$, or

$$\begin{aligned} \delta b &= -\ \pounds_{\mathbf{w}}\, b = -\,\mathbf{w}\cdot\nabla\, b\,, \\ \delta D\ d^3x &= -\ \pounds_{\mathbf{w}}\,(D\, d^3x) = -\,\nabla\cdot(D\mathbf{w})\ d^3x\,. \end{aligned} \tag{3.4}$$

Hence, Hamilton's principle (2.13) with this dependence yields

$$
\begin{aligned}
0 &= \delta \int dt\ \ell(\mathbf{u}; b, D) \\
&= \int dt \left[\frac{\delta \ell}{\delta \mathbf{u}} \cdot \delta \mathbf{u} + \frac{\delta \ell}{\delta b}\ \delta b + \frac{\delta \ell}{\delta D}\ \delta D \right] \\
&= \int dt \left[\frac{\delta \ell}{\delta \mathbf{u}} \cdot \Big(\frac{\partial \mathbf{w}}{\partial t} - \mathrm{ad}_{\mathbf{u}}\, \mathbf{w} \Big) - \frac{\delta \ell}{\delta b}\ \mathbf{w} \cdot \nabla b - \frac{\delta \ell}{\delta D} \Big(\nabla \cdot (D\mathbf{w}) \Big) \right] \\
&= \int dt\ \mathbf{w} \cdot \left[- \frac{\partial}{\partial t} \frac{\delta \ell}{\delta \mathbf{u}} - \mathrm{ad}^*_{\mathbf{u}} \frac{\delta \ell}{\delta \mathbf{u}} - \frac{\delta \ell}{\delta b}\ \nabla b + D\ \nabla \frac{\delta \ell}{\delta D} \right] \\
&= - \int dt\ \mathbf{w} \cdot \left[\Big(\frac{\partial}{\partial t} + \pounds_{\mathbf{u}} \Big) \frac{\delta \ell}{\delta \mathbf{u}} + \frac{\delta \ell}{\delta b}\ \nabla b - D\ \nabla \frac{\delta \ell}{\delta D} \right] ,
\end{aligned}
\tag{3.5}
$$

where we have consistently dropped boundary terms arising from integrations by parts, by invoking natural boundary conditions. Specifically, we may impose $\hat{\mathbf{n}} \cdot \mathbf{w} = 0$ on the boundary, where $\hat{\mathbf{n}}$ is the boundary's outward unit normal vector and $\mathbf{w} = \delta\eta_t \circ \eta_t^{-1}$ vanishes at the endpoints.

Euler-Poincaré framework for continuum GFD The Euler-Poincaré equations for continua (2.15) may now be summarized in vector form for advected Eulerian variables a in the set (3.1). We adopt the notational convention of the circulation map I in equations (2.19) and (2.22) that a one form density can be made into a one form (no longer a density) by dividing it by the mass density D and we use the Lie-derivative relation for the continuity equation $(\partial/\partial t + \pounds_{\mathbf{u}})D d^3x = 0$. Then, the Euclidean components of the Euler-Poincaré equations for continua in equation (3.5) are expressed in Kelvin theorem form (2.21) with a slight abuse of notation as

$$
\Big(\frac{\partial}{\partial t} + \pounds_{\mathbf{u}} \Big) \Big(\frac{1}{D} \frac{\delta \ell}{\delta \mathbf{u}} \cdot d\mathbf{x} \Big) + \frac{1}{D} \frac{\delta \ell}{\delta b} \nabla b \cdot d\mathbf{x} - \nabla \Big(\frac{\delta \ell}{\delta D} \Big) \cdot d\mathbf{x} = 0 , \tag{3.6}
$$

in which the variational derivatives of the Lagrangian ℓ are to be computed according to the usual physical conventions, i.e., as Fréchet derivatives. Formula (3.6) is the Kelvin-Noether form of the equation of motion for ideal continua. Hence, we have the explicit Kelvin theorem expression, cf. equations (2.19) and (2.20),

$$
\frac{d}{dt} \oint_{\gamma_t(\mathbf{u})} \frac{1}{D} \frac{\delta \ell}{\delta \mathbf{u}} \cdot d\mathbf{x} = - \oint_{\gamma_t(\mathbf{u})} \frac{1}{D} \frac{\delta \ell}{\delta b} \nabla b \cdot d\mathbf{x} , \tag{3.7}
$$

where the curve $\gamma_t(\mathbf{u})$ moves with the fluid velocity $\mathbf{u}$. Then, by Stokes' theorem, the Euler equations generate circulation of $\mathbf{v} := (D^{-1}\delta l/\delta \mathbf{u})$

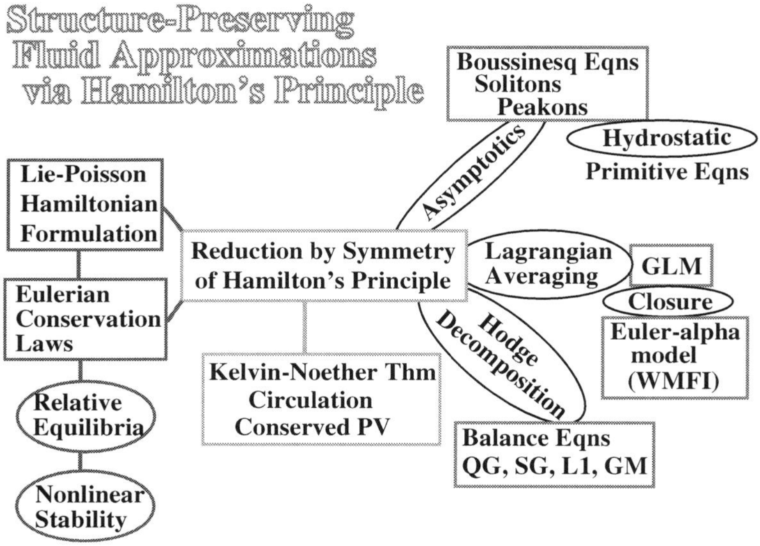

Fig. 3.1. The late Bill Burke was known as a great geometer. When he saw the pattern in this Figure a few years ago at UC Santa Cruz, he smiled during my lecture. When I asked him, "Bill, why are you smiling?" he answered, "Oh, Darryl, I was thinking this diagram would make such a great Tee-shirt."

whenever the gradients ∇b and $\nabla(D^{-1}\delta l/\delta b)$ are not collinear. The corresponding ***conservation of potential vorticity*** q on fluid parcels is given by

$$\frac{\partial q}{\partial t} + \mathbf{u}\cdot\nabla q = 0\,, \quad \text{where} \quad q = \frac{1}{D}\nabla b\cdot \text{curl}\left(\frac{1}{D}\frac{\delta \ell}{\delta \mathbf{u}}\right). \tag{3.8}$$

This is also called ***PV convection.*** Equations (3.6-3.8) embody most of the panoply of equations for GFD. The vector form of equation (3.6) is,

See [3] and [33] for detailed accounts.

3.1 Discussion: Structure-preserving fluid approximations via Hamilton's principle

The EP framework provides a unified approach: by using it, one may change the list of model GFD equations in [1] that looks rather like a page of the phonebook, into the more structured arrangement in Figure

3.1. This arrangement looks perhaps more organic, with HP at the center.

In Figure 3.1, Hamilton's Principle lies at the center. That is, the variation of an action vanishes and this creates the Euler-Poincaré equation. The Kelvin circulation theorem immediately follows from HP and, in turn, yields another central concept in oceanography and atmospherics emerges as PV convection, in equation (3.8). Here, PV is "potential vorticity" and, as we have seen, its convection follows from the Kelvin circulation theorem whenever a passive scalar (such as the buoyancy b) is available, by an application of the Stokes Theorem. One may apply various approximations, including balance, Lagrangian averaging, asymptotics, hydrostasy, etc., at any level in Hamilton's principle and still preserve the variational structure that yields Kelvin's circulation theorem, PV convection and proper energetics.

From Hamilton's Principle, one can Legendre transform to the Hamiltonian side. The EP equations are then found to be equivalent to Hamiltonian equations in terms of a Lie-Poisson bracket. (These equations will be derived later for solitons in *one* spatial dimension.) Thus, the corresponding partners of the Euler-Poincaré equations on the Hamiltonian side are the Lie-Poisson equations, which have been studied in great detail. Considerable machinery is available on the Hamiltonian side for classifying relative equilibria and obtaining their stability conditions, and so forth. See, e.g., Holm, Marsden, Ratiu and Weinstein [34] for a review of the Lie-Poisson Hamiltonian theory and its many assets. All this Hamiltonian machinery is available to us for GFD, provided we write those ocean models in the EP form. There are various ways of doing this. For example, one may perform asymptotics in Hamilton's Principle and introduce constraints as in [2], [25] and [24]. This procedure generates most of the various equations on the ABN list of balance equations for GFD, via asymptotic expansions that change the kinetic energy in the Lagrangian and impose constraints such as the geostrophic relation. The GFD equations are reviewed systematically and new GFD models are derived in the framework of the EP theorem in [3] and [33].

The strategy of this procedure is to make approximations in the Lagrangian, but still keep the same EP form of the equations that generates the fluid motion in the Eulerian representation. These model equations can then be implemented numerically. There are many physical choices available for decomposing solutions, or making guesses about how the averaging process, the course graining process, or some other approximation, or coupling, will change the kinetic energy, or smooth the solution in the kinetic energy, or constrain it. Having made the appropriate physical choices, the EP framework provides the motion equations and endows them with the required properties (energetic balance, Kelvin circulation theorem, PV convection, etc.) for properly modeling ocean and atmosphere circulation in the Eulerian representation.

4 Lagrangian reduction and EP turbulence closures

We are discussing reduction of HP by symmetries and EP approximations in the fluid setting. The EP framework provides a means of using the invariance properties of the Lagrangian or action in HP to reduce the number of variables in the equations of motion. In the previous section we explained how to use it for generating approximate ocean circulation models. The EP framework also provides a framework for introducing approximate turbulence closure models, and we shall discuss the main ideas behind the recent progress in this direction due to [32], [10], [11], [49] and [50]. We expect the EP framework will eventually also help generate climate models. However, EP climate modeling efforts are presently only in a preliminary stage. Additional degrees of freedom, such as order parameters, may also be accommodated by using the method of Lagrangian reduction by stages (LRBS). The LRBS approach allows one to obtain the equations for the motion of complex fluids such as liquid crystals, superfluids, Yang-Mills fluids and other fluids whose order parameters are defined on coset spaces by the physical process of symmetry breaking, as in [28]. However, complex fluids are not the subject of this lecture. Instead, we shall discuss the simplest of the Euler-Poincaré equations for incompressible fluids, beyond the Euler equations themselves, for the purpose of introducing the concepts used recently in making turbulence closures in the EP framework.

All of the Eulerian fluid equations we derive will take the EP form; so all these model Eulerian fluid equations will have a Kelvin circulation theorem. This is what we mentioned earlier in the spider-web diagram in Figure 3.1 that Bill Burke liked so much. At the center is Hamilton's Principle (HP) and it is equivalent to the EP equations which, in turn, imply Kelvin's circulation theorem. Thus, HP is a unified way of thinking about fluid circulation in the ocean, or in the atmosphere. And HP can potentially be a guide for creating numerical models for ocean-atmosphere circulation, in which the circulation doesn't just start up by itself for unphysical reasons. HP is also a guide for thinking about how to model other fluid equations by extending from the EP equation for incompressible flow.

The use of the EP approach in deriving GFD equations for ocean models has already been explained in [3] and [33]. Rather than repeat details from those papers, in this section we shall simplify by specializing to the EP geodesic motion equation for 3D incompressible fluid motion. This will lead us to turbulence models, when we add viscosity and forcing to the 3D incompressible fluid motion equations. These turbulence models are needed, in turn, so that one may eventually model subgrid scale effects in the computations using the GFD models of ocean circulation. One should have expected this: solving a complex problem always requires solving other complex subproblems. The subproblem of GFD that we must address is three-dimensional incompressible turbulence. Fortunately, we will be able to use the same EP framework as a guiding principle in developing models for the subproblem of turbulence closure modeling.

4.1 Incompressible 3D flows

We consider the Euler-Poincaré equation (3.6) for the following Lagrangian,

$$\ell(\mathbf{u};D) = \int \frac{1}{2} D\,\mathbf{u}\cdot Q_{op}\mathbf{u} - p(D-1)\,d^{\,3}x\,, \tag{4.1}$$

where Q_{op} is a positive-definite, symmetric, operator. (We shall also assume for simplicity that Q_{op} is translation invariant and isotropic under rotations.) For this Lagrangian, the two velocities $\mathbf{v} = [\delta\ell/\delta\mathbf{u}]_{D=1}$ and $\mathbf{u}$ are related by

$$\mathbf{v} = \frac{\delta\ell}{\delta\mathbf{u}}\bigg|_{D=1} = Q_{op}\mathbf{u}\,. \tag{4.2}$$

Conversely, the velocity $\mathbf{u} = g*\mathbf{v}$ is a smoothed, or filtered, version of the velocity $\mathbf{v}$ in (2.22).

$$\mathbf{u} = g*\mathbf{v} = \int g(|\mathbf{x}-\mathbf{y}|)\,\mathbf{v}(\mathbf{y},t)\,d^{\,3}y\,, \tag{4.3}$$

for an isotropic, translation-invariant, filter function g.[1] Expanding out the Lie derivative in the vector form (2.23) of the Euler-Poincaré equation for this Lagrangian gives

$$\frac{\partial}{\partial t}\mathbf{v} + \mathbf{u}\cdot\nabla\mathbf{v} + \nabla\mathbf{u}^T\cdot\mathbf{v} = -\nabla(p - \frac{1}{2}\,\mathbf{u}\cdot\mathbf{v})\,. \tag{4.4}$$

Here we have set $D = 1$ according to the constraint imposed by variation of the Lagrangian ℓ in (4.1) with respect to the Lagrange multiplier p (pressure). By the continuity equation (3.3), this constraint, in turn, imposes incompressibility on the transport velocity, $\nabla\cdot\mathbf{u} = 0$. Preservation of incompressibility then implies the pressure p as the solution of the Neumann problem implied by equation (4.4). If the filter function g is a delta function, so that $\mathbf{u} = \mathbf{v}$, then we may use $\nabla\mathbf{v}^T\cdot\mathbf{v} = \frac{1}{2}\nabla|\mathbf{v}|^2$, which is a gradient, and the EP equation (4.4) reduces to Euler's equation for incompressible fluids,

$$\frac{\partial}{\partial t}\mathbf{v} + \mathbf{v}\cdot\nabla\mathbf{v} = -\nabla p\,, \quad \text{with} \quad \nabla\cdot\mathbf{v} = 0\,. \tag{4.5}$$

For other filter functions, g, in the filter relation $\mathbf{u} = g * \mathbf{v}$ in (4.3), the transport velocity $\mathbf{u}$ is smoother than the transport-ed velocity $\mathbf{v}$ and the first term in equation (4.4) can be regarded as being smoothed advection. In fact, the second term in equation (4.4) also derives from smoothed advection. Kelvin's circulation theorem illustrates the effect of this smoothed advection in equation (4.4) as,

$$\frac{d}{dt}\oint_{\gamma_t(\mathbf{u})}\mathbf{v}\cdot d\mathbf{x} = \oint_{\gamma_t(\mathbf{u})}\Big[\frac{\partial}{\partial t}\mathbf{v} + \mathbf{u}\cdot\nabla\mathbf{v} + \nabla\mathbf{u}^T\cdot\mathbf{v}\Big]\cdot d\mathbf{x} = 0\,, \tag{4.6}$$

where the curve $\gamma_t(\mathbf{u})$ moves with the filtered fluid velocity $\mathbf{u} = g * \mathbf{v}$. This is the principle of ***Kelvin filtering***. We note that Kelvin filtering is a Lagrangian-averaging step, since the curve $\gamma_t(\mathbf{u})$ moves with the fluid parcels. For more details of Lagrangian-averaging in the Euler-Poincaré context, see [29], [30], [47].

Parallels of Kelvin filtering with Leray regularization Smoothing the advection velocity in the Kelvin loop $\gamma_t(\mathbf{u})$ in (4.6) is an important

[1] In principle, the filtered velocity $\mathbf{u}$ may be determined from the defiltered velocity $\mathbf{v}$, by defining the filter function $g(\mathbf{x})$ as the Green's function for the corresponding differential operator, Q_{op}, so that $Q_{op}\cdot g(\mathbf{x}) = \delta(\mathbf{x})$, where $\delta(\mathbf{x})$ is the Dirac delta and appropriate boundary conditions are supplied for $\mathbf{u}$. Because Q_{op} is positive-definite and symmetric, the quantity $\|\mathbf{u}\| = \big(\int\mathbf{u}\cdot Q_{op}\mathbf{u}\,d^3x\big)^{1/2}$ is a norm for the velocity $\mathbf{u}$. The kinetic energy in the Lagrangian (4.1) when evaluated on the constraint surface $D = 1$ is half the square of this norm.

step, because it alters the original nonlinearity $(\mathbf{v} \cdot \nabla \mathbf{v})$ in Euler's equation (4.5). This nonlinearity is the mechanism by which fluid kinetic energy cascades to smaller and smaller scales in turbulent flows. In fact, the original nonlinearity $(\mathbf{v} \cdot \nabla \mathbf{v})$ cascades the kinetic energy to smaller and smaller scales at rates that are *faster and faster*. However, Leray [44] already noticed that if the transport velocity $(\mathbf{u})$ were smoothed, relative to the transported velocity $(\mathbf{v})$, so that the nonlinearity would take the form $(\mathbf{u} \cdot \nabla \mathbf{v})$ as in the first term in equation (4.4), then one would be able to regularize the ***Leray-modified Navier-Stokes equations***,

$$\frac{\partial}{\partial t}\mathbf{v} + \mathbf{u} \cdot \nabla \mathbf{v} = -\nabla p + \nu \Delta \mathbf{v} + \mathbf{F}\,, \quad \text{with} \quad \nabla \cdot \mathbf{v} = 0\,, \tag{4.7}$$

provided the filter g in the relation (4.3) smoothes sufficiently to supply $\mathbf{u}$ with two derivatives. Here $\nu \Delta \mathbf{v}$ is Navier-Stokes viscosity and $\mathbf{F}$ is forcing.

If instead of adopting the Leray equation, we add the *same Navier-Stokes viscosity and forcing* as on the right hand side of (4.7) to the Euler-Poincaré equation (4.4), then we find the the ***EP-regularized equation***,

$$\frac{\partial}{\partial t}\mathbf{v} + \mathbf{u} \cdot \nabla \mathbf{v} + \nabla \mathbf{u}^T \cdot \mathbf{v} = -\nabla p + \nu \Delta \mathbf{v} + \mathbf{F}\,, \quad \text{with} \quad \nabla \cdot \mathbf{u} = 0\,. \tag{4.8}$$

The term $\nabla \mathbf{u}^T \cdot \mathbf{v} = v_j \nabla u^j$ in the EP-regularized Navier-Stokes equation (4.8) did not appear in the Leray theory [44]. However, this term is crucial in the Kelvin circulation theorem in equation (4.6). Namely, it is responsible for the stretching of the line element in the time derivative of the Kelvin circulation integral. This is the last term in equation (4.6).

Since Leray [44], people have been studying regularizations of the Navier-Stokes equations obtained by smoothing the transport velocity (relative to the transport-ed velocity). Leray already obtained results for existence and uniqueness of strong solutions with his smoothing approach, without addressing the line element stretching term. So the EP-regularized Navier-Stokes equation (4.8) obtained using Kelvin filtering may be regarded as a variant of the Leray [44] regularization program for Navier-Stokes that *restores* its Kevin circulation property. See Foias *et al.* [18] for discussions of the extensive literature about Leray's regularization of the Navier-Stokes equations.

4.2 Lagrangian-averaged 3D Euler-alpha equations

The *LAE-α* equations We consider the Euler-Poincaré equation (3.6) for the following Lagrangian, introduced in [32],

$$\ell(\mathbf{u}; D) = \int \frac{1}{2} D\left(|\mathbf{u}|^2 + \alpha^2|\nabla\mathbf{u}|^2\right) - p(D-1)\, d^3x\,, \tag{4.9}$$

where the quantity α is a constant length scale. On the constraint surface $D = 1$, this Lagrangian is the H^1 norm of the velocity,

$$\big[\ell(\mathbf{u}; D)\big]_{D=1} = \frac{1}{2}\|\mathbf{u}\|^2_{H^1}\,. \tag{4.10}$$

The velocity $\mathbf{v}$ in equation (2.22), when evaluated on the constraint surface $D = 1$ is given by

$$\mathbf{v} = \left[\frac{1}{D}\frac{\delta\ell}{\delta\mathbf{u}}\right]_{D=1} = \mathbf{u} - \alpha^2\Delta\mathbf{u}\,. \tag{4.11}$$

Thus, the velocities $\mathbf{u}$ and $\mathbf{v}$ are related by the Helmholtz operator in 3D, $Q_{op} = (1-\alpha^2\Delta)$. Consequently, the velocity $\mathbf{u} = g * \mathbf{v}$ is a smoothed, or filtered, version of the velocity $\mathbf{v}$, obtained by inverting the Helmholtz operator. That is, relation (4.3) is given by

$$\mathbf{u} = g * \mathbf{v} = \int g(|\mathbf{x}-\mathbf{y}|)\,\mathbf{v}(\mathbf{y},t)\,d^3y \tag{4.12}$$

in which isotropic, translation-invariant, filter function g is the Green's function for the Helmholtz operator in 3D, $Q_{op} = (1-\alpha^2\Delta)$, in the given domain. Thus, $\mathbf{u}$ is smoother than $\mathbf{v}$, by two additional derivatives, just as required in the Leray regularization of the NS equations. The Euler-Poincaré equation for the Lagrangian (4.9) gives

$$\frac{\partial}{\partial t}\mathbf{v} + \mathbf{u}\cdot\nabla\mathbf{v} + \nabla\mathbf{u}^T\cdot\mathbf{v} = -\nabla\Big(p - \frac{1}{2}|\mathbf{u}|^2 - \frac{1}{2}\alpha^2|\nabla\mathbf{u}|^2\Big)\,, \tag{4.13}$$

with $\mathbf{v} = \mathbf{u} - \alpha^2\Delta\mathbf{u}$ and $\operatorname{div}\mathbf{u} = 0$. These are called the Lagrangian-averaged Euler-alpha (*LAE-α*) equations. The *LAE-α* equations were first introduced on geometrical grounds in [32], as a three-dimensional incompressible generalization of the one-dimensional unidirectional soliton equation for shallow water waves first derived by Camassa and Holm in [9]. We shall review some of the dynamical properties of the CH equation as an Euler-Poincaré equation in section 5. The idea of applying Lagrangian averaging to the equations of GFD was pioneered by Andrews and McIntyre [4]. The relation between the present approach of Lagrangian-averaged Hamilton's principles for fluids and that of Andrews and McIntyre [4] is discussed in [29] and [30]. A wave, mean-flow closure of the Andrews and McIntyre theory for stratified rotating fluids was obtained in [24] from the viewpoint that eventually became the EP framework.

4.3 Damped, driven incompressible 3D flows and turbulence models

Equation (4.13) for the *LAE*-α model is arguably purely geometric. For example, it may be expressed in differential-geometric language as

$$(\partial_t + \pounds_{\mathbf{u}})(\mathbf{v} \cdot d\mathbf{x}) = -d\Big(p - \frac{1}{2}|\mathbf{u}|^2 - \frac{1}{2}\alpha^2|\nabla\mathbf{u}|^2\Big)\,. \tag{4.14}$$

Moreover, this is the Euler-Poincaré equation for geodesic motion on the volume-preserving diffeomorphism group with respect to the kinetic energy metric given by the H^1 norm of the velocity in (4.10). If we now add some damping and driving to this equation and think of viscosity and forcing, then we will have a damped and driven EP equation. When the operator Q_{op} is chosen to be the Helmholtz operator, the resulting damped, driven EP equation is,

$$\frac{\partial}{\partial t}\mathbf{v} + \mathbf{u}\cdot\nabla\mathbf{v} + \nabla\mathbf{u}^T\cdot\mathbf{v} = -\nabla\pi + \nu\Delta\mathbf{v} + \mathbf{F}\,, \quad \text{with} \quad \nabla\cdot\mathbf{u} = 0\,. \tag{4.15}$$

Also, $\pi = p - \frac{1}{2}|\mathbf{u}|^2 - \frac{1}{2}\alpha^2|\nabla\mathbf{u}|^2$ and $\mathbf{v} = \mathbf{u} - \alpha^2\Delta\mathbf{u}$. This is called the ***Lagrangian-averaged Navier-Stokes-alpha*** (LANS$-\alpha$) model. Smoothing of the transport velocity by inversion of the Helmholtz operator in determining $\mathbf{u}$ from $\mathbf{v}$ provides this equation with analytical properties that are considerably more regular than the Navier-Stokes equations are known to possess.

Summary of analytical properties for LANL-alpha in 3D Let's summarize the analytical results for the Navier-Stokes-alpha model in 3D. First, the LANL-alpha equations in 3D possess existence, uniqueness, and strong solutions. They also have a global attractor and moreover this global attractor is finite dimensional. That is, it has finite fractal dimension. In addition, for every sequence $\alpha_j \to 0$, the LANL-alpha solution will converge to a NS solution: so you have a shadowing theorem for finite alpha as well. All this power and control over the LANL-alpha solutions, however, gives us no information at all about solutions of the original NS equations. This is because no proof is known for the existence and uniqueness of strong solutions of NS. Without uniqueness, each of these sub-sequences could be converging to a different NS solution. So, existence and uniqueness for NS aside, LANL-alpha is a regularization of the NS equations that is obtained by strengthening the kinetic energy norm in the geodesic flow that underlies the nonlinearity in these problems. The analytical properties of the LANL-alpha model are discussed in [17], [46]. In the absence of dissipation and forcing, the LA-Euler-alpha equations (4.13) possess local existence for short time, as in the theorem of Ebin and Marsden [12] for the Euler equations. For further discussions of the analytical properties of the LA-Euler-alpha equations, see [53], [54], [46].

The closed model LAE equations for ideal incompressible flow in three dimensions were first obtained using the EP framework in Holm, Marsden and Ratiu [32]. For more discussions of how this equation set was developed into a turbulence model by using the EP framework, see papers by [10], [16] and [46]. Another, more general, variant of the LAE-α closure based on the WKB approximation was introduced for stratified and rotating incompressible fluids in [24]. This other variant was part of the background thinking for later developments of the EP theory in [27] and [47].

4.3.1 Hausdorf, or Lyapunov, dimension of the global attractor for the Navier-Stokes-alpha model of turbulence

The dimension of the global attractor for the Navier-Stokes-alpha model may be discussed in the context of counting dimensions as degrees of freedom in turbulence. Namely, one counts dimensions at a particular Reynolds number, which is the ratio of the nonlinearity to the viscous force. (The Reynolds number measures the intensity of turbulence.) Counting is done the following way. Suppose we are forcing at the low wave numbers, and we look at the log-log graph of the spectral energy density. This is the Fourier transform of the kinetic energy. We will see that it decreases with wavenumber k following an algebraic law: it decreases as $E(k) \sim k^{-5/3}$ for homogeneous isotropic Navier-Stokes, as [41] showed. This algebraic law holds until it reaches wavenumbers in the viscous dissipation range, where the length scales of the motions are small enough to be converted to heat by viscosity. The viscosity can then take over at the small scales to dissipate the flux of energy, that is put into the system by external forcing at the large scales. In the viscous dissipation range, the graph of the spectral energy density turns downward more steeply.

The wavenumber at which the viscous dissipation rate finally balances the rate of energy transport by nonlinearity and the graph of the spectral energy density turns downward defines the end of the inertial range. (The inertial range is the region in wavenumber where the transport is large compared to the dissipation.) The energy transport in the inertial range is governed by the nonlinearity. Because we are changing the nonlinearity in the LANS-alpha model, we expect to change the *balance* between the nonlinearity and dissipation. The particular change in the balance between nonlinearity and dissipation for the LANS-alpha model leads to a global attractor with finite (Hausdorf, or Lyapunov) dimension, as shown in [17].

Counting degrees of freedom for Navier-Stokes Let's first count dimensions as degrees of freedom for the Navier-Stokes equations. For this, one divides the length scale at which the dissipation takes place (called the Kolmogorov length scale) into the size of the domain, or the integral length scale of the motion. One discovers that this ratio scales like Reynolds number to the three quarters, $Re^{3/4}$. So the wavenumber at which the balance with dissipation occurs for the Navier-Stokes equations scales like Reynolds to the three quarters. The corresponding number of degrees of freedom is found by counting the number of little boxes of the size of the Kolmogorov scale that would be required to fill the domain. This number scales as the cube of the Kolmogorov wavenumber, which goes like Reynolds to the nine quarters, $Re^{9/4}$. This is Landau's classic estimate of the number of degrees of freedom in turbulence [43]. See also [20] for a slightly more refined version of this argument.

In estimating the computational complexity of turbulence as an evolutionary problem, one must arrange to take time steps that are small enough that the flow does not cross more than one of these boxes per time step. This CFL (Courant-Friedrichs-Levi) condition on the time step introduces another power of Reynolds to the three quarters ($Re^{3/4}$) into the computational complexity, which thus scales as Reynolds cubed (Re^3). This Re^3 scaling law provides insight into why direct numerical simulations (DNS) of turbulence are difficult: The task of computationally simulating all the way down to the Kolmogorov scale requires BOTH very small resolution scales and very small time steps. Hence, the scaling law Re^3 arises as a measure of the estimated computational complexity of Navier-Stokes turbulence.

Counting degrees of freedom for Navier-Stokes-alpha The spectral scaling of the energy cascade for the LANL-alpha model is different from NS. The kinetic energy spectral density for LANL-alpha obeys the same $k^{-5/3}$ scaling as NS only for wavenumbers satisfying $k\alpha < 1$. Thereafter, for higher wavenumbers satisfying $k\alpha > 1$, the spectral scaling breaks, and scales more steeply, as k^{-3} instead, as shown in [16]. This different scaling law yields a second inertial range, which ends by balancing viscous dissipation at a new dissipation wavenumber. The corresponding NS$-\alpha$ dissipation wavenumber scales as Reynolds to the one-half, $Re^{1/2}$, instead of Reynolds to the three quarters, $Re^{3/4}$, as for Navier-Stokes turbulence. Now counting the number of little boxes of size $O(Re^{-1/2})$ that would be required to fill the domain gives the scaling law $Re^{3/2}$ for the number of degrees of freedom in Navier-Stokes-alpha turbulence, with corresponding computational complexity of Reynolds squared, Re^2.

The $Re^{3/2}$ scaling law for degrees of freedom in Navier-Stokes-alpha and its Re^2 scaling for computational complexity are to be compared to the scalings of Reynolds to the three quarters, $Re^{9/4}$, for degrees of freedom, and Reynolds cubed, Re^3, for computational complexity of Navier-Stokes turbulence. See [17] for more details and for more refined discussions of the estimates of fractal dimension for the Navier-Stokes-alpha model of turbulence. The scaling law Re^2 for the Navier-Stokes-alpha equations, versus Re^3 for the Navier-Stokes equations gives a two-thirds power law in relative computational complexity.

4.3.2 Two-thirds power scaling in complexity for the NS$-\alpha$ model

The two-thirds-power scaling law $Re^{1/2} = (Re^{3/4})^{2/3}$ between the dissipation wavenumbers for the Navier-Stokes equations and the NS$-\alpha$ model is encouraging, especially in light of the convergence results proven for the NS$-\alpha$ model as $\alpha \to 0$. Provided alpha is sufficiently small, these convergence results allow one to expect that numerical computations with the NS$-\alpha$ equations will perform accurately in predicting turbulence at scales sufficiently larger than alpha. In addition, the regularization of the NS$-\alpha$ equations and their scaling law $Re^3 \rightarrow Re^2$ for computational complexity as a function of Reynolds number relative to NS shows that evolutionary turbulence computations requiring three decades of spatial resolution with the Navier-Stokes equations, should only require two decades of spatial resolution with the NS$-\alpha$ model, provided dissipation properly balances nonlinear transport at the end of the inertial range.

This two-thirds power scaling may be rigorously verified by computing the Lyapunov, or Hausdorf, dimension of the global attractor for the NS$-\alpha$ model. The result for this Hausdorf dimension is given in [17] as

$$\dim \mathcal{A} \le C \frac{L^{4/3}}{\alpha^{4/3}} Re^{3/2}$$

with a constant C that depends on the amplitude of forcing (its Grashoff number), but is independent of alpha. Here L is either the integral scale of the turbulence, or the size of the domain, which is taken to be spatially periodic in [17]. The $Re^{3/2}$ scaling in this upper bound is the same as one finds from the dimension estimates given earlier by simple box counting. Perhaps not surprisingly, this estimate for the Lyapunov, or Hausdorf, dimension of the global attractor for the the NS$-\alpha$ model is lost as $\alpha \to 0$. In this limit, one returns to the Navier-Stokes equations, for which no global attractor is known in the presence of nonzero forcing.

4.3.3 Comparison of the Leray model with LES similarity models for numerical computations of turbulence

The "similarity models" form a class of Large Eddy Simulation (LES) models for numerical computations of turbulence. The approach of the similarity models is to filter scales of wavenumber $2K$ relative to those at wavenumber K by assuming the spectral densities in the two regions scale similarly with wavenumber. This assumption yields a stress tensor for the sub-grid scales of the form that was first introduced by Bardina *et al.* in [6],

$$\tau_{SGS} = \overline{\bar{\mathbf{u}}\bar{\mathbf{u}}} - \bar{\bar{\mathbf{u}}}\bar{\bar{\mathbf{u}}}\,, \tag{4.16}$$

where bar $\overline{(\,\cdot\,)}$ indicates the filtering implied by the similarity assumption.

The stress tensor that one computes for the Leray-alpha model (4.7) is reminiscent of (4.16), but it still differs crucially from the similarity class. The Leray motion equation (4.7) may be rewritten in the present notation as,

$$\frac{\partial}{\partial t}\mathbf{u} + \bar{\mathbf{u}}\cdot\nabla\mathbf{u} = -\nabla p + \nu\Delta\mathbf{u} + \mathbf{F}\,, \quad \text{with} \quad \nabla\cdot\mathbf{u} = 0\,. \tag{4.17}$$

Taking the average $\overline{(\,\cdot\,)}$ of this equation and rearranging gives,

$$\frac{\partial}{\partial t}\bar{\mathbf{u}} + \bar{\mathbf{u}}\cdot\nabla\bar{\mathbf{u}} = -\nabla\bar{p} + \nu\Delta\bar{\mathbf{u}} + \bar{\mathbf{F}} - \operatorname{div}\tau_{SGS}\,, \quad \text{with} \quad \nabla\cdot\bar{\mathbf{u}} = 0\,, \tag{4.18}$$

where τ_{SGS} is the stress tensor for the corresponding sub-grid scales in the form advocated in Geurts and Holm [21, 22, 23], in their numerical simulations of the turbulent mixing layer,

$$\tau_{SGS} = \overline{\bar{\mathbf{u}}\,\mathbf{u}} - \bar{\mathbf{u}}\bar{\mathbf{u}}\,. \tag{4.19}$$

Thus, the choice of filter $\bar{\mathbf{u}} = L(\mathbf{u})$ and its inverse $\mathbf{u} = L^{-1}(\bar{\mathbf{u}})$ on the resolved mesh points determines the SGS stress tensor for the Leray regularization, viewed as an LES turbulence model. This crucially differs from the Bardina similarity model in requiring the defiltered velocity $\mathbf{u} = L^{-1}(\bar{\mathbf{u}})$ on the resolved mesh. See [21, 22, 23] for details of the performance of this approach in numerical simulations of turbulent mixing layers.[1]

Future LES models may benefit from introducing these filtering choices by using the EP framework, which provides the motion equations and endows them with the required properties (Kelvin circulation theorem, PV conservation, energy balance, etc.) for properly modeling ocean and atmosphere circulation in the Eulerian representation.

5 Pulsons and peakons: the EP equation for compressible geodesic motion in 1D

5.1 EP equation in 1D

We shall now simply further, by leaving the three dimensional incompressible flows and considering the one dimensional version of the Euler-alpha model called the CH equation, which possesses "peakon" solutions found by Camassa and Holm in [9]. The peakons are weak solutions that are also genuine solitons. That is, their solution profile has a jump in derivative at its peak. However, they propagate as confined pulses that interact elastically and their asymptotic speeds correspond to the discrete spectrum of an isospectral eigenvalue problem. The study of the one-dimensional evolutionary equation of the Euler-alpha model (4.13) raises questions about nonlinear aspects of balance and evolution for these equations that should be answered before one can fully understand their three-dimensional behavior as a basis for modeling turbulence.

We shall further simplify the problem by considering the EP equation for *compressible geodesic motion*, generated by the kinetic energy

[1] Other alternative choices exist for the sub-grid scale stress tensor τ_{SGS} in (4.18) that are intermediate between the the Bardina model (4.16) and the Leray model (4.19). An interesting example is $\tau_{SGS} = \overline{\bar{\mathbf{u}}\,\bar{\mathbf{u}}} - \bar{\mathbf{u}}\bar{\mathbf{u}}$. In this example, when the filter inverse is the Helmholtz operator, $L^{-1} = 1 - \alpha^2\Delta$, kinetic energy balance implies H^1 control of $\bar{\mathbf{u}}$, just as for the LANS$-\alpha$ model. However, the resulting motion equation fails to possess the proper Kelvin circulation theorem.

Lagrangian with no pressure constraint, cf. equation (4.1),

$$\ell(\mathbf{u}) = \int \frac{1}{2}\,\mathbf{u}\cdot Q_{op}\mathbf{u}\,d^{\,3}x\,, \tag{5.1}$$

where Q_{op} is again a positive-definite, symmetric, operator.

In this section, we shall discuss the simplest EP partial differential equation — the equation for compressible geodesic motion in one spatial dimension. This is the EP equation in 1D,[1]

$$\partial_t m + \mathrm{ad}_u^* m = 0\,, \quad \text{or, equivalently,} \quad \partial_t m + u m_x + 2u_x m = 0\,, \tag{5.2}$$

See Fringer and Holm [19] and references therein for discussions of the solutions of this equation. We shall consider this equation for geodesic motion on the diffeomorphism group with respect to a family of metrics for the fluid velocity $u(t,x)$, with notation,

$$m = \frac{\delta\ell}{\delta u} = Q_{op}u \quad \text{and for a Lagrangian} \quad \ell(u) = \frac{1}{2}\int u\,Q_{op}u\,dx\,. \tag{5.3}$$

In one dimension, we retain the earlier notation Q_{op} in equation (4.1) for the positive, symmetric operator that defines the kinetic energy metric for the velocity. The EP geodesic equation (5.2) is written in terms of the variable $m = \delta\ell/\delta u = Q_{op}u$, whose inverse is $u = g{*}m$ where g is the Green's function for the operator Q_{op} on the real line. It is appropriate to call this variational derivative m, because it is the momentum density associated with the fluid velocity u.

The EP equation (5.2) describes geodesic motion on the diffeomorphism group in one spatial dimension, with respect to the kinetic energy metric appearing in the reduced Lagrangian ℓ, defined for functions on the real line. Physically, the first nonlinear term in equation (5.2) is fluid transport. The coefficient 2 arises in the second nonlinear term, because, in one dimension, two of the summands in $\mathrm{ad}_u^*\, m = um_x + 2u_x m$ are the *same*, cf. equation (2.6). The momentum is expressed in terms of the velocity by $m = \delta\ell/\delta u = Q_{op}u$. Equivalently, for solutions that vanish at spatial infinity, one may think of the velocity as being obtained from the convolution, cf. equation (4.12),

$$u = g * m = \int g(x-y)m(y)\,dy\,, \tag{5.4}$$

[1] A one-form density in 1D takes the form $m\,(dx)^2$ and the EP equation is given by

$$\frac{d}{dt}\big(m\,(dx)^2\big) = \frac{dm}{dt}(dx)^2 + 2m\,(du)(dx) = 0 \quad \text{with} \quad \frac{d}{dt}dx = du = u_x dx\,,$$

where $dx/dt = u = G{*}m$ and $G{*}$ denotes convolution with G, the Green's function kernel for the operator Q_{op}.

where g is the Green's function for the operator Q_{op} on the real line. The operator Q_{op} and its Green's function g are chosen to be even under reflection, so that u and m have the same parity. Moreover, for $g(-x) = g(x)$, the EP equation (5.2) conserves the total momentum $M = \int m(y)\, dy$.

When u and m have the same parity, then equation (5.2) is reversible. That is, equation (5.2) is invariant under $t \to -t$ and $u \to -u$, $m \to -m$. Hence, the transformation $u(x,t) \to -u(x,-t)$ takes solutions into solutions, and in particular, it reverses the direction and amplitude of the traveling wave $u(x,t) = cg(x - ct)$.

We choose $g(x)$ to be an even function so that m and $u = g * m$ both have odd parity under mirror reflections. Hence, equation (5.2) is also invariant under the parity reflections $u(x,t) \to -\, u(-x,t)$, and the solutions of even and odd parity form invariant subspaces.

Equation (5.2) implies a similar reversible, parity invariant equation for the **absolute value** $|m|$,

$$|m|_t + u|m|_x + 2\, u_x|m| = 0\,, \quad \text{with} \quad u = g * m\,. \tag{5.5}$$

So, the positive and negative components $m_\pm = \frac{1}{2}(m \pm |m|)$ satisfy equation (5.2) separately. Also, if m is initially zero, it remains so. Consequently, equation (5.2) conserves the signature of m.

5.2 Pulsons

The geodesic equation (5.2) on the real line has the remarkable property that its solutions ***collectivize*** into a finite dimensional invariant manifold, on which the solution for the momentum $m = Q_{op}u$ is singular (measure-valued) and is given by a sum of delta functions,

$$m(x,t) = \sum_{i=1}^{N} p_i(t)\, \delta(x - q_i(t))\,. \tag{5.6}$$

The corresponding solutions for velocity appear as "$N-$pulsons" which were discovered for a special (integrable) form of g in Camassa and Holm [9], then were extended for any even g in Fringer and Holm [19],

$$u(x,t) = \sum_{i=1}^{N} p_i(t)\, g(x - q_i(t))\,. \tag{5.7}$$

This solution for velcoity arises from the singular solution for momentum via the convolution $u = g * m$, because $g(x)$ is the Green's function for the operator Q_{op} in the definition $m = Q_{op}u$. Thus, the time-dependent "collective coordinates" $q_i(t)$ and $p_i(t)$ are the positions and velocities of the N pulses in this solution, whose pulse shape is given by the Green's function $g(x)$. The parameters $q_i(t)$ and $p_i(t)$ satisfy the finite dimensional geodesic motion equations obtained as canonical Hamiltonian equations,

$$\dot{p}_i = -\frac{\partial H_N}{\partial q_i} = -p_i \sum_{j=1}^{N} p_j \, g'(q_i - q_j)\,, \tag{5.8}$$

$$\dot{q}_i = \frac{\partial H_N}{\partial p_i} = \sum_{j=1}^{N} p_j \, g(q_i - q_j). \tag{5.9}$$

in which the Hamiltonian is given by the quadratic form,

$$H_N = \frac{1}{2} \sum_{i,j=1}^{N} p_i \, p_j \, g(q_i - q_j)\,. \tag{5.10}$$

Thus, the canonical equations for the Hamiltonian H_N describe the nonlinear collective interactions of the $N-$pulson solutions of the geodesic equation (5.2) as finite-dimensional geodesic motion of a particle on an $N-$dimensional surface whose co-metric is

$$g^{ij}(q) = g(q_i - q_j)\,. \tag{5.11}$$

Fringer and Holm [19] showed numerically that the $N-$pulson solutions provide the dominant emergent pattern in the solution of the initial value problem for equation (5.2) with spatially confined initial conditions. This is confirmed in Figure 5.1 for the "peakon" case discussed below, in which $g = e^{-|x|/\alpha}$ and α is a length scale.

Integrability Calogero and Francoise [7], [8] found that for any finite N the Hamiltonian equations for H_N in (5.10) are completely integrable in the Liouville sense[1] for $g \equiv g_1(x) = \lambda + \mu\cos(\nu x) + \mu_1 \sin(\nu|x|)$ and $g \equiv g_2(x) = \alpha + \beta|x| + \gamma x^2$, with λ, μ, μ_1, ν, and α, β, γ being arbitrary constants, such that λ and μ are real and μ_1 and ν both real or both imaginary.[2] Particular cases of g_1 and g_2 are the peakons $g_1(x) = e^{-|x|/\alpha}$ of [9] and the compactons $g_2(x) = \max(1 - |x|, 0)$ of the Hunter-Saxton equation, [40]. The latter is the EP equation (5.2), with $m = u_{xx}$.

[1] A Hamiltonian system is integrable in the Liouville sense if the number of independent constants of motion is the same as the number of its degrees of freedom.

[2] This choice of the constants keeps H_N real in (5.10).

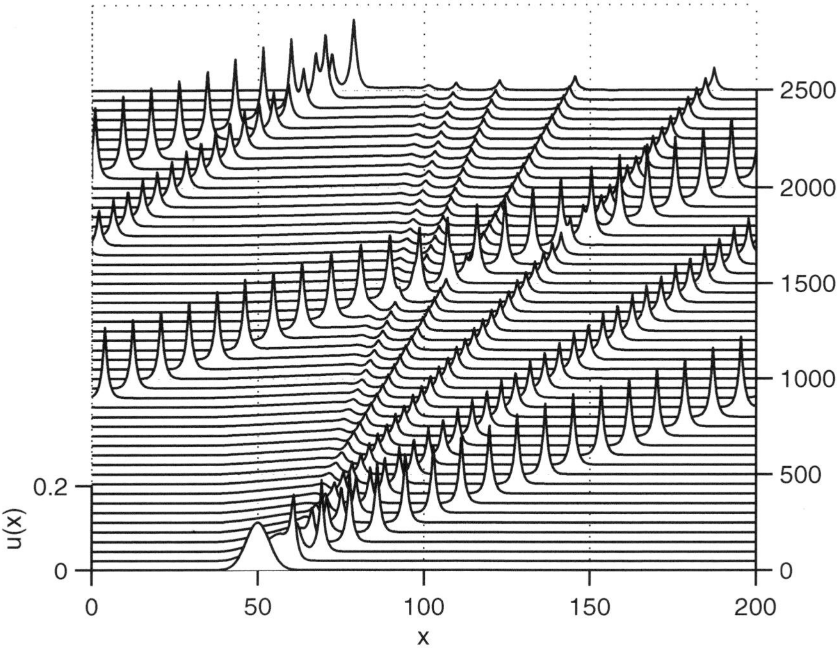

Fig. 5.1. Evolution under equation (5.2) of a Gaussian initial velocity distribution of unit area and width 5α with $\alpha = 1$. Here the filter function is the "peakon" shape $g(x) = e^{-|x|/\alpha}$, which is the Green's function for the Helmholtz operator $Q_{op} = 1 - \alpha^2\partial^2$ in one spatial dimension. Several peakons emerge from the initial Gaussian. The speed of each peakon is equal to its height and the peakons interact elastically as they cross and recross the periodic domain.

Lie-Poisson bracket In terms of m, the conserved energy Hamiltonian for the geodesic partial differential equation (5.2) is obtained by Legendre transforming the kinetic energy Lagrangian, as

$$h = \left\langle \frac{\delta\ell}{\delta u}, u \right\rangle - \ell(u) .$$

Thus, the Hamiltonian depends on m, as

$$h(m) = \frac{1}{2}\int m(x)g(x-y)m(y)\,dxdy ,$$

which also reveals the geodesic nature of the starting equation (5.2) and the role of $g(x)$ in the kinetic energy metric on the Hamiltonian side.

The corresponding Lie-Poisson bracket for this Hamiltonian evolution equation is given by,

$$\partial_t m = \{m, h\} = -\,(\partial m + m\partial)\frac{\delta h}{\delta m} \quad \text{and} \quad \frac{\delta h}{\delta m} = u\,,$$

which recovers the starting equation and explains its connection with fluid equations on the Hamiltonian side.

Collectivization One may verify that substituting the sum over delta functions for the $N-$pulson solution into the Hamiltonian h recovers the $N-$pulson Hamiltonian H_N. The $N-$pulson solution of the 1D geodesic equation (5.2) is an example of exact ***collectivization*** of the dynamics of the geodesic partial differential equation into the dynamics of a finite set of ordinary differential equations. In this case, the reduced equations also describe geodesic motion, on the reduced space. For further discussion and examples of collectivization, see Marsden and Ratiu [45]. The underlying reason for collectivization turns out to be the presence of a momentum map, as will be discussed briefly below and is explained fully in Holm and Marsden [31]. We formulate this result as a theorem,

Theorem 5.1 *The finite-dimensional invariant manifold of singular solutions of the EP geodesic equation (5.2) given by (5.6) is a momentum map.*

For the proof, see Holm and Marsden [31]. This observation is the key to understanding the generalization to higher dimensions of the peakon solutions of the geodesic equation (5.2).

5.3 Peakons

The case $g(x) = e^{-|x|/\alpha}$ with a constant lengthscale α is the Green's function for which the operator in the kinetic energy Lagrangian (5.3) is $Q_{op} = 1 - \alpha^2\partial_x^2$. For this (Helmholtz) operator Q_{op}, the Lagrangian and corresponding kinetic energy norm are given by,

$$\ell\,[u] = \frac{1}{2}\|u\|^2 = \frac{1}{2}\int u\,Q_{op}u\,dx = \frac{1}{2}\int u^2 + \alpha^2 u_x^2\,dx\,, \quad \text{for} \quad \lim_{|x|\to\infty} u = 0\,.$$

This Lagrangian is the H^1 norm of the velocity in one dimension. In this case, the EP equation (5.2) is also the zero-dispersion limit of the completely integrable CH equation for unidirectional shallow water waves derived in Camassa and Holm [9],

$$m_t + u m_x + 2m u_x = -c_0 u_x + \Gamma u_{xxx}\,, \qquad m = u - \alpha^2 u_{xx}\,. \tag{5.12}$$

This equation describes shallow water dynamics as completely integrable soliton motion at *quadratic* order in the asymptotic expansion for unidirectional shallow water waves on a free surface under gravity. The famous Korteweg-de Vries (KdV) equation provides the corresponding soliton description of shallow water waves at *linear* order in this asymptotic expansion. See Dullin, Gottwald and Holm [13, 14, 15] for more details and explanations of this asymptotic expansion for unidirectional shallow water waves to quadratic order.

In the dispersionless case $c_0 = 0 = \Gamma$, the CH equation (5.12) becomes

$$m_t + um_x + 2mu_x = 0\,, \qquad m = u - \alpha^2 u_{xx}\,. \tag{5.13}$$

The traveling wave solutions of the CH equation (5.13) in this dispersionless case are the "peakons," described by the reduced, or collective, solutions (5.7)-(5.10) for EP equation (5.2) with traveling waves

$$u(x,t) = c\, g(x - ct) = c\, e^{-|x-ct|/\alpha}\,.$$

In this case, the geodesic equation (5.2) may also be written as a conservation law for momentum,

$$\partial_t m = -\partial_x \Big(um + \frac{1}{2}u^2 - \frac{\alpha^2}{2}u_x^2 \Big)\,. \tag{5.14}$$

There is another Hamiltonian form of the CH equation (5.12), given by [9]

$$\partial_t m = \big\{ m, h_2 \big\}_2 = -\,(\partial - \alpha^2 \partial^3)\frac{\delta h_2}{\delta m}\,,$$

with a second Hamiltonian,

$$h_2 = \frac{1}{2}\int u^3 + \alpha^2 u u_x^2 \; dx\,.$$

One may verify the second Hamiltonian form of equation (5.12) by using the identity,

$$(1 - \alpha^2 \partial^2)\frac{\delta h_2}{\delta m} = \frac{\delta h_2}{\delta u} = um + \frac{1}{2}u^2 - \frac{\alpha^2}{2}u_x^2\,.$$

The two Hamiltonian operators

$$B_1 = \partial - \alpha^2 \partial^3 \quad \text{and} \quad B_2 = \partial m + m\partial$$

in the bi-Hamiltonian expression,[1]

$$\partial_t m = B_1 \frac{\delta h_2}{\delta m} = B_2 \frac{\delta h_1}{\delta m}$$

were used in combination with standard theorems for bi-Hamiltonian systems to derive the isospectral problem for this geodesic equation in [9]. Its bi-Hamiltonian property allows the nonlinear equation (5.12) to be expressed as the compatibility condition for two linear equations, namely, the *isospectral eigenvalue problem,*

$$2\lambda \left(\frac{1}{4} - \alpha^2 \partial_x^2 \right) \psi = m(x,t) \psi \,, \tag{5.15}$$

and the *evolution equation* for the eigenfunction ψ,

$$\psi_t = -(u + \lambda)\, \psi_x + \frac{1}{2} u_x \, \psi \,.$$

Compatibility of these linear equations ($\psi_{xxt} = \psi_{txx}$) and isospectrality ($d\lambda/dt = 0$) imply equation (5.12). Consequently, equation (5.12) admits the Inverse Spectral Transform (IST) method for the solution of its initial value problem, just as the KdV equation does, as discovered in [9].

This isospectral problem forms the basis for completely integrating the CH equation (5.13) as a Hamiltonian system and, thus, for finding its soliton (peakon) solutions. Remarkably, the isospectral problem (5.15) has purely discrete spectrum on the real line and the $N-$soliton solutions for this equation have the peakon form, cf. equation (5.7),

$$u(x,t) = \sum_{i=1}^{N} p_i(t) e^{-|x - q_i(t)|/\alpha} \,. \tag{5.17}$$

Here $p_i(t)$ and $q_i(t)$ satisfy the finite dimensional geodesic motion equations obtained as canonical Hamiltonian equations

$$\dot{q}_i = \frac{\partial H}{\partial p_i} \quad \text{and} \quad \dot{p}_i = -\frac{\partial H}{\partial q_i} \,,$$

when the Hamiltonian is given by, cf. equation (5.10),

$$H = \frac{1}{2} \sum_{i,j=1}^{N} p_i \, p_j \, e^{-|q_i - q_j|/\alpha} \,.$$

Thus, we have,

[1] A system is bi-Hamiltonian, if it may be expressed in Hamiltonian form in two inequivalent ways, and the sum its two Hamiltonian operators is also a Hamiltonian operator.

Theorem 5.2 *The CH peakons (5.17) are an integrable subcase of the pulsons (5.7).*

Integrability of the $N-$peakon dynamics One may verify integrability of the $N-$peakon dynamics by substituting the $N-$peakon solution (5.17) (which produces the sum of delta functions (5.6) for the momentum m) into the isospectral problem (5.15). This substitution reduces (5.15) to an $N \times N$ matrix eigenvalue problem. Isospectrality then implies that the traces of the powers of the matrix (or, equivalently, its N eigenvalues) yield N independent constants of the motion. Hence, the canonically Hamiltonian $N-$peakon dynamics is integrable.

6 Momentum filaments and surfaces: the EP equation for compressible geodesic motion in higher dimensions

As an example of the EP theory in higher dimensions, we shall generalize the one-dimensional pulson solutions of the previous section to $n-$dimensions.

6.1 $n-$dimensional Euler-Poincaré equation

Eulerian geodesic motion of a fluid in $n-$dimensions is generated as an EP equation via Hamilton's principle, when the Lagrangian is given by the kinetic energy. The kinetic energy defines a norm $\|\mathbf{u}\|^2$ for the Eulerian fluid velocity, $\mathbf{u}(\mathbf{x}, t) : R^n \times R^1 \to R^n$. The choice of the kinetic energy as a positive functional of fluid velocity $\mathbf{u}$ is a modeling step that depends upon the physics of the problem being studied. Following our earlier procedure, as in equations (4.1) and (5.3), we shall choose the Lagrangian,

$$\|\mathbf{u}\|^2 = \int \mathbf{u} \cdot Q_{op} \mathbf{u} \, d^n x = \int \mathbf{u} \cdot \mathbf{m} \, d^n x \,, \tag{6.1}$$

so that the positive-definite, symmetric, operator Q_{op} defines the norm $\|\mathbf{u}\|$, for appropriate boundary conditions. The EP equation for Eulerian geodesic motion of a fluid is given abstractly in equation (2.15) by [32, 45]

$$\frac{d}{dt} \frac{\delta \ell}{\delta \mathbf{u}} + \mathrm{ad}^*_{\mathbf{u}} \frac{\delta \ell}{\delta \mathbf{u}} = 0 \,, \quad \text{with} \quad \ell[\mathbf{u}] = \frac{1}{2} \|\mathbf{u}\|^2 \,. \tag{6.2}$$

Here ad^* is the dual of the vector-field ad-operation (the commutator) under the natural L^2 pairing $\langle \cdot \, , \cdot \rangle$ induced by the variational derivative $\delta \ell[\mathbf{u}] = \langle \frac{\delta \ell}{\delta \mathbf{u}} \, , \delta \mathbf{u} \rangle$. This pairing provides the definition of ad^*, cf.

equation (2.4),

$$\langle \mathrm{ad}^*_{\mathbf{u}}\, \mathbf{m}\,,\, \mathbf{v} \rangle = -\, \langle \mathbf{m}\,,\, \mathrm{ad}_{\mathbf{u}} \mathbf{v} \rangle\,, \tag{6.3}$$

where $\mathbf{u}$ and $\mathbf{v}$ are vector fields, $\mathrm{ad}_{\mathbf{u}}\mathbf{v} = [\mathbf{u}, \mathbf{v}]$ is the commutator and $\mathbf{m} = \delta\ell/\delta\mathbf{u}$ is the fluid momentum, a one-form density whose co-vector components are also denoted as $\mathbf{m}$. The coordinate form of $\mathrm{ad}^*_{\mathbf{u}}$ is given in equation (2.6). In this notation, the abstract EP equation (6.2) may be written explicitly in Euclidean coordinates as a partial differential equation for a co-vector function $\mathbf{m}(\mathbf{x}, t) : R^n \times R^1 \to R^n$. Namely, cf. equation (2.6),

$$\frac{\partial}{\partial t}\mathbf{m} + \mathbf{u}\cdot\nabla\mathbf{m} + \nabla\mathbf{u}^T\cdot\mathbf{m} + \mathbf{m}(\mathrm{div}\,\mathbf{u}) = 0\,, \quad \text{with} \quad \mathbf{m} = \frac{\delta\ell}{\delta\mathbf{u}}\,. \tag{6.4}$$

Equivalently, in terms of the operators div, grad and curl, in $2D$ and $3D$, the Euclidean-coordinate EP equation (6.4) becomes,

$$\frac{\partial}{\partial t}\mathbf{m} - \mathbf{u}\times\,\mathrm{curl}\,\mathbf{m} + \nabla(\mathbf{u}\cdot\mathbf{m}) + \mathbf{m}(\mathrm{div}\,\mathbf{u}) = 0\,. \tag{6.5}$$

After introducing the auxiliary variable $\mathbf{v} := \mathbf{m}/D$ and using the continuity equation (3.3) for D, equation (6.5) becomes the pressureless version of the EP equation (2.24), with right hand side set to zero. Namely, equation (6.5) is equivalent to

$$\frac{\partial}{\partial t}\mathbf{v} - \mathbf{u}\times\,\mathrm{curl}\,\mathbf{v} + \nabla(\mathbf{u}\cdot\mathbf{v}) = 0\,, \quad \text{with} \quad \frac{\partial D}{\partial t} = -\,\nabla\cdot(D\mathbf{u})\,. \tag{6.6}$$

These equations in terms of the auxiliary variables $\mathbf{v} := \mathbf{m}/D$ and D are convenient for formal manipulations. However, the fundamental solutions we shall seek for $\mathbf{m}$ will best be found in the primitive form of equation (6.5).

The scalar product of the EP equation in div-grad-curl form (6.5) with the velocity $\mathbf{u}$ shows that evolution under this equation preserves the kinetic energy norm $\langle \mathbf{m}\,,\, \mathbf{u}\rangle = \|\mathbf{u}\|^2$, as a constant of the motion. Thus, the evolution of the div-grad-curl EP equation (6.5) yields geodesic motion on the diffeomorphism group, with respect to the metric associated with the kinetic energy norm:

$$\ell[\mathbf{u}] = \frac{1}{2}\|\mathbf{u}\|^2 = \frac{1}{2}\langle \mathbf{m}\,,\, \mathbf{u}\rangle \quad \text{and} \quad \mathbf{m} = \frac{\delta\ell[\mathbf{u}]}{\delta\mathbf{u}} = Q_{op}\mathbf{u}\,. \tag{6.7}$$

Legendre transforming to the Hamiltonian side Legendre transforming the Lagrangian (6.7) yields the **Hamiltonian**,

$$H[\mathbf{m}] = \langle \mathbf{m}\,,\, \mathbf{u}\rangle - \frac{1}{2}\|\mathbf{u}\|^2 = \frac{1}{2}\int \mathbf{m}\cdot G * \mathbf{m}\; d^n x \equiv \frac{1}{2}\|\mathbf{m}\|^2\,, \tag{6.8}$$

which also defines a norm $\|\mathbf{m}\|$ via a convolution kernel G that is symmetric and positive, when the Lagrangian $\ell[\mathbf{u}]$ is a norm. By the usual rules of the Legendre transformation [32, 45], the norm $\|\mathbf{m}\|$ given by the Hamiltonian $H[\mathbf{m}]$ specifies the velocity $\mathbf{u}$ in terms of its dual momentum $\mathbf{m}$ by the variational operation,

$$\mathbf{u} = \frac{\delta H}{\delta \mathbf{m}} = G * \mathbf{m} \equiv \int G(\mathbf{x} - \mathbf{y})\, \mathbf{m}(\mathbf{y})\, d^n y\,. \tag{6.9}$$

We shall choose the kernel $G(\mathbf{x} - \mathbf{y})$ to be translation-invariant (so Noether's theorem implies that total momentum $\mathbf{M} = \int \mathbf{m}\, d^n x$ is conserved) and symmetric under spatial reflections (so that $\mathbf{u}$ and $\mathbf{m}$ have the same parity). The corresponding Legendre-dual relations are,

$$\mathbf{u} = G * \mathbf{m} \quad \text{and} \quad \mathbf{m} = Q_{op}\mathbf{u}\,, \tag{6.10}$$

where G is the **Green's function** for the operator Q_{op}, with appropriate boundary conditions (on $\mathbf{u}$) that allow inversion of the operator Q_{op} to determine $\mathbf{u}$ from $\mathbf{m}$.

After the Legendre transformation (6.8), the EP equation (6.2) appears in its equivalent **Lie-Poisson Hamiltonian form**,

$$\frac{\partial}{\partial t}\mathbf{m} = \{\mathbf{m}, H\} = -\,\mathrm{ad}^*_{\frac{\delta H}{\delta \mathbf{m}}}\,\mathbf{m}\,. \tag{6.11}$$

Here the operation $\{\cdot\,,\,\cdot\}$ denotes the Lie-Poisson bracket dual to the (right) action of vector fields amongst themselves by vector-field commutation. For more details and additional background concerning the relation of classical EP theory to Lie-Poisson Hamiltonian equations, see [32, 45].

6.2 $n-$dimensional Analogs of Pulsons for the EP equation

The momentum for the one-dimensional pulson solutions (5.6) on the real line is supported at points via the Dirac delta measures in its singular solution ansatz for momentum in Camassa and Holm [9],

$$m(x,t) = \sum_{i=1}^{N} p_i(t)\, \delta\big(x - q_i(t)\big)\,, \quad m \in R^1\,. \tag{6.12}$$

Holm and Staley [39] developed $n-$dimensional analogs of these one dimensional pulson solutions for the Euler-Poincaré equation (6.5) by generalizing this solution ansatz to allow measure-valued $n-$dimensional vector solutions $\mathbf{m} \in R^n$ for which the Euler-Poincaré momentum is supported on co-dimension$-k$ *subspaces* R^{n-k} with integer $k \in [1, n]$. In an example in section 6.2.2, we shall consider a two-dimensional vector momentum $\mathbf{m} \in R^2$ in the plane that is supported on one-dimensional curves (momentum fronts). Likewise, in three dimensions, one could consider two-dimensional momentum surfaces (sheets), one-dimensional momentum filaments, etc. The corresponding vector momentum ansatz introduced Holm and Staley [39] is the following, cf. the pulson solutions (6.12),

$$\mathbf{m}(\mathbf{x}, t) = \sum_{i=1}^{N} \int \mathbf{P}_i(s,t)\, \delta\big(\, \mathbf{Q}_i(s,t)\,, \mathbf{Q}_{j\ i}(s,t)\,\big) ds\,, \qquad \mathbf{m} \in R^n\,. \quad (6.13)$$

Here, $\mathbf{P}_i, \mathbf{Q}_i \in R^n$ for $i = 1, 2, \ldots, N$. For example, when $n - k = 1$, so that $s \in R^1$ is one-dimensional, the delta function in solution (6.13) supports an evolving family of vector-valued curves, called **momentum filaments**. (For simplicity of notation, we suppress the implied subscript i in the arclength s for each $\mathbf{P}_i$ and $\mathbf{Q}_i$.) The Legendre-dual relations (6.10) imply that the velocity corresponding to the momentum filament ansatz (6.13) is,

$$\mathbf{u}(\mathbf{x}, t) = G * \mathbf{m} = \sum_{j=1}^{N} \int \mathbf{P}_j(s', t)\, G\big(\, \mathbf{x}\,, \mathbf{Q}_{\,j}(s', t)\,\big) ds'\,. \qquad (6.14)$$

Just as for the 1D case of the pulsons, we shall show that substitution of the $n-$D solution ansatz (6.13) and (6.14) into the Euler-Poincaré equation (6.4) produces canonical geodesic Hamiltonian equations for the $n-$dimensional vector parameters $\mathbf{Q}_i(s,t)$ and $\mathbf{P}_i(s,t)$, $i = 1, 2, \ldots, N$.

The singular momentum solutions in (6.13) are vector-valued functions supported in $\mathbb{R}^n$ on a set of N surfaces (or curves) of codimension $(n-k)$ for $s \in \mathbb{R}^k$ with $k < n$. They may, for example, be supported on sets of points (vector peakons, $k = 0$), one-dimensional filaments (strings, $k = 1$), or two-dimensional surfaces (sheets, $k = 2$) in three dimensions.

One of the main results in Holm and Marsden [31] is the theorem stating that the singular solution ansatz in (6.13) is an equivariant momentum map. This result helps to organize the theory and to suggest new avenues of exploration, as they explain.

6.2.1 Canonical Hamiltonian dynamics of momentum filaments, or strings, in R^n

For definiteness in what follows, we shall consider the example of momentum filaments $\mathbf{m} \in R^n$ supported on one-dimensional space curves in R^n, so $s \in R^1$ is the arclength parameter of one of these curves. This solution ansatz is reminiscent of the Biot-Savart Law for vortex filaments, although the flow is not incompressible. The dynamics of momentum surfaces, for $s \in R^k$ with $k < n$, follow a similar analysis.

Substituting the momentum filament ansatz (6.13) for $s \in R^1$ and its corresponding velocity (6.14) into the Euler-Poincaré equation (6.4), then integrating against a smooth test function $\phi(\mathbf{x})$ implies the following canonical equations (denoting explicit summation on $i, j \in 1, 2, \dots N$),

$$\begin{aligned} \frac{\partial}{\partial t}\mathbf{Q}_i(s,t) &= \sum_{j=1}^{N} \int \mathbf{P}_j(s',t)\, G(\mathbf{Q}_i(s,t) - \mathbf{Q}_j(s',t))\,)ds' \\ &= \frac{\delta H_N}{\delta \mathbf{P}_i}\,, \end{aligned} \tag{6.15}$$

$$\begin{aligned} \frac{\partial}{\partial t}\mathbf{P}_i(s,t) &= \\ & -\sum_{j=1}^{N} \int \big(\mathbf{P}_i(s,t)\cdot\mathbf{P}_j(s',t)\big)\, \frac{\partial}{\partial \mathbf{Q}_i(s,t)} G\big(\mathbf{Q}_i(s,t)\,, \mathbf{Q}_j(s',t)\big)\, ds' \\ &= -\frac{\delta H_N}{\delta \mathbf{Q}_i}\,, \quad \text{(sum on } j\text{, no sum on } i\text{)}\,. \end{aligned} \tag{6.16}$$

The dot product $\mathbf{P}_i \cdot \mathbf{P}_j$ denotes the inner, or scalar, product of the two vectors $\mathbf{P}_i$ and $\mathbf{P}_j$ in R^n. Thus, the solution ansatz (6.13) yields a closed set of integro-partial-differential equations (IPDEs) given by (6.15) and (6.16) for the vector parameters $\mathbf{Q}_i(s,t)$ and $\mathbf{P}_i(s,t)$ with $i = 1, 2 \dots N$. These equations are generated canonically by the following Hamiltonian function $H_N : (R^n \times R^n)^{\otimes N} \to R$,

$$H_N = \frac{1}{2}\iint \sum_{i\,,\,j=1}^{N} \big(\mathbf{P}_i(s,t)\cdot\mathbf{P}_j(s',t)\big)\, G\big(\mathbf{Q}_i(s,t)\,, \mathbf{Q}_j(s',t)\big)\, ds\, ds'\,. \tag{6.17}$$

This Hamiltonian arises by substituting the momentum ansatz (6.13) into the Hamiltonian (6.8) obtained from the Legendre transformation of the Lagrangian corresponding to the kinetic energy norm of the fluid velocity. Thus, the evolutionary IPDE system (6.15) and (6.16) represents canonically Hamiltonian geodesic motion on the space of curves in

R^n with respect to the co-metric given on these curves in (6.17). The Hamiltonian $H_N = \frac{1}{2}\|\mathbf{P}\|^2$ in (6.17) defines the norm $\|\mathbf{P}\|$ in terms of this co-metric that combines convolution using the Green's function G and sum over filaments with the scalar product of momentum vectors in R^n.

Summary The singular momentum solution ansatz in (6.13) is a momentum map for the (right) action of diffeomorphisms on distributions that reduces, or ***collectivizes*** the solution of the geodesic EP PDE (6.4) in $n+1$ dimensions into the system (6.15) and (6.16) of $2N$ evolutionary IPDEs on smoothly embedded subspaces $s \in R^k$ with $k < n$ of R^n with codimension k. For more details of the use of momentum maps for singular solutions, see [31].

Potential applications of singular EP solutions

One of the potential applications of the two-dimensional version of this problem involves the internal waves on the interface between two layers of different density in the ocean.

Figs 6.1 and 6.2 show a striking agreement between two internal wave trains propagating at the interface of different density levels in the South China Sea, and the solution appearing in the simulations of the EP equation (2.23) in two dimensions. For other work on the 2D CH equation in the context of shallow water waves, see Kruse, et al. (2001) [42].

Another potential application of the two-dimensional version of this problem occurs in image processing for computational anatomy, e.g., brain mapping from PET scans. For this application, one envisions the geodesic motion as an optimization problem whose solution maps one measured two-dimensional PET scan to another, by interpolation in three dimensions along a geodesic path between them in the space of diffeomorphisms. In this situation, the measure-valued solutions of geodesic flow studied here correspond to "cartoon" outlines of PET scan images. The geodesic "evolution" in the space between them provides a three dimensional image that is optimal for the chosen norm. For a review of this imaging approach, which is called "template matching" in computational anatomy, see Miller and Younes [2002] [48]. See [37] for a discussion of the soliton paradigm of elastic collisions and momentum exchange in computational anatomy.

The momentum filaments are contact discontinuities. The canonically Hamiltonian IPDEs for momentum filaments in (6.15) and (6.16) were first considered in Holm and Staley [39], where the evolution of a single momentum filament interacting with itself was first discussed. There is a faint similarity of this system to vortex dynamics for the incompressible Euler equations. However, there are also fundamental differences. The main difference from the Hamiltonian motion of vortex filaments is that the momentum filaments possess inertia, while vortex filaments do not. Thus, N vortex filaments are described by N first-order equations, while N momentum filaments are described by $2N$ first-order equations. Holm and Staley [39] showed numerically that the momentum filament solutions represent the dominant emergent pattern in the initial value problem for the geodesic EP equation (6.4). They also showed these solutions have discontinuities in slope what move with the velocity of the flow. That is, they are *contact discontinuities.* The next subsection gives examples in which these singular solutions occur, both in nature (as internal wave fronts) and in applications (in medical imaging).

6.2.2 Zero-dispersion shallow water waves in 2D: Two interesting choices for the operator Q_{op}

The operator Q_{op} in the momentum relation $\mathbf{m} = Q_{op}\mathbf{u}$ in (6.10) corresponding to $m = u - \alpha^2 u_{xx}$ in the 1D CH equation (5.13) for zero-dispersion shallow water waves may be defined in two dimensions as *either* of two natural choices,

$$\mathbf{m} = \mathbf{u} - \alpha^2 \Delta \mathbf{u}\,, \quad \text{or} \quad \mathbf{m} = \mathbf{u} - \alpha^2 \nabla\,\mathrm{div}\,\mathbf{u}\,. \tag{6.18}$$

For the first choice of momentum definition in (6.18), the EP equation (6.4) corresponds to the (pressureless) Euler-alpha model, whose Lagrangian (6.7) is the conserved H^1 norm,[1]

$$\|\mathbf{u}\|^2_{H^1} = \int \mathbf{u}\cdot(1-\alpha^2\Delta)\mathbf{u}\,dx\,dy = \int |\mathbf{u}|^2 + \alpha^2(\mathrm{div}\,\mathbf{u})^2 + \alpha^2|\mathrm{curl}\,\mathbf{u}|^2\,dx\,dy\,.$$

The last equality assumes either homogeneous, or periodic boundary conditions, so that boundary terms may be neglected in integrating by parts.

For the second natural choice of momentum in (6.18), the conserved

[1] When incompressibility ($\mathrm{div}\,\mathbf{u} = 0$) is imposed as an additional constraint in this Lagrangian via a Lagrange multiplier (the pressure), as in equation (4.1), then the corresponding Euler-Poincaré equation (2.23) becomes the 2D *LA*-Euler equation (4.4) derived in [32]. The Lagrangians defined by the kinetic energy norms $\|\mathbf{u}\|^2_{H^1}$ and $\|\mathbf{u}\|^2_{KS}$ have no pressure constraint; that is, they allow compressible motion.

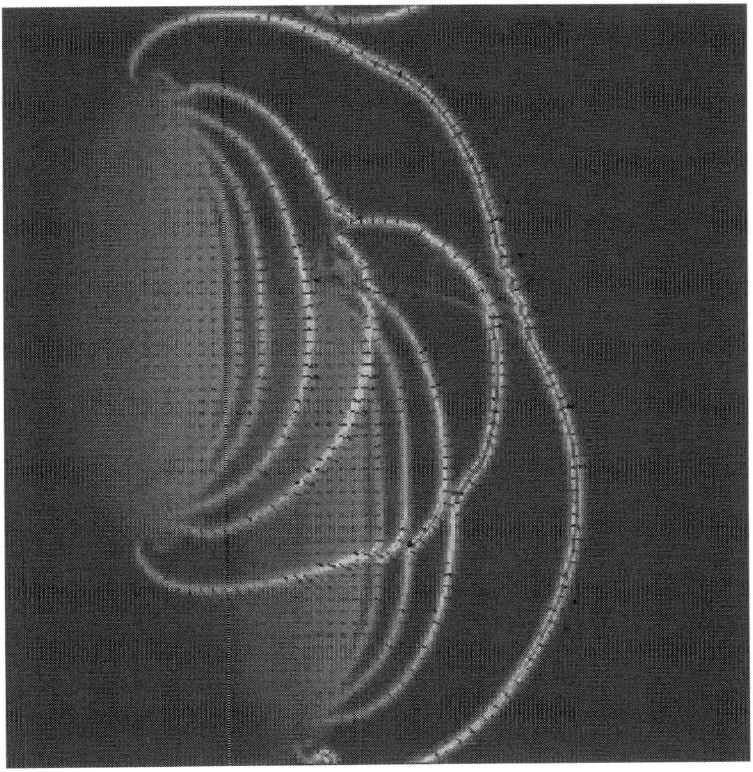

Fig. 6.1. Simulation of the full EP equation (2.23), courtesy of Martin Staley.

kinetic energy norm becomes, instead,

$$\|\mathbf{u}\|_{KS}^2 = \int \mathbf{u}\cdot(1-\alpha^2\nabla\,\mathrm{div}\,\mathbf{u})\mathbf{u}\,dx\,dy = \int |\mathbf{u}|^2 + \alpha^2(\mathrm{div}\,\mathbf{u})^2\,dx\,dy\,,$$

and kinetic energy conservation no longer controls $\mathrm{curl}\,\mathbf{u}$. This is the norm associated with vertically-averaged kinetic energy that arises when one approximates the Green-Naghdi equations for shallow water motion by neglecting variations in surface elevation in the potential energy and in the Lagrange-to-Euler Jacobian.[1] The second term proportional to α^2

[1] In this approximation for 2D shallow water waves, $\mathrm{curl}\,\mathbf{m} = \mathrm{curl}\,\mathbf{u}$ and $\mathrm{div}\,\mathbf{m} = (1-\alpha^2\Delta)\,\mathrm{div}\,\mathbf{u}$. Thus, setting $\mathbf{u} = \hat{\mathbf{z}}\times\nabla\psi - \nabla\phi$ allows one to solve for the stream function ψ and velocity potential ϕ from the momentum $\mathbf{m}$ via,

$$\hat{\mathbf{z}}\cdot\mathrm{curl}\,\mathbf{m} = -\,\Delta\psi \quad \text{and} \quad \mathrm{div}\,\mathbf{m} = -\,\Delta(1-\alpha^2\Delta)\phi\,.$$

These two relations allow one to update the potentials ψ and ϕ for the velocity

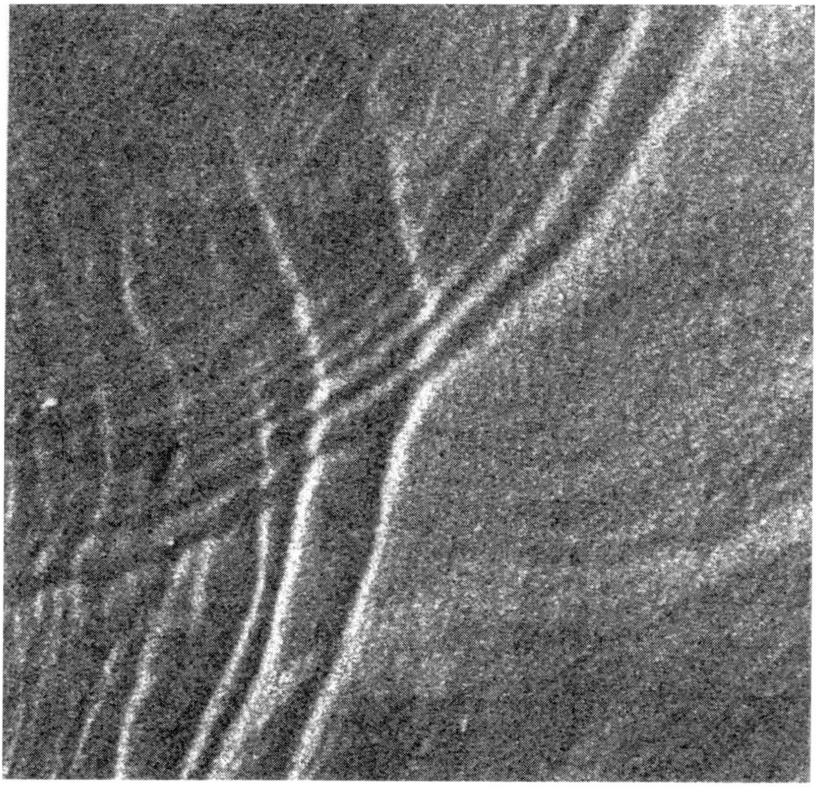

Fig. 6.2. Internal waves in the South China Sea (bottom).

approximates (twice) the vertically-averaged kinetic energy due to vertical motion. For more details of the latter shallow water approximation, see Kruse and Schreule [42].

Holm & Staley [39] integrated the div-grad-curl form (6.5) of the EP equation numerically using a difference scheme that preserved the properties of the operators div, grad and curl (divcurl=0 and curlgrad=0). The main discovery of the numerical results of Holm & Staley [39] was that the evolution of the geodesic PDE (6.5) was found to be dominated by the emergent dynamics of momentum filaments, arising from confined intial conditions for *either* choice of momentum-velocity relation

$\mathbf{u}$, given the momentum $\mathbf{m}$ at each time step, provided these potentials satisfy boundary conditions that allow inversion of the Laplacian operator for ψ and the Helmholtz-Laplace operator for ϕ. Boundary conditions must be chosen for this inversion that are consistent with the diffeomorphism group.

in equation (6.18). Thus, the momentum filament solutions in both of these cases were stable, and *no other types of solution* were created in the evolution of equation (6.5) in the periodic plane. The dynamics of the momentum filaments that emerged was quasi-one-dimensional, with greater variation of the solution in the direction transverse to the filaments. Thus, the interaction dynamics for the momentum filaments was found to be dominantly in the direction transverse to the filaments. Consequently, the filament interaction was governed primarily by elastic-scattering dynamics reminiscent of the one-dimensional solutions, as seen in soliton dynamics. In fact, the one-dimensional soliton collision rules were found to provide a good interpretation of the interactions among the momentum filaments. These interactions were found to allow reconnection of the quasi-one-dimensional momentum filaments, shown in Figure 6.1. For more information about the role of momentum maps in these singular solutions, see [31]. For numerical results illustrating their behavior, see [39]. For additional results and analysis for singular EP solutions with cylindrical and spherical symmetry, see [35] and [36].

Acknowledgements

I am grateful to the students of the MASIE Summer School for their enthusiastic questions and discussions at Peyresq, France, September 2000. I am especially grateful for the helpful feedback prepared by H.R. Dullin and G.A. Gottwald for the CH equation which resulted in papers [13, 14, 15]. Some of these lectures were also delivered to post-graduate students at Imperial College London in January 2004. I am grateful to R.C. Malone and R.D. Smith for the use of their numerical results in Figure 1.1. Figures 5.1, 6.1 and 6.2 were kindly produced by Martin Staley, see [38, 39] for more details. This work was supported by the US Department of Energy, Los Alamos National Laboratory and US DoE, Office of Science.

Bibliography

[1] ALLEN, J. S., BARTH, J. A. AND NEWBERGER, P. A. 1990 On intermediate models for barotropic continental shelf and slope flow fields: Part I, Formulation and comparison of exact solutions. *J. Phys. Oceanogr.*, **20**, 1017-1042.

[2] ALLEN, J. S., AND HOLM, D. D. 1996 Extended-geostrophic Hamiltonian models for rotating shallow water motion. *Physica D*, **98** 229-248.

[3] ALLEN, J. S., HOLM, D. D., AND NEWBERGER, P. A. 2002 Toward an extended-geostrophic Euler-Poincaré model for mesoscale oceanographic flow. In *Large-Scale Atmosphere-Ocean Dynamics 1: Analytical Methods and Numerical Models.* Edited by J. Norbury & I. Roulstone, Cambridge University Press: Cambridge, pp. 101-125.

[4] ANDREWS, D. G. AND MCINTYRE, M. E. 1978 An exact theory of nonlinear waves on a Lagrangian-mean flow. *J. Fluid Mech.* **89**, 609-646.

[5] ARNOLD, V. I AND KHESIN, B. A. 1998 *Topological Methods in Hydrodynamics.* Springer: New York.

[6] BARDINA, J., FERZIGER, J. H. AND REYNOLDS, W. C. 1984 Improved turbulence models based on large eddy simulations of homogeneous incompressible turbulence. Stanford University, Report TF-19.

[7] CALOGERO, F. 1995 An integrable hamiltonian system. *Phys. Lett. A* **201**, 306-310.

[8] CALOGERO, F. AND FRANCOISE, J.-P. 1996 An integrable hamiltonian system. *J. Math. Phys.* **37**, 2863-2871.

[9] CAMASSA, R. AND HOLM, D. D. 1993 An integrable shallow water equation with peaked solitons. *Phys. Rev. Lett.* **71**, 1661-64.

[10] CHEN, S. Y., FOIAS, C., HOLM, D. D., OLSON, E. J., TITI, E. S. AND WYNNE, S. 1998 The Camassa-Holm equations as a closure model for turbulent channel and pipe flows. *Phys. Rev. Lett.* **81**, 5338-5341;*Ibid* 1999a The Camassa-Holm equations and turbulence in pipes and channels. *Physica D* **133**, 49-65; *Ibid* 1999b A connection between the Camassa-Holm equations and turbulence in pipes and channels. *Phys. Fluids* **11**, 2343-2353.

[11] CHEN, S. Y., FOIAS, C., HOLM, D. D., MARGOLIN, L. G. AND ZHANG,

R. 1999 Direct numerical simulations of the Navier-Stokes alpha model. *Physica D* **133**, 66-83.

[12] EBIN, D. G. AND MARSDEN, J. E. 1970 Groups of diffeomorphisms and the motion of an incompressible fluid. *Ann.of Math.* **92**, 102-163.

[13] DULLIN, H., GOTTWALD, G. AND HOLM, D. D. 2001 An integrable shallow water equation with linear and nonlinear dispersion. *Phys. Rev. Lett.* **87** 194501-04.

[14] DULLIN, H., GOTTWALD, G. AND HOLM, D. D. 2003 Camassa-Holm, Korteweg-de Vries-5 and other asymptotically equivalent equations for shallow water waves. *Fluid Dyn. Res.* **33**, 7395.

[15] DULLIN, H., GOTTWALD, G. AND HOLM, D. D. 2004 On asymptotically equivalent shallow water wave equations. *Physica D* **190** 1-14.

[16] FOIAS, C., HOLM, D. D. AND TITI, E. S. 2001 The Navier-Stokes-alpha model of fluid turbulence. *Physica D* **152**, 505-519.

[17] FOIAS, C., HOLM, D. D. AND TITI, E. S. 2002 The three dimensional viscous Camassa-Holm equations, and their relation to the Navier-Stokes equations and turbulence theory. *J. Dyn. and Diff. Eqns.* **14**, 1-35.

[18] FOIAS, C., MANLEY, O., TEMAM, R. AND ROSA, R. 2001 *Navier-Stokes Equations and Turbulence* (Encyclopedia of Mathematics and Its Applications, Vol 83) Cambridge University Press: Cambridge.

[19] FRINGER, O. B. AND HOLM, D. D. 2001 Integrable vs nonintegrable geodesic soliton behavior. *Physica D* **150** 237-263.

[20] FRISCH, U. 1995 *Turbulence: The Legacy of A. N. Kolmogorov.* Cambridge University Press.

[21] GEURTS, B. J. AND HOLM, D. D. 2002a Alpha-modeling strategy for LES of turbulent mixing. In *Turbulent Flow Computation* (ed. D. Drikakis & B. G. Geurts). Kluwer: London.

[22] GEURTS, B. J. AND HOLM, D. D. 2002b Leray simulation of turbulent shear layers. In *Advances in Turbulence IX: Proceedings of the Ninth European Turbulence conference.* (Ed. J. P. Castro and P. E. Hancock) CIMNE:Barcelona, pp 337-340 (2002). ArXiv:nlin.CD/0202062.

[23] GEURTS, B. J. AND HOLM, D. D. 2003 Regularization modeling for large-eddy simulation. *Phys. Fluids* **15**, L13-L16.

[24] GJAJA, I. AND HOLM, D. D. 1996 Self-consistent wave-mean flow interaction dynamics and its Hamiltonian formulation for a rotating stratified incompressible fluid. *Physica D*, **98**, 343-378.

[25] HOLM, D. D. 1996 Hamiltonian balance equations. *Physica D*, **98** 379-414.

[26] HOLM, D. D. 1996 The ideal Craik-Leibovich equations. *Physica D*, **98** (1996) 415-441.

[27] HOLM, D. D. 1999 Fluctuation effects on 3D Lagrangian mean and Eulerian mean fluid motion. *Physica D* **133**, 215-269; *Ibid* 2002 Averaged Lagrangians and the mean dynamical effects of fluctuations in continuum mechanics. *Physica D* **170**, 253-286.

[28] HOLM, D. D. 2002 Euler-Poincaré dynamics of perfect complex fluids. In *Geometry, Mechanics, and Dynamics: in honor of the 60th birthday of Jerrold E. Marsden* edited by P. Newton, P. Holmes and A. Weinstein.

Springer, pp. 113-167.

[29] Holm, D. D. 2002 Variational principles for Lagrangian averaged fluid dynamics. *J. Phys. A: Math. Gen.* **35**, 1-10.

[30] Holm, D. D. 2002 Lagrangian averages, averaged Lagrangians, and the mean effects of fluctuations in fluid dynamics. *Chaos* **12**, 518-530.

[31] Holm, D. D. and Marsden, J. E. 2004 Momentum maps and measure valued solutions (peakons, filaments, and sheets) of the Euler-Poincare equations for the diffeomorphism group. To appear, *In The Breadth of Symplectic and Poisson Geometry*, (Marsden, J. E. and T. S. Ratiu, eds), Festshrift for Alan Weinstein (to appear). Birkhäuser Boston, ArXiv:nlin.CD/0312048.

[32] Holm, D. D., Marsden, J. E. and Ratiu, T. S. 1998 Euler-Poincaré models of ideal fluids with nonlinear dispersion. *Phys. Rev. Lett.* **80**, 4173-4177; *Ibid* The Euler-Poincaré equations and semidirect products with applications to continuum theories. *Adv. in Math.* **137**, 1-81.

[33] Holm, D. D., Marsden, J. E. and Ratiu, T. S. 2002 The Euler-Poincaré equations in geophysical fluid dynamics. In *Large-Scale Atmosphere-Ocean Dynamics 2: Geometric Methods and Models*. Edited by J. Norbury and I. Roulstone, Cambridge University Press: Cambridge, pp. 251-299.

[34] Holm, D. D., Marsden, J. E., Ratiu, T. S. and Weinstein, A. 1985 Nonlinear stability of fluid and plasma equilibria. *Physics Reports* **123**, 1-116.

[35] Holm, D. D., Munn, J. and Stechmann, S. N. 2004 Reduced singular solutions of EPDiff equations on manifolds with symmetry. Submitted to *Nonlinearity*, ArXiv:nlin.CD/0402044.

[36] Holm, D. D., Putkaradze, V. and Stechmann, S. N. 2004 Rotating concentric peakons. Submitted to *Nonlinearity*, ArXiv:nlin.SI/0312012.

[37] Holm, D. D., Rananather, J. T., Younes, L. and Trouvé, A. 2004 Soliton dynamics in computational anatomy. *NeuroImage*, to appear.

[38] Holm, D. D. and Staley, M. F. 2002 Wave structures and nonlinear Balances in a family of 1+1 evolutionary PDEs. ArXiv:nlin.CD/0202059. Submitted to *SIADS*.

[39] Holm, D. D. and Staley, M. F. 2004 Interaction dynamics of singular wave fronts in nonlinear evolutionary fluid equations. Submitted to *SIAM J. Appl. Dyn. Syst.*

[40] Hunter, J. K. and Zheng, Y. 1994 On a completely integrable nonlinear hyperbolic variational equation. *Physica D* **79**, 361-386.

[41] Kolmogorov, A. N. 1941 Dissipation of energy in locally isotropic turbulence. *Dokl. Akad. Nauk SSSR* **32**, 16-18 (reprinted in *Proc. Roy. Soc. London A* **434** (1991) 15-17).

[42] Kruse, H. P., Schreule, J. and Du, W. 2001 A two-dimensional version of the CH equation. In *Symmetry and Perturbation Theory: SPT 2001* Edited by D. Bambusi, G. Gaeta and M. Cadoni. World Scientific: New York, pp 120-127.

[43] Landau, L. D. and Lifschitz, E. M. 1987 *Fluid Mechanics.* Pergamon Press: New York, 2nd Edition.

[44] Leray, J. 1934 Sur le mouvement d'un liquide visqueux emplissant

l'espace. *Acta Math.* **63**, 193-248 Reviewed, e.g., in CONSTANTIN, P., FOIAS, C., NICOLAENKO, B. AND TEMAM, R. *Integral Manifolds and Inertial Manifolds for Dissipative Partial Differential Equations.* Applied Mathematical Sciences, **70**, (Springer-Verlag, New York-Berlin, 1989).

[45] MARSDEN, J. E. AND RATIU, T. S. 1999 *Intro Mechanics and Symmetry* Springer.

[46] MARSDEN, J. E. AND SHKOLLER, S. 2001 Global well-posedness for the Lagrangian averaged Navier-Stokes (LANS$-\alpha$) equations on bounded domains. *Proc. Roy. Soc. London A* **359**, 1449-1468.

[47] MARSDEN, J. E. AND SHKOLLER, S. 2002 The anisotropic Lagrangian-averaged Euler and Navier-Stokes equations. *Arch. Rat. Mech. Anal.* **166**, 27-46.

[48] MILLER, M. I. AND YOUNES, L. 2001 Group actions, homeomorphisms, and matching: a general framework. *International Journal of Computer Vision* **41**, 61-84.

[49] MOHSENI, K., SHKOLLER, S., KOSOVIĆ, B., MARSDEN, J. E., CARATI, D., WRAY, A. AND ROGALLO, R. 2000 Numerical simulations of homogeneous turbulence using the Lagrangian averaged Navier-Stokes equations. Center for Turbulence Research, Proceedings of the Summer Program, 2000, 271-283.

[50] MOHSENI, K., KOSOVIĆ, B., SHKOLLER, S. AND MARSDEN, J. E. 2002 Numerical simulations of the Lagrangian averaged Navier-Stokes (*LANS*-α) equations for homogeneous isotropic turbulence. *Phys. Fluids* **15**, 524-544.

[51] NORBURY, J. S. AND ROULSTONE, I. 2002 Editors, *Large-Scale Atmosphere-Ocean Dynamics, Vol. 1: Analytical Methods and Numerical Models* and *Vol. 2: Geometric Methods and Models.* Cambridge University Press: Cambridge.

[52] POINCARÉ, H. 1901 Sur une forme nouvelle des équations de la méchanique *C.R. Acad. Sci.* **132**, 369-371.

[53] SHKOLLER, S. 1998 Geometry and curvature of diffeomorphism groups with H^1 metric and mean hydrodynamics *J. Funct. Anal.,* **160**, 337-365.

[54] SHKOLLER, S. 2000 Analysis on groups of diffeomorphisms of manifolds with boundary and the averaged motion of a fluid *J. Differential Geometry,* **55**, 145-191.

IV

No Polar Coordinates

Konstantinos Efstathiou and Dmitrií Sadovskií

Based on lectures by Richard Cushman

From the lecturer

R.C. would like to thank the organizer Dr. James Montaldi of the Mechanics and Symmetry Euro Summer School which was held in Peyresq, France in September 2–16, 2000 for inviting him to give several lectures on integrable Hamiltonian systems.[1]

R.C. declines to take any credit for the contents of these notes even though he helped write and typeset them, contributed his scratchy notes to the preparers, and wrote up his homework (sec. D).

The principal printed source which should be read along with these notes is the "blue book" [5].

Contents

[1] This work was partially supported by the EU project MASIE contract HPCFCT-1999-00067.

Avant propos

It is a great pleasure for me to introduce these lecture notes. In the last few years after our first meeting in 1997 I have been constantly learning from Richard Cushman and am glad to be one of his co-workers. What I appreciate the most in Richard's lectures and in his work is that he presents and studies modern mathematics based on examples of concrete dynamical systems which he considers in great detail. As such his approach is very accessible to physicists and practitioners.

The five lectures are presented in the way they were given, except for interchanging Lecture IV and V back to their intended logical order. Some comments and discussion are added separately at the end of each lecture. The idea of these lecture notes is to give a very informal introduction to Richard's work. We even tried to preserve some of the style of Richard's presentation peppered with such phrases as "don't be caught dead in the water", "you will eat crow", "sneaky gadget", "dirty trick", "this is deep", etc. We feel that this personalization is inseparable from the imprint his lectures left in our heads. We therefore attempted to take full advantage of the lecture note format, especially since a more conventional and detailed presentation of the subject can be found in [5] (known to connoisseurs as "the blue book").

The title of these lectures originates from a comment of Hans Duistermaat on the blue book [5]. He remarked that Lagrange was proud to have stated in the front matter to *Mécanique Analytique* (Paris, 1788) that his book had no pictures. Hans suggested that in the dedication to the blue book should appear the statement: "This book uses no polar coordinates".

Initially the number of volunteers to write up these notes was larger and each could choose which lecture was closest to his work and interests. In the end, Konstantinos worked on the harmonic oscillator and I was left with practically everything else. Fortunately, Richard himself came to my rescue. He wrote up the Euler top (I should confess that he used this as an occasion to further polish his presentation which deviates a bit from the original lecture) and an appendix which contains his solution of his "homework" problem. I concentrated on the spherical pendulum (one of Richard's favorites) and on the editorial work. Richard Morrison volunteered as a technical editor and I like to end by citing him :

> I have to confess that my understanding of Cushman's lectures was not maximal, and indeed I volunteered for this project to motivate my working through his notes at a more steady pace than had I been doing it purely for fun. ... given the speed that Cushman lectures and my unfamiliarity with the mathematics at the time, I don't really have what could be described as a complete set of accurate figures in my notes from the lectures. On the bright side, my (somewhat nominal) involvement with this project has meant I have been motivated to work a little on the notes and increase my understanding (no doubt they would have been relegated to gathering dust along with many other things that I have little time for academically).

It seems to me that much of this applies equally to the rest of us.

D. Sadovskiĭ, Boulogne-sur-Mer

1 Lectures I and II. The two-dimensional harmonic oscillator

We will be dealing with 'simple' integrable systems: 2D harmonic oscillator, Euler top, and spherical pendulum. People often ask why Cushman still works on these examples. He replies: because they are tricky.

1.1 The harmonic oscillator

1.1.1 Preliminaries

Configuration space. The configuration space of the two-dimensional harmonic oscillator is $\mathbf{R}^2$ with coordinates $x = (x_1, x_2)$.

Phase space. The phase space is $T^*\mathbf{R}^2 \cong \mathbf{R}^4$ with coordinates $(x, y) = (x_1, x_2, y_1, y_2)$.

Canonical 1-form. On $T^*\mathbf{R}^2$ the canonical 1-form is

$$\Theta = y_1 \, \mathrm{d}x_1 + y_2 \, \mathrm{d}x_2 = \langle y, \mathrm{d}x \rangle.$$

Symplectic form. The symplectic form on $T^*\mathbf{R}^2$ is the closed nondegenerate 2-form

$$\omega = -\,\mathrm{d}\Theta = \mathrm{d}x_1 \wedge \mathrm{d}y_1 + \mathrm{d}x_2 \wedge \mathrm{d}y_2 \tag{1.1}$$

with the matrix representation

$$\omega = \begin{pmatrix} \mathrm{d}x \\ \mathrm{d}y \end{pmatrix}^t \begin{pmatrix} 0 & I_2 \\ -I_2 & 0 \end{pmatrix} \begin{pmatrix} \mathrm{d}x \\ \mathrm{d}y \end{pmatrix}. \tag{1.2}$$

Hamiltonian function. The Hamiltonian function of the two dimensional harmonic oscillator is

$$H : T^*\mathbf{R}^2 \to \mathbf{R} : (x, y) \mapsto \tfrac{1}{2}\,(x_1^2 + y_1^2) + \tfrac{1}{2}\,(x_2^2 + y_2^2). \tag{1.3}$$

Vector field. The corresponding Hamiltonian vector field

$$X_H = \langle X_1, \frac{\partial}{\partial x} \rangle + \langle X_2, \frac{\partial}{\partial y} \rangle$$

is computed using $X_H \,\lrcorner\, \omega = \mathrm{d}H = -X_2 \, \mathrm{d}x + X_1 \, \mathrm{d}y$. We get

$$X_1 = \frac{\partial H}{\partial y} \quad \text{and} \quad X_2 = -\frac{\partial H}{\partial x}. \tag{1.4}$$

Therefore the equations of motion of the harmonic oscillator are

$$\dot{x} = y \quad \text{and} \quad \dot{y} = -x. \tag{1.5}$$

Flow. The solution to the above equations is the one parameter family of transformations

$$\phi_t^H(x,y) = A(t) = \begin{pmatrix} (\cos t)\, I_2 & -(\sin t)\, I_2 \\ (\sin t)\, I_2 & (\cos t)\, I_2 \end{pmatrix} \begin{pmatrix} x \\ y \end{pmatrix} \tag{1.6}$$

This defines an S^1 action on $T^*\mathbf{R}^2$, which is a map from $\mathbf{R}$ to $\mathrm{Sp}(4,\mathbf{R})$ that sends t to the 4×4 symplectic matrix $A(t)$, which is periodic of period 2π.

As Poincaré liked to say, formulae are not the answer. This means that we have not solved our problem yet.

Symplectic group. A real $2n \times 2n$ matrix is symplectic if it satisfies the relation

$$A^t J A = J. \tag{1.7}$$

These matrices form a Lie group denoted by $\mathrm{Sp}(2n,\mathbf{R})$. The Lie algebra $\mathfrak{sp}(2n,\mathbf{R})$ of the symplectic group $\mathrm{Sp}(2n,\mathbf{R})$ consists of the *Hamiltonian* matrices X that satisfy $X^t J + JX = 0$.

Conservation of energy. We calculate

$$L_{X_H} H = \langle y, \frac{\partial H}{\partial x}\rangle - \langle x, \frac{\partial H}{\partial y}\rangle = \langle y, x\rangle - \langle x, y\rangle = 0 \tag{1.8}$$

This shows that H is constant along the integral curves of X_H.

Invariant manifold. Therefore the manifold

$$H^{-1}(h) = \{(x,y) \in \mathbf{R}^4 \big| x^2 + y^2 = 2h, h > 0\} \cong \mathbf{S}^3_{\sqrt{2h}} \tag{1.9}$$

is invariant under the flow of X_H.

1.1.2 S^1 symmetry.

S^1 action on $\mathbf{R}^2$. The configuration space $\mathbf{R}^2$ is invariant under the S^1 action

$$S^1 \times \mathbf{R}^2 \to \mathbf{R}^2 : (t,x) \mapsto R_t x,$$

where R_t is the matrix $\begin{pmatrix} \cos t & -\sin t \\ \sin t & \cos t \end{pmatrix}$.

Lift. This map lifts to a symplectic action Φ_t of S^1 on phase space $T^*\mathbf{R}^2$ that sends (x,y) to $\Phi_t(x,y) = (R_t x, R_t y)$.

Generator. This infinitesimal generator of this action is

$$Y(x,y) = \frac{d}{dt}\bigg|_{t=0} \Phi_t(x,y) = (-x_2, x_1, -y_2, y_1). \tag{1.10}$$

Conservation of angular momentum. The vector field Y is Hamiltonian corresponding to the Hamiltonian function

$$L(x,y) = \langle y, (x_2, -x_1)\rangle = x_1 y_2 - x_2 y_1,$$

that is, $Y = X_L$. L is readily recognized as the angular momentum. The hamiltonian of the harmonic oscillator is an integral of X_L.

Check. Since $H(\Phi_t(x,y)) = H(x,y)$, we find that

$$0 = L_{X_L} H = \{H, L\} = -\{L, H\} = -L_{X_H} L. \tag{1.11}$$

Invariant manifold. Since both H and L are integrals of X_H, the manifold

$$\mathcal{M}_{h,\ell} = H^{-1}(h) \cap L^{-1}(\ell) \tag{1.12}$$

is invariant.

What is it? I claim that in most cases it is a 2-torus. How do we derive this?

We have to diagonalize the flow.

1.1.3 The geometry of $\mathcal{M}_{h,\ell}$.

Diagonalize L. In order to understand the geometry of the set $\mathcal{M}_{h,\ell}$, we want to find a transformation in $\mathrm{Sp}(4,\mathbf{R}) \cap \mathrm{O}(4,\mathbf{R})$ that diagonalizes L. Such a transformation P is given by

$$\begin{pmatrix} x \\ y \end{pmatrix} = \begin{pmatrix} A & -B \\ B & A \end{pmatrix} \begin{pmatrix} \xi \\ \eta \end{pmatrix}, \tag{1.13}$$

where $A = \left(\begin{smallmatrix} 0 & 0 \\ 1 & -1 \end{smallmatrix}\right)$ and $B = \left(\begin{smallmatrix} -1 & -1 \\ 0 & 0 \end{smallmatrix}\right)$.

Transform. In the new coordinates (ξ, η) the functions H and L become

$$\widetilde{H}(\xi,\eta) = (H \circ P)(\xi,\eta) = \tfrac{1}{2}\left(\eta_1^2 + \xi_1^2 + \eta_2^2 + \xi_2^2\right) \tag{1.14}$$

and

$$\widetilde{L}(\xi,\eta) = (L \circ P)(\xi,\eta) = \tfrac{1}{2}\left(-\eta_1^2 - \xi_1^2 + \eta_2^2 + \xi_2^2\right), \tag{1.15}$$

respectively.

Level sets. We have

$$\mathcal{M}_{h,\ell} = H^{-1}(h) \cap L^{-1}(\ell) = \widetilde{H}^{-1}(h) \cap \widetilde{L}^{-1}(\ell). \qquad (1.16)$$

The level set $\mathcal{M}_{h,\ell}$ is determined by

$$\begin{aligned} \eta_2^2 + \xi_2^2 &= h + \ell \\ \eta_1^2 + \xi_1^2 &= h - \ell. \end{aligned}$$

Therefore

$$\mathcal{M}_{h,\ell} = \begin{cases} \emptyset, & \text{if } |\ell| > h \\ 0, & \text{if } h = \ell = 0 \\ \mathbf{S}^1, & \text{if } |\ell| = h,\ h > 0 \\ \mathbf{T}^2, & \text{if } |\ell| < h. \end{cases} \qquad (1.17)$$

Energy-momentum mapping. Define

$$\mathcal{EM} : \mathbf{R}^4 \to \mathbf{R}^2 : (x, y) \mapsto (H(x,y), L(x,y)). \qquad (1.18)$$

Obviously $\mathcal{EM}^{-1}(h,\ell) = \mathcal{M}_{h,\ell}$.

Bifurcation diagram. We summarize the preceding discussion of the level sets of the energy-momentum mapping in the *bifurcation diagram* that shows the change of topological type of $\mathcal{EM}^{-1}(h,\ell)$ as (h,ℓ) changes, see figure 1.1.

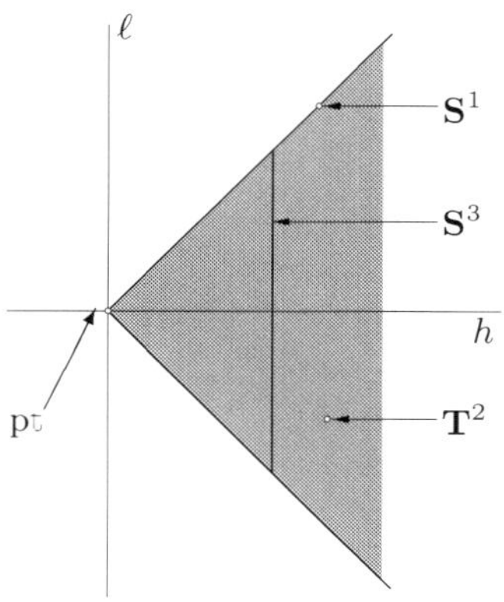

Fig. 1.1. The bifurcation diagram.

Regular values of $\mathcal{EM}$. $\mathcal{R} = \{(h,\ell) \in \mathbf{R}^2 \big| |\ell| < h, h > 0\}$ is the set of regular values of the energy-momentum map. For $(h,\ell) \in \mathcal{R}$ the fibers $\mathcal{EM}^{-1}(h,\ell)$ are 2-tori. This is a consequence of the *Arnol'd-Liouville* theorem, since $\mathrm{d}H$ and $\mathrm{d}L$ are linearly independent on $\mathcal{EM}^{-1}(\mathcal{R})$, X_H and X_L are complete, and $\mathcal{EM}^{-1}(\mathcal{R})$ is an open dense subset of $T^*\mathbf{R}^2$.

The Arnol'd-Liouville theorem. *(A very powerful result, Avez and others are somewhere in here too.)* We consider a symplectic manifold (M^{2n}, ω) and a Hamiltonian function $H : M^{2n} \to \mathbf{R}$. Consider a collection of n functions $F_1 = H, F_2, \ldots, F_n$ such that

1. $F_1, \ldots, F_n$ are integrals of X_H and the corresponding vector fields X_{F_i} have flows which are defined for all time.
2. $\{F_i, F_j\} = 0$ for all i, j.
3. $dF_1 \wedge \cdots \wedge dF_n \neq 0$ on an open dense subset of M^{2n}.

Define the momentum map $\mathcal{EM} : M^{2n} \to \mathbf{R}^n : x \to (F_1(x), \ldots, F_n(x))$. If

4. the set of regular values $\mathcal{R}$ of $\mathcal{EM}$ is a nonempty open subset of $\mathbf{R}^n$, and
5. for $c \in \mathcal{R}$, the set $\mathcal{EM}^{-1}(c)$ is compact[1] and connected,

then $\mathcal{EM}^{-1}(c)$ is an n-torus.

The Arnol'd-Liouville theorem is boring, because it tells us everything there is to know about connected components of fibers of the energy momentum mapping corresponding to regular values. But what about the singular values? Knowing all individual fibers is not going to finish the problem either. We should also understand how these fibers fit together.

At this point we know the topological type of each fiber of the energy-momentum mapping of the harmonic oscillator, but we can not say anything about the way that $\mathbf{S}^3$ is made up from 2 circles and a bunch of 2-tori. This is the question that we study next.

1.2 U(2) momentum map

Quadratic integrals. We now find all the quadratic integrals of X_H. Any quadratic function on $\mathbf{R}^4$ can be expressed as

$$F(x,y) = \tfrac{1}{2} \begin{pmatrix} x \\ y \end{pmatrix}^t \begin{pmatrix} -B & A^t \\ A & C \end{pmatrix} \begin{pmatrix} x \\ y \end{pmatrix} \tag{1.19}$$

[1] Compactness is needed to make sure that near a given torus we should find other tori on which the motion is also quasi-periodic.

where A, B and C are 2×2 matrices with $B = B^t$ and $C = C^t$.

Hamiltonian vector field. The corresponding hamiltonian vector field is

$$X_F(x, y) = \begin{pmatrix} A & B \\ C & -A^t \end{pmatrix} \begin{pmatrix} x \\ y \end{pmatrix} \tag{1.20}$$

where $X_F \in \mathfrak{sp}(4, \mathbf{R})$.

Statement. For any two quadratic functions F and H

$$0 = L_{X_H} F = \{F, H\} \Leftrightarrow [X_H, X_F] = 0. \tag{1.21}$$

Application. When $X_H = \begin{pmatrix} 0 & -I \\ I & 0 \end{pmatrix}$, $X_F = \begin{pmatrix} A & C \\ B & -A^t \end{pmatrix}$ and $0 = [X_H, X_F]$, then

$$X_F = \begin{pmatrix} A & -B \\ B & A \end{pmatrix}, \tag{1.22}$$

where A, B are 2×2 real matrices such that $A = -A^t$ and $B = B^t$.

The Lie algebra $\mathfrak{u}(2)$. By definition

$$\mathfrak{u}(2) = \{\mathcal{A} \in \mathrm{gl}(2, \mathbf{C}) \,|\, \bar{\mathcal{A}}^t + \mathcal{A} = 0\} \tag{1.23}$$

Setting $\mathcal{A} = A + \mathrm{i}B$ we see that the set of solutions (1.22) is isomorphic to $\mathfrak{u}(2)$ (the Lie algebra of U(2)).

Hamiltonian. Consider the linear vector field $X_v(z) = v(z)$ with $v \in \mathfrak{sp}(4, \mathbf{R})$, then X_v is hamiltonian with hamiltonian function

$$F_v(z) = \tfrac{1}{2}\, \omega(v(z), z) \tag{1.24}$$

Therefore if v is of the form (1.22), then F_v is an integral of X_H. Let Q be the set of quadratic integrals of X_H.

A basis for Q. Since Q is isomorphic to $\mathfrak{u}(2)$ we can find a basis for Q by taking a basis of $\mathfrak{u}(2)$ and then tranforming it. We select the following basis for $\mathfrak{u}(2)$:

$$\epsilon_1 = \begin{pmatrix} 0 & \mathrm{i} \\ \mathrm{i} & 0 \end{pmatrix} \quad \epsilon_2 = \begin{pmatrix} 0 & -1 \\ 1 & 0 \end{pmatrix} \quad \epsilon_3 = \begin{pmatrix} \mathrm{i} & 0 \\ 0 & -\mathrm{i} \end{pmatrix} \quad \epsilon_4 = \begin{pmatrix} \mathrm{i} & 0 \\ 0 & \mathrm{i} \end{pmatrix}. \tag{1.25}$$

and then take the corresponding Hamiltonian matrices in $\mathfrak{sp}(4,\mathbf{R})$ and their Hamiltonian functions. This way we get the basis

$$\begin{aligned} w_1(x,y) &= x_1x_2 + y_1y_2 \\ w_2(x,y) &= x_1y_2 - x_2y_1 \\ w_3(x,y) &= 1/2(x_1^2 + y_1^2 - x_2^2 - y_2^2) \\ w_4(x,y) &= 1/2(x_1^2 + x_2^2 + y_1^2 + y_2^2), \end{aligned} \tag{1.26}$$

for Q. We see that Q with the usual Poisson bracket is a Lie algebra isomorphic to $\mathfrak{u}(2)$. The commutation relations between the four basis functions w_i are given in the following table

$\{w_i, w_j\}$	w_1	w_2	w_3	w_4
w_1	0	$2w_3$	$-2w_1$	0
w_2	$-2w_3$	0	$2w_2$	0
w_3	$2w_1$	$-2w_2$	0	0
w_4	0	0	0	0

The Lie group U(2). By definition

$$\begin{aligned} \mathrm{U}(2) &= \{u \in \mathrm{GL}(2,\mathbf{C}) \,|\, \bar{u}^t u = I\} \\ &= \{\left(\begin{smallmatrix} a & -b \\ b & a \end{smallmatrix}\right) \in \mathrm{GL}(4,\mathbf{R}) \,|\, a^t a + b^t b = I,\ a^t b = b^t a, a, b \in \mathrm{GL}(2,\mathbf{R})\} \\ &\cong \mathrm{Sp}(4,\mathbf{R}) \cap \mathrm{O}(4,\mathbf{R}) \end{aligned} \tag{1.27}$$

Consider the linear action $\Phi : \mathrm{U}(2) \times \mathbf{R}^4 \to \mathbf{R}^4 : (u, z = (x,y)) \mapsto u(z)$. Φ_u is a linear symplectic map on $(\mathbf{R}^4, \omega)$.

Flow. If $u \in \mathfrak{u}(2)$, then $\Phi_{\exp tu}$ is the flow of the linear hamiltonian vector field

$$X^u(z) = \left.\frac{d}{dt}\right|_{t=0} \Phi_{\exp tu} z = u(z). \tag{1.28}$$

Associated to X^u is the hamiltonian function

$$J^u(z) = \tfrac{1}{2}\, \omega(u(z), z). \tag{1.29}$$

The function J^u depends *linearly* on u.

Momentum map. We define the U(2) momentum map $J : \mathbf{R}^4 \rightarrow \mathfrak{u}(2)^*$ of the U(2) action Φ to be $J(x,y)(u) = J^u(x,y)$. If E_j^* is the dual basis of $\mathfrak{u}(2)^*$ then

$$J(x,y) = \sum_j w_j(x,y) E_j^*. \tag{1.30}$$

J intertwines the linear action of U(2) on $\mathbf{R}^4$ with the coadjoint action of U(2) on $\mathfrak{u}(2)^*$, that is,

$$J(Uz) = \mathrm{Ad}^t_{U^{-1}} J(z). \tag{1.31}$$

Some people call the coadjoint action the dual action. Actually it is the contragradient action.

Check.

$$\begin{aligned} J(Uz)u &= J^u(Uz) = \tfrac{1}{2}\,\omega(u(Uz), Uz) \\ &= \omega(U^{-1}uUz, z) = J^{U^{-1}uU}(z) = J(z)(U^{-1}uU) \\ &= J(z)(\mathrm{Ad}_{U^{-1}}u) = \mathrm{Ad}^t_{U^{-1}}(J(z)u). \qquad \square \end{aligned}$$

In the original problem we saw only S^1 symmetry which acted on the configuration space. Now we found a larger symmetry which acts on phase space.

Killing form. By definition the Killing form on $\mathfrak{u}(2)$ is

$$\mathrm{k} : \mathfrak{u}(2) \times \mathfrak{u}(2) \rightarrow \mathbf{C} : (u,v) \mapsto -\tfrac{1}{2}\,\mathrm{trace}(u\bar{v}^t). \tag{1.32}$$

Let $\widetilde{J} : \mathbf{R}^4 \rightarrow \mathfrak{u}(2) : z \mapsto \mathrm{k}^\flat \circ J(z)$. Then

$$\widetilde{J}(z) = (w_1(z), w_2(z), w_3(z), w_4(z))$$

because the ϵ_j's form an orthonormal basis with respect to k.

1.3 Hopf fibration

Hopf map. Define the *Hopf map* (formerly known as $\widetilde{J}$) by

$$\mathcal{H} : \mathbf{R}^4 \rightarrow \mathbf{R}^4 : z = (x,y) \mapsto (w_1(z), w_2(z), w_3(z), w_4(z)). \tag{1.33}$$

Obviously

$$w_1^2 + w_2^2 + w_3^2 = w_4^2. \tag{1.34}$$

Hopf fibration. Restricting $\mathcal{H}$ to the sphere

$$\mathbf{S}^3_{\sqrt{2h}} = \{(x,y) \in \mathbf{R}^4 \big| x_1^2 + x_2^2 + y_1^2 + y_2^2 = 2h\} \tag{1.35}$$

we get the *Hopf fibration*

$$\mathcal{F} : \mathbf{S}^3_{\sqrt{2h}} \to \mathbf{S}^2_h : z = (x,y) \mapsto w = (w_1(z), w_2(z), w_3(z)) \tag{1.36}$$

where

$$\mathbf{S}^2_h = \{(w_1, w_2, w_3) \in \mathbf{R}^3 \big| w_1^2 + w_2^2 + w_3^2 = h^2\}. \tag{1.37}$$

1.3.1 Properties of the Hopf fibration

Property 1. Let $w \in \mathbf{S}^2_h$. Then $\mathcal{F}^{-1}(w)$ is a great circle on $\mathbf{S}^3_{\sqrt{2h}}$.

Proof.
CASE 1. $w \in \mathbf{S}^2_h - \{(0,0,-h)\}$. Suppose $(x,y) \in \mathcal{F}^{-1}(w)$. Since $x_1^2 + y_1^2 + x_2^2 + y_2^2 = 2h$ and $x_1^2 + y_1^2 - x_2^2 - y_2^2 = 2w_3$ it follows that $x_1^2 + y_1^2 = h + w_3 > 0$. Therefore we may solve the linear equations

$$\begin{pmatrix} x_1 & y_1 \\ -y_1 & x_1 \end{pmatrix} \begin{pmatrix} x_2 \\ y_2 \end{pmatrix} = \begin{pmatrix} w_1 \\ w_2 \end{pmatrix} \tag{1.38}$$

to obtain

$$\begin{aligned} w_1 x_1 - w_2 y_1 + (h + w_3) x_2 &= 0 \\ w_2 x_1 + w_1 y_1 + (h + w_3) y_2 &= 0. \end{aligned}$$

The above equations define a 2-plane Π^w in $\mathbf{R}^4$, since $\left(\begin{smallmatrix} w_1 & -w_2 & h+w_3 & 0 \\ w_2 & w_1 & 0 & h+w_3 \end{smallmatrix}\right)$ has rank 2. Hence $\mathcal{F}^{-1}(w) \subseteq \Pi^w \cap \mathbf{S}^3_{\sqrt{2h}}$. Reversing the argument shows that $\Pi^w \cap \mathbf{S}^3_{\sqrt{2h}} \subseteq \mathcal{F}^{-1}(w)$.
CASE 2. $w = (0,0,-h)$. Then $x_1^2 + y_1^2 = 0$ which implies $x_1 = y_1 = 0$. Thus

$$\mathcal{F}^{-1}(w) = \{(0, x_2, 0, y_2) \in \mathbf{R}^4 \big| x_2^2 + y_2^2 = 2h\}, \tag{1.39}$$

which is a great circle and it is $\mathbf{S}^3_{\sqrt{2h}} \cap \{x_1 = y_1 = 0\}$. □

Consequence 1. Each fiber of the Hopf fibration is a single orbit of the harmonic oscillator of energy h. In other words, the orbit space $H^{-1}(h)/S^1$ of the harmonic oscillator of energy h is $\mathbf{S}^2_h$.

Property 2. Let $w, v \in \mathbf{S}^2_h$ with $w \neq v$. Then $\mathcal{F}^{-1}(w)$ and $\mathcal{F}^{-1}(v)$ are linked once in $\mathbf{S}^3_{\sqrt{2h}}$.

Proof. Since $v \neq w$, the 2-planes Π^v and Π^w are transverse, that is, $\Pi^v \cap \Pi^w = \{0\}$. Let Π be any 3-plane containing Π^w. Then $\Pi^v \not\subseteq \Pi$, so $\Pi^v \cap \Pi = \ell^v$ which is a line through the origin. $\Pi \cap \mathbf{S}^3_{\sqrt{2h}}$ is a great 2-sphere $\mathbf{S}^2_{\sqrt{2h}}$ with equator $\Pi^w \cap \mathbf{S}^3_{\sqrt{2h}}$. Let H^+ be an open hemisphere of $\mathbf{S}^2_{\sqrt{2h}}$ whose closure has boundary $\Pi^w \cap \mathbf{S}^3_{\sqrt{2h}}$. Since $\ell^v \cap \Pi^w = \{0\}$, ℓ^v intersects H^+ at one point p. Hence the great circle $\Pi^v \cap \mathbf{S}^3_{\sqrt{2h}}$ intersects H^+ only at p. Thus the fibers $\mathcal{F}^{-1}(v)$ and $\mathcal{F}^{-1}(w)$ are linked once in $\mathbf{S}^3_{\sqrt{2h}}$. □

Consequence 2. There is *no global* Poincaré section for the flow of X_H on $H^{-1}(h)$.

Proof. Suppose that a 2-disc $\mathbf{D}^2 \subseteq \mathbf{S}^3_{\sqrt{2h}}$ is a global cross section. Since every orbit of X_H on $\mathbf{S}^3_{\sqrt{2h}}$ is a circle, it would follow that $\mathbf{S}^3_{\sqrt{2h}}$ is homeomorphic to $\mathbf{D}^2 \times \mathbf{S}^1$. But two distinct orbits of X_H, are two distinct fibers of the Hopf fibration. Therefore they are linked in $\mathbf{S}^3_{\sqrt{2h}}$ but they would be unlinked in $\mathbf{D}^2 \times \mathbf{S}^1$. This is impossible if these sets are topologically the same. The same argument works for any topological 2-manifold in $\mathbf{S}^3_{\sqrt{2h}}$. □

Consequence 3. The orbit space $H^{-1}(h)/S^1$ is *not* a submanifold of $H^{-1}(h)$.

Proof. See the last sentence of the proof of consequence 2. □

Consequence 4. We need at least two local Poincaré sections. (In fact, two are enough. We will see that the orbit space is a 2-sphere. Any $\mathbf{S}^2$ requires two charts.)

Visualization. We visualize $\mathbf{S}^3$ using stereographic projection. In figure 1.2 we have drawn the level sets of w_1 (the angular momentum). Each level set is a 2-torus. The two critical points of the energy-momentum map correspond to the two thick black curves in figure 1.2, given by $(0, 0, t)$ and $(\cos t, \sin t, 0)$. Both curves are circles, thinking of $\mathbf{S}^3$ as $\mathbf{R}^3$ together with a point at infinity.

Proof of consequence 4. Select any level set ℓ of w_1 with $|\ell| < h$ and consider the two open disks A and B in figure 1.3. Each open disk is a local Poincaré section, since any orbit that begins on one of the disks crosses the same disk again. □

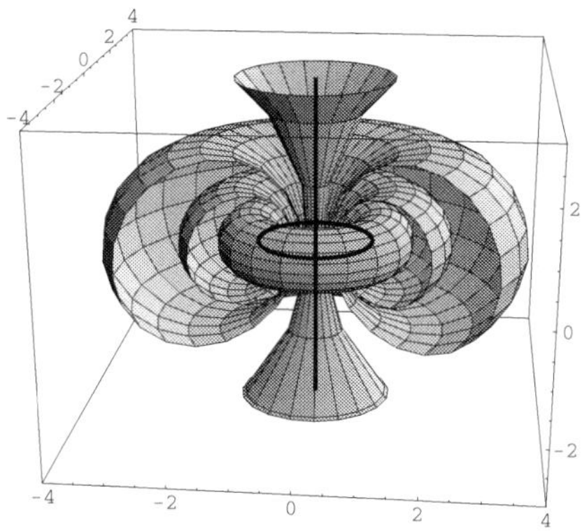

Fig. 1.2. Visualization of $\mathbf{S}^3$

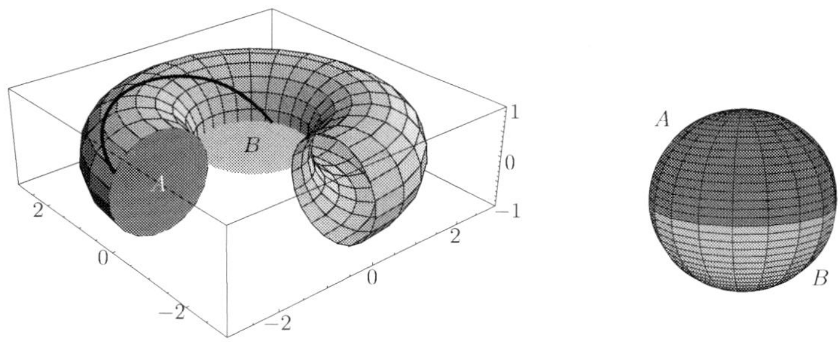

Fig. 1.3. Poincaré disks used to construct the orbit space (left). The orbit space $H^{-1}(h)/S^1 = \mathbf{S}^2_h$ (right).

The orbit space. *(For all except Boris Zhilinskií: an orbifold is an orbit space of a locally free action.)* Our local Poincaré sections (figure 1.3, left) are charts of the orbit space. every orbit intersects at least once one of the two disks. We glue them together and obtain a 2-sphere. Indeed, as shown in figure 1.3, left, an orbit that begins on a point $q \in \partial A$ will cross ∂B at a point p exactly once before returning to its initial point. Identifying q and p gives a 2-sphere, which is the orbit space $H^{-1}(h)/S^1$, see figure 1.3. The orbit space $\mathbf{S}^2$ is *not* sitting in the energy level $\mathbf{S}^3$ (shown in figure 1.2); thus $\mathbf{S}^2$ is an *abstract* manifold (as Hopf showed in 1935). Indeed, every orbit intersects the orbit space $\mathbf{S}^2$ transversally. So if our $\mathbf{S}^2$ *were* a submanifold of $\mathbf{S}^3$ then $\mathbf{S}^3$ would decompose as $\mathbf{S}^1 \times \mathbf{S}^2$. But it does not. Thus $\mathbf{S}^3$ is a nontrivial $\mathbf{S}^1$ bundle over $\mathbf{S}^2$.

We have completed *regular reduction* *(= Marsden-Weinstein reduction).*

1.4 Normalization

1.4.1 Dynamics on the orbit space

Harmonic oscillator symmetry. Consider a map $K : T^*\mathbf{R}^2 \to \mathbf{R}$ that factors through the $\mathfrak{u}(2)$ momentum map $\widetilde{\mathcal{J}} : T^*\mathbf{R}^2 \to \mathbf{R}^4$, that is, there is a smooth function $\widetilde{K} : \mathbf{R}^4 \to \mathbf{R}$ such that $K(x,y) = \widetilde{\mathcal{J}}^*\widetilde{K}(x,y)$. In other words

$$K(x,y) = \widetilde{K}(w_1(x,y), w_2(x,y), w_3(x,y), w_4(x,y)) \tag{1.40}$$

Integral. K is an integral of X_H, that is, $L_{X_H}K = 0$.

Induced equations of motion on $\mathbf{R}^4$. Since $\{w_j, w_4\} = 0$ for $j = 1, \ldots, 4$, we obtain

$$\begin{aligned} \dot{w}_j &= \{w_j, \widetilde{K}\} = \sum_{k=1}^{3} \{w_j, w_k\} \frac{\partial \widetilde{K}}{\partial w_k} \\ &= \sum_{k=1}^{3}\sum_{l=1}^{3} 2\varepsilon_{jkl} w_l \frac{\partial \widetilde{K}}{\partial w_k} = 2(\nabla \widetilde{K} \times w)_j \end{aligned}$$

for $j = 1, 2, 3$ and $\dot{w}_4 = 0$.

Restrict to $H^{-1}(h)$. Restricting $\widetilde{K}$ to $H^{-1}(h)$ gives

$$\widetilde{K}_h(w_1, w_2, w_3) = \widetilde{K}(w_1, w_2, w_3, h). \tag{1.41}$$

Induced equations of motion on $\mathbf{R}^3$. Set $w = (w_1, w_2, w_3)$. Then

$$\dot{w} = 2(\nabla \widetilde{K}_h \times w) \tag{1.42}$$

is satisfied by integral curves of a vector field X on $\mathbf{R}^3$.

Invariant manifold. The sphere $\mathbf{S}^2_h$ is invariant under the flow of the vector field X.

Check.

$$L_X\langle w, w\rangle = 2\langle w, \dot{w}\rangle = 4\langle w, \nabla \widetilde{K}_h(w) \times w\rangle = 0. \tag{1.43}$$

X is hamiltonian. Consider the matrix of the Poisson structure

$$W(w) = (\{w_j, w_k\}) = 2\begin{pmatrix} 0 & -w_3 & w_2 \\ w_3 & 0 & -w_1 \\ -w_2 & w_1 & 0 \end{pmatrix} \tag{1.44}$$

Since $\ker W(w) = \mathrm{span}\{w\}$ and $T_w\mathbf{S}^2_h = \mathrm{span}\{w\}^{\perp}$, the matrix $W(w)\big|_{T_w\mathbf{S}^2_h}$ is invertible. On $\mathbf{S}^2_h$ define the symplectic form

$$\omega(w)_h(u,v) = \langle W^t(w)^{-1}u, v\rangle.$$

Since $W^t(w)y = -2w \times y = z$ we have

$$\begin{aligned} w \times z = w \times (-2w \times y) &= -2w \times (w \times y) = -2(w\,\langle w, y\rangle - y\,\langle w, w\rangle) \\ &= 2y\,\langle w, w\rangle = 2h^2\,y, \end{aligned}$$

which implies $y = \frac{1}{2h^2} w \times z$. Therefore

$$\omega_h(w)(u,v) = \frac{1}{2h^2}\,\langle w \times u, v\rangle = \frac{1}{2h^2}\,\langle w, u \times v\rangle. \tag{1.45}$$

The vector field X is hamiltonian with respect to ω_h with hamiltonian function $\widetilde{K}_h$, because

$$\begin{aligned} \omega_h(w)(X(w), u) &= \frac{1}{h^2}\,\langle w, (\nabla\widetilde{K}_h \times w) \times u\rangle = \frac{1}{h^2}\,\langle w \times (\nabla\widetilde{K}_h \times w), u\rangle \\ &= \frac{1}{h^2}\,\langle -w\,\langle\nabla\widetilde{K}_h, w\rangle + \nabla\widetilde{K}_h\,\langle w, w\rangle, u\rangle = \langle\nabla\widetilde{K}_h, u\rangle = \mathrm{d}\widetilde{K}_h(w)u, \end{aligned}$$

where $u, v \in T_w\mathbf{S}^2_h$.

Complex variables. On $\mathbf{R}^4$ introduce complex variables

$$z_j = x_j + \mathrm{i}y_j \quad \bar{z}_j = x_j - \mathrm{i}y_j.$$

Hamiltonian. The Hamiltonian of the harmonic oscillator becomes

$$\widetilde{H}(z_1, z_2) = \tfrac{1}{2}\,(z_1\bar{z}_1 + z_2\bar{z}_2).$$

Symplectic form. The symplectic form ω becomes

$$\Omega = \frac{1}{2\mathrm{i}}(dz_1 \wedge d\bar{z}_1 + dz_2 \wedge d\bar{z}_2).$$

Vector field. The hamiltonian vector field corresponding to $\widetilde{H}$ satisfies

$$X_{\widetilde{H}} \lrcorner \, \Omega = \mathrm{d}\widetilde{H}, \tag{1.46}$$

where

$$\widetilde{H} = \tfrac{1}{2}\,(z_1\,\mathrm{d}\bar{z}_1 + z_2\,\mathrm{d}\bar{z}_2 + \bar{z}_1\,\mathrm{d}z_1 + \bar{z}_2\,\mathrm{d}z_2).$$

Using equation (1.46) a calculations shows that

$$X_{\widetilde{H}} = i\left(z_1\,\frac{\partial}{\partial z_1} + z_2\,\frac{\partial}{\partial z_2} - \bar{z}_1\,\frac{\partial}{\partial \bar{z}_1} - \bar{z}_2\,\frac{\partial}{\partial \bar{z}_2}\right).$$

whose flow is

$$\phi_t^{\widetilde{H}}(z_1, z_2, \bar{z}_1, \bar{z}_2) = (e^{\mathrm{i}t} z_1, e^{\mathrm{i}t} z_2, e^{-\mathrm{i}t}\bar{z}_1, e^{-\mathrm{i}t}\bar{z}_2).$$

Quadratic integrals. In complex coordinates

$$\begin{array}{llll} w_1 & = \operatorname{Im} z_1\bar{z}_2 & w_2 & = \operatorname{Re} z_1\bar{z}_2 \\ w_3 & = \tfrac{1}{2}\,(z_1\bar{z}_1 - z_2\bar{z}_2) & w_4 & = \tfrac{1}{2}\,(z_1\bar{z}_1 + z_2\bar{z}_2). \end{array} \tag{1.47}$$

Assertion. The integrals w_1, w_2, w_3, w_4 generate the algebra of polynomials invariant under the flow of the harmonic oscillator vector field X_H.

Proof. Consider a monomial $M = z_1^{j_1} z_2^{j_2} \bar{z}_1^{k_1} \bar{z}_2^{k_2}$ such that

$$0 = L_{X_{\widetilde{H}}} M = \mathrm{i}(j_1 + j_2 - k_1 - k_2)M.$$

Then M is invariant under the flow $\phi_t^{\widetilde{H}}$ if and only if $j_1 + j_2 = k_1 + k_2$. We write the factors of M in two lists

$$\overbrace{z_1 \cdots\cdots z_1}^{j_1}\ \overbrace{z_2 \cdots z_2}^{j_2}$$

$$\underbrace{\bar{z}_1 \cdots \bar{z}_1}_{k_1}\ \underbrace{\bar{z}_2 \cdots\cdots \bar{z}_2}_{k_2}$$

Since these lists have equal length, their entries can be paired off. This expresses M as a product of $z_1\bar{z}_1$, $z_1\bar{z}_2$, $z_2\bar{z}_1$ and $z_2\bar{z}_2$. □

Consequence. By a theorem of Schwarz (*a heavy theorem about smooth invariant functions on [37]*) every *smooth* integral of X_H factors through $\widetilde{J}$.

1.5 Normalization of the Hénon-Heiles hamiltonian

Hamiltonian. The Hénon-Heiles hamiltonian is

$$H : \mathbf{R}^4 \to \mathbf{R} : (x, y) \mapsto \tfrac{1}{2}\,(y_1^2 + y_2^2 + x_1^2 + x_2^2) + \frac{\varepsilon}{3}(x_1^3 - 3x_1x_2^2). \quad (1.48)$$

Normalization. Normalizing H means that we find a symplectic change of coordinates so that the new Hamiltonian has a two dimensional harmonic symmetry, that is, it commutes with w_4 up to a certain order, see [11] for more details.

Normalized Hénon-Heiles. The normalized Hénon-Heiles hamiltonian up to sixth order is

$$\mathcal{H} = \mathcal{H}^{(2)} + \varepsilon^2\mathcal{H}^{(4)} + \varepsilon^4\mathcal{H}^{(6)} + \cdots \quad (1.49)$$

where

$$\begin{aligned} \mathcal{H}^{(2)} &= \frac{1}{2}w_4 \\ \mathcal{H}^{(4)} &= \frac{1}{48}(7w_2^2 - 5w_4^2) \\ \mathcal{H}^{(6)} &= \frac{1}{64}\Big(-\frac{67}{54}w_4^3 - \frac{7}{8}w_2^2w_4 - \frac{28}{9}w_3^3 + \frac{28}{3}w_1^2w_3\Big). \end{aligned} \quad (1.50)$$

1.5.1 The Hénon-Heiles hamiltonian normalized up to 4th order

Restrict. We sit on the constant energy surface $w_4 = h$, that is, on the set $\mathbf{S}_h^2$. The fourth order normalized hamiltonian restricted to $\mathbf{S}_h^2$ is

$$\mathcal{H}_h = \mathcal{H}\big|_{\mathbf{S}_h^2} = \frac{h}{2} + \frac{\varepsilon^2}{48}(7w_2^2 - 5h^2). \quad (1.51)$$

Simplification. Simplify the hamiltonian $\mathcal{H}_h$ by removing the additive constants and rescaling time. We get

$$\mathcal{H}_h = w_2^2. \quad (1.52)$$

Critical points of $\mathcal{H}_h$ on $\mathbf{S}_h^2$. In order to find the critical points of $\mathcal{H}_h$ on the surface $\mathbf{S}_h^2$ we solve

$$(0, 0, 0) = D\mathcal{H}_h(w) + \lambda DG(w) \quad \text{and} \quad G(w) = 0, \quad (1.53)$$

where

$$G : \mathbf{R}^3 \to \mathbf{R} : (w_1, w_2, w_3) \mapsto w_1^2 + w_2^2 + w_3^2 - h^2.$$

Solutions. The system of equations for the critical points is

$$\begin{aligned} 2\lambda\, w_1 &= 0 \\ 2\, w_2 + 2\lambda\, w_2 &= 0 \\ 2\lambda\, w_3 &= 0 \end{aligned}$$

and the constraint $w_1^2 + w_2^2 + w_3^2 = h^2$. For $\lambda \neq 0$ the solutions of these equations are $w_1 = 0$, $w_3 = 0$, $w_2 = \pm h$, $\lambda = -1$. These correspond to two critical points $p_\pm = (0, \pm h, 0)$. For $\lambda = 0$ the solution is $w_2 = 0$, $w_1^2 + w_3^2 = h^2$. This is a critical submanifold of $\mathbf{S}_h^2$, which is the heavy darkened circle in figure 1.4.

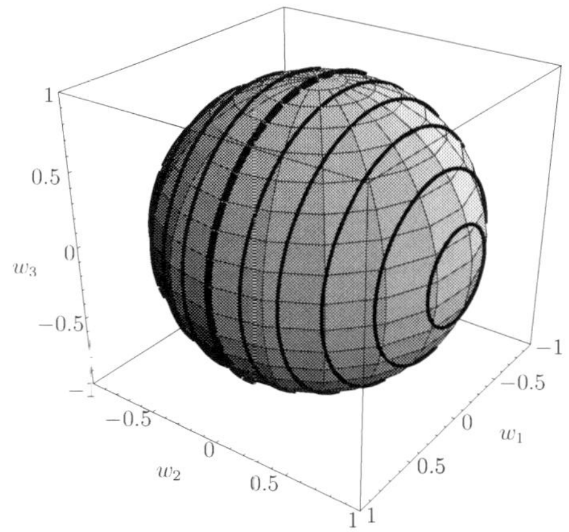

Fig. 1.4. Constant level sets of the reduced Hénon-Heiles Hamiltonian $\mathcal{H}_h$ of order 4 on the reduced phase space $\mathbf{S}_h^2$; heavy darkened circle is the critical set at $w_2 = 0$.

Hessian. The Hessian of $\mathcal{H}_h\big|_{\mathbf{S}_h^2}$ at the critical points $p_\pm$ is

$$D^2\mathcal{H}_h{}_{\,S_h^2}(p_\pm) = (D^2\mathcal{H}_h - D^2G)\big|_{T_{p_\pm}\mathbf{S}_h^2}, \tag{1.54}$$

where $T_w\mathbf{S}_h^2 = \ker DG(w)$. Since $DG(p_\pm) = (0, \pm 2h, 0)$, we see that $\ker DG(w) = \operatorname{span}\{e_1, e_3\}$. Therefore

$$D^2\mathcal{H}\big|_{S_h^2}(p_\pm) = \left(\begin{pmatrix} 0 & & \\ & 2 & \\ & & 0 \end{pmatrix} - \begin{pmatrix} 2 & & \\ & 2 & \\ & & 2 \end{pmatrix} \right)\Big|_{\operatorname{span}\{e_1, e_3\}} = -2I_2 \tag{1.55}$$

and the critical points $p_\pm$ are maxima of $\mathcal{H}_h$ on $\mathbf{S}_h^2$.

Until 1982 physicists did not know how to carry on since the 4th order system remained degenerate.

Our function behaves like a Morse function at the two critical points with $w_2 = \pm 1$. These points remain in place and will remain stationary due to symmetry, a small perturbation will not destroy them. On the contrary, the critical circle is a nondegenerate Bott-Morse critical submanifold. It will nearly completely disappear due to the small perturbation of higher order.

1.5.2 The Hénon-Heiles hamiltonian normalized up to 6th order

Restrict We sit on the constant energy surface $\mathbf{S}_h^2$. The sixth order normalized hamiltonian restricted to $\mathbf{S}_h^2$ is

$$\begin{aligned}\mathcal{H}_h &= \mathcal{H}^{(2)} + \varepsilon^2\mathcal{H}^{(4)} + \varepsilon^6\mathcal{H}^{(6)} \\ &= \frac{7}{48}w_2^2 + \frac{\varepsilon^2}{64}\left(-\frac{7h}{8}w_2^2 - \frac{28}{9}w_3^3 + \frac{28}{3}w_1^2 w_3\right)\end{aligned}$$

Critical points. To find the critical points of $\mathcal{H}_h|\mathbf{S}_h^2$ we solve

$$(0,0,0) = D\mathcal{H}_h(w) + \lambda DG(w) \quad \text{and} \quad G(w) = 0 \tag{1.56}$$

where

$$G : \mathbf{R}^3 \to \mathbf{R} : (w_1, w_2, w_3) \mapsto w_1^2 + w_2^2 + w_3^2 - h^2$$

as before.

Solutions. The system of equations for the critical points is

$$\begin{aligned}&\frac{7\varepsilon^2}{24}w_1 w_3 + 2\lambda w_1 = 0 \\ &\frac{7}{24}w_2 - \frac{7\varepsilon^2 h}{256}w_2 + 2\lambda w_2 = 0 \\ &\frac{7\varepsilon^2}{48}w_1^2 - \frac{7\varepsilon^2}{48}w_3^2 + 2\lambda w_3 = 0.\end{aligned}$$

together with the constraint $w_1^2 + w_2^2 + w_3^2 = h^2$. We search for solutions where $w_2 = 0$. Set $\lambda = 7\varepsilon^2\mu/24$. Then the above system of equations becomes

$$\begin{aligned}&w_1 w_2 + 2\mu w_1 = 0 \\ &w_1^2 - w_3^2 + 4\mu w_3 = 0 \\ &w_1^2 + w_3^2 = h^2\end{aligned}$$

The solutions of this system are $w_1 = 0$, $w_3 = \pm h$, $\mu = \pm\frac{1}{2}h$ and $w_1 = \pm\frac{1}{2}\sqrt{3}h$, $w_3 = \pm\frac{1}{2}h$, $\mu = \mp\frac{1}{2}h$.

Geometry. Because of the introduction of the sixth order term, the nondegenerate critical manifold that we had for the fourth-order hamiltonian breaks up into 6 critical points: three of them stable (elliptic) and three unstable (hyperbolic) that are connected by their stable and unstable manifolds (figure 1.5). For a geometric explanation of this bifurcation see [4]. This picture does not change qualitatively if we add higher order terms to the hamiltonian.

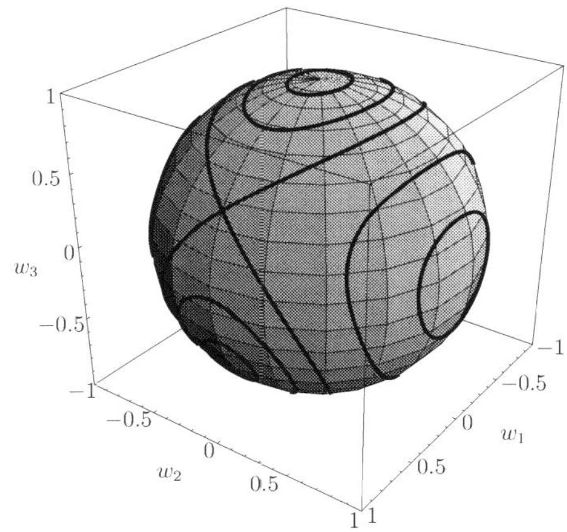

Fig. 1.5. Constant level sets of the reduced Hénon-Heiles hamiltonian $\mathcal{H}_h$ of order 6 on the reduced phase space $\mathbf{S}^2_h$.

Reconstruction. In the case of the 4th order normalized Hénon-Heiles hamiltonian, we found two critical points $p_\pm = (0, \pm h, 0)$ and a critical manifold $w_2 = 0$. After reconstruction, the critical points become periodic orbits in phase space, while the critical manifold becomes a 2-torus on which the flow of the normalized hamiltonian has rotation number 0.

In the case of the 6th order normalized Hénon-Heiles hamiltonian, we found eight critical points. Six of the critical points have $w_2 = 0$ while the other two are again $p_\pm = (0, \pm h, 0)$. Three of the critical points with $w_2 = 0$ are elliptic, while the other three are hyperbolic. The hyperbolic critical points are connected by their stable and unstable manifolds. After reconstruction these manifolds form a 2-torus in phase space that intersects itself three times cleanly along three circles.

A Comments on lecture I. Hénon-Heiles system

A great number of papers on this system has appeared since the first publication by Hénon and Heiles in 1964 [22], see [33] for a brief review. It has served both as a model of a nonintegrable (chaotic) system and as a test bed for various normalization techniques. Although originating in astronomy, the Hénon-Heiles system is quite popular in molecular physics where it has many analogues, such as doubly degenerate vibrations of a triatomic molecule A_3 (for example H_3^+ [18]) or of a tetrahedral molecule AB_4. Here we discuss aspects of the Hénon-Heiles system related to its finite symmetry which simplify significantly the analysis in [3]. Despite extensive work, this paper has been overlooked.

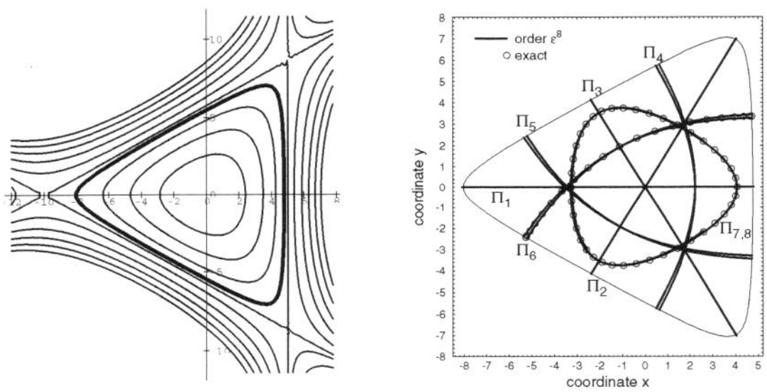

Fig. A.1. Hénon-Heiles potential $V(x)$ calculated for $\epsilon = 0.1$ and $E/E_{\text{saddle}} = 0.2, 0.45, 0.7, \mathbf{0.9}, 1, 1.2, \ldots$ (left); Relative equilibria (nonlinear normal modes) of the Hénon-Heiles system reconstructed from the ϵ^8 normal form at the energy $E/E_{\text{saddle}} = 0.9$ (right).

The spatial symmetry group of (1.48) is a dihedral group D_3. The full symmetry group is $D_3 \times \mathcal{T}$ where $\mathcal{T}$ is a Z_2 symmetry of the kind $(q,p) \to (q,-p)$ or equivalently $z \to \bar{z}$, which is often called *time reversal* or *momentum reversal*. Operations of the spatial group D_3 commute with the oscillator symmetry S^1. Operations which involve $\mathcal{T}$ are anti-symplectic and do not commute with S^1.

A.1 Invariants and integrity basis

Dynamical invariants. As in the lecture, we consider quadratic polynomial invariants of the oscillator symmetry. For obscure historical reasons[1], our definition differs by a factor 2, namely,

$$H_0 = 2J = \tfrac{1}{2}(z_2\bar{z}_2 + z_1\bar{z}_1) = w_4,$$

$$\mathbf{J} = \begin{pmatrix} J_1 \\ J_2 \\ J_3 \end{pmatrix} = \frac{1}{4}\begin{pmatrix} z_2\bar{z}_1 + z_1\bar{z}_2 \\ iz_2\bar{z}_1 - iz_1\bar{z}_2 \\ z_2\bar{z}_2 - z_1\bar{z}_1 \end{pmatrix} = \frac{1}{2}\begin{pmatrix} w_2 \\ w_1 \\ -w_3 \end{pmatrix}.$$

The Poisson algebra generated by the components of the 3-vector $\mathbf{J}$ is the standard so(3) with bracket $\{J_a, J_b\} = \epsilon_{abc} J_c$ and the Casimir $J = |\mathbf{J}|$ (or H_0).

Action of $D_3 \times \mathcal{T}$. The action of the symmetry group $D_3 \times \mathcal{T}$ on the components of $\mathbf{J}$ is equivalent to the action of the point group D_{3h} of transformations of $\mathbf{R}^3$ with coordinates (J_1, J_2, J_3) [20]. Since J_2 is invariant with respect to any rotation of the initial coordinate plane (x_1, x_2), it is convenient to choose J_2 along the vertical axis in $\mathbf{R}^3$. Then time reversal $\mathcal{T}$, which sends (J_1, J_2, J_3) to $(J_1, -J_2, J_3)$ (as can be verified directly), acts as the horizontal reflection plane of D_{3h}.

Integrity basis. Due to the relation

$$J_1^2 + J_2^2 + J_3^2 = J^2 = \tfrac{1}{4}h^2. \tag{A.1}$$

the ring $\mathcal{R}$ of invariant polynomials generated by J and (J_1, J_2, J_3) is not free. To analyze the normalized system we should have the way to express the normal form H_{nf} uniquely in terms of (J_1, J_2, J_3, J). The standard recipe for this is a Gröbner basis. We use a slightly more sophisticated integrity basis which (when it works) has certain advantages. An *integrity basis* consists of *principal* and *auxiliary* polynomials. The ring $\mathcal{R}$ decomposes as $\mathbf{R}[J, J_a, J_b] \oplus J_c\,\mathbf{R}[J, J_a, J_b]$ meaning that any member of $\mathcal{R}$ can be written uniquely as a real polynomial in the principal polynomials $\{J, J_a, J_b\}$ and J_c times another polynomial in $\{J, J_a, J_b\}$. Using (A.1) any power of J_3 can be represented this way.

In general, the number of principal polynomials equals the dimension of the reduced phase space (which is 2 for $\mathbf{S}^2_h$) plus the number of integrals of motion (we have one such integral J). Since the values of principal polynomials distinguish orbits of the action of the dynamical symmetry, they can serve as coordinates for charts of the reduced phase space, while auxiliary polynomials can be used to distinguish different charts. Thus for the reduced space $\mathbf{S}^2_J$ we need two charts $J_c > 0$ and $J_c < 0$ with coordinates (J_a, J_b).

[1] Our factors correspond to the quantum mechanical analogue of (A.1) called Schwinger [38] or boson representation of the angular momentum system.

Molien function. An explicit construction of an integrity basis is aided by knowing the Molien generating function. The generating function for the polynomials in four initial phase space variables $(z_1, z_2, \bar{z}_1, \bar{z}_2)$ invariant with respect to the S^1 oscillator symmetry is

$$g(\lambda) = (1+\lambda^2)/(1-\lambda^2)^3. \tag{A.2}$$

Here the formal variable λ represents one of $\{z, \bar{z}\}$. This function can be computed directly from Molien's theorem in representation theory. It indicates that there are three principal invariants represented by terms $1-\lambda^k$ in the denominator and one nontrivial auxiliary invariant represented by terms λ^k in the numerator. Since $k=2$, all invariants are of degree 2 in $\{z, \bar{z}\}$. This kind of information is invaluable in high dimensional situations.

Fully symmetrized integrity basis. Our polynomials $\{J_1, J_2, J_3\}$ are *not* symmetric with respect to $D_3 \times \mathcal{T}$. The group $D_3 \times \mathcal{T}$ acts on (J_3, J_1, J_2) in the same way as D_{3h} acts on (X, Y, Z) in 3-space, that is, (J_3, J_1, J_2) span the $E \oplus A_2$ representation of D_{3h}. The Molien generating function for the $D_3 \times \mathcal{T}$ invariant polynomials in (J_1, J_2, J_3) is

$$g(E \oplus A_2 \to A_1; \lambda) = \frac{1}{(1-\lambda^2)(1-\lambda^3)}.$$

This can be obtained straightforwardly from the action of the finite group $D_3 \times \mathcal{T}$ on (J_1, J_2, J_3). Note that here λ stands for any one of $\{J_1, J_2, J_3\}$. We conclude that the ring of all polynomial invariants of the combined action of $D_3 \times \mathcal{T}$ and oscillator symmetry S^1 is freely generated by (n, μ, ξ), where n is the main oscillator invariant (see (A.1)), and μ and ξ are polynomials in $\{J_1, J_2, J_3\}$ of degree 2 and 3 respectively. The invariants $\{n, \mu, \xi\}$ can be chosen explicitly as follows

$$n = 2J, \quad \mu = J_2^2, \quad \xi = \tfrac{1}{2} J_3(3J_1^2 - J_3^2). \tag{A.3}$$

This means that the normalized Hénon-Heiles Hamiltonian is a function $H_{\mathrm{nf}}(n, \mu, \xi)$ with n later relegated as a parameter. This basic result of invariant theory has not been appreciated in the numerous studies on the Hénon-Heiles system, including Cushman's early work in [7] and his lecture in Peyresq. Yet, this observation along with the rest of our comment is entirely in the spirit of Cushman's contemporary approach to the analysis of reduced systems [5].

A.2 Qualitative analysis of the reduced system

Most of the qualitative information on the Hénon-Heiles system presented in the lecture can be obtained simply from the $\mathbf{S}^2_h$ topology of the reduced phase space and the full use of the symmetry group action on it.

Table A.1. *Critical orbits of the $D_3 \times \mathcal{T} \sim D_{3h}$ action on the reduced phase space $\mathbf{S}^2_h$ of the Hénon-Heiles system. The $C_3 \wedge \mathcal{T}_2$ subgroup of $D_3 \times \mathcal{T}$ is generated by C_3 and $\mathcal{T}_2 = C_2 \circ \mathcal{T}$; the groups $C_3 \wedge \mathcal{T}_2$, D_3, and C_{3v} are isomorphic as abstract groups. "Historic" labels Π_k were introduced for the nonlinear normal modes in [3, 29] and used in [18].*

orbit	stabilizer	ξ/J^3	μ/J^2	$(J_3, J_1, J_2), J$
$\Pi_{7,8}$	$C_3 \wedge \mathcal{T}_2$	0	1	$(0, 0, \pm 1)$
$\Pi_{4,5,6}$	$C_2 \times \mathcal{T}$	$-1/2$	0	$(1, 0, 0)$, $(\cos\frac{2\pi}{3}, \pm\sin\frac{2\pi}{3}, 0)$
$\Pi_{1,2,3}$	$C_2' \times \mathcal{T}$	$1/2$	0	$(-1, 0, 0)$, $(\cos\frac{\pi}{3}, \pm\sin\frac{\pi}{3}, 0)$

Stratification of the reduced phase space. The action of the symmetry group $D_3 \times \mathcal{T}$ on $\mathbf{S}^2_h \subseteq \mathbf{R}^3$ follows from the action of the point group D_{3h} of transformations acting on $\mathbf{R}^3$ with coordinates $(X, Y, Z) = (J_3, J_1, J_2)$. This action has 8 fixed points which form three critical orbits characterized in table A.1. Note the immediate advantage of fully symmetrized main invariants $\{\xi, \mu\}$ over the coordinates (J_3, J_1, J_2). The values of (ξ, μ) represent entire orbits of the group action. This amounts to introducing polar coordinates by the back door, which is a no no, according to Cushman.

Orbit space. As shown in figure A.2, right, the orbit space $\mathcal{O}$ of the $D_3 \times \mathcal{T}$ action on $\mathbf{S}^2_h$ is the semialgebraic variety in $\mathbf{R}^3$ with coordinates (ξ, η, t) defined by

$$0 \leq \frac{\mu}{J^2} \leq 1 - t^2, \quad \frac{|\xi|}{J^3} \leq \tfrac{1}{2}\, t^3, \quad t \in [0, 1].$$

Each point in the interior of $\mathcal{O}$ represents a 12-point generic orbit of the $D_3 \times \mathcal{T}$ group action. Its three singular boundary points correspond to critical orbits in table A.1. The other boundary points represent 6-point orbits with nontrivial stabilizers $\mathcal{T}$ or $\mathcal{T}_s$. Knowing how D_{3h} acts on

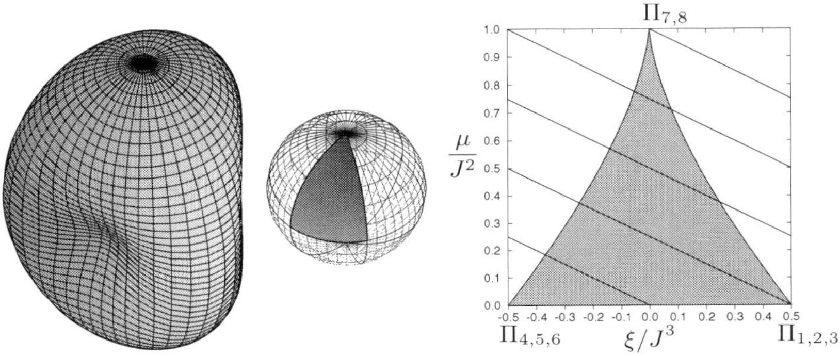

Fig. A.2. Relative equilibria of the Hénon-Heiles system as stationary points of the reduced Hamiltonian H_{nf}^{J} on the reduced phase space $\mathbf{S}_J^2$. On the left we show H_{nf}^{J} as a function on $\mathbf{S}_J^2$. The shaded area on the right and central panel represents the orbit space (orbifold) $\mathcal{O}$ of the $D_3 \times \mathcal{T}$ action on $\mathbf{S}^2$; straight lines in the right panel are constant level sets of the simplest $D_3 \times \mathcal{T}$-invariant Morse Hamiltonian $\mathcal{H} = \mu + \epsilon\xi$.

the $\mathbf{S}_h^2$ (figure A.2, centre), we see that $\mathcal{O}$ is the image of the triangular petal on $\mathbf{S}_h^2$ cut out by the three symmetry planes: two vertical planes intersecting at the angle $\pi/3$ and the horizontal plane. Those who prefer using "pure algebra" (and avoid any scent of polar coordinates) would do better by considering

$$\mathcal{J} = \det\left[\frac{\partial(\mu, \xi, J)}{\partial(J_1, J_2, J_3)}\right] = -6J_1J_2(3J_3^2 - J_1^2) = 0$$

and observing that the boundary and singular points of $\mathcal{O}$, that is, its 1 and 0-dimensional strata on $\mathbf{S}_h^2$, correspond to simple and double zeroes of $\mathcal{J}$.

Symmetric Morse functions. We now ask the question: what is a typical $D_3 \times \mathcal{T}$ symmetric function $\mathcal{H}$ on $\mathbf{S}_h^2$? We characterize $\mathcal{H}$ primarily by finding its set of critical points which in our case correspond to *relative equilibria* of our system. Points on the critical orbits are isolated and *must* be critical points of $\mathcal{H}$. Points in the same orbit are *equivalent* and therefore have the same stability. Furthermore, the two equivalent points $\Pi_{7,8}$ must be elliptic (stable) because of their high local symmetry. If we further assume that $\mathcal{H}$ is a Morse function, that is, a function with only nondegenerate critical points, and remember that

Morse's relation for the Euler characteristic of the 2-sphere is

$$c_0 - c_1 + c_2 = 2,$$

where c_0, c_1, and c_2 are the number of maxima, saddle (hyperbolic), and minima of $\mathcal{H}$, we can conclude that a function $\mathcal{H}$ with minimum possible number of critical points has three equivalent elliptic points and three equivalent saddle points in addition to $\Pi_{7,8}$. One possible such simplest Morse function is drawn in figure A.2, left.[1] It has maxima at $\Pi_{7,8}$, minima at $\Pi_{1,2,3}$ and saddle points at $\Pi_{4,5,6}$. The other possibility is to have an oblate shape with two minima at $\Pi_{7,8}$ and three maxima. Trajectories of the reduced system shown in figure 1.5 are constant level sets of $\mathcal{H}$ which can be obtained as intersections of the surface in figure A.2 left, and spheres of different radii.[2]

Simplest polynomial Morse Hamiltonian. The most natural way to construct the simplest Morse Hamiltonian $\mathcal{H}$ explicitly is to consider $\mathcal{H}$ as a polynomial in (μ, ξ) defined on the orbit space $\mathcal{O}$. It can be seen that a linear function $\mathcal{H}(\mu, \xi) = a\mu + b\xi$ with nonzero a and b is generic. Indeed, while μ alone is too symmetric (it has axial symmetry S^1), together with the cubic invariant ξ it reproduces all symmetry properties of $D_3 \times \mathcal{T}$ correctly. The absence of auxiliary integrity basis invariants also indicates that we need no other terms in $\mathcal{H}$. In general, coefficients in $\mathcal{H}(\mu, \xi)$ are functions of the parameter J. Since ξ is of higher degree in $(z, \bar{z})$ than μ, the contribution $b(J)$ is likely to be smaller (at least for low values of J) than $a(J)$. Therefore, the reduced Hénon-Heiles system at low J should be qualitatively correctly represented by the level sets of $\mathcal{H} = \mu + \varepsilon\xi$ where $0 < \varepsilon \ll 1$. As shown in figure A.2, right, the family of constant level sets of such $\mathcal{H}$ on the orbifold $\mathcal{O}$ has three exceptional (critical) levels which pass at $\Pi_{1,2,3}$, $\Pi_{4,5,6}$, and $\Pi_{7,8}$. The extremal levels correspond necessarily to stable relative equilibria, the critical level at the intermediate energy $\mathcal{H}_{\Pi_{4,5,6}}$ contains unstable relative equilibria and their stable/unstable manifold (separatrix).

[1] Plots of this kind are used in molecular physics to represent effective rotational energy of nonrigid molecules as function of the orientation of the total angular momentum **J** (orientation of the rotation axis), they are called *rotational energy surfaces* [21]

[2] Note that vertical axis in figure A.2 corresponds to the horizontal axis w_2 in figure 1.5.

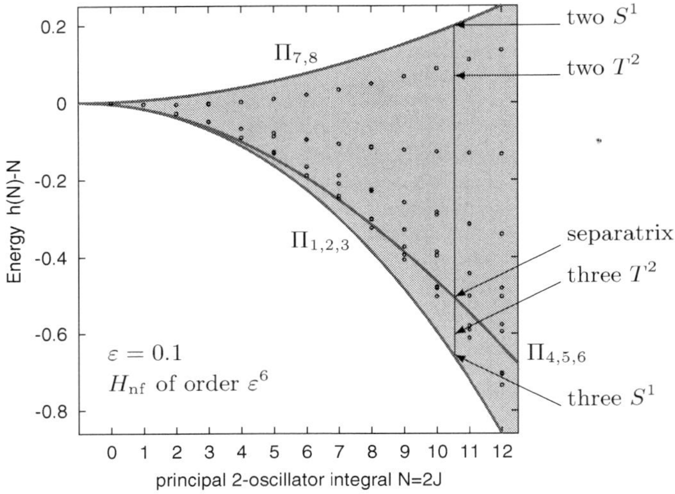

Fig. A.3. Image of the energy-momentum map (shaded area), energies of relative equilibria (solid lines) and quantum energies (circles) of the Hénon-Heiles system with $\varepsilon = 0.1$ obtained using order ε^6 normal form $H_{\rm nf}$. The classical action (momentum) $2J = n$ equals $N + 1$ where N is the polyad quantum number.

A.3 Normal form and remarks on further analysis

Now, after the Hénon-Heiles system has been understood qualitatively, we compute the normal form

$$H_{\rm nf} = n - \varepsilon^2 \left(\tfrac{5}{12}n^2 - \tfrac{7}{3}\mu\right) - \varepsilon^4 \left(\tfrac{67}{432}n^3 + \tfrac{7}{36}\mu n - \tfrac{56}{9}\xi\right) + \dots, \tag{A.4}$$

where the coefficients in the higher orders are listed below.

order	1	μn^{-2}	ξn^{-3}	$\mu^2 n^{-4}$	$\mu\xi n^{-5}$
$\varepsilon^6 n^4$	$-\frac{42229}{155520}$	$-\frac{76447}{6480}$	$\frac{2093}{135}$	$\frac{115171}{1944}$	
$\varepsilon^8 n^5$	$-\frac{15624833}{18662400}$	$-\frac{11656729}{2332800}$	$\frac{353843}{8100}$	$\frac{2217943}{233280}$	$\frac{6701639}{4050}$

It comes as little surprise that $H_{\rm nf}$ is a function of (μ, ξ) and parameter n. We can use table A.1 to find the energy of $H_{\rm nf}$ at the critical points Π_k. The results plotted against n give the image of the energy-momentum map $\mathcal{EM}$ of the system, see figure A.3. Note that the $\mathcal{EM}$ map of $\mathcal{H} = \mu + \varepsilon\xi$ has qualitatively the same image.

We now look at reconstruction. In other words, we lift constant energy sets of H_{nf} on the orbifold $\mathcal{O}$ first to the reduced phase space $\mathbf{S}_h^2$ and then all the way back to the initial phase space $\mathbf{R}^4$. This process is a good exercise for those who like to understand the role of the $D_3 \times \mathcal{T}$ symmetry of the system in detail. From the same point of view, it is helpful to compare the image of the $\mathcal{EM}$ map in figure A.3 to that in figure 1.1 and to reconstruct $\mathcal{EM}^{-1}(h, n)$. In the simple case of relative equilibria Π_k, we can describe qualitatively the corresponding periodic orbits S^1 in $\mathbf{R}^4$ entirely on the basis of their local symmetry properties (stabilizers) listed in table A.1. This reproduces the results of [3, 29]. Figure A.1, right, demonstrates how these periodic orbits can be reconstructed analytically using the inverse normal form transformation. Finally we can consider quantum analogue of our system on the basis of the EBK torus quantization, see figure A.3.

We conclude with one more remark. We have seen that much of the analysis of the normalized Hénon-Heiles system can be simplified, if not avoided entirely, after we take discrete symmetries into account. Of course this does not reflect the general situation. In certain cases, typically when symmetries are low and dimensions are high, the critical point analysis of the kind presented in the first two lectures becomes necessary. Rather the general conclusion should be that analyzing symmetries helps to distinguish specific properties of the system from more common dynamical behaviour.

2 Lectures III and V. The Euler top

Physically the Euler top is a rigid body which is spinning around its (fixed) center of mass with no other forces acting upon it.

2.1 Preliminaries on the rotation group

Rotation group. The group of rotations in $\mathbf{R}^3$ is

$$\mathrm{SO}(3) = \{A \in \mathrm{GL}(3, \mathbf{R}) \big| A^t A = I \text{ and } \det A = 1\}.$$

Lie algebra. The Lie algebra of SO(3) is

$$\mathrm{so}(3) = \{X \in \mathrm{gl}(3, \mathbf{R}) \big| X + X^t = 0\}$$

with Lie bracket $[X, Y] = XY - YX$.

Isomorphism. $(\mathbf{R}^3, \times) \simeq (\mathrm{so}(3), [\,,\,])$. The isomorphism is

$$x = \begin{pmatrix} x_1 \\ x_2 \\ x_3 \end{pmatrix} \rightarrow \widehat{x} = \begin{pmatrix} 0 & -x_3 & x_2 \\ x_3 & 0 & -x_1 \\ -x_2 & x_1 & 0 \end{pmatrix} = X.$$

Properties of this isomorphism. $\widehat{x}(y) = Xy = x \times y$, and $[\widehat{x}, \widehat{y}] = \widehat{x \times y}$; for $A \in \mathrm{SO}(3)$, $A\widehat{x}A^{-1} = \widehat{Ax}$.

Inner product on so(3)**.** Define an inner product on so(3) as

$$\mathrm{k}(X,Y) = -\tfrac{1}{2}\,\mathrm{tr}XY^t = \langle \widehat{x}, \widehat{y} \rangle,$$

where $\langle\,,\,\rangle$ is the Euclidean inner product on $\mathbf{R}^3$.

2.2 Traditional derivation of the equations of motion

Here we derive the equations of motion of the Euler top in the traditional nonhamiltonian manner. We use Coriolis' theorem and the conservation of angular momentum.

2.2.1 Reference frames

Let V be a three dimensional real vector space with Euclidean inner product $\langle\,,\,\rangle$. A *frame of reference* $\mathcal{F}$ is a positively oriented orthonormal basis $\{f_1, f_2, f_3\}$ of V. A vector $v \in V$ *looks like* the vector $x \in \mathbf{R}^3$ *in the frame* $\mathcal{F}$ means $v = \sum_{i=1}^3 x_i\, f_i$. Corresponding to the frame $\mathcal{F}$ is its *coframe* $\mathcal{F}^* = \{f_1^*, f_2^*, f_3^*\}$, where $f_i^*(f_j) = \delta_{ij}$. Suppose that $\mathcal{A} = \{a_1, a_2, a_3\}$ is another reference frame such that the vector $v \in V$ looks like the vector $X \in \mathbf{R}^3$, that is, $v = \sum_{i=1}^3 X_i\, a_i$. Let A be the 3×3 matrix whose ijth entry is $f_i^*(a_j)$, that is, a_j looks like the jth column of A in the frame $\mathcal{F}$. Then

$$x = AX, \tag{2.1}$$

because

$$x_i = f_i^*(\sum_{j=1}^3 x_j\, f_j) = f_i^*(\sum_{j=1}^3 X_j\, a_j) = \sum_{j=1}^3 f_i^*(a_j) X_j.$$

In other words, the vector $v \in V$, which in the frame $\mathcal{F}$ looks like the vector $x \in \mathbf{R}^3$, in the frame $\mathcal{A}$ looks like the vector X.

2.2.2 Rotating frame

Let

$$A : \mathbf{R} \to \mathrm{SO}(3) : t \mapsto A(t) = \mathrm{col}(a_1(t), a_2(t), a_3(t)).$$

Then $\mathcal{A} = \{a_1(t), a_2(t), a_3(t)\}$ is a frame for V whose j^{th} member $a_j(t)$ looks like the jth column of $A(t)$ with respect to the fixed frame $\mathcal{F}$. We say that $\mathcal{A}$ is a *frame which rotates with respect to the fixed frame* $\mathcal{F}$.

2.2.3 Coriolis' theorem

Before starting the derivation Richard points out that "Physicists don't know how to prove this theorem". Before a fight between the mathematicians and physicists in the audience has time to break out Daryl Holm replies "Don't mock the alligator until you've crossed the river safely." The lecture continues.

Let $x : \mathbf{R} \to \mathbf{R}^3 : t \mapsto x(t)$ be a differentiable function. Suppose that $\Xi : \mathbf{R} \to V : t \mapsto \Xi(t)$ is a motion in V so that its position $\Xi(t)$ at time t in the fixed frame $\mathcal{F}$ looks like $x(t)$, while its position in the rotating frame $\mathcal{A}$ looks like $X(t)$. Then from (2.1) we obtain

$$x(t) = A(t)X(t). \tag{2.2}$$

Differentiating (2.2) gives

$$\frac{dx}{dt} = A'(t)X + A(t)\frac{dX}{dt} = A'(t)A^{-1}(t)x + A(t)\frac{dX}{dt}. \tag{2.3}$$

The velocity of $t \mapsto \Xi(t)$ at time t with respect to the fixed frame $\mathcal{F}$ is a vector in V which looks like $\frac{dx}{dt}$, while with respect to the rotating frame $\mathcal{A}$ it is a vector in V which looks like $\frac{dX}{dt}$. The skew symmetric matrix $A'(t)A^{-1}(t)$ is an *infinitesimal motion* in the fixed frame. The corresponding vector $\omega(t) \in \mathbf{R}^3$, where

$$\widehat{\omega(t)} = A'(t)A^{-1}(t),$$

is the *angular velocity in the fixed frame at time t of the rotating frame.* We can rewrite (2.3) as

$$\frac{dx}{dt} - \omega(t) \times x(t) = A(t)\frac{dX}{dt}, \tag{2.4}$$

which is a form of Coriolis' theorem (in the fixed frame). Define $\Omega(t)$ to be the vector in $\mathbf{R}^3$ which looks like $\omega(t)$ in the rotating frame, that is,

$$\omega(t) = A(t)\Omega(t). \tag{2.5}$$

Using the definition of $\omega(t)$ we find that

$$\widehat{\omega(t)} = A'(t)A^{-1}(t) = A(t)(A^{-1}(t)A'(t))A^{-1}(t).$$

Taking the hat of both sides of (2.5) gives $\widehat{\omega(t)} = A(t)\widehat{\Omega(t)}A^{-1}(t)$. Thus

$$\widehat{\Omega(t)} = A^{-1}(t)A'(t).$$

Thus we may rewrite (2.3) as

$$\begin{aligned}\frac{dx}{dt} &= A(t)[A^{-1}(t)A'(t)X + \frac{dX}{dt}] = A(t)[\widehat{\Omega(t)}X + \frac{dX}{dt}] \\ &= A(t)[\Omega(t) \times X + \frac{dX}{dt}], \qquad (2.6)\end{aligned}$$

which is another form of Coriolis' theorem (in the rotating frame).

Richard says that he has now crossed the river safely.

2.2.4 Constant angular momentum

Suppose we have a rigid body $\mathcal{B}$ in $\mathbf{R}^3$ made up of a finite number of point masses m_i at position r_i, not all on a single line through the origin. Suppose that the center of mass of $\mathcal{B}$ lies at the origin O of $\mathbf{R}^3$ and that $\mathcal{B}$ is subjected to no external forces.

Fix the frame $\mathcal{E} = \{e_1, e_2, e_3\}$ consisting of the standard basis vectors in $\mathbf{R}^3$. We call $\mathcal{E}$ the *space frame*. The angular momentum of $\mathcal{B}$ with respect to the space frame is given by

$$\ell = \sum_i m_i \, r_i \times v_i, \qquad (2.7)$$

where $v_i = \frac{dr_i}{dt}$ is the velocity of the ith point mass in $\mathcal{B}$ with respect to the space frame. ℓ is constant throughout the motion of $\mathcal{B}$.

Proof. Differentiating (2.7) gives

$$\frac{d\ell}{dt} = \sum_i m_i \frac{dr_i}{dt} \times v_i + \sum_i m_i r_i \times \frac{dv_i}{dt} = \sum_i r_i \times \frac{d(m_i v_i)}{dt} = \sum_i r_i \times F_i.$$

F_i is the total force exerted on the ith point mass.

$$F_i = F_i^{\text{int}} + F_i^{\text{ext}},$$

where F_i^{int} and F_i^{ext} is the internal and external forces, respectively. $F_i^{\text{int}} = \sum_{j \neq i} F_{ij}^{\text{int}}$, where F_{ij}^{int} is the force exerted on the ith particle by the jth particle of the body. But action and reaction are equal and lie along a line joining the ith and jth particle, that is,

$$0 = r_i \times F_{ij}^{\text{int}} + r_j \times F_{ji}^{\text{int}} = (r_i - r_j) \times F_{ij}^{\text{int}}.$$

Consequently,

$$\begin{aligned}\sum_i r_i \times F_i^{\text{int}} &= \sum_{\substack{i,j\\ i\neq j}} r_i \times F_{ij}^{\text{int}} = \sum_{i<j} r_i \times F_{ij}^{\text{int}} + \sum_{j<i} r_i \times F_{ij}^{\text{int}} \\ &= \sum_{i<j} (r_i \times F_{ij}^{\text{int}} - r_j \times F_{ij}^{\text{int}}) = 0.\end{aligned}$$

Thus

$$\frac{d\ell}{dt} = \sum_i r_i \times F_i = \sum_i r_i \times (F_i^{\text{int}} + F_i^{\text{ext}}) = \sum_i r_i \times F_i^{\text{ext}} = 0. \quad \square$$

2.2.5 Euler's equations

Attach an orthonormal frame to $\mathcal{B}$ with origin at O in $\mathbf{R}^3$. As $\mathcal{B}$ rotates, the attached frame rotates with it and thus defines a differentiable curve $\mathbf{R} \to \mathrm{SO}(3) : t \mapsto A(t)$. The column vectors of $A(t)$ define the *body frame.* Let $L = A(t)^{-1}\ell$ be the *angular momentum in the body frame.* Coriolis' formula (2.6) applied to ℓ gives

$$\frac{d\ell}{dt} = A(t)[\Omega(t) \times L + \frac{dL}{dt}], \tag{2.8}$$

Since $\frac{d\ell}{dt} = 0$, we obtain

$$\frac{dL}{dt} = L \times \Omega, \tag{2.9}$$

where L is the angular momentum in the body frame and Ω is the angular velocity of the body in the body frame. Now

$$L = I(\Omega). \tag{2.10}$$

I is the *moment of inertia tensor* of $\mathcal{B}$ in the body frame. I does not depend on t as the body is rigid, which means that the positions and the magnitudes of the masses are constant in the body frame. Thus (2.9) can be written as

$$I(\dot{\Omega}) = I(\Omega) \times \Omega, \tag{2.11}$$

which are called *Euler's equations.* We may choose the body frame so that $\{e_1, e_2, e_3\}$ are the *principal axes* of $\mathcal{B}$, that is, $I(e_j) = I_j\, e_j$ for $j = 1, 2, 3$. From now on we assume that $0 < I_1 < I_2 < I_3$. In components (2.11) reads

$$\begin{aligned} I_1\, \dot{\Omega}_1 &= (I_2 - I_3)\, \Omega_2 \Omega_3 \\ I_2\, \dot{\Omega}_2 &= (I_3 - I_1)\, \Omega_1 \Omega_3 \\ I_3\, \dot{\Omega}_3 &= (I_1 - I_2)\, \Omega_1 \Omega_2. \end{aligned} \tag{2.12}$$

Let $a = I_1^{-1}$, $b = I_2^{-1}$, $c = I_3^{-1}$ (so $0 < c < b < a$) and let $p_j = I_j \Omega_j$. Then (2.12) becomes

$$\begin{aligned} \dot{p}_1 &= -(b-c)\, p_2 p_3 \\ \dot{p}_2 &= (a-c)\, p_1 p_3 \\ \dot{p}_3 &= -(a-b)\, p_1 p_2. \end{aligned} \tag{2.13}$$

2.3 Qualitative behavior of solutions of Euler's equations

To describe the qualitative behavior of the solutions of Euler's equations (2.11), we note that the functions

$$E = \tfrac{1}{2}\langle I\Omega, \Omega\rangle = \tfrac{1}{2}\langle I^{-1}(p), p\rangle \tag{2.14}$$

$$L = \langle I\Omega, I\Omega\rangle = \langle p, p\rangle \tag{2.15}$$

are constant on the solutions of (2.11) and thus are constant on the solutions of (2.13).

Check.

$$\dot{E} = \langle I(\dot{\Omega}), \Omega\rangle = \langle I(\Omega) \times \Omega, \Omega\rangle = 0$$

and

$$\dot{L} = 2\,\langle I(\dot{\Omega}), I(\Omega)\rangle = \langle I(\Omega) \times \Omega, I(\Omega)\rangle = 0. \qquad \square$$

The function E is a Morse function on the 2-sphere $\mathbf{S}^2_{|\ell|}$ defined by $\langle p, p\rangle = |\ell|^2$. It has six nondegenerate critical points: 2 of Morse index 0, 2 of index 1, and 2 of index 0.

Check. If p^0 is a critical point of E on $\mathbf{S}^2_{|\ell|}$, then

$$0 = \mathrm{d}E(p^0) - \lambda^0\, \mathrm{d}L(p^0) = (I^{-1} - \lambda^0\, \mathrm{id})p^0 \quad \text{and} \quad \langle p^0, p^0\rangle = |\ell|^2.$$

Then p^0 is an eigenvector of length $|\ell|$ of $I^{-1} = \mathrm{diag}(a, b, c)$ corresponding to the eigenvalue λ^0. Thus

$$p^0 = \begin{cases} \pm|\ell| e_1, & \text{when } \lambda^0 = a \\ \pm|\ell| e_2, & \text{when } \lambda^0 = b \\ \pm|\ell| e_3, & \text{when } \lambda^0 = c. \end{cases}$$

The Hessian of $E\big|_{\mathbf{S}^2_{|\ell|}}$ at the critical point p^0 is

$$\begin{aligned} D^2(E\big|_{\mathbf{S}^2_{|\ell|}})(p^0) &= (D^2E(p^0) - \lambda^0 \, D^2L(p^0))\big|_{T_{p^0}\mathbf{S}^2_{|\ell|}} \\ &= (I^{-1} - \lambda^0 \, \mathrm{id})\big|_{T_{p^0}\mathbf{S}^2_{|\ell|}} \\ &= \begin{cases} \mathrm{diag}(b-a, c-a), & p^0 = \pm|\ell|e_1 \\ \mathrm{diag}(a-b, c-b), & p^0 = \pm|\ell|e_2 \\ \mathrm{diag}(a-c, b-c), & p^0 = \pm|\ell|e_3. \end{cases} \end{aligned}$$

Its Morse index is 2, 1, 0, if p^0 is $\pm|\ell|e_1$, $\pm|\ell|e_2$, and $\pm|\ell|e_3$, respectively. □

According to the Morse lemma, the level sets of $E\big|_{\mathbf{S}^2_{|\ell|}}$ near $p^0 = \pm|\ell|e_1$ or $\pm|\ell|e_3$ are circles, whereas those near $p^0 = \pm|\ell|e_2$ are hyperbolas. In fact the $\frac{1}{2}\,b|\ell|^2$-level set of $E\big|_{\mathbf{S}^2_{|\ell|}}$ is

$$\begin{aligned} \tfrac{1}{2}\,(a\,p_1^2 + b\,p_2^2 + c\,p_3^2) &= \tfrac{1}{2}\,b \\ p_1^2 + p_2^2 + p_3^2 &= |\ell|^2. \end{aligned}$$

Multiplying the first equation above by $|\ell|^2$ and subtracting $\frac{1}{2}\,b$ times the second equation gives

$$\begin{aligned} 0 =& \tfrac{1}{2}\,(a-b)p_1^2 - \tfrac{1}{2}\,(b-c)p_2^2 \\ =& \tfrac{1}{2}\,(\sqrt{a-b}\,p_1 + \sqrt{b-c}\,p_2)(\sqrt{a-b}\,p_1 - \sqrt{b-c}\,p_2). \end{aligned}$$

This means that the $\frac{1}{2}\,b|\ell|^2$-level set of E on $\mathbf{S}^2_{|\ell|}$ is the intersection of $\mathbf{S}^2_{|\ell|}$ with two transverse 2-planes (which intersect along the p_3-axis). Thus the $\frac{1}{2}\,b|\ell|^2$-level set of $E\big|_{\mathbf{S}^2_{|\ell|}}$ is the union of two great circles. All other level sets are diffeomorphic to *two* circles, except for $\frac{1}{2}\,a|\ell|^2$ and $\frac{1}{2}\,c|\ell|^2$, which are two distinct points.

2.4 Quantitative behavior of solutions of Euler's equations

2.4.1 A crash course in Jacobi elliptic functions

In order to solve Euler's equations quantitatively, we need Jacobi elliptic functions. Consider the system

$$\begin{aligned} \dot{x} &= yz \\ \dot{y} &= -xz \\ \dot{z} &= -k^2xy, \end{aligned} \tag{2.16}$$

on $\mathbf{R}^3$, where the parameter k lies in $(0,1)$. Define the *Jacobi elliptic functions* sn, cn, and dn as the solution

$$t \to (x(t), y(t), z(t)) = (\mathrm{sn}(t;k), \mathrm{cn}(t;k), \mathrm{dn}(t;k)) \tag{2.17}$$

of (2.16) with initial condition $(0,1,1)$. The functions

$$x^2 + y^2 \quad \text{and} \quad k^2x^2 + z^2$$

are integrals of (2.16). Hence

$$\begin{aligned} \mathrm{sn}^2(t;k) + \mathrm{cn}^2(t;k) &= 1 \\ k^2\mathrm{sn}^2(t;k) + \mathrm{dn}^2(t;k) &= 1, \end{aligned}$$

which implies that for all $t \in \mathbf{R}$

$$|\mathrm{sn}(t;k)| \le 1, \ |\mathrm{cn}(t;k)| \le 1 \text{ and } k' = \sqrt{1-k^2} \le \mathrm{dn}(t;k) \le 1. \tag{2.18}$$

Since $x^2 + y^2 = 1$ and $k^2x^2 + z^2 = 1$, we may drop the equations for $\frac{dy}{dt}$ and $\frac{dz}{dt}$ from (2.16) and obtain

$$\frac{dx}{dt} = \sqrt{(1-x^2)(1-k^2x^2)}. \tag{2.19}$$

Since the right hand side of (2.19) is positive when $x \in (-1,1)$, we find that

$$x \mapsto t(x) = \int_0^x \frac{dx}{\sqrt{(1-x^2)(1-k^2x^2)}} \tag{2.20}$$

is a smooth inverse to the function

$$x : \mathbf{R} \to (-1,1) : t \mapsto x(t) = \mathrm{sn}(t;k).$$

Because $t(\pm 1) = \pm K(k) = \pm K$ is finite, the function x is continuous on $[-1,1]$. Thus $\mathrm{sn}(K;k) = 1$, which implies that $\mathrm{cn}(K;k) = 0$ and $\mathrm{dn}(K;k) = k'$. From the definition of $t(x)$ it follows that for $k=0$ and 1 the Jacobi elliptic functions degenerate to trigonometric functions. Explicitly, for $k=0$ we have

$$\mathrm{sn}(t;0) = \sin t, \ \mathrm{cn}(t;0) = \cos t, \text{ and } \mathrm{dn}(t;0) = 1;$$

while for $k=1$ we have

$$\mathrm{sn}(t;1) = \tanh t, \ \mathrm{cn}(t;1) = \mathrm{sech}\, t, \text{ and } \mathrm{dn}(t;1) = \mathrm{sech}\, t.$$

We now show that sn, cn, and dn are periodic. Let

$$\xi(t) = \frac{\mathrm{cn}(t;k)}{\mathrm{dn}(t;k)}, \ \eta(t) = -k'\frac{\mathrm{sn}(t;k)}{\mathrm{dn}(t;k)}, \text{ and } \zeta(t) = k'\frac{1}{\mathrm{dn}(t;k)}.$$

Then $t \longrightarrow (\xi(t), \eta(t), \zeta(t))$ is an integral curve of (2.16) with initial condition $(1, 0, k')$. But so is $t \longrightarrow (\mathrm{sn}(t+K;k), \mathrm{cn}(t+K;k), \mathrm{dn}(t+K;k))$. Hence

$$\begin{aligned}
\mathrm{sn}(t+K;k) &= \frac{\mathrm{cn}(t;k)}{\mathrm{dn}(t;k)} \\
\mathrm{cn}(t+K;k) &= -k'\frac{\mathrm{sn}(t;k)}{\mathrm{dn}(t;k)} \\
\mathrm{dn}(t+K;k) &= k'\frac{1}{\mathrm{dn}(t;k)}.
\end{aligned}$$

This implies that $\mathrm{sn}(t;k)$ and $\mathrm{cn}(t;k)$ are periodic of period $4K(k)$, while $\mathrm{dn}(t;k)$ is periodic of period $2K(k)$.

2.4.2 Explicit solutions of Euler's equations

Using Jacobi elliptic functions we find explicit solutions of Euler's equations. There are two cases which correspond to the two types of stable relative equilibria, see figure 2.5. This is a bit messy.

CASE 1. $|\ell|^2 b \ge 2h \ge |\ell|^2 c$.

Solving

$$\begin{aligned}
a\,p_1^2 + b\,p_2^2 + c\,p_3^2 &= 2h \\
p_1^2 + p_2^2 + p_3^2 &= |\ell|^2
\end{aligned}$$

for p_1^2 and p_3^2 we obtain

$$\begin{aligned}
p_1^2 &= \frac{1}{a-c}(2h - |\ell|^2 c - (b-c)p_2^2) \\
p_3^2 &= \frac{1}{a-c}(-2h + |\ell|^2 a - (a-b)p_2^2).
\end{aligned} \tag{2.21}$$

Thus the equation $\dot{p}_2 = (a-c)\,p_1 p_3$ becomes

$$\frac{dp_2}{dt} = \sqrt{(2h - |\ell|^2 c - (b-c)p_2^2)(-2h + |\ell|^2 a - (a-b)p_2^2)}. \tag{2.22}$$

We now transform (2.22) into (2.19). Let

$$\begin{aligned}
\tau &= t\,n, \quad \text{where } n = \sqrt{(b-c)(a|\ell|^2 - 2h)} \\
x &= p_2\sqrt{\frac{b-c}{2h - |\ell|^2 c}},
\end{aligned}$$

and

$$k^2 = \frac{(a-b)(2h - |\ell|^2 c)}{(b-c)(|\ell|^2 a - 2h)}.$$

Then

$$\begin{aligned}
\frac{d\tau}{dx} &= \frac{d\tau}{dt} \Big/ \left(\frac{dx}{dp_2} \frac{dp_2}{dt} \right) \\
&= \frac{\sqrt{(b-c)(|\ell|^2 a - 2h)}}{\sqrt{\frac{b-c}{2h-|\ell|^2 c}} \sqrt{2h - |\ell|^2 c - (b-c)p_2^2)(|\ell|^2 a - 2h - (a-b)p_2^2)}} \\
&= \frac{1}{\sqrt{(1-x^2)(1-k^2x^2)}}. \qquad (2.23)
\end{aligned}$$

Consequently, $x(\tau) = \mathrm{sn}(\tau; k) = \mathrm{sn}(nt; k)$. From (2.23) and (2.21) we obtain

$$\begin{aligned}
p_1(t) &= A\,\mathrm{cn}(nt; k) \\
p_2(t) &= B\,\mathrm{sn}(nt; k) \\
p_3(t) &= C\,\mathrm{dn}(nt; k),
\end{aligned}$$

where

$$A^2 = \frac{2h - |\ell|^2 c}{a-c}, \quad B^2 = \frac{2h - |\ell|^2 c}{b-c}, \quad \text{and} \quad C^2 = \frac{|\ell|^2 a - 2h}{a-c}.$$

The signs of the square roots are chosen so that $t \mapsto (p_1(t), p_2(t), p_3(t))$ sweeps out a connected component of $E^{-1}(h) \cap L^{-1}(|\ell|^2)$.

Case 2. $|\ell|^2 a \ge 2h \ge |\ell|^2 b$.

A similar argument gives

$$\begin{aligned}
p_1(t) &= A\,\mathrm{dn}(nt; k) \\
p_2(t) &= B\,\mathrm{sn}(nt; k) \\
p_3(t) &= C\,\mathrm{cn}(nt; k),
\end{aligned}$$

where

$$n = \sqrt{(a-b)(2h - c|\ell|^2)}, \quad k^2 = \frac{(b-c)(a|\ell|^2 - 2h)}{(a-b)(2h - |\ell|^2 c)},$$

and

$$A^2 = \frac{2h - |\ell|^2 c}{a-c}, \quad B^2 = \frac{a|\ell|^2 - 2h}{a-b}, \quad C^2 = \frac{a|\ell|^2 - 2h}{a-c}.$$

Exercise: if two of the moments of inertia are equal you get the case of precession, which can be integrated with sines and cosines.

2.5 The Euler-Arnol'd equations

We know all the solutions of Euler's equations and seem to know everything, yet we still have to integrate $\dot{A}(t)$ to obtain the motion of the top in space.

To find the motion of the Euler top $\mathcal{B}$ in the *space frame*, assuming we know the solution $t \mapsto \Omega(t) = I^{-1}(p(t))$ of Euler's equations (2.12), we first convert the curve $\Omega : \mathbf{R} \to \mathbf{R}^3 : t \mapsto \Omega(t)$ of angular velocities into a curve $\xi : \mathbf{R} \to \mathrm{so}(3) : t \mapsto \xi(t)$ of infinitesimal motions. To do this we choose $\xi(t)$ so that $\xi(t) = \widehat{\Omega(t)}$. Since $\xi(t) = A(t)^{-1} \frac{dA}{dt}$ we have

$$\frac{dA}{dt} = A(t)\, \xi(t), \tag{2.24}$$

which is a system of linear differential equations with time dependent coefficients. Here $A : \mathbf{R} \to \mathrm{SO}(3) : t \mapsto A(t)$ is the curve we would like to find, as it describes the motion of the body in the space frame. Equations (2.24) and (2.16) are the *Euler-Arnol'd* equations of a rigid body with respect to the space frame. To find a particular solution of the Euler-Arnol'd equations of course initial conditions need to be imposed. Note that if $t \mapsto (\xi(t), A(t))$ is a solution of the Euler-Arnol'd equations and if A_0 is a fixed rotation, then $t \mapsto (\xi(t), A_0(A(t)))$ is also a solution of the Euler-Arnol'd equations.

Euler's equations (2.13) only describe the motion of the angular velocity vector $\Omega(t)$ (or the angular momentum vector $L(t) = I(\Omega(t)))$ in a frame corotating with the body. Note that this corotating frame is *not* an inertial frame. With respect to the space frame the angular momentum vector ℓ is actually constant through out the motion of the body. The center of mass of the body is fixed at the origin. The body does not necessarily come back to the same position even if the motion of the angular momentum vector L in the body frame is periodic. If the angular momentum vector has returned after time t to the same position in the body frame, all one can conclude is that the body frame has rotated in space around the angular momentum vector ℓ. To describe the motion of $\mathcal{B}$ in space we have to determine how much the body frame has rotated about the angular momentum vector ℓ after time t. First we choose a better space frame, namely a frame $\widetilde{\mathcal{F}} = \{\widetilde{f}_1, \widetilde{f}_2, \widetilde{f}_3\}$ so that the angular momentum of the body lies along the positive $\widetilde{f}_3$-axis. Let A_0 be the matrix whose j^{th} column looks like e_j in the new space frame. Then the angular momentum vector looks like $\widetilde{\ell} = A_0\ell = |\ell|\widetilde{f}_3$ in the new space frame $\{\widetilde{f}_1, \widetilde{f}_2, \widetilde{f}_3\}$. Write $\widetilde{A}(t) = (A_0\, A)(t)$. The jth column of $\widetilde{A}(t)$ describes the jth member of the body frame in the new space frame.

Physicists are a tough bunch of people. They like the old stuff. They still think that Poinsot solves everything, even though Poincaré did not. Now you think that Cushman's going to fall on his nose — but he didn't!

2.5.1 Qualitative Poinsot description

Since

$$I^{-1}(L(t)) = \Omega(t) = (\widetilde{A}(t))^{-1}\omega(t),$$

in order to describe the motion of the body in space it suffices to know the angular velocity vector $\omega(t)$ in the new space frame. We now give a geometric interpretation, due to Poinsot [31], of the curve $t \mapsto -\omega(t)$, see figure 2.1.

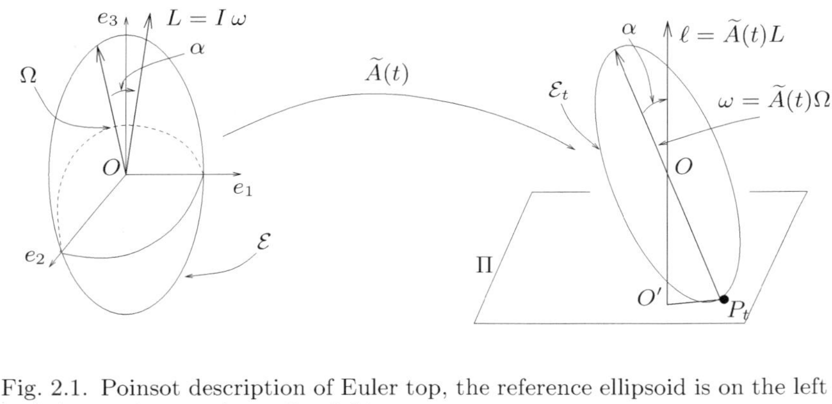

Fig. 2.1. Poinsot description of Euler top, the reference ellipsoid is on the left and the moving one is on the right.

I had a lot of trouble reading Goldstein [19] about all this, and usually this means that he is wrong (at least in my experience). Your best reference remains Whittaker [42].

Recall $\Omega(t)$ lies on the reference ellipsoid $\mathcal{E} = \{\Omega \in \mathbf{R}^3 \,|\, \langle I\Omega, \Omega\rangle = 2h\}$. Since $\omega(t) = \widetilde{A}(t)\Omega(t)$, the vector $\omega(t)$ lies on the ellipsoid $\mathcal{E}_t$ which is obtained by applying the rotation $\widetilde{A}(t)$ to the reference ellipsoid.

That is what rotating coordinates is all about guys!

We give more details. If $\widetilde{I}(t) = \widetilde{A}(t)\, I(\widetilde{A}(t))^{-1}$ denotes the matrix of the moment of inertia tensor with respect to the space frame $\{\widetilde{f}_1, \widetilde{f}_2, \widetilde{f}_3\}$, then $\mathcal{E}_t = \{\omega \in \mathbf{R}^3 \,\big|\, \langle \widetilde{I}(t)\omega, \omega\rangle = 2h\}$. We may think of $\mathcal{E}_t$ as an ellipsoid, which moves with respect to the space frame $\{\widetilde{f}_1, \widetilde{f}_2, \widetilde{f}_3\}$, with center of mass fixed at O. The inner product between the angular momentum vector $\widetilde{\ell} = \widetilde{I}(t)\omega(t)$ in the space frame and the angular velocity vector $\omega(t)$ in the space frame is $2h$ and $\widetilde{\ell}$ is constant. Hence $-\omega(t)$ lies on a fixed affine plane Π, which is perpendicular to $\widetilde{\ell}$ and consists of those vectors whose inner product with $-\widetilde{\ell}$ is $2h$. Let $-\omega(t)$ be the vector $\overrightarrow{OP_t}$. The point P_t lies on the plane Π as well as on the moving ellipsoid $\mathcal{E}_t$.

Fix $t = t_0$. Since the normal to $\mathcal{E}_t$ at P_t is

$$\mathrm{grad}_{-\omega}\langle I(t)\omega, \omega\rangle = -2\, I(t)\omega \;=\; -2\,\widetilde{\ell},$$

which is parallel to $\widetilde{\ell}$, the plane Π is tangent to $\mathcal{E}_t$ at P_t. Thus P_t is the point of contact of the ellipsoid $\mathcal{E}_t$ with Π. Consider the point P_0 on the reference ellipsoid whose image under $\widetilde{A}(t_0)$ is P_{t_0}. The velocity of the image of P_0 under $\widetilde{A}(t)$ with respect to the space frame at $t = t_0$ is

$$\omega \times \overrightarrow{OP_t} = \omega \times (-\omega) \;=\; 0.$$

This means that the moving ellipsoid $\mathcal{E}_t$ rolls without slipping on the plane Π. Its center of mass is fixed at O, which is a constant height $\frac{2h}{|\ell|}$ above Π. Thus $t \to -\omega(t)$ is the curve traced out on the invariant plane Π by the point of contact P_t of the rolling ellipsoid $\mathcal{E}_t$.

Many people – including Arnol'd – stop here without showing how to find the point of contact P_t. Thus Poinsot is not a quantitative solution as it should be. We have to find P_t and show what the body is doing.

2.5.2 Integration of the Euler-Arnol'd equations

Given a solution $t \mapsto \Omega(t)$ of Euler's equations, we now find a formula for the position $\widetilde{A}(t)$ of the body frame with respect to the space frame $\{\widetilde{f}_1, \widetilde{f}_2, \widetilde{f}_3\}$.

Now, what is the most friendly parametrization of the rotation group? Some people will say Euler angles — without even thinking. I use a different one, namely, two orthonormal vectors.

Write $x(t)$ for the first column of $\widetilde{A}(t)$ and $y(t)$ for the second. Then

$$\widetilde{A}(t) = (A_0\, A)(t) \;=\; \mathrm{col}(x(t), y(t), x(t) \times y(t)), \tag{2.25}$$

with

$$\langle x(t), x(t)\rangle = 1, \quad \langle y(t), y(t)\rangle = 1, \text{ and } \langle x(t), y(t)\rangle = 0. \tag{2.26}$$

Note that $\dfrac{d\widetilde{A}}{dt} = \widetilde{A}(t)\xi(t)$, so equation (2.24) takes the form

$$\mathrm{col}\big(\dot{x}(t), \dot{y}(t), (x \dot{\times} y)(t)\big) = \mathrm{col}\big(x(t), y(t), (x \times y)(t)\big)\begin{pmatrix} 0 & -\Omega_3 & \Omega_2 \\ \Omega_3 & 0 & -\Omega_1 \\ -\Omega_2 & \Omega_1 & 0 \end{pmatrix}.$$

Because the third column in the above equation is redudant, we see that the Euler-Arnol'd equations (2.24) and (2.16) are equivalent to the following vector equations

$$\dot{x} = \Omega_3\, y - \Omega_2(x \times y), \tag{2.27a}$$
$$\dot{y} = -\Omega_3\, x + \Omega_1(x \times y), \tag{2.27b}$$
$$I(\dot{\Omega}) = I(\Omega) \times \Omega. \tag{2.27c}$$

subject to the constraints (which follow from (2.26))

$$x_1^2 + x_2^2 + x_3^2 = 1, \tag{2.28a}$$
$$y_1^2 + y_2^2 + y_3^2 = 1, \tag{2.28b}$$
$$x_1y_1 + x_2y_2 + x_3y_3 = 0. \tag{2.28c}$$

With no choice of chart of any kind we have reduced the motion of the Euler top in space to six equations (plus restrictions). These equations are Hamiltonian even though the symplectic form is a mess. This is in the blue book [5]. There is no other reference. Numerically these equations are incredibly stable near the unstable manifold where the motion of Euler's top is the most interesting.

Since we have chosen the body frame so that the matrix of the moment of inertia tensor I is $\mathrm{diag}(I_1, I_2, I_3)$ and since

$$\begin{aligned} I(\Omega) =& L = (\widetilde{A}(t))^{-1}\widetilde{\ell} = \widetilde{A}(t)^t\,\widetilde{\ell} = |\ell|\,\widetilde{A}(t)^t\,\widetilde{f}_3 \\ &= |\ell|\ \mathrm{row}(x(t), y(t), (x \times y)(t))\widetilde{f}_3, \end{aligned} \tag{2.29}$$

we obtain

$$x_3 = |\ell|^{-1} I_1\, \Omega_1 = M_1 \tag{2.30a}$$
$$y_3 = |\ell|^{-1} I_2\, \Omega_2 = M_2 \tag{2.30b}$$
$$x_1y_2 - x_2y_1 = |\ell|^{-1}\, I_3\Omega_3 = M_3. \tag{2.30c}$$

Suppose that we know a solution $t \mapsto \Omega(t) = (\Omega_1(t), \Omega_2(t), \Omega_3(t))$ of Euler's equations (2.16) whose energy is h and whose angular momentum has magnitude $|\ell|$. The rotating frame $\{x, y, x \times y\}$ gives the position of the body with respect to the space frame $\{\widetilde{f}_1, \widetilde{f}_2, \widetilde{f}_3\}$. We want to find how much the $\widetilde{f}_1$-$\widetilde{f}_2$-component of the vector x has rotated about the $\widetilde{f}_3$-axis after time t. More precisely, we seek a differential equation for the angle θ that the projection of x on the $\widetilde{f}_1$-$\widetilde{f}_2$-plane makes with the $\widetilde{f}_1$-axis.

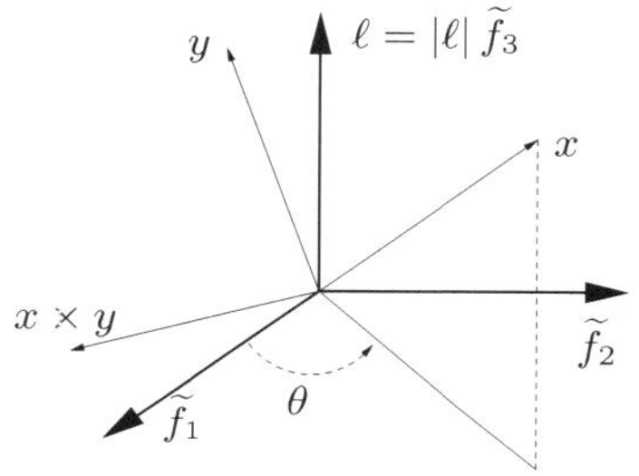

Fig. 2.2. Definition of the angle θ.

Because $\Omega = (\Omega_1, \Omega_2, \Omega_3)$ are assumed to be known functions of time, from (2.30a) and (2.30b) we see that x_3 and y_3 are also known. Eliminating x_3 and y_3 from (2.27a), (2.27b), (2.28a), (2.28b), (2.28c) and using (2.30a) and (2.30b) gives

$$\begin{aligned} \frac{d}{dt}\begin{pmatrix} x_1 \\ x_2 \end{pmatrix} &= |\ell| I_3^{-1} M_3 \begin{pmatrix} y_1 \\ y_2 \end{pmatrix} - |\ell| I_2^{-1} M_2 \begin{pmatrix} -M_1\, y_2 + M_2\, x_2 \\ M_1\, y_1 - M_2\, x_1 \end{pmatrix} \\ \frac{d}{dt}\begin{pmatrix} y_1 \\ y_2 \end{pmatrix} &= -|\ell| I_3^{-1} M_3 \begin{pmatrix} x_1 \\ x_2 \end{pmatrix} + |\ell| I_1^{-1} M_1 \begin{pmatrix} -M_1\, y_2 + M_2\, x_2 \\ M_1\, y_1 - M_2\, x_1 \end{pmatrix} \end{aligned}$$

$$\begin{aligned} x_1^2 + x_2^2 &= 1 - |\ell|^{-2} I_1^2\, \Omega_1^2 = 1 - M_1^2 && \text{(2.31a)} \\ y_1^2 + y_2^2 &= 1 - |\ell|^{-2} I_2^2\, \Omega_2^2 = 1 - M_2^2 && \text{(2.31b)} \\ x_1 y_1 + x_2 y_2 &= -|\ell|^{-2} I_1 I_2\, \Omega_1 \Omega_2 = -M_1 M_2. && \text{(2.31c)} \end{aligned}$$

Suppose that $\Omega \neq \left(\pm\frac{|\ell|}{I_1}, 0, 0\right)$. This is equivalent to assuming that the solution $t \to \Omega(t)$ of Euler's equations of energy h and magnitude of the angular momentum $|\ell|$ does not correspond to either one of the equilibrium points $\pm\frac{1}{I_1} e_1$. Consequently, the right hand side of (2.31a) is never zero. Writing (2.31c) and (2.30c) as

$$\begin{pmatrix} x_1 & x_2 \\ -x_2 & x_1 \end{pmatrix} \begin{pmatrix} y_1 \\ y_2 \end{pmatrix} = \begin{pmatrix} -M_1 M_2 \\ M_3 \end{pmatrix}$$

We may solve this equation for y_1 and y_2 obtaining

$$y_1 = \frac{-1}{1-M_1^2}\Big(M_1M_2\,x_1 + M_3\,x_2\Big) \tag{2.32a}$$

$$y_2 = \frac{1}{1-M_1^2}\Big(M_3\,x_1 - M_1M_2\,x_2\Big). \tag{2.32b}$$

In studying the Hopf fibration in the 2-dimensional harmonic oscillator (section 1.3), we have had to solve similar linear *equations.*

We now obtain the world's simplest differential equation (linear with time dependent coefficients)! Substituting (2.32a) and (2.32b) into the second equation above (2.31a) gives

$$\begin{aligned} \frac{dx_1}{dt} &= \alpha\,x_1 - \beta\,x_2 \\ \frac{dx_2}{dt} &= \beta\,x_1 + \alpha\,x_2, \end{aligned} \tag{2.33}$$

where

$$\begin{aligned} \alpha &= |\ell|(I_2^{-1} - I_3^{-1})\,\frac{M_1M_2M_3}{1-M_1^2} = I_1(I_3 - I_2)\,\frac{\Omega_1(t)\Omega_2(t)\Omega_3(t)}{|\ell|^2 - I_1^2\Omega_1^2(t)} \\ \beta &= |\ell|\,\frac{I_2^{-1}M_2^2 + I_3^{-1}M_3^2}{1-M_1^2} = |\ell|\,\frac{I_2\Omega_2^2(t) + I_3\Omega_3^2(t)}{|\ell|^2 - I_1^2\Omega_1^2(t)}. \end{aligned} \tag{2.34}$$

It cannot be that ridiculously simple — but it is. After all we are on a circle.

The angle θ that the projection of the vector x on the $\widetilde{f}_1$-$\widetilde{f}_2$ plane makes with the $\widetilde{f}_1$ axis is $\tan^{-1}(x_2/x_1)$. Therefore, using (2.33),

$$\dot\theta = \frac{x_1\dot x_2 - x_2\dot x_1}{x_1^2 + x_2^2} = \beta. \tag{2.35}$$

Oops, polar coordinates are trying to stick their ugly puss in here — but θ is an angle parametrizing a circle, so we are OK.

Integrating (2.35) gives

$$\theta(t) = \theta(0) + |\ell| \int_0^t \frac{I_2\,\Omega_2^2(s) + I_3\,\Omega_3^2(s)}{|\ell|^2 - I_1^2\,\Omega_1^2(s)}\,ds. \tag{2.36}$$

I am not a master of Weierstrass' theory of elliptic functions, so I won't do this integral. But Whittaker [42] is and he does it. See also [1].

θ is the rotation angle (a physical parameter) of the flow of the Euler-Arnol'd equations on a connected component of $E^{-1}(h) \cap L^{-1}(\ell)$, which is a 2-dimensional torus.

Montgomery [28] has found the rotation angle by hard *work.*

Knowing $t \mapsto \theta(t)$ and the $t \mapsto \Omega_i(t)$ we will now find the curve of rotations

$$t \mapsto \widetilde{A}(t) = \mathrm{col}(x(t), y(t), (x \times y)(t)),$$

which determines the position of the body with respect to the space frame $\{\widetilde{f}_1, \widetilde{f}_2, \widetilde{f}_3\}$. From the definition of θ and (2.31a) we find that

$$\begin{aligned} x_1(t) &= \sqrt{x_1^2 + x_2^2}\,\cos\theta = \sqrt{1 - M_1^2}\,\cos\theta && (2.37\mathrm{a}) \\ x_2(t) &= \sqrt{1 - M_1^2}\,\sin\theta && (2.37\mathrm{b}) \\ x_3(t) &= M_1, \quad \text{using (2.30a)}. \end{aligned}$$

Substituting (2.37a) and (2.37b) into (2.32a) and (2.32b) gives

$$\begin{aligned} y_1(t) &= \frac{-1}{\sqrt{1 - M_1^2}}\Big[M_1M_2\,\cos\theta + M_3\sin\theta\Big] && (2.38\mathrm{a}) \\ y_2(t) &= \frac{1}{\sqrt{1 - M_1^2}}\Big[M_3\cos\theta - M_1M_2\,\sin\theta\Big] && (2.38\mathrm{b}) \\ y_3(t) &= M_2, \quad \text{using (2.30b)}. && (2.38\mathrm{c}) \end{aligned}$$

Therefore

$$\begin{aligned} (x \times y)_1(t) &= \frac{-1}{\sqrt{1 - M_1^2}}\Big[M_1M_3\,\cos\theta - M_2\sin\theta\Big] \\ (x \times y)_2(t) &= \frac{-1}{\sqrt{1 - M_1^2}}\Big[M_2\cos\theta + M_1M_3\,\sin\theta\Big] \\ (x \times y)_3(t) &= M_3, \quad \text{using (2.30c)}. \end{aligned}$$

Thus we have found the curve $t \to \widetilde{A}(t)$ of motion of the body in space under the assumption that we know $t \to \Omega(t)$ and $t \to \theta(t)$.

So this is the solution of the Euler-Arnol'd equations – complete, straightforward, explicit, except for one quadrature. You can grumble, but not very much.

2.5.3 Quantitative Poinsot description

Using the curve $t \mapsto \widetilde{A}(t)$, which describes the motion of the body in space, we give an explicit parametrization of the curve $t \mapsto -\omega(t)$ traced out by the point of contact of the rolling moment of inertia ellipsoid on the invariant plane Π. This makes Poinsot's description of the motion of the Euler top in space *quantitative.*

We now find the instantaneous angular velocity $\omega(t)$ of the body $\mathcal{B}$

at time t with respect to the new space frame. By definition $\omega(t) = \widetilde{A}(t)\,\Omega(t)$. We compute the components of $\omega(t)$ as follows. From the construction of $\widetilde{A}(t)$ we have

$$\begin{aligned}
\omega_1(t) &= x_1\,\Omega_1 + y_1\,\Omega_2 + (x\times y)_1\Omega_3 \\
&= \frac{1}{\sqrt{1-M_1^2}}\Big\{\Big[\Omega_1 - M_1(M_1\Omega_1 + M_2\Omega_2 + M_3\Omega_3)\Big]\cos\theta + \\
&\qquad\qquad + \Big[\Omega_3 M_2 - \Omega_2 M_3\Big]\sin\theta\Big\} \\
&= \frac{1}{\sqrt{1-M_1^2}}\Big\{\Big[\Omega_1 - 2|\ell|^{-1}h\,M_1\Big]\cos\theta - \Big[\Omega_2 M_3 - \Omega_3 M_2\Big]\sin\theta\Big\}.
\end{aligned}$$

A similar argument gives

$$\omega_2(t) = \frac{1}{\sqrt{1-M_1^2}}\Big\{\Big[\Omega_2 M_3 - \Omega_3 M_2\Big]\cos\theta + \Big[\Omega_1 - 2|\ell|^{-1}h\,M_1\Big]\sin\theta\Big\}.$$

Also

$$\omega_3(t) = M_1\Omega_1 + M_2\Omega_2 + M_3\Omega_3 \;=\; 2|\ell|^{-1}h.$$

Note that this confirms that the inner product of the angular velocity vector with the angular momentum vector $|\ell|\widetilde{f}_3$ is constant. The above results may be written in matrix form as

$$\begin{aligned}
\begin{pmatrix}\omega_1\\ \omega_2\\ \omega_3\end{pmatrix} &= \begin{pmatrix}\Omega_1 - 2|\ell|^{-1}h\,M_1 & -(\Omega_2 M_3 - \Omega_3 M_2) & 0\\ \Omega_2 M_3 - \Omega_3 M_2 & \Omega_1 - 2|\ell|^{-1}h\,M_1 & 0\\ 0 & 0 & 1\end{pmatrix}\begin{pmatrix}\frac{\cos\theta}{\sqrt{1-M_1^2}}\\ \frac{\sin\theta}{\sqrt{1-M_1^2}}\\ 2|\ell|^{-1}h\end{pmatrix} \\
&= \begin{pmatrix}\cos u & -\sin u & 0\\ \sin u & \cos u & 0\\ 0 & 0 & 1\end{pmatrix}\begin{pmatrix}R\cos\theta\\ R\sin\theta\\ 2|\ell|^{-1}h\end{pmatrix},
\end{aligned}$$

where

$$\tan u(t) = \frac{\Omega_2 M_3 - \Omega_3 M_2}{\Omega_1 - 2|\ell|^{-1}h\,M_1} = \frac{|\ell|(I_3 - I_2)\Omega_2(t)\Omega_3(t)}{(|\ell|^2 - 2I_1 h)\Omega_1(t)} \tag{2.39a}$$

and

$$\begin{aligned}
R(t) &= \sqrt{\frac{(\Omega_1 - 2|\ell|^{-1}h\,M_1)^2 + (\Omega_2 M_3 - \Omega_3 M_2)^2}{1-M_1^2}} \\
&= \frac{1}{|\ell|}\sqrt{\frac{(|\ell|^2 - 2I_1 h)^2\,\Omega_1^2(t) + |\ell|^2(I_3 - I_2)\,\Omega_2^2(t)\Omega_3^2(t)}{|\ell|^2 - I_1^2\Omega_1^2(t)}}.
\end{aligned} \tag{2.39b}$$

Therefore we obtain

$$t \to \omega(t) = \begin{pmatrix} R(t)\cos(\theta(t) + u(t)) \\ R(t)\sin(\theta(t) + u(t)) \\ 2|\ell|^{-1} h \end{pmatrix}. \tag{2.40}$$

Remember that $u(t)$ is obtained from our solution to the Euler's equations. For the curve Γ traced out on the invariant plane Π by the point of contact P_t of the rolling ellipsoid $\mathcal{E}_t$ we get

$$t \to -\omega(t) = \begin{pmatrix} R(t)\cos(\theta(t) + u(t) + \pi) \\ R(t)\sin(\theta(t) + u(t) + \pi) \\ -2|\ell|^{-1} h \end{pmatrix}. \tag{2.41}$$

So we have parameterized Γ at no additional cost. Γ lies in an annulus

$$\mathcal{A} = \{(\varphi, R) \in \Pi \,\big|\, 0 < R_{\min} \le R(t) \le R_{\max}\}$$

and is alternately tangent to a different component of the boundary of $\mathcal{A}$. The rotation angle θ of the flow of the Euler-Arnol'd equations on a connected component of $E^{-1}(h) \cap L^{-1}(\ell)$ is the angle between every *second* (and not every) point of tangency on the same boundary component, if $a\ell^2 > 2h > b\ell^2$ and this angle *plus* 2π, if $c\ell^2 < 2h < b\ell^2$. For more details see [1] and [5].

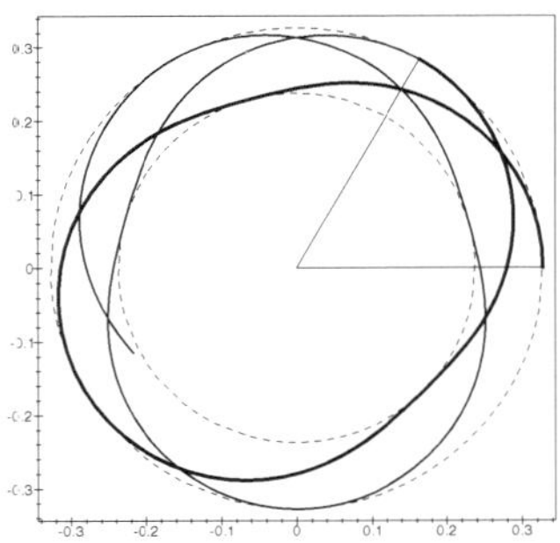

Fig. 2.3. The rotation angle of the flow of the Euler-Arnol'd equations on a 2-torus in $E^{-1}(h) \cap L^{-1}(\ell)$ as determined by the Poinsot description. The moments of inertia in this example are $I_1 = 1$, $I_2 = 2$, and $I_3 = 2.9$; $|\ell|$ is set equal to 1, initial point $(x, y) = \left(\frac{1}{\sqrt{2}}, 0, \frac{1}{\sqrt{2}}, 0, 1, 0\right)$, the initial point for Euler's equations is $(\Omega_1, \Omega_2, \Omega_3) = \left(\frac{1}{\sqrt{2}}, 0, \frac{1}{2.9\sqrt{2}}\right)$ with Euler energy $h = .344$, the Euler period is 17, and rotation number equals 1.1666.

2.6 Abstract derivation of equations of motion

In this section we give a Hamiltonian derivation of the Euler–Arnol'd equations.

2.6.1 Geodesic equations on a Lie group

Let G be a Lie group with the algebra $\mathfrak{g}$. On phase space T^*G with its canonical symplectic form $\widetilde{\omega}$ suppose that we have a hamiltonian $\widetilde{\mathcal{H}} : T^*G \to \mathbf{R}$. Consider the map *left translation* by $g \in G$, namely,

$$L_g : G \to G : h \to g \cdot h.$$

We have the *left trivialization*

$$\widetilde{\lambda} : G \times \mathfrak{g}^* \to T^*G : (g, \alpha) \mapsto (T_g L_{g^{-1}})^t \alpha = \alpha_g.$$

Warning. This trivialization does *not* give coordinates, because for $\xi \in \mathfrak{g}$ the coordinate vector fields $X^{\xi}(g) = \left.\frac{d}{dt}\right|_{t=0} L_{\exp t\xi}(g)$ on G, which are dual to the 1-forms α^{ξ}, do *not* commute.

2.6.2 Hamilton's equations on a Lie group

On $G \times \mathfrak{g}^*$ we have the 2-form $\widetilde{\Omega}$, which is the pull back by $\widetilde{\lambda}$ of the canonical 2-form $\widetilde{\omega}$. Thus

$$\widetilde{\Omega}(g, \alpha)\big((T_e L_g \xi, \beta), (T_e L_g \eta, \gamma)\big) = -\beta(\eta) + \gamma(\xi) + \alpha([\xi, \eta]),$$

where $\xi, \eta \in \mathfrak{g}$ and $\alpha, \beta, \gamma \in \mathfrak{g}^*$, see [5]appendix A. Pulling $\widetilde{\mathcal{H}}$ back by $\widetilde{\lambda}$ gives the Hamiltonian

$$\widetilde{H} = \widetilde{\lambda}^* \widetilde{\mathcal{H}} : G \times \mathfrak{g}^* \to \mathbf{R} : (g, \alpha) \mapsto \widetilde{H}(g, \alpha).$$

By definition, the hamiltonian vector field $X_{\widetilde{H}}$ satisfies $X_{\widetilde{H}} \mathbin{\lrcorner} \widetilde{\Omega} = \mathrm{d}\widetilde{H}$. The integral curves of $X_{\widetilde{H}}$ are the solutions of the Euler-Arnol'd equations

$$\begin{aligned} \dot{g} &= T_e L_g D_2 \widetilde{H}(g, \alpha) \\ \dot{\alpha} &= -(T_e L_g)^t D_1 \widetilde{H}(g, \alpha) + \mathrm{ad}^t_{D_2 \widetilde{H}} \alpha \end{aligned} \tag{2.42}$$

These are *Hamilton's equations* on the Lie group G.

Suppose that $\mathcal{H}$ is *left invariant*, that is, $\widetilde{\mathcal{H}}(\alpha_{gh}) = \widetilde{\mathcal{H}}(\alpha_h)$, Then

$$\widetilde{H}(g \cdot h, \alpha) = \widetilde{H}(h, \alpha).$$

Since $D_1\widetilde{H} = 0$, the Euler–Arnol'd equations become

$$\begin{aligned}\dot{g} &= T_eL_g D_2\widetilde{H}(g,\alpha)\\ \dot{\alpha} &= \mathrm{ad}^t_{D_2\widetilde{H}}\alpha.\end{aligned} \tag{2.43}$$

2.6.3 Special case

Let $\mathcal{H}^*(\alpha_g) = \frac{1}{2}\, k^*(g)(\alpha_g, \alpha_g)$ be a hamiltonian on T^*G, where k^* is a left invariant dual metric on G. In other words,

$$k(g)(v_g, w_g) = k^*(g)\big(k^*(g)^\flat(v_g), k^*(g)^\flat(w_g)\big)$$

is a left invariant metric on G. The hamiltonian $\mathcal{H}^*$ is purely kinetic. We now show that the solutions of the Euler-Arnol'd equations (2.43) for $\mathcal{H}^*$ give the geodesic flow on the Lie group G.

Pull back the Hamiltonian system $(\mathcal{H}^*, T^*G, \widetilde{\omega})$ by the map $k^\sharp : TG \to T^*G$. The resulting Hamiltonian on $(TG, \omega = k^\flat\widetilde{\omega})$ is

$$\mathcal{H}(v_g) = \tfrac{1}{2}\, k(g)(v_g, v_g)$$

. Pulling the symplectic form ω back by the left trivialization of TG

$$\lambda : G \times \mathfrak{g} \to TG : (g, v) \mapsto v_g = T_eL_g v$$

gives a 2-form Ω on $G \times \mathfrak{g}$ on TG. Explicitly,

$$\Omega(g,v)((T_eL_g\xi, v), (T_eL_g\eta, w)) = -\mathrm{k}(v,\eta) + \mathrm{k}(w,\xi) + \mathrm{k}(v, [\xi,\eta]), \tag{2.44}$$

where $\mathrm{k} = \lambda^* k(e)$. The Hamiltonian $\mathcal{H}$ becomes

$$H = \lambda^*\mathcal{H} : G \times \mathfrak{g} \to \mathbf{R} : (g,v) \to \tfrac{1}{2}\,\mathrm{k}(v,v),$$

which is a left invariant metric on $G \times \mathfrak{g}$. By definition the hamiltonian vector field $X_H(g,v) = (T_eL_gX_1, X_2)$ satisfies

$$X_H \;\lrcorner\; \Omega = \mathrm{d}H. \tag{2.45}$$

We compute X_H as follows. From (2.45) and (2.44) we obtain

$$\begin{aligned}-\mathrm{k}(X_2,\eta) + \mathrm{k}(X_1,w) + \mathrm{k}(v,[X_1,\eta]) &= \Omega(g,v)((T_eL_gX_1, X_2), (T_eL_g\eta, w))\\ &= \mathrm{d}H(g,v)(T_eL_g\eta, w) = D_1H(g,v)T_eL_g\eta + D_2H(g,v)w\\ &= \mathrm{k}(v,w),\end{aligned} \tag{2.46}$$

for every $(w,\eta) \in \mathfrak{g}$. In (2.46) set $\eta = 0$. Then $\mathrm{k}(X_1,w) = \mathrm{k}(v,w)$ for every $w \in \mathfrak{g}$. Since k is nondegenerate, we have $X_1 = v$. In (2.46) set $w = 0$. Then

$$\mathrm{k}(X_2,\eta) = \mathrm{k}(v,[v,\eta]) = \mathrm{k}(B(v),\eta),$$

for every $\eta \in g$. This implies $X_2 = B(v)$. Consequently, the Euler-Arnol'd equations are

$$\begin{aligned} \dot{g} &= T_e L_g v \\ \dot{v} &= B(v). \end{aligned} \tag{2.47}$$

These are equations for geodesics on G of the left invariant metric k.

2.6.4 An even more special case

Now we restrict ourselves to the case when G is semisimple. Then $\mathfrak{g}$ has an Ad-invariant nondegenerate inner product k called the Killing metric.

Using a k-symmetric invertible linear mapping $I : \mathfrak{g} \rightarrow \mathfrak{g}$, called the *generalized moment of inertia tensor*, we can write every metric k on $\mathfrak{g}$ as

$$k(v, w) = \mathrm{k}(I(v), w).$$

Thus for every $\eta \in g$

$$k(I(B(v)), \eta) = \mathrm{k}(B(v), \eta) = \mathrm{k}(v, [v, \eta]) = \mathrm{k}(I(v), [v, \eta]) = k([I(v), v], \eta),$$

since k is Ad-invariant. Consequently, $B(v) = I^{-1}([I(v), v])$. Hence the Euler-Arnol'd equations for geodesics of a left invariant metric on a semisimple Lie group are

$$\begin{aligned} \dot{g} &= T_e L_g v \\ \dot{v} &= I^{-1}[I(v), v]. \end{aligned} \tag{2.48}$$

Now suppose that $G = \mathrm{SO}(3)$ and $\mathfrak{g} = \mathrm{so}(3) \simeq (\mathbf{R}^3, \times)$. In addition, suppose that k is the Euclidean inner product on $\mathbf{R}^3$ and that the moment of inertia tensor I is $\mathrm{diag}(I_1, I_2, I_3)$. Then the solutions of the Euler-Arnol'd equations (2.48) give integral curves of a vector field V on $\mathrm{SO}(3) \times \mathbf{R}^3$ which satisfy

$$\begin{aligned} \dot{A} &= A\widehat{\Omega} \\ I(\dot{\Omega}) &= I(\Omega) \times \Omega, \end{aligned} \tag{2.49}$$

The above equations are the Euler-Arnol'd equations for the Euler top. They are *geodesic equations* and are Hamiltonian even though they do not look like it.

2.6.5 Integrals and reduction

The vector field V on $\mathrm{SO}(3) \times \mathbf{R}^3$ has two integrals:

Energy. $E(A,\Omega) = \frac{1}{2}\langle I(\Omega),\Omega\rangle$.

Check.

$$\dot{E} = \langle I(\dot{\Omega}),\Omega\rangle = \langle I(\Omega)\times\Omega,\Omega\rangle = 0.$$

and

Angular momentum. $L(A,\Omega) = AI\Omega$.

Check.

$$\begin{aligned}\dot{L} &= \dot{A}I(\Omega) + AI(\dot{\Omega}) = A\big(\widehat{\Omega}(I(\Omega))\big) + A(I(\Omega)\times\Omega)\\ &= A(\Omega\times I(\Omega) + I(\Omega)\times\Omega) = 0. \quad \square\end{aligned}$$

Angular momentum L comes from the lift of the action of left translation of SO(3) on itself to $T\,\mathrm{SO}(3)$. This action is Hamiltonian and commutes with the Hamiltonian H, which is a left invariant metric on SO(3). We have arranged that on $\mathrm{SO}(3)\times\mathbf{R}^3$ we have $L(A,\Omega) = \ell = |\ell|\, e_3$. Thus L is constant on the integral curves of V. Consequently,

$$L^{-1}(\ell) = \{(A, I^{-1}A^{-1}\ell) \in \mathrm{SO}(3)\times\mathbf{R}^3 \,\big|\, A \in \mathrm{SO}(3)\}$$

is an invariant manifold of the vector field V, which is diffeomorphic to SO(3).

Solid ball model of SO(3). Consider an open ball $\mathbf{B}^3_\pi \subseteq \mathbf{R}^3$, whose closure has boundary which is a 2-sphere $\mathbf{S}^2_\pi$ of radius π (see fig. 2.4). Every point in $\mathbf{B}^3_\pi$ is a vector ℓ of length less than π and defines a unique rotation about axis ℓ by angle $|\ell| < \pi$. The vectors $\pm\ell$ define rotation about $\pm\ell$ by angle $\pm|\ell|$. We should be more careful with the points on $\mathbf{S}^2_\pi$ where $|\ell| = \pi$. In this case, $\pm\ell$ define the *same* rotation. Therefore, we should identify diametrically opposite points of $\mathbf{S}^2_\pi$, so that it becomes real projective 2-space $\mathbf{RP}^2$. To obtain all of SO(3) we add $\mathbf{B}^3_\pi$ and obtain real projective 3-space $\mathbf{RP}^3$.[1] We obtain the same picture of

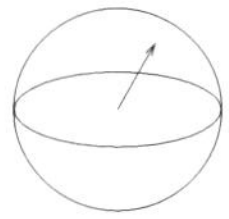

Fig. 2.4. Solid ball model of SO(3).

SO(3) ([5], p. 95) when we define the covering map $\mathbf{S}^3 \to \mathrm{SO}(3) \simeq \mathbf{RP}^3$.

[1] We have already seen that $\mathrm{so}(3) \simeq \mathbf{R}^3$. Correspondingly, SO(3) is locally an $\mathbf{R}^3$. Globally it is an $\mathbf{RP}^3$.

The 3-sphere $\mathbf{S}^3$ of unit radius is the space of all quaternions[2] q of unit length, $q\bar{q} = 1$. There is a one to one correspondence between these q s and the matrices

$$\begin{pmatrix} \alpha & \beta \\ -\overline{\beta} & \overline{\alpha} \end{pmatrix} \in \mathrm{SU}(2), \quad \alpha, \beta \in \mathbf{C}, \quad |\alpha|^2 + |\beta|^2 = 1,$$

which in turn correspond to rotations.[1] The exact correspondence is given by the two to one covering map $\rho : \mathrm{SU}(2) \to \mathrm{SO}(3); q \mapsto R_q$, where R_q is a real linear map

$$R_q : \mathbf{R}^3 \subseteq \mathbf{H} \to \mathbf{R}^3 \subseteq \mathbf{H} : x \to q \cdot x \cdot \bar{q},$$

and $x = x_1 i + x_2 j + x_3 k \in \mathbf{R}^3 \subseteq \mathbf{H}$. It can be shown that R_q is a rotation of $\mathbf{R}^3$ with its standard Euclidean inner product. It it is easy to see that q and $-q$ correspond to the same rotation R_q. Thus $\ker \rho = \mathbf{Z}_2$. Geometrically, this means that if we identify antipodal points q and $-q$ on the 3-sphere $\mathbf{S}^3 \subseteq \mathbf{R}^4$ we obtain real projective 3-space $\mathbf{RP}^3$. (This is similar to identifiying antipodal points on $\mathbf{S}^2$, which produces $\mathbf{RP}^2$.) After identifying antipodal points, the covering map ρ is a diffeomorphism. Hence SO(3) is $\mathbf{RP}^3$.

2.6.6 Reduction

Reduction on the ℓ level set of L. Consider the isotropy group

$$\mathrm{SO}(3)_\ell = \{B \in \mathrm{SO}(3) \big| B\ell = \ell\} \simeq S^1$$

of the action

$$\mathrm{SO}(3) \times (\mathrm{SO}(3) \times \mathrm{so}(3)) \to \mathrm{SO}(3) \times \mathrm{so}(3) : (B, (A, \widehat{\Omega}) \to (BA, B\widehat{\Omega}B^{-1}),$$

which comes from the lift of the action of inverse of *right* multiplication by SO(3) on itself to the left trivialization of the tangent bundle $T\,\mathrm{SO}(3)$ of SO(3). We have an action of $\mathrm{SO}(3)_\ell$ on $L^{-1}(\ell)$ defined by

$$\Phi : \mathrm{SO}(3)_\ell \times L^{-1}(\ell) \to L^{-1}(\ell) : (B, (A, \Omega)) \mapsto (BA, \Omega),$$

Check.

$$L(BA, \Omega) = B(AI\Omega) = BL(A, \Omega) = B\ell = \ell.$$

We now show that Φ is a proper free action with the orbit space $\mathbf{S}^2_{|\ell|}$.

[2] Recall that quaternions $\mathbf{H}$ are real linear combinations of $(1, i, j, k)$, where $i^2 = j^2 = k^2 = -1$ and $ij = k$ together with its cyclic permutations.

[1] Recall that $\mathrm{SU}(2) \to \mathrm{SO}(3)$ is a two to one homomorphism of groups. The variables α and β are known as Caley–Klein parameters for SU(2).

Proof. If $(A, \Omega) \in L^{-1}(\ell)$ and $B \in \mathrm{SO}(3)_\ell$ then $\Phi_B(A, \Omega) = (A, \Omega)$ implies that $BA = A$. Thus $B = e$, the identity element of SO(3). Hence the action Φ is free. Furthermore, since $\mathrm{SO}(3)_\ell$ is compact, Φ is a proper action. Consequently, the orbit space $L^{-1}(\ell)/\mathrm{SO}(3)_\ell$ is a smooth manifold.

Reduction map. Recall that $\ell = |\ell| e_3$. The reduction map removing the $\mathrm{SO}(3)_\ell$ symmetry on $L^{-1}(\ell)$ is

$$\pi_\ell : L^{-1}(\ell) \to \mathbf{S}^2_{|\ell|} : (A, I^{-1}A^{-1}\ell) \mapsto z = A^{-1}\ell = |\ell|\, A^{-1}e_3.$$

Check. If $\pi_\ell(A, \Omega) = \pi_\ell(A', \Omega')$, then $A^{-1}\ell = (A')^{-1}\ell$ implies $A'A^{-1} = B \in \mathrm{SO}(3)_\ell$. Thus

$$\begin{aligned}(A', \Omega') &= (A', I^{-1}(A')^{-1}\ell) = \big(BA, I^{-1}(A^{-1}B^{-1}\ell)\big) \\ &= (BA, I^{-1}A^{-1}\ell) = \Phi_B(A, \Omega).\end{aligned}$$

Therefore, π_ℓ^{-1} is a *unique* Φ-orbit in $L^{-1}(\ell)$ and the orbit space $L^{-1}(\ell)/\mathrm{SO}(3)_\ell$ is $\mathbf{S}^2_{|\ell|}$. □

Precomposing the reduction map π_ℓ with the two to one covering map $\mathbf{S}^3 \to \mathrm{SO}(3)$ gives the Hopf fibration. In other words, the double covering of the reduction map π_ℓ is the Hopf fibration.

Note that $\mathrm{SO}(3) \simeq \mathbf{RP}^3$ is not simply connected. Consequently, linking numbers *cannot* be defined for two closed curves in SO(3). That is why we need S^3. The double cover of the integral curves of the Euler top on a level set of angular momentum lie in S^3 and have linking number 1.

Reduced equations on $\mathbf{S}^2$. From $\ell = AI(\Omega)$ and $z = A^{-1}\ell$ it follows that $z = I(\Omega)$. Thus the reduced equations of motion are Euler's equations on $\mathbf{S}^2_{|\ell|}$, namely,

$$\dot{z} = I(\dot{\Omega}) = I(\Omega) \times \Omega = z \times I^{-1}(z).$$

Euler's equations are Hamilton's equations with respect to the symplectic form

$$\omega_{|\ell|}(z)(u, v) = \frac{1}{2|\ell|^2}\langle z, u \times v\rangle$$

and correspond to the reduced Hamiltonian

$$H_{|\ell|}(z) = \tfrac{1}{2}\,\langle I^{-1}(z), z\rangle.$$

$H_{|\ell|}$ is a Morse function on $\mathbf{S}^2_{|\ell|}$. $H_{|\ell|}$ is a *Morse function* with 6 nondegenerate critical points: 2 of index 0, 2 of index 1, and 2 of index 2 which are maxima, hyperbolic, and minima, respectively (see figure 2.5).[1] For each regular value $|\ell|$ the level set of $H_{|\ell|}$ on $\mathbf{S}^2_{|\ell|}$ consists of two equivalent disconnected circles. The nontrivial level set where $H_{|\ell|} = h_s = \frac{1}{2}\, b|\ell|^2$ is a heteroclinic connection and consists of two unstable relative equilibria connected by two great circles – their stable and unstable manifolds.

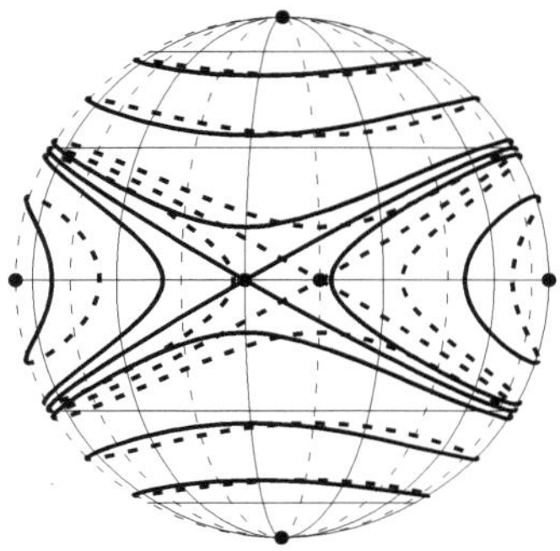

Fig. 2.5. Trajectories of the reduced Euler top on $\mathbf{S}^2_{|\ell|}$.

Qualitative reconstruction. We now reconstruct the geometry of the level sets of the reduced Hamiltonian $H_{|\ell|}$ on $\mathbf{S}^2_{|\ell|}$. This will describe how the level sets of the energy foliate a level set of angular momentum. Because the reduced Hamiltonian $H_{|\ell|}$ is a Morse function on $\mathbf{S}^2_{|\ell|}$ and because the Hamiltonian

$$H\big|_{L^{-1}(\ell)} : L^{-1}(\ell) \to \mathbf{R}$$

is $\mathrm{SO}(3)_\ell$ invariant, it follows that $H\big|_{L^{-1}(\ell)}$ is a *Bott-Morse function* on $\mathrm{SO}(3)$ with 6 nondegenerate critical circles: two of index 0, two of index 1 and two of index 2. Each critical point of $H_{|\ell|}$ lifts under the reduction map π_ℓ to a critical circle of $H\big|_{L^{-1}(\ell)}$ on $L^{-1}(\ell)$. A regular level set of $H_{|\ell|}$ lifts to two smooth 2-tori $\mathbf{T}^2$. The singular level set of $H^{-1}_{|\ell|}(h_s)$ corresponding to the hyperbolic critical points and their heteroclinic stable and unstable manifolds lifts to two 2-tori which intersect cleanly along two circles.

[1] Each pair of critical points with the same index lie on the same orbit of a nontrivial finite symmetry group $D_2 \times Z_2$ of the Euler top. They are said to be *equivalent*.

We now show how these level sets of $H_{|\ell|}$ fit together to form $L^{-1}(\ell)$. Remove a small open 2-disk $\mathbf{D}$ about the north pole of $\mathbf{S}^2_{|\ell|}$ and use stereographic projection to map $\mathbf{S}^2_{|\ell|} - \mathbf{D}$ onto a region $\mathcal{E} \subseteq \mathbf{R}^2$. The relative equilibria (except the one at the north pole) and the stable and unstable manifold are mapped into three points corresponding to elliptic relative equilibria and two circles which intersect transversely at point corresponding to the hyperbolic relative equilibria, respectively. The three elliptic points each lie in a compact region $\mathcal{D}_1 \cup \mathcal{D}_2$ bounded by the circles, see figure 2.6.

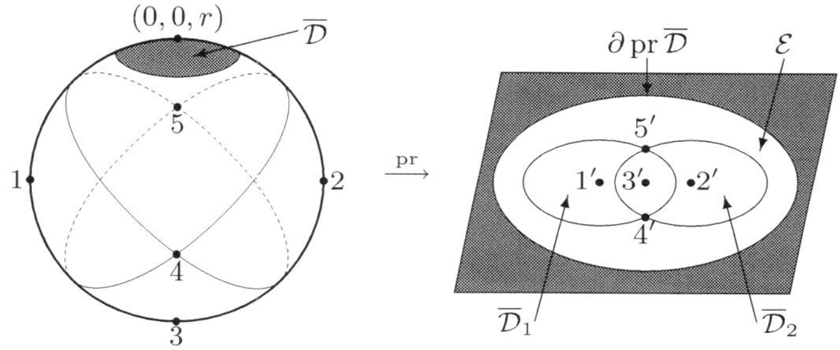

Fig. 2.6. The level sets of the reduced Hamiltonian flattened out. Stereographic projection of $\mathbf{S}^2$ without the north pole $(0, 0, r)$.

Remove the solid torus $\mathcal{E} \times \mathbf{S}^1$ from $\mathrm{SO}(3)$. We have to replace $(\mathcal{D}_1 \cup \mathcal{D}_2) \times \mathbf{S}^1$ (which is homeomorphic to $\mathcal{E} \times \mathbf{S}^1$) in the cored apple $\mathrm{SO}(3) - (\mathcal{E} \times \mathbf{S}^1)$, see figure 2.7.

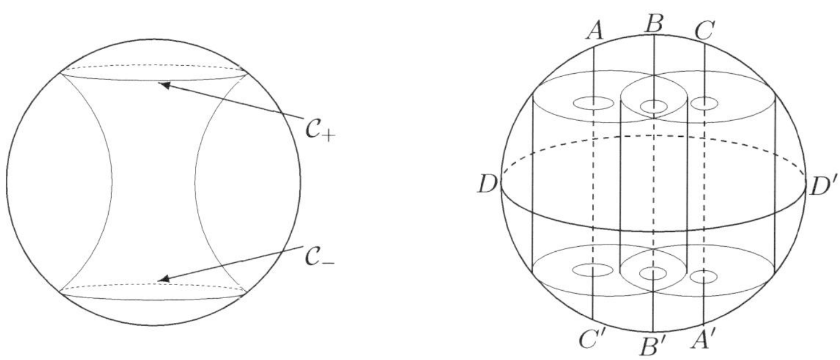

Fig. 2.7. The "cored apple" $\mathrm{SO}(3) - \mathcal{C}$ and the solid cylinder core $\mathcal{C} = (\mathcal{D}_1 \cup \mathcal{D}_2) \times [0, 1]$ (left). Replacement of core with no twists (right).

The problem is: how many twists do we give the core, which is the solid cylinder $\mathcal{C} = (\mathcal{D}_1 \cup \mathcal{D}_2) \times [0, 1]$, (whose ends are identified antipodally in the solid ball model of $\mathrm{SO}(3)$ and give $(\mathcal{D}_1 \cup \mathcal{D}_2) \times \mathbf{S}^1$) before replacing it in the cored apple $\mathrm{SO}(3) - (\mathcal{E} \times \mathbf{S}^1)$? No twists does not work, because then $H\big|_{L^{-1}(\ell)}$ would have two critical circles of elliptic type instead of three, see figure 2.7. One and one half twists do not work either,

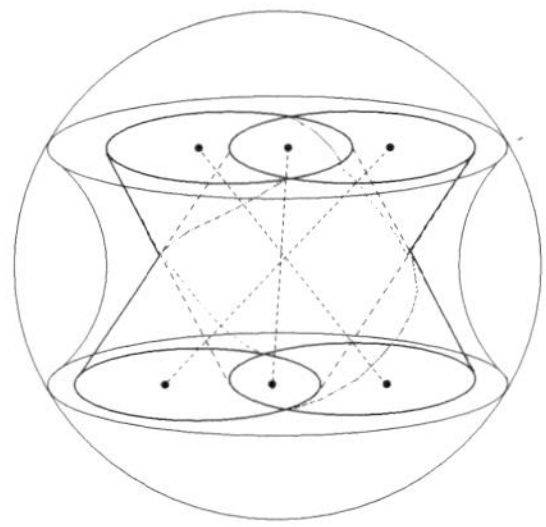

Fig. 2.8. Foliation of $L^{-1}(\ell)$ by the level sets of $H|L^{-1}(\ell)$.

because then $H\big|_{L^{-1}(\ell)}$ would have three critical circles of elliptic type whose two fold cover would have linking number more than one in $\mathbf{S}^3$. Generalizing this shows that placing the core $\mathcal{C}$ back in the cored apple with *one half a twist* is the only one possible. Thus we have obtained a *global* qualitatively accurate picture of how the level sets of $H\big|_{L^{-1}(\ell)}$ fit together to form $L^{-1}(\ell)$, see figure 2.8.

B Comments on lecture III.

B.1 The herpolhode

The curve traced out by the point P_t of contact of the moment of inertia ellipsoid rolling on the invariant plane was called the *herpolhode* by Poinsot [31]. It comes from the Greek word *herpes* meaning snake. Poinsot drew a picture of a snakelike herpolhode. On §9 page 472 Routh [34] gives a proof that the herpolhode has no inflection points for a physically realizable Euler top, that is, one in which the principal moments of inertia satisfy the inequalities

$$I_1 \le I_2 + I_3, \;\; I_2 \le I_1 + I_3 \text{ and } I_3 \le I_1 + I_2.$$

Thus the herpolhode is not snakelike at all. Routh says that Darboux [14] was the first to show this. Whittaker [42] leaves this as an exercise (# 29 on page 174) and refers to Lecornu [24] for a short proof.

In figure B.1 below we see that the herpolhode can indeed be snakelike for nonphysical Euler tops.

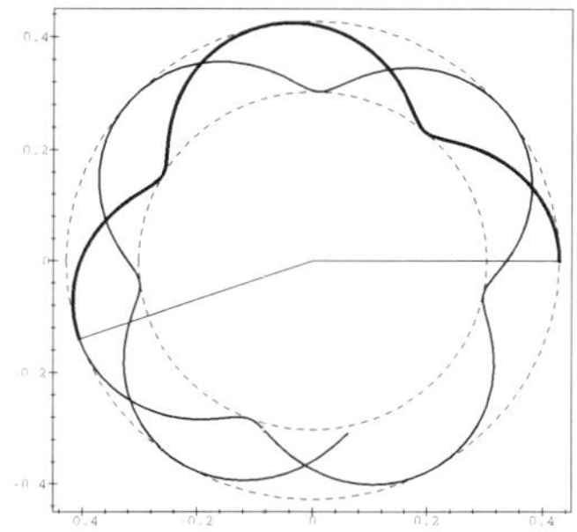

Fig. B.1. The herpolhode for an unphysical Euler top with moments of inertia $I_1 = 1$, $I_2 = 2.8$, and $I_3 = 7$. Angular momentum has magnitude equal to 1; the Euler period is 13.21, and rotation number equals 0.5532.

B.2 Finite symmetries of the reduced Euler top

We saw in the lectures III and V that there is a great deal of similarity between the harmonic oscillator and the Euler top. The obvious difference between these systems is the remaining finite symmetry group.

For the Hénon-Heiles system (section A) this group is $D_3 \times \mathcal{T}$. From its action on the reduced phase space $\mathbf{S}^2_h$ we found all critical points (relative equilibria) of the simplest reduced Hamiltonian for low energies. What is the symmetry of the rigid rotor? The reduced Hamiltonian

$$H_{|\ell|} = \tfrac{1}{2}\left(I_1^{-1}L_1^2 + I_2^{-1}L_2^2 + I_3^{-1}L_3^2\right)$$

is invariant with respect to any changes of the signs of L_1, L_2, and L_3. These operations form the group of order 8 with the structure $D_2 \times \mathbf{Z}_2$. Taking into account that $\mathbf{L} = (L_1, L_2, L_3)$ is an axial 3-vector which changes sign under time reversal $\mathcal{T} \sim \mathbf{Z}_2 : \mathbf{L} \mapsto -\mathbf{L}$, we can readily come up with the physical realization of this group in terms of transformations of the 3-space $\mathbf{R}^3$ with coordinates (L_1, L_2, L_3). Rotations $C_2^{(a)}$ by angle π about any of the principal axes of inertia 1, 2, or 3 constitute the abelian group D_2, which is extended by time reversal $\mathcal{T}$. Since the latter is equivalent to an inversion of $\mathbf{R}^3$, the group $D_2 \times \mathcal{T}$ corresponds to the Schoenflies point group D_{2h}; the three mutually orthogonal reflection planes of D_{2h} correspond, of course, to combinations $C_2 \circ \mathcal{T}$. The action of $D_2 \times \mathcal{T}$ on $\mathbf{S}^2$ is the action of the spatial group D_{2h} on a sphere in $\mathbf{R}^3$ (see figure B.2). It has *six* fixed points, which are grouped into three pairs of equivalent points (two-point orbits) with stabilizers $C_2^{(a)}$,

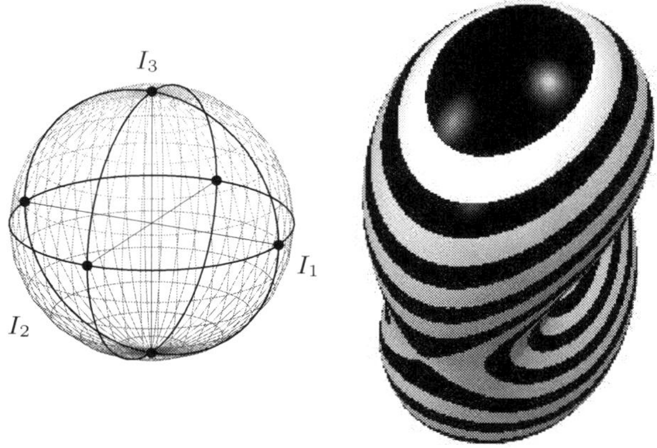

Fig. B.2. Action of the $D_2 \times \mathcal{T}$ group on the sphere $\mathbf{S}^2$ (left). Reduced rigid body Hamiltonian $H_{|\ell|}$ for $|\ell| = 1$, $\frac{1}{2} I_1^{-1} = 0.4$, $\frac{1}{2} I_2^{-1} = 0.8$, and $\frac{1}{2} I_3^{-1} = 1.25$ as a simplest $D_2 \times \mathcal{T}$-invariant Morse function on $\mathbf{S}^2$ (right). Black and white stripes on the surface correspond to constant h (energy) levels drawn equidistantly, the width is ≈ 0.05.

$a = 1, 2, 3$. The two equivalent points are mapped into each other by operations $C_2^{(b)}$, $b \neq a$ and correspond to a rotation about a stationary axis a in two opposite directions ($\langle \mathbf{L}, \mathbf{e}_a \rangle = \pm |\ell|$).

A generic Morse function on $\mathbf{S}^2$ is required by the topology of this space to have two stationary points, a maximum and a minimum. In the presence of symmetry all critical points of the group action are necessarily stationary. The simplest $D_2 \times \mathcal{T}$-symmetric Morse function on $\mathbf{S}^2$ has *six* points situated on the critical orbits of the $D_2 \times \mathcal{T}$ action. Two points are minima, two are maxima, and two are hyperbolic (saddles), so that Euler's equation for the sphere remains satisfied. If $I_1 > I_2 > I_3$ then the values of $H_{|\ell|}$ lie in the interval $\frac{1}{2} I_1^{-1} \leq h \leq \frac{1}{2} I_3^{-1}$, and the value at the hyperbolic critical point (= unstable relative equilibrium) equals $\frac{1}{2} I_2^{-1}$. The Hamiltonian $H_{|\ell|}$ is the simplest $D_2 \times \mathcal{T}$ symmetric Morse function on $\mathbf{S}^2$. We have plotted $H_{|\ell|}$ in figure B.2, right, as a surface function defined over the sphere $\mathbf{S}^2$. This representation is a familiar sight to many molecular physicists, who call it a "rotational energy surface" [21]. The constant energy levels painted on this surface show the trajectories of the reduced system. Many *asymmetric top* molecules, such as H_2O, O_3, have the zero-order rotational Hamiltonian of the type

$H_{|\ell|}$. Higher order molecular terms emerge because the molecule is not rigid, that is, because there are interactions with vibrations.

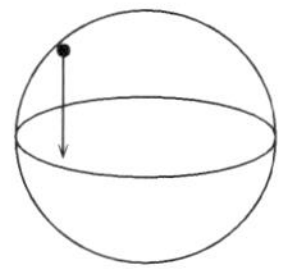

Fig. 2.3. Spherical pendulum.

3 Lecture IV. The spherical pendulum and monodromy

Spherical pendulum was discovered by Dutchman Christiaan Huygens about twenty years before Newton's *Principia*.[1]

This is immediately followed by confusion concerning the correct Dutch pronunciation of the name Huygens. A heated exchange with a Dutch grad student Bob Rink follows. The lecture continues.

3.1 The unconstrained system

Phase space. Give $T\mathbf{R}^3$ coordinates (x, y). We can confuse it with $T^*\mathbf{R}^3$ because we have the standard Euclidean inner product $\langle \, , \, \rangle$ on $\mathbf{R}^3$. In fact $T\mathbf{R}^3 \simeq \mathbf{R}^3 \times \mathbf{R}^3$.

Symplectic form. The standard symplectic form on $T\mathbf{R}^3$ is

$$\widetilde{\omega} = \sum \mathrm{d}x_j \wedge \mathrm{d}y_j.$$

The unconstrained hamiltonian on $T\mathbf{R}^3$ is

$$\widetilde{H}(x, y) = \tfrac{1}{2} \langle y, y \rangle + \langle x, e_3 \rangle.$$

S^1 **symmetry.** The unconstrained system $(\widetilde{H}, T\mathbf{R}^3, \widetilde{\omega})$ has an S^1 symmetry given by

$$S^1 \times T\mathbf{R}^3 \to T\mathbf{R}^3 : (t, (x, y)) \mapsto (R_t x, R_t y), \tag{3.1}$$

where $R_t = \begin{pmatrix} \cos t & -\sin t & 0 \\ \sin t & \cos t & 0 \\ 0 & 0 & 1 \end{pmatrix}$ is the matrix of rotation about axis e_3 by angle t.

[1] The work by Huygens (1629–1695) appeared in 1673, thirty years after Newton was born (1642). There are indications that Huygens understood this system earlier, see more in [5] on p. 402 and [40]. *Principia* [30] was published in 1687.

The unconstrained S^1-momentum map associated to the S^1-action (3.1) is

$$\widetilde{L}(x,y) = x_1y_2 - x_2y_1.$$

3.2 Constrained system

Constrained phase space. The phase space of the constrained system (= the spherical pendulum) is

$$T\mathbf{S}^2 = \{(x,y) \in T\mathbf{R}^3 \mid \langle x,x\rangle - 1 = 0, \quad \langle x,y\rangle = 0\}.$$

The constrained symplectic form on $T\mathbf{S}^2$ is $\omega = \widetilde{\omega}\big|_{T\mathbf{S}^2}$.

Hamiltonian. On $(T\mathbf{S}^2, \omega)$ the spherical pendulum hamiltonian is

$$H(x,y) = \widetilde{H}(x,y)\big|_{T\mathbf{S}^2}.$$

The equations of motion of the spherical pendulum are

$$\begin{aligned} \dot{x} &= y \\ \dot{y} &= -e_3 + \big(\langle x, e_3\rangle - \langle y,y\rangle\big)\, x. \end{aligned} \tag{3.2}$$

They determine the integral curves of the Hamiltonain vector field X_H on $(T\mathbf{S}^2, \omega)$, see p. 148 and 296 in [5]. Actually, equations (3.2) define a vector field on $T\mathbf{R}^3$ whose restriction to $T\mathbf{S}^2$ is X_H. To see this we need to verify that $T\mathbf{S}^2$ is an invariant manifold of (3.2).

Check.

$$\frac{d}{dt}\langle x,x\rangle = 2\langle x,\dot{x}\rangle = 2\langle x,y\rangle = 0,$$

$$\begin{aligned} \frac{d}{dt}\langle x,y\rangle &= \langle \dot{x},y\rangle + \langle x,\dot{y}\rangle = \langle y,y\rangle - \langle x,e_3\rangle + \big(\langle x,e_3\rangle - \langle y,y\rangle\big)\langle x,x\rangle \\ &= 0 \end{aligned}$$

on $T\mathbf{S}^2$. □

Now, those people who use polar coordinates will give you only one system of equations for the spherical pendulum. In fact, they need at least two, since $T\mathbf{S}^2$ requires at least two charts.

S^1 symmetry. The S^1 symmetry of the spherical pendulum is

$$S^1 \times T\mathbf{S}^2 \to T\mathbf{S}^2 : \big(t,(x,y)\big) \mapsto (R_t x, R_t y). \tag{3.3}$$

The S^1-momentum map of the spherical pendulum is $L = \widetilde{L}\big|_{T\mathbf{S}^2}$.

Conserved quantity. Since S^1 action (3.3) preserves the constrained Hamiltonian H, we find that the Lie derivative of the momentum L with respect to the Hamiltonian vector field X_H of the spherical pendulum vanishes identically. Thus the spherical pendulum is Liouville integrable.

3.3 Reduction of S^1 symmetry

To remove the S^1 symmetry of the spherical pendulum we use invariants, because the S^1-action (3.3) is *not* free.

When you have compact group actions, invariant theory is the way to go. For general proper actions the situation is more complicated.

Algebra of invariants. The algebra of polynomials on $T\mathbf{R}^3$ which are *invariant* under the S^1 action (3.1) is generated by

$$\begin{array}{lll} \sigma_1 = x_3 & \sigma_3 = y_1^2 + y_2^2 + y_3^2 & \sigma_5 = x_1 y_1 + x_2 y_2 \\ \sigma_2 = y_3 & \sigma_4 = x_1^2 + x_2^2 & \sigma_6 = x_1 y_2 - x_2 y_1. \end{array}$$

Relation. The above invariants satisfy the relation

$$\sigma_5^2 + \sigma_6^2 = \sigma_4(\sigma_3 - \sigma_2^2), \text{ where } \sigma_4 \geq 0,\ \sigma_3 \geq \sigma_2^2. \tag{3.4}$$

This relation defines the orbit space $T\mathbf{R}^3/S^1$, which is a connected, *irreducible* semialgebraic variety in $\mathbf{R}^6$.

Orbit map. The orbit map is

$$\pi : T\mathbf{R}^3 \to \mathbf{R}^6 : (x, y) \to \big(\sigma_1(x, y), \ldots, \sigma_6(x, y)\big).$$

In other words, each of the fibers of π is a unique S^1 orbit of the action (3.1). The orbit space $T\mathbf{R}^3/S^1$ is *singular*, because the S^1 action on $T\mathbf{R}^3$ leaves the points $(0, 0, x_3, 0, 0, y_3)$ fixed.

Another orbit space. What we really want is the orbit space of the S^1 action (3.1) restricted to $T\mathbf{S}^2$. This is the orbit space $T\mathbf{S}^2/S^1$ of the S^1 action (3.3). We obtain the defining equations of $T\mathbf{S}^2/S^1$ if we add two more relations to (3.4) (which come from the equations defining $T\mathbf{S}^2$), namely,

$$\sigma_4 + \sigma_1^2 = 1 \quad \text{and} \quad \sigma_5 + \sigma_1\sigma_2 = 0. \tag{3.5}$$

The gadget defined by (3.4) and (3.5) has *no chance* to be smooth because the action is *not free*. Singularities of $T\mathbf{S}^2/S^1$ contain dynamical information. We can use (3.5) to get rid of σ_4 and σ_5 in (3.4). We obtain

$$\sigma_1^2\sigma_2^2 + \sigma_6^2 = (\sigma_3 - \sigma_2^2)(1 - \sigma_1^2) = \sigma_3(1 - \sigma_1^2) - \sigma_2^2 + \sigma_1^2\sigma_2^2.$$

Simplifying gives the following description of $T\mathbf{S}^2/S^1$ as a semialgebraic variety in $\mathbf{R}^4$ (with coordinates $(\sigma_1, \sigma_2, \sigma_3, \sigma_6)$).

$$\sigma_2^2 + \sigma_6^2 = \sigma_3(1 - \sigma_1^2), \quad \text{where} \quad |\sigma_1| \le 1 \text{ and } \sigma_3 \ge 0. \tag{3.6}$$

The reduced phase space. The (singular) reduced phase space P_ℓ of the spherical pendulum is the orbit space $L^{-1}(\ell)/S^1$, where ℓ is the value of the momentum L. In terms of invariants L is σ_6. Thus P_ℓ is defined by adding the relation $\sigma_6 = \ell$ to equation (3.6). Thus as a subvariety of $\mathbf{R}^3$ (with coordinates $(\sigma_1, \sigma_2, \sigma_3)$), the singular reduced space P_ℓ is defined by

$$\sigma_2^2 + \ell^2 = \sigma_3(1 - \sigma_1^2), \quad |\sigma_1| \le 1, \quad \sigma_3 \ge 0.$$

In other words, P_ℓ is a $\sigma_6 = \ell$ *slice* of the orbit space $T\mathbf{S}^2/S^1$. When $\ell \ne 0$, P_ℓ is

$$\sigma_3 = \frac{\sigma_2^2 + \ell^2}{1 - \sigma_1^2} \quad \text{and} \quad |\sigma_1| < 1,$$

which is smooth and is diffeomorphic to $\mathbf{R}^2$. When $\ell = 0$ we have a "canoe" (see figure 3.1), whose two singular points $(\pm 1, 0, 0)$ are the fixed points of the S^1 action on $T\mathbf{S}^2$.

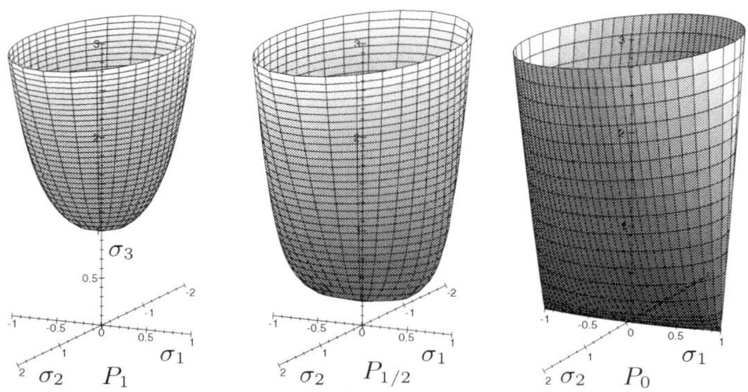

Fig. 3.1. Reduced phase spaces P_ℓ.

Note that all these guys P_ℓ have Poisson structure and there is reduced dynamics but I am not going to talk about it. The map $(x, y) \mapsto (\sigma_1, \sigma_2, \sigma_3)$ looks very much like a Hopf fibration but it is not and I will not discuss this either. I have completed reduction for the spherical pendulum.

3.4 Analysis of the reduced system

Reduced Hamiltonian. On the singular reduced space P_ℓ we have the reduced Hamiltonian

$$H_\ell : P_\ell \subseteq \mathbf{R}^3 \to \mathbf{R} : (\sigma_1, \sigma_2, \sigma_3) \mapsto \tfrac{1}{2}\,\sigma_3 + \sigma_1.$$

Critical values of H_ℓ on P_ℓ. How does $H_\ell = h$ intersect P_ℓ?

Calculus is difficult to use when P_ℓ has singularities. So we use a little bit of algebra instead. Computations proceed at maximum speed so that no one in the audience can follow what is happening.

Consider a family of 2-planes $\pi_h : \frac{1}{2}\,\sigma_3 + \sigma_1 = h$. Look for values of h where the intersection of π_h with P_ℓ has a point with multiplicity greater than 1. In other words, we look for those values (h, ℓ) for which polynomial

$$P(\sigma_1, \sigma_2) = \sigma_2^2 + \ell^2 - 2(h - \sigma_1)(1 - \sigma_1^2), \quad |\sigma_1| \le 1,$$

in (σ_1, σ_2) has a zero of multiplicity greater than 1 and σ_1 lies in $[-1, 1]$. (The polynomial P is obtained by eliminating σ_3 from the defining equation of P_ℓ using $\sigma_3 = 2(h - \sigma_1)$. This is equivalent to $\sigma_2 = 0$ and finding those values of (h, ℓ) where the cubic polynomial

$$p(\sigma_1) = (h - \sigma_1)(1 - \sigma_1^2) - \tfrac{1}{2}\,\ell^2, \tag{3.7}$$

in σ_1 has a zero of multiplicity greater than 1 in $[-1, 1]$. Let $s \in [-1, 1]$ be a zero of p of multiplicity at least 2 and let $t \in \mathbf{R}$ be zero of multiplicity at least 1. Then necessarily

$$\begin{aligned} p(\sigma_1) &= \sigma_1^3 - h\sigma_1^2 - \sigma_1 + h - \tfrac{1}{2}\,\ell^2 = (\sigma_1 - s)^2(\sigma_1 - t) \\ &= \sigma_1^3 - (2s + t)\sigma_1^2 + s(s + 2t)\sigma_1 - s^2 t. \end{aligned}$$

Comparing coefficients of powers of σ_1 gives

$$2s + t = h, \quad s(s + 2t) = -1, \quad \text{and} \quad -s^2 t = h - \tfrac{1}{2}\,\ell^2.$$

Solving the above equations for $\{h, \ell\}$ gives

$$\begin{aligned} h &= \tfrac{1}{2}\,(3s^2 - 1)/s \\ \ell^2 &= -(1 - s^2)^2/s, \end{aligned} \tag{3.8}$$

where $s \in [-1, 0) \cup \{1\}$. The first two equations in (3.8) give a parametrization of the *discriminant locus* Δ of p. When s varies between -1 and 0, the parametrization traces out two 1-smooth branches of the discriminant locus Δ, which join together when $s = -1$ and form an angle. When $s = 1$, $(h, \ell) = (1, 0)$ is an *isolated* point of the discriminant locus Δ. Thus 1 is a critical value of the reduced Hamiltonian H_1 on the reduced space P_0 corresponding to the critical point $(1, 0, 0)$, which is a singular point of P_0, see figure 3.2.

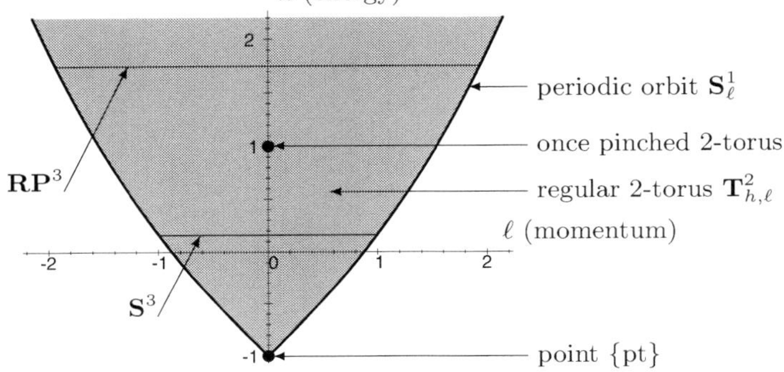

Fig. 3.2. Image and fibers of the energy-momentum map $\mathcal{EM}$ of the spherical pendulum.

Energy-momentum map $\mathcal{EM}$. The energy momentum map of the spherical pendulum is

$$\mathcal{EM} : T\mathbf{S}^2 \rightarrow \mathbf{R}^2 : p \mapsto (H(p), L(p)).$$

The region bounded by the black curves in figure 3.2 is the image of $\mathcal{EM}$. Its set of regular values is the grey shaded region; the black curves are the singular values of $\mathcal{EM}$ (= critical values of H_ℓ) together with the big dots, which mark the critical values $(1, 0)$ and $(-1, 0)$. Recall that critical values of H_ℓ are the same as critical values of $H\big|_{L^{-1}(\ell)}$ which in turn are the same as critical values of the energy-momentum map $\mathcal{EM}$.

3.5 Reconstruction

What is the topology of individual fibers $\mathcal{EM}^{-1}(h, \ell)$ of the energy-momentum map?

As S. Smale used to say: "whatever you do, don't lose geometry."

To determine the topology of fibers look at the projection of P_ℓ on the $\{\sigma_2 = 0\}$ plane. Remember that each regular point of P_ℓ lifts to a circle $\mathbf{S}^1$ and each singular point (of P_0) lifts to a point.

Regular fibers. If (h, ℓ) is a regular value of $\mathcal{EM}$, then the fiber $\mathcal{EM}^{-1}(h, \ell)$ is a smooth 2-torus. (The Liouville–Arnol'd theorem only shows that a *connected* component of $\mathcal{EM}^{-1}(h, \ell)$ is a 2-torus). To verify this, note that figure 3.3 shows that the h-level set of the reduced Hamiltonian H_ℓ on the reduced space P_ℓ is a circle C. Each point on C under the reduction map π lifts to a single $\mathbf{S}^1$ orbit of the S^1-action (3.5). Thus the lift of the circle C is the product $\mathbf{S}^1 \times \mathbf{S}^1$, which is the 2-torus $\mathcal{EM}^{-1}(h, \ell)$. Regular level sets of the reduced Hamiltonian H_0 on

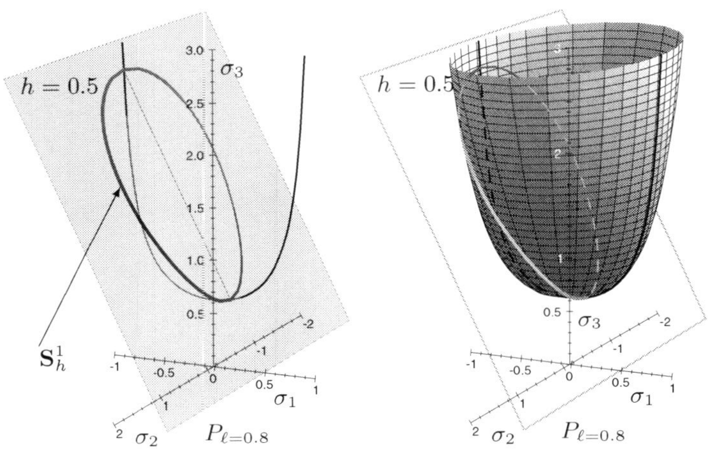

Fig. 3.3. h-level sets of H_ℓ on the regular reduced phase space P_l.

the singular reduced phase space P_0 are also circles which lift to 2-tori (see figure 3.4). Dynamically, there are two different types of regular level sets: those *below* the critical energy $h = 1$ and those *above*. The image under the tangent bundle projection $\tau : T\mathbf{S}^2 \rightarrow \mathbf{S}^2 : (x, y) \mapsto x$ of an integral curve of the spherical pendulum on $\mathcal{EM}^{-1}(h, \ell)$, when (h, ℓ) is a regular value of $\mathcal{EM}$, is shown in figure 3.5.

Energy levels. How do regular tori fit together? The level $H^{-1}(h)$ is $\mathbf{RP}^3$ when $h > 1$ and $\mathbf{S}^3$ when $-1 < h < 1$. In more detail, using the solid ball model of SO(3), we see that SO(3) ($\simeq \mathbf{RP}^3$) is the union of two solid 2-tori, which are in turn the union of 2-tori with core a circle.

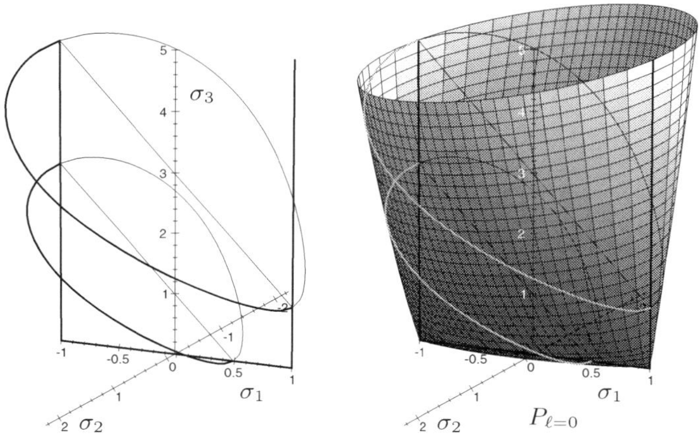

Fig. 3.4. Regular h-level sets of H_ℓ on the singular reduced phase space P_0.

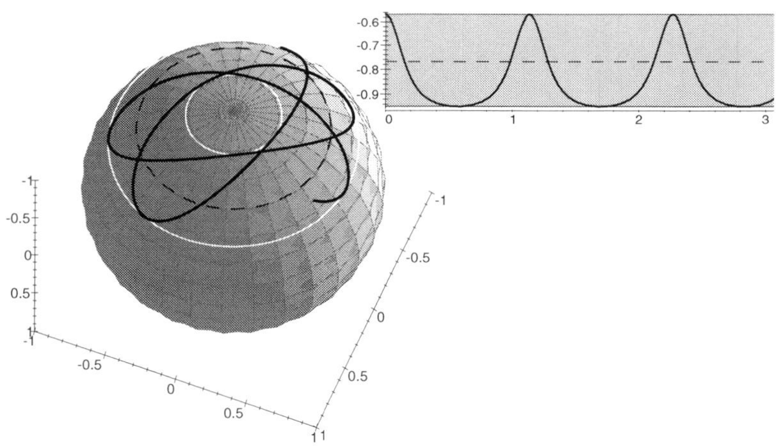

Fig. 3.5. Integral curve of the spherical pendulum on the configuration space $\mathbf{S}^2$ (view from the bottom, cf. figure 2.3) corresponding to the regular h-level set of H_ℓ. White circles indicate maximum and minimum elevation x_3^+ and x_3^-. The insert (top right) demonstrates the rotation angle $\theta > \pi$ by showing elevation $x_3(t)$ as function of longitude $\varphi(t)/\pi$. In this example $\ell = 0.3$, $h = -0.5$, rotation angle θ equals 1.1327π, and the period of $x_3(t)$ is 3.3453.

This describes the foliation of $H^{-1}(h)$ by level sets of L when $h > 1$. When $-1 < h < 1$, the foliation is the same as that given by the Hopf fibration.

Singular fibers. For critical values $(h, \ell \neq 0)$ of $\mathcal{EM}$, which form the boundary of the discriminant locus Δ, the h-level set of the reduced Hamiltonian H_ℓ is a point on P_l (see figure 3.6). The reconstructed fiber $\mathcal{EM}^{-1}(h, \ell)$ is a periodic orbit of the Hamiltonian vector field X_H which is also an orbit of the S^1-action (3.5). In other words, it is a *relative equilibrium* of X_H. On the singular reduced phase space P_0

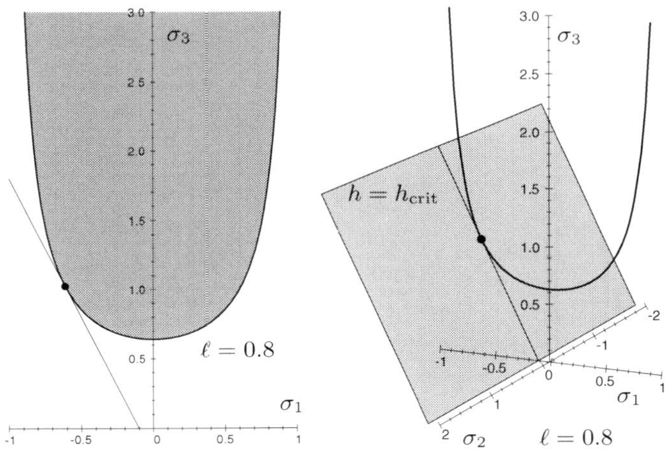

Fig. 3.6. Singular h-level sets of H_ℓ (relative equilibria) on the regular reduced phase space.

we have two kinds of critical slices (see figure 3.7). The level $h = -1$ is a point $(\sigma_1, \sigma_2, \sigma_3) = (-1, 0, 0)$ and the level $h = 1$ is a circle with one singular point $(\sigma_1, \sigma_2, \sigma_3) = (1, 0, 0)$. The point $\sigma = (-1, 0, 0)$ on P_0 corresponds to the stable equilibrium point $(0, 0, -1, 0, 0, 0)$ in $T\mathbf{S}^2$ of the spherical pendulum, because $h = -1$ is an absolute minimum energy. $\mathcal{EM}^{-1}(-1, 0)$ is, of course, a point. The point $\sigma = (1, 0, 0)$ in P_0 corresponds to the unstable equilibrium point $(0, 0, 1, 0, 0, 0)$ in $T\mathbf{S}^2$. Because this equilbrium point is a fixed point of the S^1 action (3.5), it lifts to a point in $T\mathbf{S}^2$. The rest of the points on $H_0^{-1}(1)$ lift to $\mathbf{S}^1$ orbits. From this it follows that $\mathcal{EM}^{-1}(1, 0)$ is a once pinched 2-torus (see figure 3.6.2 on page 280). This once pinched 2-torus is a homoclinic connection of the stable and unstable manifolds of the hyperbolic equilibrum of X_H at $(0, 0, 1, 0, 0, 0) \in T\mathbf{S}^2$.

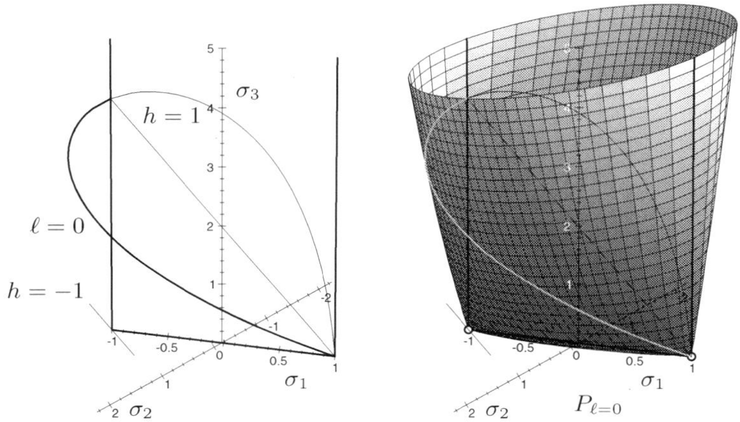

Fig. 3.7. Two singular h-level sets of H_ℓ on the singular reduced phase space P_0. The $h = 0$ level set lifts to the point (stable equilibrium), the $h = 1$ level set lifts to the pinched torus $\mathcal{EM}^{-1}(1,0)$.

3.6 Monodromy

What we want to do now is to describe how the 2-torus fibers $\mathcal{EM}^{-1}(h,\ell)$ fit together as (h,ℓ) runs over a parameterized subset of the set of regular values of $\mathcal{EM}$. Now suppose that this set of regular values is a small open punctured disc $\mathbf{D}^* = \mathbf{D} - \{(h,\ell)_{\mathrm{crit}}\}$, which lies in the image of $\mathcal{EM}$. In other words, the critical value $(h,\ell)_{\mathrm{crit}}$ is *isolated*. Let Γ be a circle $\mathbf{S}^1 \subseteq \mathbf{D}^*$, which cannot be contracted to a point in $\mathbf{D}^*$, that is, $\mathbf{S}^1$ goes around the puncture $(h,\ell)_{\mathrm{crit}}$ as shown below

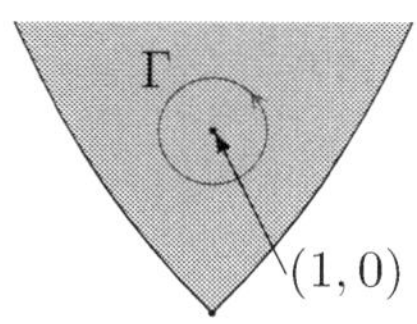

What is the topology of the 2-torus bundle $\Pi : \mathcal{EM}^{-1}(\Gamma) \to \Gamma$ over the circle Γ? (Recall the reconstruction of energy momentum level sets for regular values (h,ℓ)). Answer: The bundle Π is *non-trivial*. In other words, topologically $\mathcal{EM}^{-1}(\Gamma)$ is *not* a product $\mathbf{S}^1 \times \mathbf{T}^2$.

Monodromy map. How do we describe the topology of the 2-torus bundle Π over the circle Γ? Take the circle Γ and cut it at a point $\mathrm{pt} = (h,\ell)$.

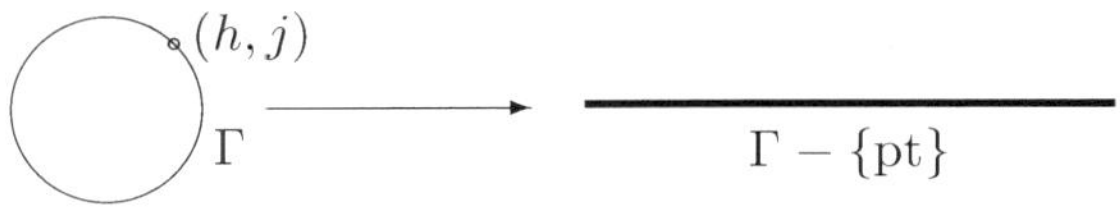

The resulting 2-torus bundle over the interval $\Gamma - \{\mathrm{pt}\}$ is *trivial* because the interval is can be contracted to a point. To obtain $\mathcal{EM}^{-1}(\Gamma)$ we glue the two tori over the end points of $\Gamma - \{\mathrm{pt}\}$ together. This is tricky business.

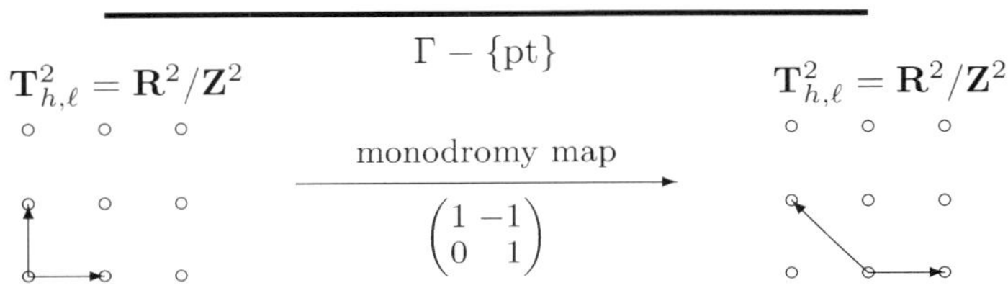

The map which identifies these tori is called the *monodromy map.* How do we glue the end 2-tori together? Since the tori of a Liouville integrable system are affine, such a 2-torus is $\mathbf{R}^2/\mathbf{Z}^2$, which is the 2-plane with points whose coordinates both differ by an integer being identified. The map which identifies the end tori is given by a 2×2 matrix M with integer entries having determinant 1, since M preserves $\mathbf{Z}^2$. If M is not conjugate by an integer 2×2 matrix with determinant 1 to the identity matrix, as in the figure above, then the bundle Π is nontrivial. When this is the case we say that the integrable system has monodromy.

Consequences. You may say – so what if the system has monodromy? Wait a minute. If the Liouville integrable Hamiltonian system with two degrees of freedom has monodromy, it *does not* have globally defined action-angle coordintes.

3.6.1 Analytical description of mondromy. Rotation angle

To compute monodromy analytically we need to use the *rotation angle* $\Theta_{(h,\ell)}$ of the flow of the Hamiltonian vector field X_H on a smooth 2-torus $\mathcal{EM}^{-1}(h,\ell)$, where (h,ℓ) is a regular value of $\mathcal{EM}$. The X_H and X_L on $\mathcal{EM}^{-1}(h,\ell)$ are linearly independent. The flow $\varphi^L_{t''}$ of X_L is periodic on the torus $\mathbf{T}^2_{(h,\ell)}$. We define the rotation angle $\Theta_{(h,\ell)}$ so that for *any* initial condition p

$$\varphi^L_{\Theta_{(h,\ell)}}(p) = \varphi^H_T(p),$$

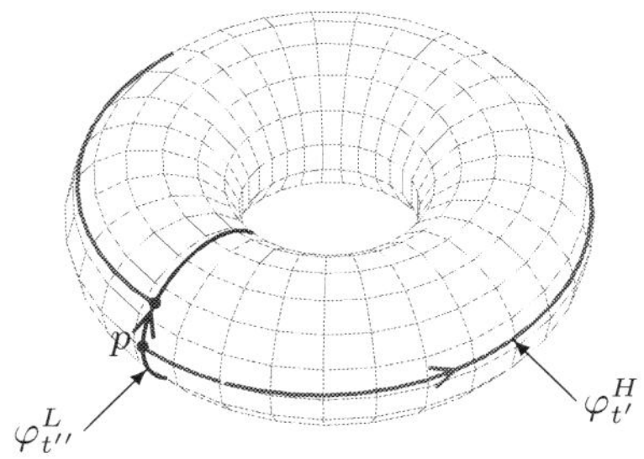

Fig. 3.8. Flow of two linearly independent vector fields on the torus $\mathbf{T}^2_{(h,\ell)}$ and definition of the rotation angle.

where T is the period of the flow $\varphi_t^{H_\ell}$ of the reduced vector field X_{H_ℓ} on $H_\ell^{-1}(h)$, see figure 3.8. The vector field $T(h,\ell)X_H + \Theta_{(h,\ell)}X_L$ has periodic flow on $T^2_{(h,\ell)}$. Projecting the tangent bundle $T\mathbf{S}^2$ on the configuration space $\mathbf{S}^2$ we can define $\Theta_{(h,\ell)}$ as the angle by which the pendulum turns about axis x_3 during the period of oscillation in x_3 (height)[1], see figure 3.5.
A standard classical argument shows that

$$\Theta_{(h,\ell)} = 2\sqrt{2}\ell \int_{\sigma_1^-}^{\sigma_1^+} \frac{d\sigma_1}{(1-\sigma_1^2)\sqrt{p(\sigma_1)}},$$

where p is the polynomial (3.7) and $\sigma_1^{\pm}$ are its real zeroes in $[-1,1]$ with $\sigma_1^- \le \sigma_1^+$. The function $\Theta_{(h,\ell)}$ is a *multivalued* real analytic function on the set of regular values of $\mathcal{EM}$, for more details see [5]. When we let (h,ℓ) run over the curve Γ, the value of $\Theta_{(h,\ell)/2\pi}$ does not return to its original value. Instead it decreases by 1. Hence the variation of the lattice spanned by periodic Hamiltonian vector fields X_L and $X_{T(h,\ell)H+\Theta_{(h,\ell)}L}$ on $\mathcal{EM}^{-1}(h,\ell)$ corresponding to the local actions is $\begin{pmatrix} 1 & -1 \\ 0 & 1 \end{pmatrix}$.

3.6.2 Geometric monodromy theorem

We now state the geometric monodromy theorem, which allows us to compute the monodromy using only geometry. Consider a two degree of freedom Liouville integrable Hamiltonian system with phase space $(\mathcal{M},\omega)$, which is a 4-dimensional symplectic manifold and Poisson com-

[1] Rotation angle equals π in the planar pendulum limit $\varphi^L \equiv 0$. For all regular (h,ℓ) $\pi < \Theta_{(h,\ell)} < 2\pi$, see chapter IV.6.3 of [5].

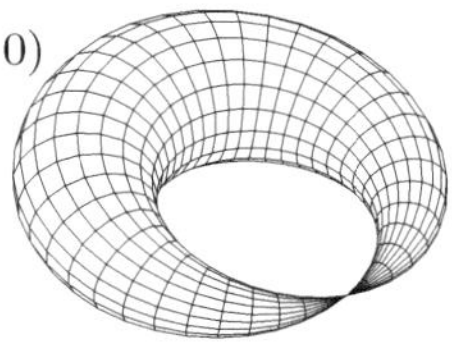

Fig. 3.9. A once pinched 2-torus.

muting integrals (F_1, F_2). Suppose that the integral map

$$\mathcal{I} : \mathcal{M} \to \mathbf{R}^2 : p \mapsto (F_1(p), F_2(p)).$$

has an isolated critical value of $(0,0)$ and that $\mathbf{D}^* = \mathbf{D} - \{(0,0)\}$ is contained its set of regular values. Suppose that for every c in $\mathbf{D}^*$ the preimage $\mathcal{I}^{-1}(c)$ is a smooth 2-torus, while $\mathcal{I}^{-1}(0,0)$ is a 2-torus which is once pinched at the point x. In other words, x is the only singular point of $\mathcal{I}^{-1}(0,0)$ and is a hyperbolic equilibrium point of X_{F_1}, that is, the linearization of X_{F_1} at x has two nonreal complex eigenvalues with positive real part and two nonreal complex eigenvalues with negative real part. Moreover, the whole of $\mathcal{I}^{-1}(0,0)$ is a homoclinic connection of the stable and unstable manifolds of x. If Γ is a smooth positively oriented circle in $\mathbf{D}^*$, then the 2-torus bundle $\mathcal{I}^{-1}(\Gamma)$ has monodromy $\left(\begin{smallmatrix} 1 & -1 \\ 0 & 1 \end{smallmatrix}\right)$.

If the singular fiber $\mathcal{I}^{-1}(0,0)$ is a k-pinched 2-torus, that is, is a heteroclinic k-cycle, then the monodromy is $\left(\begin{smallmatrix} 1 & -k \\ 0 & 1 \end{smallmatrix}\right)$.

C Comments on lecture IV

The concept of monodromy for two degree of freedom integrable Hamiltonian systems was first formulated by Duistermaat [15]. The spherical pendulum was historically the first example of a system with monodromy [15, 6]. It is discussed at length in chapt. IV of [5].

Reduction of the S^1 symmetry in the spherical pendulum is an example of *singular reduction* using invariants which was pioneered by Cushman. In his lecture Cushman relies on the simple elegant method of qualitative analysis (of functions defined on algebraic varieties, such as reduced Hamiltonians H defined on singular reduced phase spaces P) which can be called the "method of slices". Having both P and h-level sets of H described as surfaces in an ambient space with invariant polynomial coordinates, he studies the topology of their intersections D_h

which form a continuous one-parameter family. Critical sections $D_{h_{\rm crit}}$ have special topology and are isolated in this family, the values $h_{\rm crit}$ are critical values of the function H on P. A similar method (for functions defined on orbit spaces) was independently used by Zhilinskií (see section 5.6.1 of [26], appendix A in [27], and [36]).

The presentation of the spherical pendulum gives a convincing illustration of Cushman's leitmotiv "no polar coordinates", that is, of an analysis based on polynomial invariants. The reader who is still (uh) not converted to the faith should certainly enjoy drawing a picture of the singular reduced phase space P_0 in polar (cylindrical) coordinates.

Initially monodromy was uncovered analytically in terms of local angle-action variables and the variation of the period lattice [5]. In this approach the monodromy matrix was calculated directly after an explicit relation between the different local angle action variable charts was established. The geometric monodromy theorem was formulated later in [10]. According to this theorem and the results of [25] and [43] we can determine whether the system has monodromy and even find the monodromy matrix on the basis of the geometric reconstruction of the fibers of the energy-momentum map $\mathcal{EM}$.

All figures in this section were prepared from numerical computations.

C.1 Discrete symmetries

The spherical pendulum has a number of discrete symmetries in addition to the S^1 symmetry discussed in the lecture. To find these symmetries we consider operations which leave invariant the unconstrained Hamiltonian $\widetilde{H}$ *and* the phase space $T\mathbf{S}^2$. In this way we see that our system is invariant with respect to any reflection σ_v in a plane containing the e_3 axis, for example,

$$(x_1, x_2, x_3, y_1, y_2, y_3) \mapsto (x_1, -x_2, x_3, y_1, -y_2, y_3).$$

The resulting symmetry group has two classes of conjugate elements: one which contain rotations (S^1 symmetry) and the other the reflection σ_v. In the Schoenflies classification this group is called $C_{\infty v}$. The action of $C_{\infty v}$ on $T\mathbf{S}^2$ is symplectic. The S^1-invariant polynomials $\sigma_1, \ldots, \sigma_6$ are invariant under a larger group $C_{\infty v}$. When the action of $C_{\infty v}$ is reduced, its orbit space is $T\mathbf{S}^2/C_{\infty v}$.

The full symmetry group of the spherical pendulum is $C_{\infty v} \times \mathbf{Z}_2$. The nontrivial operation of the order two group $\mathbf{Z}_2 = \{\pm 1\}$ is the anti-symplectic symmetry $(x, y) \to (x, -y)$, which is sometimes called *time reversal*. After reduction this symmetry induces a nontrivial symmetry on $\mathbf{R}^4$ defined by

$$\mathbf{Z}_2 \times \mathbf{R}^4 \to \mathbf{R}^4 : (-1, (\sigma_1, \sigma_2, \sigma_3, \sigma_6)) \mapsto (\sigma_1, -\sigma_2, \sigma_3, -\sigma_6).$$

This has two important consequences:

(i) Points on $T\mathbf{S}^2$ with $L = \ell \neq 0$ which differ by the direction of rotation about axis e_3 are *equivalent*. Therefore it suffices to study only the case $\ell \geq 0$.

(ii) The reduced phase space P_ℓ can be "flattened" into the *fully reduced phase space* $P_\ell/\mathbf{Z}_2$ which is a projection of P_ℓ onto the plane with coordinates (σ_1, σ_3). The use of $P_\ell/\mathbf{Z}_2$ makes the geometric analysis of the level sets of $H_\ell = \frac{1}{2}\sigma_3 + \sigma_1$ particularly simple.

C.2 Geometric analysis on $P_\ell/\mathbf{Z}_2$

Analysis of the level sets of H_ℓ on P_ℓ can be done using level sets of H_ℓ on $P_\ell/\mathbf{Z}_2$ for $\ell \geq 0$. The h-level set of H_ℓ is an intersection of the line

$$\sigma_3 = 2(h - \sigma_1)$$

and the region of the coordinate plane (σ_1, σ_3) defined by the inequality

$$\sigma_3 \geq \ell^2/(1 - \sigma_1^2).$$

Note that on the boundary of $P_\ell/\mathbf{Z}_2$ the value of σ_2 is 0, while the value of H_ℓ is

$$H_\ell = \tfrac{1}{2}\sigma_3 + \sigma_1 = \frac{\ell^2}{2(1 - \sigma_1^2)} + \sigma_1.$$

Regular level sets of H_ℓ on $P_\ell/\mathbf{Z}_2$ are closed intervals. Its critical level sets are points (see figures in the lecture) on the boundary of $P_\ell/\mathbf{Z}_2$. To find the critical set for $\ell > 0$ we find those values of σ_1 where the slope of the boundary of $P_\ell/\mathbf{Z}_2$ equals that of the h-level set, that is,

$$\ell^2 \frac{d}{d\sigma_1}(1 - \sigma_1^2)^{-1} = -2, \quad \sigma_1 < 0.$$

Solving the above equation for ℓ gives

$$\ell = \pm(1 - s^2)/\sqrt{s}, \quad \text{where } s = -\sigma_1.$$

This yields the parametrization

$$\big(\ell(s), h(\ell(s), s)\big) = \left(\pm\frac{1 - s^2}{\sqrt{s}}, \frac{1 - s^2}{2\,s} - s\right)$$

of the discriminant locus Δ used in the lecture.

C.3 Quantum monodromy

Quantum mechanics provides a very clear interpretation of monodromy [8] which we think is worth mentioning here. We recall that in the quantized spherical pendulum it is the energy h and momentum ℓ of the classical system which are quantized. According to the Einstein-Brilluoin-Kramers (EBK) quantization principle, we should find regular tori $\mathbf{T}_{h,\ell}$ for which the actions are an integer value times an overall scale factor $2\pi\hbar$.

For the spherical pendulum the action ℓ, which corresponds to the S^1 symmetry, is quantized so that

$$\frac{\ell}{2\pi\hbar} = 0, 1, 2, \dots \ .$$

The action corresponding to the second vector field on $\mathbf{T}_{h,\ell}$ with periodic flow should be computed locally in (h, ℓ) and then quantized. The result of such computation is shown in figure C.1, which was kindly provided to us by Igor Kozin [23]. Note that the value of $\hbar$ was "adjusted" so that there are enough quantum states in the region near the isolated critical value $(h, \ell) = (1, 0)$.

All quantum states (EBK tori) form a two-dimensional lattice of points in the image of the $\mathcal{EM}$ map (inside the discriminant locus). We can choose many regular one-dimensional sequences of nodes starting locally at any given node in this lattice; the distance between the neighbors in such sequences is a "quantum", and the distance between the last state and the border of the classical locus (measured along the direction of the sequence) is half of this quantum. In other words, we can define local quantum numbers which correspond to local action variables.

Boris Zhilinskií proposed to make manifest that the monodromy of the quantum system is a defect of the lattice formed by quantum states in the image of the $\mathcal{EM}$-map. This improved the original picture of Cushman and Duistermaat [8] so that anyone who sees one of Boris' pictures (including Cushman and Duistermaat) immediately begins to play around and try to "perfect" it. Therefore we call such a picture as figure C.1 a *Zhilinskií diagram.*

We begin Zhilinskií's diagram (figure C.1) by drawing an "elementary cell", which is a small quadrangle defined by one choice of local quantum numbers. We then propogate this cell along the contour Γ.

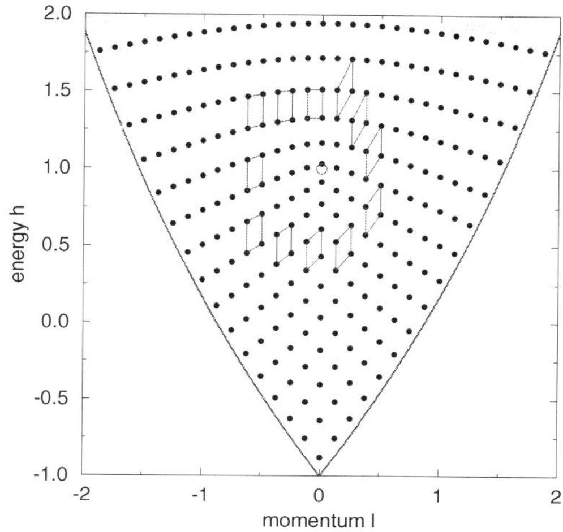

Fig. C.1. Image of the classical and quantum $\mathcal{EM}$ map of the spherical pendulum. Hollow circle shows the isolated critical value, black circles represent quantum levels (= EBK tori). The family of quadrilaterals (= Zhilinskiǐ's diagram) shows how local quantum numbers (= local action variables) are chosen along the circle Γ.

At each sufficiently small step along Γ, the choice of the new neighboring cell preserves the same choice of the quantum numbers and thus is unambiguous. However, after making the whole circuit of Γ, the cell does not come to itself because a global choice of quantum numbers is impossible. The map from the original cell to the final cell is of course the monodromy map. More precisely, the matrix which gives the transformation of the cell is the inverse transpose of the classical monodromy matrix. A rigorous mathematical formulation of quantum monodromy was given by Vu Ngoc in [39]; relation between quantum and classical monodromy matrices is also discussed in the appendix of [17].

If you like playing with little puzzles as much as Boris does [44], here are a couple of provocative conjectures which can be formulated on the basis of the Zhilinskiǐ diagrams similar to the one in figure C.1.

(i) The "sign conjecture" states that the monodromy matrix is $\left(\begin{smallmatrix} 1 & -k \\ 0 & 1 \end{smallmatrix}\right)$ where k is always a *positive* integer. Indeed, try changing the direction of your contour Γ or try flipping the lattice in figure C.1 any way you can imagine (help yourself with scissors and glue, if you like). The monodromy transformation will always be $\left(\begin{smallmatrix} 1 & -k \\ 0 & 1 \end{smallmatrix}\right)$.

(ii) The "additivity conjecture" states that monodromies of several isolated critical values of the $\mathcal{EM}$-map, which all lie in the interior of the closure of a connected component of the set of regular values, can *only add up*. In other words, the distortion of the original elementary cell can only increase. For example, the monodromy matrix computed along a contour which encircles two isolated critical values with monodromy matrices $\begin{pmatrix} 1 & -k' \\ 0 & 1 \end{pmatrix}$ and $\begin{pmatrix} 1 & -k'' \\ 0 & 1 \end{pmatrix}$, respectively, should be $\begin{pmatrix} 1 & -k'-k'' \\ 0 & 1 \end{pmatrix}$.

Both of these conjectures have been recently proved by Cushman and Vu Ngoc [12]. They should help in the study of bifurcations of pinched tori associated to isolated critical values.

Among many examples of quantum systems with monodromy which have been found since 1980, we can add the textbook system of two (or more) coupled angular momenta [36]. It has also been suggested in [35] that nonintegrable systems with most of their KAM tori remaining intact can also have monodromy. This conjecture has been recently proven in [32]. In particular, the hydrogen atom in perpendicular electric and magnetic fields [35], an atomic realization of a particular perturbation of the Kepler system, has monodromy. Monodromy has been recently found in many molecular systems, notably in H_2^+ [41], a molecular realization of the two-center Kepler system, in floppy triatomic molecules, such as LiCN/CNLi [17], which are distant relatives of the spherical pendulum, and in Fermi resonant molecular elastic pendula, such as CO_2 [13]. See [17] for a brief review.

C.4 Finding monodromy by deformation argument

Monodromy is a global topological property of an integrable fibration. Thus regular fibers (tori) lying far away from the singularity, which is at the origin of monodromy (such as the pinched torus), continue to fit together in the way prescribed by the monodromy of the fibration. Furthermore, in a sufficiently small neighbourhood a small local continuous deformation, which changes qualitatively only the singularity, does not change the monodromy of the fibration. The far away tori will fit together in the same fashion as for the undeformed fibration. This simple deformation argument allows to find monodromy of many systems.

For example, consider a "quadratic spherical pendulum", which is obtained by a generic quadratic deformation $V_{a,b} = \frac{1}{2}ax_3^2 + bx_3$ of the original linear potential $V_{0,1} = x_3 = \sigma_1$ of the spherical pendulum. By analyzing the corresponding continuously deformed reduced system (an exercise which the reader is invited to do following the approach in sec. 3.4) we find that deformed system differs qualitatively from the original spherical pendulum only when $|a| > |b|$. Neglecting overall energy scaling, we can distinguish three robust deformations $V_{0,1}$, $V_{-1,\epsilon}$, and $V_{1,\epsilon}$ (here $0 < |\epsilon| < 1$), two special systems with $V_{\pm 1,0}$, and two transitional systems with $V_{\pm 1,1}$.

As shown in fig. C.2, the $\mathcal{EM}$ map of the $V_{1,\epsilon}$ system has *two* isolated critical values at $\ell = 0$ which correspond to two pinched tori. The continuous deformation $V_{1,\epsilon} \to V_{1,0}$ merges these tori into one doubly pinched torus. The degenerate system $V_{1,0}$ was studied in [2] where it was shown to have monodromy $\left(\begin{smallmatrix} 1 & -2 \\ 0 & 1 \end{smallmatrix}\right)$. This analysis was extended in [12] to $V_{1,\epsilon}$ where the authors show that the $V_{1,\epsilon}$ system also has monodromy 2 for the contour Γ which encircles both critical values. This assertion follows from a deformation argument for sufficiently small ϵ. A different continuous deformation $V_{1,\epsilon} \to V_{0,1}$ merges one of the critical values into the lower boundary of the image of $\mathcal{EM}$, while the other becomes the isolated critical value of the $V_{0,1}$. It follows that monodromy for a contour which goes around this remaining isolated critical value is 1. Replacing ϵ by $-\epsilon$, we see that monodromy for a contour around any of the two isolated critical values of the $V_{1,\epsilon}$ system is 1.

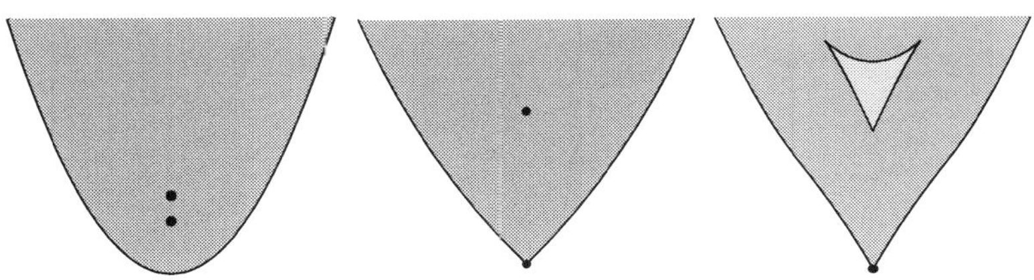

Fig. C.2. Image of the $\mathcal{EM}$ map of the spherical pendulum (center) the deformed quadratic spherical pendulum of the $V_{1,\epsilon}$ kind (left) and of the $V_{-1,\epsilon}$ kind (right, cf. [17] and [41]).

The $V_{-1,\epsilon}$ case was analyzed recently in [17] as a model of triatomic floppy molecules. The image of the $\mathcal{EM}$ map has two leaves A and B (fig. C.2, right). The "main" leaf A is unbounded from above and has a small cut $\mathcal{C}$ along the top edge of the curvilinear triangle B. The second "small" leaf B is glued to A along $\mathcal{C}$. The $\mathcal{EM}$ map is two-valued on $\mathcal{EM}^{-1}(B \setminus C)$. This system has *nonlocal monodromy* which we can observe for a contour $\Gamma \subset A$ which goes around $\mathcal{C}$. The continuous deformation $V_{-1,\epsilon} \to V_{0,1}$ shrinks B (and the cut $\mathcal{C}$) to a point, which becomes the isolated critical point of $V_{0,1}$. In the full space, the union of singular fibers which form $\mathcal{EM}^{-1}(\mathcal{C})$ shrinks to a pinch torus. Since this deformation does not involve the regular tori in $\mathcal{EM}^{-1}(\Gamma)$, monodromy of the $V_{-1,\epsilon}$ system along Γ is 1.

The transitional case $V_{1,1}$ corresponds to the moment of the bifurcation of the lower equilibrium of the system. When we move continuously in the parameter space so that $V_{0,1} \to V_{1,1} \to V_{1,\epsilon}$ (that is, going from center to left in fig. C.2) the initially stable equilibrium detaches from the two families of relative equilibria and becomes an isolated focus-focus point whose stable and unstable manifolds connect and form a pinched torus. It has been proven in [16] that this is a supercritical Hamiltonian Hopf bifrucation.

The case $V_{-1,1}$ corresponds to the moment of the subcritical Hamiltonian Hopf bifurcation of the upper equilibrium of the system [16]. When $V_{0,1} \to V_{-1,1} \to V_{-1,\epsilon}$ (going from center to right in fig. C.2) this initially isolated unstable equilibrium becomes stable (lower vertex of leaf B) and new families of relative equilibria are born (the rest of the boundary ∂B). In sec. D Cushman shows that the monodromy of the $V_{-1,1}$ system is the same as that of the original spherical pendulum $V_{0,1}$. Of course we can now anticipate this result from our deformation argument, but we invite the reader to appreciate the hard way of computing directly the monodromy of $V_{-1,1}$.

D Homework problem. Monodromy about a degenerately pinched 2-torus

This appendix is contributed entirely by Cushman after a question by Boris Zhilinskií and Dmitrií Sadovskií at the end of his lecture on the spherical pendulum. This question turned into an interesting homework problem for the lecturer.

D.1 Introduction

In this section we construct a two degree of freedom Liouville integrable Hamiltonian system on the tangent bundle $T\mathbf{S}^2$ of the 2-sphere $\mathbf{S}^2$ whose energy momentum map $\mathcal{EM}$ has the following properties.

(i) $(0,0)$ is an isolated critical value, that is, there is an open disc $\mathbf{D}$ in $\mathbf{R}^2$ containing $(0,0)$ such that $\mathbf{D}^* = \mathbf{D} \setminus \{(0,0)\}$ consists of regular values of $\mathcal{EM}$ and $\mathbf{D}^*$ lies in its image.

(ii) For every $(h,\ell) \in \mathbf{D}^*$ the (h,ℓ)-level set

$$\mathcal{EM}^{-1}(h,\ell) = \{p \in T\mathbf{S}^2 \mid \mathcal{EM}(p) = (h,\ell)\}$$

is a smooth 2-dimensional torus $T^2_{h,\ell}$.

(iii) The singular fiber $\mathcal{EM}^{-1}(0,0)$ is an immersed 2-dimensional submanifold of $T\mathbf{S}^2$ which is smooth except at two pinch points $p_\pm$ where it has a self intersection. At p_- the pinch is transverse, that is, the tangent spaces to $\mathcal{EM}^{-1}(0,0)$ at p_- are transverse; whereas at p_+ the pinch is degenerate because the tangent spaces are *not* transverse.

Our calculations show that the global monodromy of the smooth 2-torus bundle $\mathcal{EM}^{-1}(\Gamma) \rightarrow \Gamma$ over a smooth positively oriented circle Γ in $\mathbf{D}^*$, which has winding number 1 about $(0,0)$ is $\left(\begin{smallmatrix} 1 & -2 \\ 0 & 1 \end{smallmatrix}\right)$. For more backgound on monodromy in Liouville integrable Hamiltonian systems see [15] or [5]. Our calculations show that the local monodromy around the degenerate pinch point p_+ is $\left(\begin{smallmatrix} 1 & -1 \\ 0 & 1 \end{smallmatrix}\right)$, which is the *same* as the local monodromy about the transversal pinch point p_-. This result is well known if both pinch points are transverse, see [43, 10].

To determine the local monodromy around p_+, we find the variation of the rotation angle of the flow of the Hamiltonian vector field on $T^2_{h,\ell}$ as (h,ℓ) traces out the curve Γ. This uses residues and follows closely the idea of the calculation of the local monodromy for the spherical pendulum given in Chapt. V of [5].

D.2 A model system

Consider the following model Hamiltonian system. On $T\mathbf{R}^3$ with coordinates (x,y) and symplectic form $\widetilde{\omega} = \sum_{i=1}^3 \mathrm{d}x_i \wedge \mathrm{d}y_i$ let

$$\widetilde{H}(x,y) = \tfrac{1}{2}\langle y,y\rangle + V(x_3), \tag{D.1}$$

be the unconstrained Hamiltonian. Here $\langle\, ,\, \rangle$ is the Euclidean metric on $\mathbf{R}^3$. Constrain the Hamiltonian system $(\widetilde{H}, T\mathbf{R}^3, \widetilde{\omega})$ so that the motion takes place on the tangent bundle

$$T\mathbf{S}^2 = \{(x,y) \in T\mathbf{R}^2 \mid \langle x,x\rangle = 1,\ \langle x,y\rangle = 0\}$$

of the 2-sphere $\mathbf{S}^2$. The model problem we consider is the constrained system $(H, T\mathbf{S}^2, \omega)$, where $H = \widetilde{H}|T\mathbf{S}^2$ and $\omega = \widetilde{\omega}|T\mathbf{S}^2$. This system has an S^1 symmetry

$$S^1 \times T\mathbf{S}^2 \rightarrow T\mathbf{S}^2 : (t,(x,y)) \mapsto (R_t x, R_t y), \tag{D.2}$$

where $R_t = \begin{pmatrix} \cos t & -\sin t & 0 \\ \sin t & \cos t & 0 \\ 0 & 0 & 1 \end{pmatrix}$, whose momentum is

$$L : T\mathbf{S}^2 \rightarrow T\mathbf{S}^2 : (x,y) \mapsto x_1 y_2 - x_2 y_1. \tag{D.3}$$

Since the S^1 symmetry preserves the Hamiltonian H, the function L is an integral of the Hamiltonian vector field X_H. In other words, the functions H and L Poisson commute, that is, $\{H, L\} = 0$. Thus $(H, T\mathbf{S}^2, \omega)$ is Liouville integrable.

To remove the S^1 symmetry we apply singular reduction [5], which uses invariant theory. First we note that the algebra of S^1-invariant polynomials on $T\mathbf{S}^2$ is generated by

$$\begin{aligned} \rho_1 &= x_3 & \rho_3 &= y_1^2 + y_2^2 + y_3^2 & \rho_5 &= x_1 y_1 + x_2 y_2 \\ \rho_2 &= y_3 & \rho_4 &= x_1^2 + x_2^2 & \rho_6 &= x_1 y_2 - x_2 y_1 \end{aligned} \tag{D.4}$$

subject to the relations

$$\begin{aligned} \rho_5^2 + \rho_6^2 &= \rho_4(\rho_3 - \rho_2^2), \quad \rho_4 \geq 0, \ \rho_3 \geq \rho_2^2 \\ \rho_1^2 + \rho_4 &= 1 \\ \rho_1\rho_2 + \rho_5 &= 0. \end{aligned} \tag{D.5}$$

The above relations (D.5) define the space $T\mathbf{S}^2/S^1$ of S^1-orbits on $T\mathbf{S}^2$. The singular reduced space $P_\ell = L^{-1}(\ell)/S^1$ is the space of S^1-orbits on the ℓ-level set of the momentum L and is defined by equation (D.5) together with the relation $\rho_6 = \ell$. Eliminating ρ_4, ρ_5, and ρ_6 from these equations, we find that P_ℓ is the semialgebraic variety in $\mathbf{R}^3$ (with coordinates (ρ_1, ρ_2, ρ_3))

$$\rho_2^2 + \ell^2 = \rho_3(1 - \rho_1^2), \quad |\rho_1| \leq 1, \ \rho_3 \geq 0. \tag{D.6}$$

Since the Hamiltonian is invariant under the S^1 symmetry (D.2), it induces the reduced Hamiltonian

$$H_\ell : P_\ell \subseteq \mathbf{R}^3 \to \mathbf{R} : \rho = (\rho_1, \rho_2, \rho_3) \mapsto \tfrac{1}{2}\,\rho_3 + V(\rho_1). \tag{D.7}$$

D.3 A special case

We now look at the special case of the model problem when

$$V(x_3) = -(1 - x_3)^2.$$

Then the reduced Hamiltonian on the reduced space P_ℓ (D.6) is

$$H_\ell(\rho) = \tfrac{1}{2}\,\rho_3 - (1 - \rho_1)^2. \tag{D.8}$$

Its 0-level set on P_ℓ when $\ell = 0$ is illustrated in figure D.1. After reconstruction $H_0^{-1}(0)$ becomes a 2-torus in the 0-level set of the momentum L with a nontransverse pinch point at $p_+ = (0, 0, 1, 0, 0, 0)$. To see this recall that over each smooth point of $H_0^{-1}(0)$ (that is, except for $\rho = (1,0,0)$) the fiber of the S^1-reduction map

$$\pi : L^{-1}(0) \subseteq T\mathbf{S}^2 \rightarrow P_0 \subseteq \mathbf{R}^3 : (x,y) \mapsto (\rho_1(x,y), \rho_2(x,y), \rho_3(x,y)) \tag{D.9}$$

is a unique circular orbit of the S^1 symmetry (D.2); whereas the fiber over the point $(1,0,0)$ is the point p_+, since it is a fixed point of (D.2). The pinch point p_+ is nontransverse because the 0-level set of the reduced Hamiltonian H_0 has second order contact with the reduced space P_0 at $(1,0,0)$.

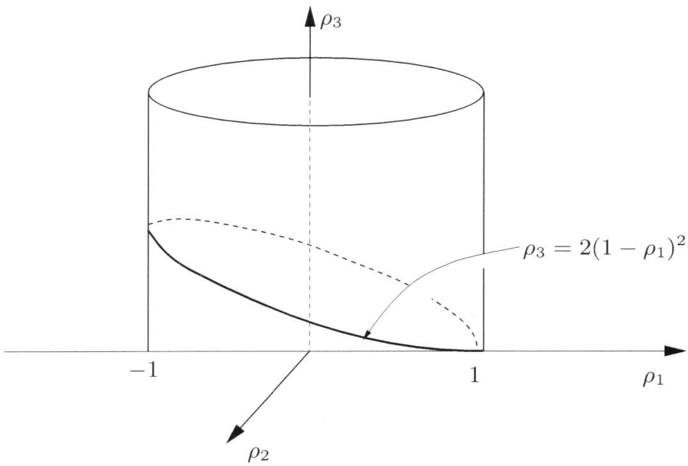

Fig. D.1. The 0-level set of the reduced Hamiltonian $H_0(\rho) = \frac{1}{2}\,\rho_3 - (1-\rho_1)^2$ on the reduced space $P_0 : \rho_2^2 = \rho_3(1-\rho_1^2), \quad |\rho_1| \le 1, \;\; \rho_3 \ge 0$.

Before we compute the local monodromy around p_+ we need to show that $(0,0)$ is an isolated critical value of the energy momentum mapping

$$\mathcal{EM} : T\mathbf{S}^2 \rightarrow \mathbf{R}^2 : (x,y) \mapsto \left(\tfrac{1}{2}\,(y_1^2 + y_2^2 + y_3^2) - (1-x_3)^2, x_1y_2 - x_2y_1\right).$$

This is equivalent to showing that $(h,\ell) = (0,0)$ is an isolated point of the discriminant locus Δ of the polynomial

$$Q(\rho_1,\rho_2) = -\rho_2^2 + 2(h + (1-\rho_1)^2)(1-\rho_1^2) - \ell^2.$$

In other words, Δ is the set of $(h,\ell) \in \mathbf{R}^2$ where Q has a multiple root in $\{(\rho_1,\rho_2) \in \mathbf{R}^2 \mid |\rho_1| \le 1, \; \rho_2 \ge 0\}$. Thus $(h,\ell) \in \Delta$ if and only if $\rho_2 = 0$ and

$$P_{h,\ell}(\rho_1) = 2(h + (1-\rho_1)^2)(1-\rho_1^2) - \ell^2 \tag{D.10}$$

has a multiple root in $[-1,1]$. A straightforward calculation shows that Δ is parmetrized by

$$\begin{cases} h &= -\dfrac{(r-1)^2(2r+1)}{r} \\ \ell^2 &= 2\,\dfrac{(r-1)^3(r+1)^2}{r}, \end{cases} \tag{D.11}$$

where $r \in [-1,0) \cup \{1\}$. Hence $(0,0)$ (which corresponds to $r=1$) is an isolated point of Δ.

Thus we may choose a disc $\mathbf{D}$ in $\mathbf{R}^2$ containing $(0,0)$, which does not intersect Δ and lies in the image of $\mathcal{EM}$. Every $(h,\ell) \in \mathbf{D}^*$ is a regular value of $\mathcal{EM}$. For each $(h,\ell) \in \mathbf{D}^*$, the h-level set of the reduced Hamiltonian H_ℓ on P_ℓ is a smooth $\mathbf{S}^1$. Hence after reconstruction, the (h,ℓ)-level set of $\mathcal{EM}$ is a smooth 2-torus $T^2_{h,\ell}$.

We now turn to analyzing the rotation angle. Let $\mathcal{R}$ be the set of regular values of $\mathcal{EM}$ which lie in its image. For every $(h,\ell) \in \mathcal{R}$, the rotation angle of the flow of the Hamiltonian vector field X_H on the 2-torus $T^2_{h,\ell}$ is

$$\theta(h,\ell) = 2\ell \int_{x_-}^{x_+} \frac{dx}{(1-x^2)\sqrt{P_{h,\ell}(x)}}, \tag{D.12}$$

where $x_\pm = x_\pm(h,\ell)$ are simple zeroes of $P_{h,\ell}$ (D.10) in $[-1,1]$.

The following lemma is the key fact needed to compute the local monodromy of the 2-torus bundle $\mathcal{EM}^{-1}(\Gamma) \to \Gamma$ about p_+.

Lemma. Suppose that $\ell > 0$. Then

$$\lim_{\ell \to 0^+} \theta(h,\ell) = \begin{cases} \pi, & \text{if } h < 0 \\ 2\pi, & \text{if } h > 0. \end{cases} \tag{D.13}$$

Proof. We use complex analysis.
CASE 1. $h < 0$.

Consider the extended complex plane with cuts as indicated in figure D.2 To see that the polynomial $P_{h,0}$ has exactly two simple roots $x_\pm$ in $(-1,1)$ when ℓ is small, first note that $P_{h,0}$ has four simple real roots ± 1 and $s_\pm = 1 \pm \sqrt{-h}$. Since $P_{h,\ell}(\pm 1) = -\ell^2$ and $P_{h,0}(0) = 2(h+1) - \ell^2$ is positive when $|h|$ and ℓ are small, $P_{h,\ell}$ has two simple roots $x_\pm$ in $(-1,1)$ and two simple roots x_0, x_1 outside of $[-1,1]$.

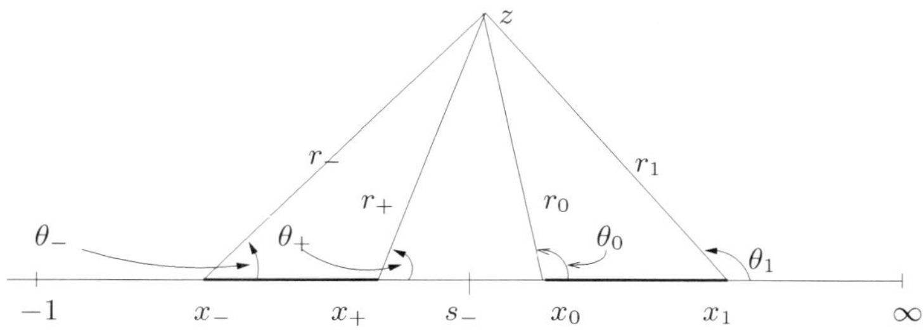

Fig. D.2. The extended complex plane is cut at $[x_-, x_+]$ and $[x_0, x_1]$.

Consider the extended complex plane with cuts as indicated in figure D.2. Write $z - x_\pm = r_\pm e^{i\,\theta_\pm}$ and $z - x_{0,1} = r_{0,1} e^{i\,\theta_{0,1}}$. Then

$$P_{h,\ell}(z) = -2(z - x_-)(z - x_+)(z - x_0)(z - x_1).$$

Define

$$\sqrt{P_{h,\ell}(z)} = i\,\sqrt{|P_{h,\ell}(z)|}\,e^{i(\theta_- + \theta_+ + \theta_0 + \theta_1)/2}.$$

On the upper part of the cut $[x_-, x_+]$ the sign of $\sqrt{P_{h,\ell}(z)}$ is $+$, because $i\,e^{(0+\pi+\pi+\pi)/2} = i(-i) = 1$. Now $z \mapsto \sqrt{P_{h,\ell}(z)}$ is a single valued holomorphic function on the cut extended complex plane, because it is single valued on a loop around each cut. Note that

$$\sqrt{P_{h,\ell}(-1)} = i\,\sqrt{|-\ell^2|}\,e^{i(\pi+\pi+\pi+\pi)/2} = i\,\ell.$$

Consider the contours in figure D.4a The contour C is homologous to $C' + C''$. Hence

$$\int_C \omega = \int_{C'} \omega + \int_{C''} \omega.$$

Here

$$\omega = \frac{1}{(1-z^2)\sqrt{P_{h,\ell}(z)}}$$

is a meromorphic 1-form on the cut extended complex plane, which has first order poles at $z = \pm 1$ and is holomorphic elsewhere. When the contour C'' shrinks to the cut $[x_-, x_+]$ by Cauchy's theorem we obtain

$$\begin{aligned} \ell \int_{C''} \omega &= 2\ell \int_{x_+}^{x_-} \frac{dx}{(1-x^2)\sqrt{P_{h,\ell}(x)}} = -2\ell \int_{x_-}^{x_+} \frac{dx}{(1-x^2)\sqrt{P_{h,\ell}(x)}} \\ &= -\theta(h, \ell). \end{aligned}$$

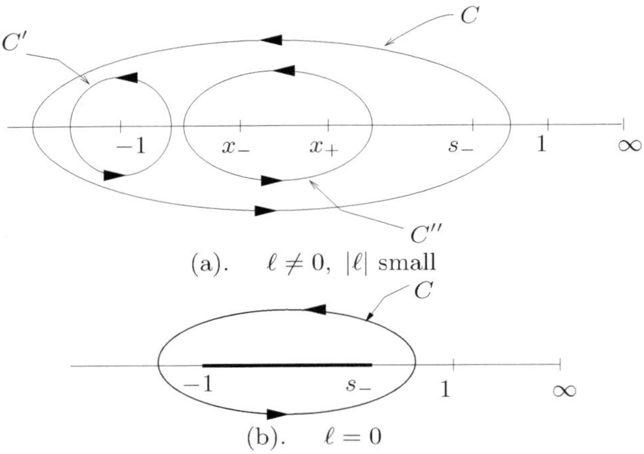

Fig. D.3. Contours on the extended complex plane when $h < 0$.

By the residue theorem we find

$$\begin{aligned}\ell \int_{C'} \omega &= 2\pi i\ell \operatorname*{Res}_{z=-1} \omega = 2\pi i\ell \lim_{z\to -1} \frac{z+1}{(1-z)(1+z)\sqrt{P_{h,\ell}(z)}} \\ &= 2\pi i\ell \frac{1}{2\sqrt{P_{h,\ell}(-1)}} = 2\pi i\ell \frac{1}{2i\ell} = \pi.\end{aligned}$$

We now show that

$$\lim_{\ell\to 0^+} (\ell \int_C \omega) = 0.$$

Thinking of $(h,\ell) \in \mathcal{R}$ as complex variables in $\mathcal{R}^{\mathbf{C}}$, the function $(h,\ell) \to \int_C \omega$ is holomorphic on $\mathcal{R}^{\mathbf{C}}$. Since the contour C does not depend of ℓ,

$$\lim_{\ell\to 0^+} \int_C \omega = \int_C \lim_{\ell\to 0^+} \omega = \int_C \widetilde{\omega},$$

where

$$\widetilde{\omega} = \frac{dz}{(1-z^2)\sqrt{2(h+(1-z)^2)(1-z^2)}}.$$

$\widetilde{\omega}$ is meromorphic in the extended complex plane as cut in figure D.3b with poles only at $z = \pm 1$. At $z = -1$, the 1-form $\widetilde{\omega}$ has a second order pole with zero residue. To see this, let $u^2 = 1 - z$. Then

$$\widetilde{\omega} = \frac{-2u\,du}{u^2(2-u^2)\sqrt{2(h+u^4)u^2(2-u^2)}} = \left(\frac{-1}{2\sqrt{h}}\frac{1}{u^2} + \mathrm{O}(1)\right) du.$$

At $z = s_- = 1 - \sqrt{-h}$ the 1-form $\widetilde{\omega}$ is holomorphic, because s_- is a simple zero of $P_{h,0}$, which is not equal to ± 1. Hence by the residue theorem

$$\int_C \widetilde{\omega} = 2\pi i \operatorname*{Res}_{z=-1} \widetilde{\omega} = 0.$$

Consequently,

$$\lim_{\ell \to 0^+} (\ell \int_C \omega) = \lim_{\ell \to 0^+} (\ell \int_C \widetilde{\omega}) = 0.$$

Taking the limit as $\ell \to 0^+$ of both sides of the equation

$$\ell \int_C \omega = \ell \int_{C'} \omega + \int_{C''} \omega = \pi - \theta(h, \ell)$$

gives

$$\lim_{\ell \to 0^+} \theta(h, \ell) = \pi,$$

when $h < 0$. This proves case 1 of the lemma.

CASE 2. $h > 0$.

The polynomial $P_{h,0}$ has four simple complex roots: ± 1, $1 \pm i\sqrt{h}$. It is positive when $x \in (-1, 1)$ and negative elsewhere. When $|\ell|$ is small and positive, the polynomial $P_{h,\ell}$ has four simple complex roots: $x_\pm$, x_0, x_1 with $x_\pm \in (-1, 1)$ and x_1 is the complex conjugate of x_0. The extended complex plane is cut along the real axis between x_- and x_+ and along a semicircle lying to the right of the line $\operatorname{Re} x = 1$ of radius $\frac{1}{2}|x_0 - x_1|$ with center at $\operatorname{Re} x_0 = 1 + O(|\ell|)$, see figure D.4

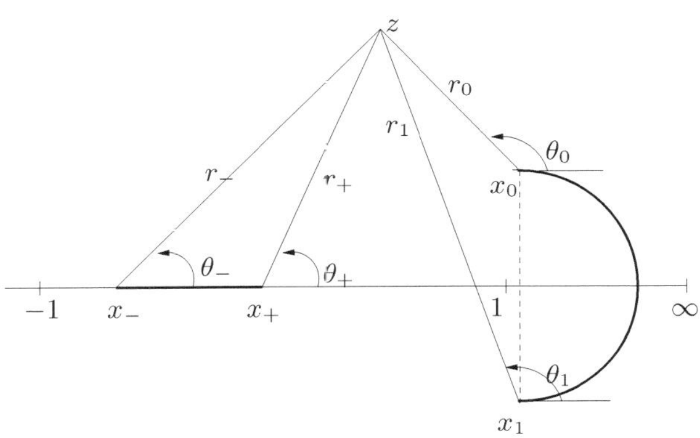

Fig. D.4. Contours on the cut extended complex plane when $h > 0$.

Consider the extended complex plane with cuts as indicated in figure D.4. As in case 1 write $z - x_\pm = r_\pm e^{i\theta_\pm}$, $z - x_{0,1} = r_{0,1}e^{i\theta_{0,1}}$ and

$$P_{h,\ell}(z) = -2(z - x_-)(z - x_+)(z - x_0)(z - x_1).$$

Define

$$\sqrt{P_{h,\ell}(z)} = i\sqrt{|P_{h,\ell}(z)|}\, e^{i(\theta_- + \theta_+ + \theta_0 + \theta_1)/2}.$$

From the fact that $\theta_0 + \theta_1 = 2\pi$ when z is real and less than 1, it follows that the sign of $\sqrt{P_{h,\ell}(z)}$ on the upper part of the cut $[x_-, x_+]$ is + (because $ie^{(0+\pi+2\pi)/2} = i(-i) = 1$). Since the square root is single valued along a loop about each cut, it is holomorphic on the extended cut complex plane. Note that

$$\sqrt{P_{h,\ell}(-1)} = i\sqrt{|-\ell^2|}\, e^{i(\pi+\pi+2\pi)/2} = i\,\ell$$

and

$$\sqrt{P_{h,\ell}(1)} = i\sqrt{|-\ell^2|}\, e^{i(0+0+2\pi)/2} = -i\,\ell.$$

Consider figure D.5a Then $\widetilde{C}$ is homologous to $C + \widetilde{C}' + \widetilde{C}''$. When the contour C shrinks to the cut $[x_-, x_+]$, by Cauchy's theorem we obtain

$$\begin{aligned}\ell \int_C \omega &= 2\ell \int_{x_+}^{x_-} \frac{dx}{(1-x^2)\sqrt{P_{h,\ell}(x)}} = -2\ell \int_{x_-}^{x_+} \frac{dx}{(1-x^2)\sqrt{P_{h,\ell}(x)}} \\ &= -\theta(h,\ell).\end{aligned}$$

By the residue theorem, we have

$$\begin{aligned}\ell \int_{\widetilde{C}'} \omega &= 2\pi i\ell \operatorname*{Res}_{z=-1} \omega = 2\pi i\ell \lim_{z\to -1}(z+1)\frac{1}{(1-z)(1+z)\sqrt{P_{h,\ell}(z)}} \\ &= 2\pi i\,\ell \frac{1}{2\sqrt{P_{h,\ell}(-1)}} = 2\pi i\,\ell\frac{1}{(2i\ell)} = \pi.\end{aligned}$$

and

$$\begin{aligned}\ell \int_{\widetilde{C}''} \omega &= 2\pi i\ell \operatorname*{Res}_{z=1} \omega = 2\pi i\ell \lim_{z\to 1}(z-1)\frac{1}{(1-z)(1+z)\sqrt{P_{h,\ell}(z)}} \\ &= 2\pi i\ell \frac{1}{-2\sqrt{P_{h,\ell}(1)}} = -2\pi i\ell\frac{1}{(-2i\ell)} = \pi.\end{aligned}$$

Hence

$$\begin{aligned}\ell \int_{\widetilde{C}} \omega &= \ell \int_C \omega + \ell \int_{\widetilde{C}'} \omega + \ell \int_{\widetilde{C}''} \omega \\ &= -\theta(h,\ell) + \ell\pi + \ell\pi = -\theta(h,\ell) + 2\pi. \qquad \text{(D.14)}\end{aligned}$$

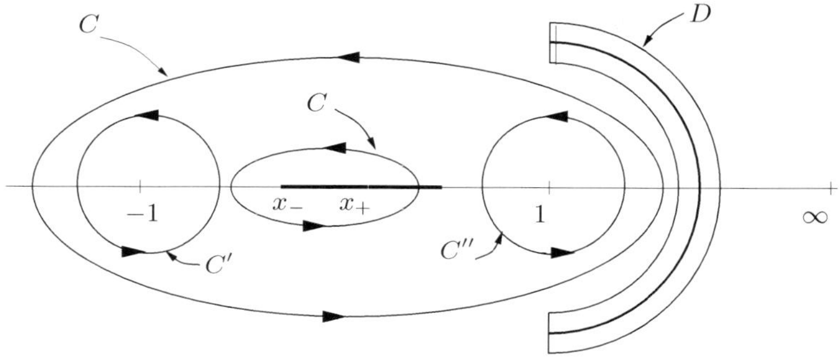

(a). $\ell \neq 0,\ |\ell| \le \ell_C,\ \ell_0$ small, when $h > 0$.

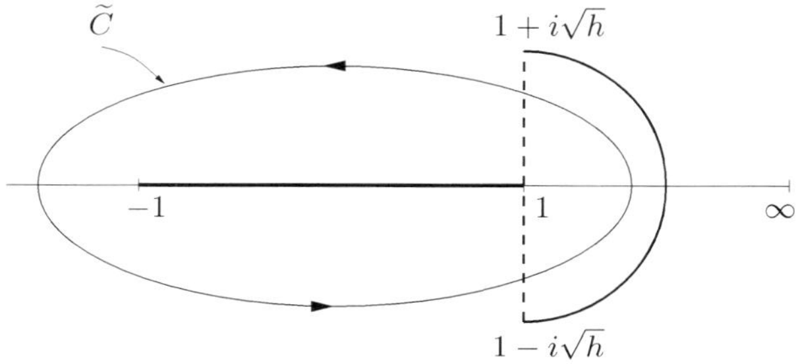

(b). $\ell = 0$ when $h > 0$

Fig. D.5. Contours on the extended complex plane when $h > 0$. The contour $\widetilde{C}$ in figure D.5a semicircular cut from x_0 to x_1 ranges when $|\ell| \le \ell_0$.

When $\ell = 0$ the contour $\widetilde{C}$ has not changed but the cut has, see fig. D.5b. As in case 1 an argument shows that $\widetilde{\omega} = \lim_{\ell \to 0^+} \omega$ is meromorphic on the cut complex plane with a second order pole at ± 1. Since $\int_{\widetilde{C}} \omega$ is a continuous function of h and ℓ when $|\ell| \le \ell_0$ with ℓ_0 small and the contour $\widetilde{C}$ does not depend on ℓ, it follows that

$$\lim_{\ell \to 0^+} (\ell \int_C \omega) = \lim_{\ell \to 0^+} (\ell \int_C \widetilde{\omega}) = 0.$$

Taking the limit as $\ell \to 0^+$ of both sides of equation (D.14) gives

$$0 = \lim_{\ell \to 0^+} \theta(h, \ell).$$

This proves case 2 of the lemma. □

From the lemma and the fact that $\theta(h, -\ell) = -\theta(h, \ell)$ when $\ell \neq 0$ we conclude that the variation of θ is -2π along the positively oriented piecewise smooth curve γ, which is a rectangle in the (h, ℓ) plane, containing $(0,0)$ in the interior of the domain it bounds, with sides parallel to the axes and the sides parallel to the ℓ axis small, see figure D.6.

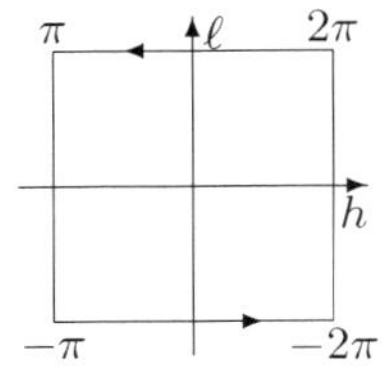

Fig. D.6. The contour γ.

Since the variation of θ depends only on the homotopy class of the curve γ in the set $\mathcal{R}$ of regular values of the energy momentum mapping, it is the equal to the variation along the curve Γ which lies in the punctured disc $\mathbf{D}^*$. Thus the variation in the period lattice associated to the 2-torus bundle $\mathcal{EM}^{-1}(\Gamma) \rightarrow \Gamma$ is $\begin{pmatrix} 0 & -1 \\ 0 & 0 \end{pmatrix}$ as Γ is traversed once in the counterclockwise fashion. Hence the local monodromy of the bundle $\mathcal{EM}^{-1}(\Gamma) \rightarrow \Gamma$ is $\begin{pmatrix} 1 & -1 \\ 0 & 1 \end{pmatrix}$.

D.4 An example of a degenerate heteroclinic cycle

A doubly pinched 2-torus bundle with a transversal pinch at $p_- = (0, 0, -1, 0, 0, 0) \in T\mathbf{S}^2$ and a degenerate pinch at $p_+ = (0, 0, 1, 0, 0, 0)$ is realized by taking the special case of the model problem with $V(x_3) = (x_3 - 1)^2(x_3 + 1)$. Note that $\mathcal{EM}(p_-) = \mathcal{EM}(p_+) = (0, 0)$ is an isolated critical value and consider a positively oriented closed curve Γ in the set of regular values of the $\mathcal{EM}$ map having winding number 1 about $(0, 0)$. By calculation in the preceding section the local monodromy about p_+ is $\begin{pmatrix} 1 & -1 \\ 0 & 1 \end{pmatrix}$. The local monodromy about p_- is $\begin{pmatrix} 1 & -1 \\ 0 & 1 \end{pmatrix}$. The global monodromy is the composition of the local monodromies. Hence the monodromy of the 2-torus bundle $\mathcal{EM}^{-1}(\Gamma) \rightarrow \Gamma$ is $\begin{pmatrix} 1 & -2 \\ 0 & 1 \end{pmatrix}$.

Bibliography

[1] L. Bates and R. Cushman, The rotation number and the herpolhode angle in Euler's top, preprint University of Calgary, September 2003; to appear in ZAMP.

[2] L. Bates and M. Zou, Degeneration of Hamiltonian monodromy cycles, Nonlinearity **6** (1993) 313–335.

[3] R. C. Churchill, M. Kummer, and D. L. Rod, On averaging, reduction and symmetry in Hamiltonian systems, J. Diff. Eqns. **71** (1983) 359–414.

[4] C. Cotter, Ph.D. Thesis, University of California at Santa Cruz, 1985.

[5] R. Cushman and L. Bates, *Global aspects of classical integrable systems*, Birkhauser, Basel, 1997.

[6] R. H. Cushman, Geometry of the energy-momentum mapping of the spherical pendulum, Centrum voor Wiskunde en Informatica Newsletter **1** (1983) 4–18.

[7] R. H. Cushman, Geometry of the bifurcations of the Henon-Heiles family, Proc. R. Soc. London A **382** (1982) 361–371.

[8] R. Cushman and J. J. Duistermaat, The quantum mechanical spherical pendulum, Bull. Am. Math. Soc. **19** (1988) 475–79.

[9] R. Cushman and D. L. Rod, Reduction of the semisimple 1:1 resonance, Physica D **6** (1982) 105–112.

[10] R. Cushman and J. J. Duistermaat, Non-Hamiltonian monodromy, J. Diff. Equations **172** (2001) 42–58.

[11] R. Cushman, The geometry of perturbation theory, in: *Deterministic Chaos and General Relativity*, ed. D. Hobill *et al*, p. 89–101, Plenum Press, New York, 1994.

[12] R. Cushman and S. Vũ Ngọc, The sign of the monodromy for Liouville integrable systems, Ann. Inst. H. Poincaré **3** (2002) 883–894.

[13] R. H. Cushman, H. R. Dullin, A. Giacobbe, D. D. Holm, M. Joyeux, P. Lynch, D. A. Sadovskií, and B. I. Zhilinskií, CO_2 molecule as a quantum realization of the 1:1:2 resonant swing-spring with monodromy, Phys. Rev. Lett. accepted in May (2004).

[14] G. Darboux, *Cours de Méchanique*, 1886.

[15] J. J. Duistermaat, On global action angle coordinates, Comm. Pure Appl. Math. **33** (1980) 687–706.

[16] K. Efstathiou, *Metamorphoses of Hamiltonian systems*, Ph.D. thesis, Université du Littoral, Dunkerque, 2004.

[17] K. Efstathiou, M. Joyeux, and D. A. Sadovskií, Global bending quantum number and the absence of monodromy in the HCN↔CNH molecule, Phys. Rev. A **69**(3) (2004) 032504-1–15.

[18] N. Fulton, J. Tennyson, D. A. Sadovskií, and B. I. Zhilinskií, Nonlinear normal modes and local bending vibrations of H_3^+ and D_3^+, J. Chem. Phys. **99** (1993) 906–918.

[19] H. Goldstein, *Classical Mechanics*, Addison Wesley, Reading MA, 1951.

[20] M. Hamermesh, *Group theory and its application to physical problems* Addison Wesley, Reading, MA, 1994.

[21] W. G. Harter and C. W. Patterson, Rotational energy surfaces and high-J eigenvalue structure of polyatomic molecules, J. Chem. Phys. **80** (1984) 4241–4261.

[22] M. Hénon and C. Heiles, The applicability of the third integral of motion: Some numberical experiments, Astronom. J. **69** (1964) 73–79.

[23] I. N. Kozin and R. M. Roberts, Monodromy in the spectrum of a rigid symmetric top molecule in an electric field, preprint, Warwick 2000; J. Chem. Phys. **118** (2003) 10523-10533.

[24] M. L. Lecornu, Sur l'herpolhodie, Bull. Soc. Math. France **XXXIV** (1906) 40–41.

[25] V. Matveev, Integrable Hamiltonian systems with two degrees of freedom. Topological structure of saturated neighborhoods of saddle-saddle and focus points, Mat. Sb. **187** (1996) 29–58.

[26] L. Michel and B. I. Zhilinskií, Symmetry, invariants, topology: Basic tools, Phys. Rep. **341** (2001) 11–84.

[27] L. Michel and B. I. Zhilinskií, Rydberg states of atoms and molecules: Basic group-theoretical and topological analysis, Phys. Rep. **341** (2001) 173–264.

[28] J. Marsden, R. Montgomery, and T. Ratiu, *Reduction, symmetry, and phases in mechanics*, Mem. Amer. Math Soc. **88**, Providence, RI, 1990.

[29] J. Montaldi, R. M. Roberts, and I. Stewart, Periodic solutions near equilibria of symmetric Hamiltonian systems, Phil. Trans. R. Soc. London Ser. A, **325** (1988) 237–293.

[30] I. Newton, *Philosophiae Naturalis Principia Mathematica*, London 1687 = English translation: *The Principia: Mathematical Principles of Natural Philosophy*, translated by B.I. Cohen and A. Whitman, University of California Press, Berkeley, 1999.

[31] L. Poinsot, *Théorie Nouvelle de la Rotation des Corps*, Paris, 1834.

[32] Bob Rink, A Cantor set of tori with monodromy near a focus-focus singularity, Nonlinearity **17** (2004) 347–356.

[33] D. L. Rod and R. C. Churchill, A guide to the Hénon-Heiles Hamiltonian, in: *Singularities and Dynamical Systems*, ed. S.N. Pnevatikos, pp. 385–395, Elsevier, New York, 1985.

[34] E. J. Routh, *Advanced Dynamics of a system of rigid bodies*, 6th ed., MacMillian, London, 1905; reprint Dover, New York, 1955.

[35] D. A. Sadovskií and R. H. Cushman, Monodromy in the hydrogen atom in crossed fields, Physica D **142** (2000) 166–96.

[36] D. A. Sadovskií and B. I. Zhilinskií, Monodromy, diabolic points, and angular momentum coupling, Phys. Lett. A **256**(4) (1999) 235–44.

[37] G. Schwarz, Smooth functions invariant under the action of a compact Lie group, Topology **14** (1975) 63–68.

[38] J. Schwinger, On angular momentum, in: *Quantum theory of angular momentum*, eds. L. C. Biedenharn and H. V. Dam, p.229–279, Academic Press, New York, 1975.

[39] S. Vũ Ngọc, Quantum monodromy in integrable systems, Commun. Math. Phys. **203** (1999) 465–479.

[40] S. Vũ Ngọc, *Systèmes integrables semiclassiques: du local au global*, thèse d'habilitation, Institut Fourier, Grenoble, 2003.

[41] H. Waalkens, A. Junge, and H. R. Dullin, Quantum monodromy in simple molecules, J. Phys. A. Math. Gen. **36** (2003) L307–L314.

[42] E. T. Whittaker, *Analytical dynamics of particles and rigid bodies*, 4th ed., Cambridge University Press, Cambridge, UK, 1937.

[43] N. Tien Zung, A note on focus-focus singularities, Differential Geom. Appl. **7** (1997) 123–130.

[44] B. I. Zhilinskií, Hamiltonian monodromy as lattice defect, arXiv:quant-ph/0303181 (2003).

V

Survey on dissipative KAM theory including quasi-periodic bifurcation theory

Maria-Cristina Ciocci, Anna Litvak-Hinenzon & Henk Broer

Based on lectures by Henk Broer

ABSTRACT

Kolmogorov-Arnol'd-Moser Theory classically was mainly developed for conservative systems, establishing persistence results for quasi-periodic invariant tori in nearly integrable systems. In this survey we focus on dissipative systems, where similar results hold. In non-conservative settings often parameters are needed for the persistence of invariant tori. When considering families of such dynamical systems bifurcations of quasi-periodic tori may occur. As an example we discuss the quasi-periodic Hopf bifurcation.

Contents

1 Introduction

1.1 Motivation

Kolmogorov-Arnold-Moser Theory is concerned with the occurrence of multi- or quasi-periodic invariant tori in nearly integrable systems. Integrable systems by definition have a toroidal symmetry which produces invariant tori as orbits under the corresponding torus action. The central problem of KAM Theory, is the continuation of such tori to nearly integrable perturbations of the system.

Initially this part of perturbation theory was developed for conservative, i.e., Hamiltonian, systems that model the frictionless dynamics of classical mechanics. Related physical questions are concerned with the perpetual stability of the solar system, of tokamak accelerators, etc. Initiated by Poincaré at the end of the 19th century, the theory was further developed by Birkhoff and Siegel and later established by Kolmogorov, Arnold, Moser and others from the 1950s on. For a historical overview and further reference, see [36]. As pointed out in [66, 65] and later in [48, 24, 23], the conservative approach can be extended to many other settings, like to general dissipative systems, to volume preserving systems and to various equivariant or reversible settings. A unifying Lie algebra approach enables to reach all these results at once [66, 24].

In many cases the systems need to depend on parameters in order to ensure the persistent occurrence of quasi-periodicity. In the general dissipative formulation we therefore encounter families of quasi-periodic attractors, parametrized over a nowhere dense set of positive measure. The parametrization is smooth in the sense of Whitney [73, 24]. As an example of the physical relevance of this phenomenon we mention the Ruelle-Takens scenario for the 'onset' of turbulence [76, 75]. In this scenario a fluid dynamical system depends on parameters, where upon changes of the parameters transitions from laminar to more complicated and even turbulent dynamics are described at a low dimensional level.[1] Here chaotic dynamics plays a special role, related to the onset of turbulence and quasi-periodicity is an intermediate, pre-chaotic stage of this.

The present material on KAM Theory focuses on results in this dissipative setting. In this way the mathematical difficulties are largely decoupled from the symplectic and Hamiltonian formalism. On the one hand, this clarifies the mathematics of KAM Theory, which is difficult enough to master by itself. On the other hand we illustrate the analogy of our approach in the conservative and other contexts by quite a few remarks and excercises, particularly in the Appendix. To this end we also include a number of references for further reading.

Moreover, we included elements from quasi-periodic bifurcation theory, focussing on the Hopf case, where from n-dimensional quasi-periodic attractors $(n+1)$-dimensional quasi-periodic attractors branch off. This example fits very well in the Ruelle-Takens scenario for the onset of turbulence as sketched

[1] For example in a center manifold [45].

before, and its relationship to the more classical Landau-Hopf-Lifschitz scenario [57, 47, 58].

To a great extent the material of this course is contained in [23], to which we often shall refer for background material, for details and further reference. We thank Heinz Hanßmann, George Huitema, Jun Hoo, Vincent Naudot, Khairul Saleh, Floris Takens, Renato Vitolo and Florian Wagener for their help during the preparation of these notes.

1.2 Preliminaries

We introduce a few basic concepts, for simplicity restricting to the world of smooth dynamical systems with continuous time. Such systems are generated by vector fields, locally given as systems of ordinary differential equation.[2] For background information regarding see, e.g., Arnol'd [1], Moser [66, 65], Broer *et al.* [7, 24, 23]. Also see Broer, Dumortier, Van Strien and Takens [14], Chs. 4 and 9.

First we recall the notion of smooth conjugacy between vector fields. Let two vector fields $\dot{x} = F(x)$ and $\dot{y} = G(y)$, $x, y \in \mathbb{R}^m$, be given and a diffeomorphism $y = \Phi(x)$ of $\mathbb{R}^m$ that takes solutions of the former to the latter vector field in a time preserving way. Then, if $x(t)$ and $y(t) = \Phi(x(t))$ are such solutions, it follows by the Chain Rule that

$$G(y) = \dot{y} = D_x\Phi \cdot \dot{x} = D_x\Phi \cdot F(x),$$

where we took matrix products. Such a map Φ is called a conjugacy between the two vector fields. It follows that the condition

$$D_x\Phi \cdot F(x) \equiv G(y), \tag{1.1}$$

where $y = \Phi(x)$, is necessary for Φ being a conjugacy. By the Existence and Uniqueness Theorem [39, 46] for solutions of ordinary differential equations, condition (1.1) also is sufficient for Φ to be a conjugacy. In tensorial shorthand we often rewrite (1.1) as $G = \Phi_* F$.

Remark There exists many variations on the definition of conjugacy, compare with [1, 72]. One variation relaxes the preservation of time-parametrization; the corresponding map Φ then is called an equivalence. Another variation concerns the smoothness, which often is replaced by continuity. Indeed, if Φ is just a homeomorphism we speak of a topological conjugacy or equivalence. With respect to all such equivalence relations the notion of structural

[2] Similar considerations hold for systems with discrete time generated by diffeomorphisms.

stability can be introduced, meaning that with all nearby vector fields (in an appropriate topology) the corresponding relation holds. This is a strong way of formulating persistence of various dynamical properties, like the existence of equilibria, periodic solutions or invariant tori. It turns out that in the present setting of KAM Theory we can use smooth conjugacies or equivalences and corresponding forms of structural stability to express persistence results.

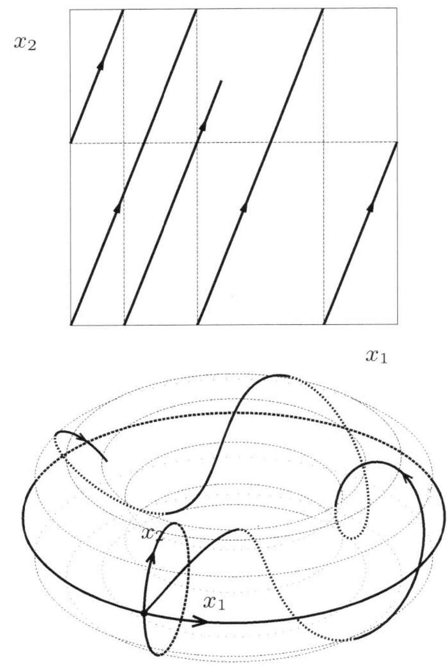

Fig. 1.1. Evolution curve of a constant vector field on the 2-torus $\mathbb{T}^2$.

Now we come to a central subject of KAM Theory, namely the quasi-periodic invariant torus. Let n be a natural number and denote by $\mathbb{T}^n = \mathbb{R}^n/(2\pi\mathbb{Z})^n$ the standard n-torus, equipped with angular coordinates $(x_1, x_2, \ldots, x_n)$ (mod 2π).

Definition 1.1 *A vector field X on a smooth manifold*[3] *M with an invariant*

[3] For all purposes one can take $M = \mathbb{R}^N$ for $N \in \mathbb{N}$ sufficiently large.

submanifold $T \subseteq M$ *is said to have parallel flow on* T, *if the restriction* $X|_T$ *is smoothly conjugate to the constant vector field*

$$\begin{array}{ccc} \dot{x}_1 & = & \omega_1 \\ \dot{x}_2 & = & \omega_2 \\ \dots & \dots & \dots \\ \dot{x}_n & = & \omega_n \end{array}$$

on $\mathbb{T}^n$.

The conjugacy is a diffeomorphism between T and $\mathbb{T}^n$ and the evolution curves of a parallel flow on T, in the coordinates associated to the conjugacy, are given by $t \mapsto (x_1 + t\omega_1, \dots, x_n + t\omega_n)$, where the addition is mod 2π, see Figure 1.1. We say that T is a parallel torus of X, with frequencies $\omega_1, \omega_2, \dots, \omega_n$. The nature of the flow on T depends on the arithmetical properties of its frequencies, where we distinguish between dependence or independence over the rationals.

Definition 1.2 *A parallel torus* T *is quasi-periodic (or non-resonant) if* ω_1, $\omega_2, \dots, \omega_n$ *are rationally independent; that is, for all* $k \in \mathbb{Z}^n \backslash \{0\}$ *one has* $\langle \omega, k \rangle \neq 0$.

Here we abbreviate $\langle \omega, k \rangle = \sum_{i=1}^n \omega_i k_i$. Quasi-periodic tori are densely filled by each of the evolution curves contained in them, see Exercise 1. The parallel torus is called resonant, if its frequencies are rationally dependent, meaning that an integer relation $\langle \omega, k \rangle = 0$ exists for some $k \in \mathbb{Z}^n \setminus \{0\}$. Resonant tori are foliated by lower dimensional tori [2].

Remarks

- Consider the case $n = 2$. For a given vector field X on $\mathbb{T}^2$ we can study the Poincaré return map P of the circle $x_1 = 0$. In the present setting where X is constant, this map is a well-defined diffeomorphism of the circle $\mathbb{T}^1$, provided that $\omega_2 \neq 0$. Indeed, the evolution curve of the vector field X starting at the point $(0, x_2)$ passes through $(2\pi, P(x_2))$, see Figure 1.2.
- Note that when X is constant, the circle map P is a rigid rotation. We quote the following result on rigid rotations, which is quite well-known, e.g., see [35]:

  Lemma *If* $\beta \in \mathbb{R}$ *is irrational, then each orbit of the rigid rotation*

  $$R_\beta : \mathbb{T}^1 \to \mathbb{T}^1, \quad x \mapsto x + 2\pi\beta \,(\text{mod}\, 2\pi),$$

  is dense in $\mathbb{T}^1$.

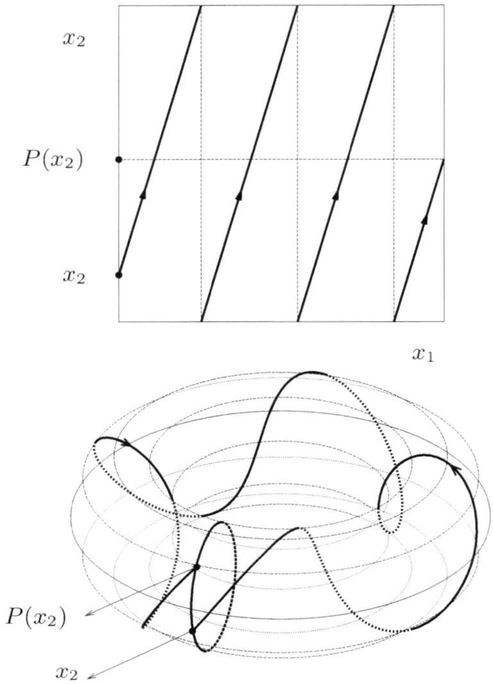

Fig. 1.2. Poincaré map associated to the section $x_1 = 0$ of $\mathbb{T}^2$.

Exercise 1 (Parallel vector fields on $\mathbb{T}^2$) On $\mathbb{T}^2$, with coordinates (x_1, x_2), both counted mod 2π, consider the constant vector field X, given by

$$\begin{aligned} \dot{x}_1 &= \omega_1 \\ \dot{x}_2 &= \omega_2. \end{aligned} \tag{1.2}$$

(i) Suppose that ω_1 and ω_2 are rationally independent, then show that any integral curve of X is dense in $\mathbb{T}^2$.
(ii) Suppose that $\omega_1/\omega_2 = p/q$ with gcd $(p, q) = 1$. Show that $\mathbb{T}^2$ is foliated by periodic solutions, all of period q.

Exercise 2 (How intrinsic is the frequency vector?) If we define $\mathbb{T}^n = \mathbb{R}^n/\mathbb{Z}^n$ and count the angles modulo 1, consider $\mathbb{T}^n$-automorphisms of the affine form $x \mapsto a + Sx$, where $a \in \mathbb{R}^n$ and $S \in \mathrm{SL}(n, \mathbb{Z})$: the $n \times n$ matrices of integers with determinant 1 [2].

(i) Show that the frequency vector of a parallel torus is well-defined up to the lattice $\mathbb{Z}^n$.
(ii) How does this translate to the present situation where $\mathbb{T}^n = \mathbb{R}^n/(2\pi\mathbb{Z}^n)$?
(iii) What can you say of the frequency vectors of X and $\Phi_* X$ when Φ is sufficiently close to the identity map?
(iv) Show that an individual vector field of the form (1.2) can never be structurally stable.

2 Quasi-periodic attractors

In what dynamical systems does the phenomenon of quasi-periodicity occur? Our answer to this question should have the following two properties:

- The occurrence is visible in the 'physical' sense, meaning that the set of initial values showing quasi-periodicity or asymptotic quasi-periodicity has positive measure.
- The occurrence is persistent for small perturbations of the system, which means that we are not interested in pathological examples of this phenomenon.

Colloquially the above often is rephrased as 'for typical systems quasi-periodicity occurs in a visible way'. Both in the conservative and in the dissipative setting many concrete examples can be given in this sense [24, 23]. Since we focus on the dissipative setting, we present a few examples with quasi-periodic attractors. For conservative analogues we refer to the Appendix. The first property of 'physical' visibility is easily met: the attractors have an open basin of attraction (occupying almost the entire phase space). The second property of persistence is more problematic and belongs to the domain of KAM Theory, which is the subject of this course. As announced before, persistence requires that the systems depend on parameters, compare with the Exercises 1 and 2. The quasi-periodic attractors will be isolated in phase space and occur on parameter sets that are nowhere dense and of positive measure, compare with [75].

2.1 Forcing or coupling of nonlinear oscillators

Basic ingredient of our examples is a nonlinear oscillator with equation of motion

$$\ddot{y} + c\dot{y} + ay + f(y, \dot{y}) = 0, \tag{2.1}$$

where $y \in \mathbb{R}$ and $\dot{y} = dy/dt$, which is assumed to have a hyperbolic periodic attractor, i.e., a periodic solution with a negative Floquet-exponent. For the

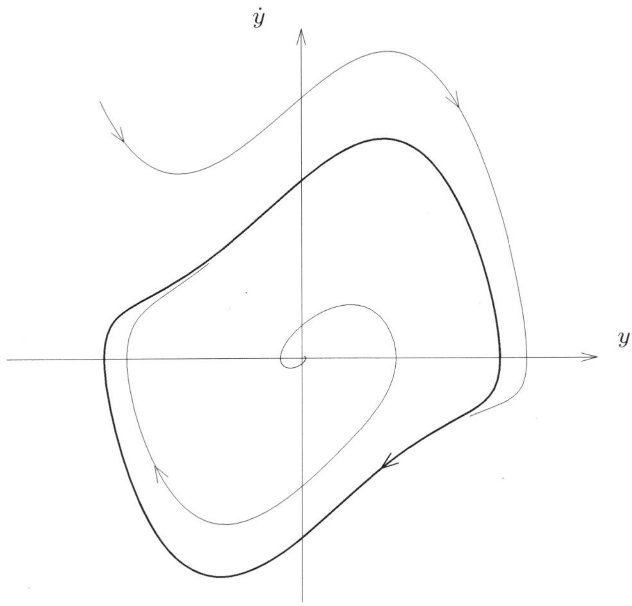

Fig. 2.1. Phase portrait of the free Van der Pol oscillator (2.1).

moment we consider coefficients like a and c as positive constants, but later on they also may be regarded as parameters. A classical example of such a system is the Van der Pol oscillator, where the nonlinearity is given by $f(y, \dot{y}) = by^2\dot{y}$, with b a real constant. For a phase portrait see Figure 2.1.

2.1.1 A nonlinear oscillators with periodic forcing

As a first example consider the oscillator (2.1) subject to a weak time-periodic forcing:

$$\ddot{y} + c\dot{y} + ay + f(y, \dot{y}) = \varepsilon g(y, \dot{y}, t), \tag{2.2}$$

$y, t \in \mathbb{R}$, where $g(y, \dot{y}, t + 2\pi/\Omega) \equiv g(y, \dot{y}, t)$, and where ε is a small perturbation parameter. As usual we take the time t as an extra state-variable, introducing the 3-dimensional (generalized) phase space $\mathbb{R}^2 \times \mathbb{T}^1$ with coordinates $(y, \dot{y}) \in \mathbb{R}^2$ and $t \in \mathbb{T}^1 = \mathbb{R}/(2\pi\mathbb{Z})$. Here the non-autonomous oscillator (2.2) defines the vector field X_ε given by

$$\begin{aligned} \dot{y} &= z \\ \dot{z} &= -ay - cz - f(y,z) + \varepsilon g(y,z,t) \\ \dot{t} &= \Omega. \end{aligned} \tag{2.3}$$

It is our aim to get this example in the format of Section 1.2, corresponding to the definition of quasi-periodicity given there. From now on we set $x_2 = t$ and $\omega_2 = \Omega$.

We start considering the unperturbed case $\varepsilon = 0$. Here the oscillator is free and decouples from the third equation $\dot{x}_2 = \omega_2$. Combining the periodic attractor of the free oscillator (2.1) with this third equation gives rise to an invariant 2-torus to be denoted by T_0. Here we use certain elements of the theory of ordinary diffenential equations, compare [1, 39, 46]. The time-parameterization of the periodic attractor provides a coordinate $x_1 \pmod{2\pi}$, such that the corresponding evolution is generated by an equation $\dot{x}_1 = \omega_1$. It can be shown that the angular coordinates $x_1 \pmod{2\pi}$ and $x_2 \pmod{2\pi}$ give a conjugacy of the restriction $X_{\varepsilon=0}|_{T_0}$ to the constant vector field

$$\begin{aligned} \dot{x}_1 &= \omega_1 \\ \dot{x}_2 &= \omega_2 \end{aligned} \tag{2.4}$$

on the standard 2-torus $\mathbb{T}^2$. We conclude that for $\varepsilon = 0$ the system (2.3) has an attracting parallel invariant 2-torus.

What happens to such an invariant torus for $\varepsilon \neq 0$? Note that by hyperbolicity of the periodic orbit, the unperturbed torus T_0 is normally hyperbolic. Thus, according to the Center Manifold Theorem, compare [45, 81], T_0 is persistent as an invariant manifold. This means that, for $|\varepsilon| \ll 1$, the vector field X_ε has a smooth invariant 2-torus T_ε, also depending smoothly on ε. Here smooth means finitely differentiable. In particular the degree of differentiability tends to ∞ as $\varepsilon \to 0$.

The remaining question then concerns the persistence of the dynamics inside T_ε. In howfar is the parallellity or quasi-periodicity of the dynamics persistent under small perturbation? This question will be answered in the next subsection. Before that we give another example, which easily can be generalized to higher dimension.

2.1.2 Coupled nonlinear oscillators

As a second example consider two nonlinear oscillators with a weak coupling

$$\begin{aligned}\ddot{y}_1 + c_1\dot{y}_1 + a_1 y_1 + f_1(y_1, \dot{y}_1) &= \varepsilon g_1(y_1, y_2, \dot{y}_1, \dot{y}_2)\\ \ddot{y}_2 + c_2\dot{y}_2 + a_2 y_2 + f_2(y_2, \dot{y}_2) &= \varepsilon g_2(y_1, y_2, \dot{y}_1, \dot{y}_2),\end{aligned}$$

$y_1, y_2 \in \mathbb{R}$. This yields the following vector field X_ε on the 4-dimensional phase space $\mathbb{R}^2 \times \mathbb{R}^2 = \{(y_1, z_1), (y_2, z_2)\}$:

$$\begin{aligned}\dot{y}_j &= z_j\\ \dot{z}_j &= -a_j y_j - c_j z_j - f_j(y_j, z_j) + \varepsilon g_j(y_1, y_2, z_1, z_2),\end{aligned} \tag{2.5}$$

$j = 1, 2$. Note that for $\varepsilon = 0$ the system decouples to a system of two independent oscillators and has an attractor in the form of a two dimensional torus T_0. This torus arises as the product of two (topological) circles, along which each of the oscillators has its periodic solution. The circles lie in the two-dimensional planes given by $z_2 = y_2 = 0$ and $z_1 = y_1 = 0$ respectively.

In the plane $z_2 = y_2 = 0$ we see the evolutions of the first oscillator only. The time-parameterization of its periodic attractor gives a coordinate $x_1(\text{mod}2\pi)$ such that the corresponding evolution is generated by $\dot{x}_1 = \omega_1$. Similarly, one finds $\dot{x}_2 = \omega_2$ for the second oscillator in the plane $z_1 = y_1 = 0$. We see that in this way the restriction $X_0|T_0$ is conjugate to the constant vector field (2.4) which, as before, lives on the standard 2-torus $\mathbb{T}^2$. The conclusion is that the system (2.5), for $\varepsilon = 0$, has an attracting parallel invariant 2-torus. It may be clear that a similar coupling of n nonlinear oscillators gives rise to an attracting parallel n-torus inside $\mathbb{R}^{2n}$.

Regarding the fate of T_0 for $|\varepsilon| \ll 1$ we can repeat the above discussion verbatim. This means that by the Center Manifold Theorem [45, 81] we find a smooth invariant 2-torus T_ε, but that regarding the parallel and the quasi-periodic dynamics on T_ε further discussion is needed.

2.2 Reduction to KAM Theory of circle maps

We now discuss the persistence of the dynamics in the invariant 2-tori T_ε as this came up in both of the above examples. After a first center manifold reduction to the standard 2-torus $\mathbb{T}^2$, we reduce to maps of the circle $\mathbb{T}^1$ by taking a Poincaré section.

2.2.1 Preliminaries

As said before, the tori T_ε are smooth (highly differentiable) center manifolds due to normal hyperbolicity [45, 81]. Indeed, for $|\varepsilon| \ll 1$, all T_ε are diffeomorphic to the standard 2-torus $\mathbb{T}^2$, where the degree of differentiability increases to ∞ as ε tends to 0. To study the dynamics inside the torus, from now on we restrict to this center manifold and reduce the perturbation problem to $\mathbb{T}^2$.

The KAM Theorem we are dealing with allows for formulations in the world of C^k-systems for k sufficiently large, including C^∞, compare [73, 24]. These versions are variations on the simpler setting where the systems are real analytic. For simplicity we therefore restrict to the case where our entire perturbation problem is real analytic.

The present version of the KAM perturbation problem looks for smooth conjugacies or smooth equivalences between X_0 and X_ε, both living on the 2-torus $\mathbb{T}^2$, where $|\varepsilon| \ll 1$. Considering a simple example like

$$\begin{aligned} \dot{x}_1 &= \omega_1 + \varepsilon_1 \\ \dot{x}_2 &= \omega_2 + \varepsilon_2, \end{aligned}$$

we observe that both quasi-periodicity and resonance have dense occurrence, compare Section 1.2. These two cases cannot be equivalent, since in the latter case all evolutions are compact, which they are not in the former. To be more precise, when classifying parallel 2-tori under (smooth) conjugacy the frequencies ω_1 and ω_2 are invariants, while under (smooth) equivalence the frequency ratio $\omega_1 : \omega_2$ is an invariant. Compare with the Exercises 1 and 2.

So for a systematic study of the persistence problem it is necessary to introduce parameters. In the examples concerning the forced oscillator (2.2) or the coupled oscillators (2.5) we may consider the coefficients a, c or a_1, a_2, c_1, c_2 respectively, as parameters. The frequencies ω_1 and ω_2 of the unperturbed vector field X_0, see (2.4), then are regarded as functions of these parameters. We claim that in our examples the following nondegeneracy condition holds, namely that the frequency ratio ω_1/ω_2 varies as a function of the relevant multi-parameter: locally this correspondence is submersive. To simplify things further we consider $\beta = \omega_1/\omega_2$ itself as a parameter.

This leaves us with a family of vector fields $X = X_{\beta,\varepsilon}(x_1, x_2)$ on the standard 2-torus $\mathbb{T}^2$, which is assumed real analytic in all the variables. We study this family by the Poincaré return map $P_{\beta,\varepsilon}$ of the circle $x_1 = 0$, considered as a circle diffeomorphism $\mathbb{T}^1 \to \mathbb{T}^1$, compare a remark following Definition 1.2. The map $P_{\beta,\varepsilon}$ has the form

$$x_2 \mapsto x_2 + 2\pi\beta + \varepsilon a(x_2, \beta, \varepsilon). \tag{2.6}$$

Note that for $\varepsilon = 0$ the unperturbed map $P_{\beta,0}$ is just the rigid rotation $R_\beta :$ $x_2 \mapsto x_2 + 2\pi\beta$ of the circle $\mathbb{T}^1$. For a proper formulation of our problem it is convenient to regard the family of circle maps as a 'vertical' map of the cylinder by considering $P_\varepsilon : \mathbb{T}^1 \times [0,1] \to \mathbb{T}^1 \times [0,1]$ defined as

$$P_\varepsilon : (x_2, \beta) \mapsto (x_2 + 2\pi\beta + \varepsilon a(x_2, \beta, \varepsilon), \beta).$$

The persistence problem now is further formalized as follows. We start looking for a diffeomorphism $\Phi_\varepsilon : \mathbb{T}^1 \times [0,1] \to \mathbb{T}^1 \times [0,1]$ conjugating the unperturbed family P_0 to the perturbed family P_ε, i.e., such that

$$P_\varepsilon \circ \Phi_\varepsilon = \Phi_\varepsilon \circ P_0. \tag{2.7}$$

The conjugacy equation (2.7) also is expressed by commutativity of the following diagram.

$$\begin{array}{ccc} \mathbb{T}^1 \times [0,1] & \xrightarrow{P_\varepsilon} & \mathbb{T}^1 \times [0,1] \\ \uparrow \Phi_\varepsilon & & \uparrow \Phi_\varepsilon \\ \mathbb{T}^1 \times [0,1] & \xrightarrow{P_0} & \mathbb{T}^1 \times [0,1] \end{array}$$

Remarks

- Conjugacies between return maps directly translate to equivalences between the corresponding vector fields, see [1, 72]. In the case of the first example (2.2) these equivalences can even be made conjugacies.
- For orientation preserving circle homeomorphisms, the rotation number is an invariant under (topological) conjugacy, e.g., compare with [1, 35]. The rotation number of such homeomorphism is the average amount of rotation effected by the homeomorphism, which for the unperturbed map $P_{\beta,0}$ exactly coincides with the frequency ratio β.
- The Denjoy Theorem [1, 35] asserts that for irrational β, whenever the rotation number of $P_{\beta',\varepsilon}$ equals β, a topological conjugacy exists between $P_{\beta,0}$ and $P_{\beta',\varepsilon}$.

2.2.2 Formal considerations and small divisors

We now study equation (2.7) for the conjugacy Φ_ε to some detail. To simplify the notation first set $x = x_2$. Assuming that $\Phi_\varepsilon : \mathbb{T}^1 \times [0,1] \to \mathbb{T}^1 \times [0,1]$ has the general form

$$\Phi_\varepsilon(x,\beta) = (x + \varepsilon U(x,\beta,\varepsilon), \beta + \varepsilon\sigma(\beta,\varepsilon)),$$

we get the following nonlinear equation for the function U and the parameter shift σ

$$\begin{aligned} U(x + 2\pi\beta, \beta, \varepsilon) - U(x,\beta,\varepsilon) &= \\ 2\pi\sigma(\beta,\varepsilon) + a\left(x + \varepsilon U(x,\beta,\varepsilon), \beta + \varepsilon\sigma(\beta,\varepsilon), \varepsilon\right). \end{aligned} \tag{2.8}$$

As is common in classical perturbation theory,[4] we expand a, U and σ as formal power series in ε and solve (2.8) by comparing coefficients. We only consider the coefficients of power zero in ε, not only because asymptotically these coefficients have the strongest effect, but also since the coefficients of higher ε-powers satisfy similar equations. So, writing

$$a(x,\beta,\varepsilon) = a_0(x,\beta) + O(\varepsilon), \quad U(x,\beta,\varepsilon) = U_0(x,\beta) + O(\varepsilon),$$

$$\sigma(\beta,\varepsilon) = \sigma_0(\beta) + O(\varepsilon),$$

substitution in equation (2.8) leads to the following, so-called homological, equation

$$U_0(x + 2\pi\beta, \beta) - U_0(x,\beta) = 2\pi\sigma_0(\beta) + a_0(x,\beta), \tag{2.9}$$

[4] Compare with Poincaré-Lindstedt series.

which has to be solved for U_0, and σ_0. Equation (2.9) is linear and therefore can be directly solved by Fourier series. Indeed, introducing [5]

$$a_0(x,\beta) = \sum_{k\in\mathbb{Z}} a_{0k}(\beta)e^{ikx} \text{ and } U_0(x,\beta) = \sum_{k\in\mathbb{Z}} U_{0k}(\beta)e^{ikx}$$

and comparing coefficients in (2.9), directly yields that

$$\sigma_0 = -\frac{1}{2\pi}a_{00}, \qquad U_{0k}(\beta) = \frac{a_{0k}(\beta)}{\mathrm{e}^{2\pi ik\beta}-1}, k \in \mathbb{Z}\backslash\{0\}, \tag{2.10}$$

while U_{00}, which corresponds to the position of the origin 0 on T, remains arbitrary. We conclude that in general a formal solution exists if and only if β is irrational. Even then one meets the problem of small divisors, caused by the accumulation of the denominators in (2.10) on 0, which makes the convergence of the Fourier series of U_0 problematic. This problem can be solved by a further restriction of β by so-called Diophantine conditions.

Definition 2.1 *Let $\tau > 2$ and $\gamma > 0$ be given. We say that $\beta \in [0,1]$ is Diophantine if for all $p, q \in \mathbb{Z}$ with $q > 0$ we have that*

$$\left|\beta - \frac{p}{q}\right| \geq \frac{\gamma}{q^\tau}. \tag{2.11}$$

Let us denote the set of β satisfying (2.11) by $[0,1]_{\tau,\gamma} \subseteq [0,1]$. It is easily seen that $[0,1]_{\tau,\gamma}$ is a closed set. From this, by the Cantor-Bendixson Theorem [42] it follows that $[0,1]_{\tau,\gamma}$ is the union of a perfect set and a countable set. The perfect set, for sufficiently small $\gamma > 0$, has to be a Cantor set, since it is compact and totally disconnected.[6] The latter means that every point of $[0,1]_{\tau,\gamma}$ has arbitrarily small neighbourhoods with empty boundary, which directly follows from the fact that the dense set of rationals is in its complement. Note that, since $[0,1]_{\tau,\gamma} \subseteq [0,1]$, so as a subset of the real line, the property of being totally disconnected is equivalent to being nowhere dense. Anyhow, the set $[0,1]_{\tau,\gamma}$ is small in the topological sense. In this case, however, the Lebesgue measure of $[0,1]_{\tau,\gamma}$ is not small, since

$$\text{measure}\left([0,1] \setminus [0,1]_{\tau,\gamma}\right) \leq 2\gamma \sum_{q\geq 2} q^{-(\tau-1)} = O(\gamma), \tag{2.12}$$

as $\gamma \downarrow 0$, by our assumption that $\tau > 2$, e.g., compare [1, 24, 23], also for further reference. Note that the estimate (2.12) implies that the union

$$\bigcup_{\gamma>0} [0,1]_{\tau,\gamma},$$

5 In [23] this is called a '1 bite small divisor problem'.
6 Or zero-dimensional.

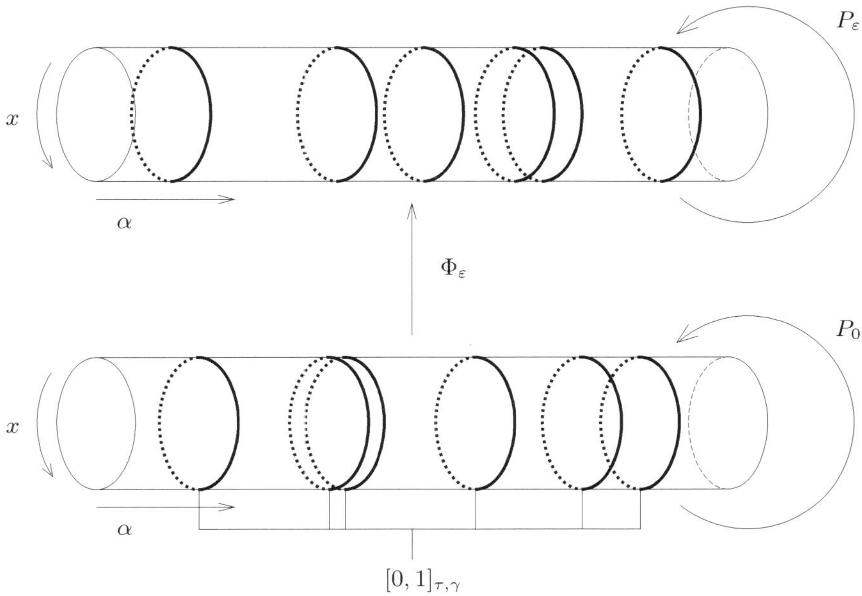

Fig. 2.2. Conjugacy between the Poincaré maps P_0 and P_ε on $\mathbb{T}^1 \times [0,1]_{\tau,\gamma}$.

for any fixed $\tau > 2$, is of full measure. For a thorough discussion of the discrepancy between the measure theoretical and the topological notion of the size of number sets, see Oxtoby [71].

Returning to (2.10), we conclude the following on the convergence of the Fourier series. First we recall that for a real analytic function a_0 the Fourier coefficients a_{0k} decay exponentially as $|k| \to \infty$. This is implied by the Paley-Wiener estimate, which, for completeness, is included in the Appendix. Also see Exercise 15. Second, a brief calculation reveals that for $\beta \in [0,1]_{\tau,\gamma}$ it follows that for all $k \in \mathbb{Z}\setminus\{0\}$ we have

$$|\mathrm{e}^{2\pi i k\beta} - 1| \geq 4\gamma |k|^{-\tau}.$$

We conclude that the coefficients $U_{0,k}$ still have exponential decay as $|k| \to \infty$, which implies that the sum U_0 again is a real analytic function. Of course this does not yet prove that the function $U = U(x, \beta, \varepsilon)$ exists in one way or another for $|\varepsilon| \ll 1$. Yet we do have the following.

Theorem 2.2 *For $|\varepsilon|$ and γ sufficiently small, there exists a C^∞-diffeomorphism $\Phi_\varepsilon : \mathbb{T}^1 \times [0,1] \to \mathbb{T}^1 \times [0,1]$, conjugating the restriction $P_0|_{[0,1]_{\tau,\gamma}}$ to a subsystem of P_ε.*

Observe that Theorem 2.2 is independent of the oscillator background provided by the above examples. Indeed, it is a general result for circle diffeomorphisms and also it forms the first KAM Theorem of this course. There exist several more or less independent proofs of Theorem 2.2. We refer to [24, 23] for further discussion and a large bibliography. The proofs of the KAM Theorems in these notes usually are not based on extending the above formal argument by showing that the power series in ε converges. Indeed, an appropriate 'Newtonian' iteration process is set up, based on a linearization related to the above homological equation (2.9). For an earlier version also see [1]. Note that by more global methods of Herman-Yoccoz [43, 88] versions of Theorem 2.2 have been obtained for large values of $|\varepsilon|$. The present formulation is related to the conservative analogue by Pöschel [73], who follows Zehnder [92, 93] and to Moser [66]. This method uses the concept of Whitney differentiable functions defined on Diophantine Cantor sets. Several aspects of this theory will be treated in more detail below, also see the Appendix.

Figure 2.2 illustrates Theorem 2.2. The invariant subsystem of the perturbed family P_ε mentioned in Theorem 2.2 consists of a collection of parallel (quasi-periodic) circles, smoothly parametrized over a Cantor set of positive measure. In the context of our examples (2.3) and (2.5) this corresponds to a similar family of quasi-periodic invariant and attracting 2-tori. By a straightforward construction, the conjugacy Φ_ε can be extended as an equivalence between the corresponding families of vector fields, restricted to the 2-tori. Compare with [72, 24].

These results all can be phrased in terms of an appropriate form of structural stability, for the occasion called quasi-periodic stability, cf. [24, 23]. Compare with the notion of Ω-stability where structural stability is restricted to the non-wandering set or Ω-set.

Remarks

- It is known that in the gaps of the Cantor set $[0,1]_{\tau,\gamma}$ in general we meet periodicity, which in the oscillator context often is called phase lock. An example of this is given by the Arnold family of circle maps, where $P_{\beta,\varepsilon}(x) = x + 2\pi\beta + \varepsilon \sin x$, compare [1, 35]. In these examples the periodicity in the (β,ε)-plane is organized in a open dense family of resonance tongues. For results and an overview on circle maps see [90, 91, 63], for references also see [23].

- The smoothness of the conjugacies Φ_ε implies the following:

  Corollary *For typical families of dynamical systems, quasi-periodic* 2-*tori occur on a set of positive measure in the parameter space.*

  By the perfectness of the Cantor sets, typically quasi-periodicity is not isolated in the parameter space. One can even show that typically each parameter point of quasi-periodicity is a Lebesgue density point of quasi-periodicity, in the sense that the relative Lebesgue measure tends to full measure as the volume of the neighborhood tends to zero, [23] pp. 131-134.

  Note that such measure theoretic results could not be obtained if Φ_ε were only known to be continuous, again see Oxtoby [71].
- A result similar to Theorem 2.2, including the above remarks on the measure theoretical and topological consequences, generally holds for the existence of quasi-periodic n-tori, again we refer to the set-up of the next section. In the seventies of the last century a 'paradox' arose in this respect, involving the names of R. Thom and V.I. Arnol'd, among others. However, while quasi-periodicity is not persistent for individual systems, for families of systems it is generally persistent on a set of positive measure in the parameter space. Again see [5, 24, 23].
- For the measure theoretic aspect it would also have sufficed to use the somewhat weaker concept of Lipschitz continuity for the conjugacy Φ_ε. However, the Whitney smoothness beyond this keeps a fine track of the geometry of the foliations that are generated by the Diophantine conditions in the product of phase space and parameter space. This is especially of importance when studying quasi-periodic bifurcations.

Exercise 3 (An equivalence turned into a conjugacy) Show that in the case of the forced nonlinear oscillator (2.3) the conjugacy Φ_ε between the return maps P_0 and P_ε can be extended to a (smooth) conjugacy between the corresponding vector fields.

3 Towards a KAM Theory of vector fields

One of the main aims of this course is to sketch set-up and proof of the general KAM Theorem as developed in [24, 23]. We like to point out here that a completely similar theory exists for diffeomorphisms, compare the examples in Section 2.2 and the Appendix. For simplicity we stay in the context of quasi-periodic attractors, so with the standard n-torus $\mathbb{T}^n$ as phase-space, the 'center manifold'. This set-up is very close to that of the KAM Theorem for Lagrangean invariant tori in Hamiltonian mechanics, compare [73].

3.1 Formulation of the Main Theorem

Let $P \subseteq \mathbb{R}^s$ be an open set of parameters and consider families of vector fields $X = X_\mu(x)$, with $X \in \mathbb{T}^n = \mathbb{R}^n/(2\pi\mathbb{Z})^n$ and $\mu \in P$. Often such a family is considered as a 'vertical' vector field on the product $\mathbb{T}^n \times P$. Throughout we assume a real analytic dependence of all vector fields in both x and μ. Also we often use vector field notation, writing

$$f(x,\mu)\partial_x \text{ instead of } \dot{x} = f(x,\mu).$$

Starting point is an integrable family

$$X_\mu(x) = \omega(\mu)\partial_x, \tag{3.1}$$

$x \in \mathbb{T}^n, \mu \in P$, where integrability amounts to x-independence, which expresses symmetry (equivariance) with respect to a natural $\mathbb{T}^n$-action. Our interest is with the family of X-invariant n-tori $\mathbb{T}^n \times \{\mu\}$, where $\mu \in P$. For obvious reasons, the analytic map $\omega : P \to \mathbb{R}^n$ is called the frequency map. The family X is said to be nondegenerate at $\mu_0 \in P$ if the derivative $D_{\mu_0}\omega$ is surjective. As before, our interest is with the fate of the X-invariant tori $\mathbb{T}^n \times \{\mu\}$, $\mu \in P$, under real analytic perturbations

$$\tilde{X}_\mu(x) = X_\mu(x) + \tilde{f}(x,\mu)\partial_x, \tag{3.2}$$

where the size of $\tilde{X} - X$ is small in the compact-open topology on holomorphic extensions.[7] The main question of KAM Theory concerns the fate of the tori $\mathbb{T}^n \times \{\mu\}$, $\mu \in P$, when perturbing from X to $\tilde{X}$. Again as before, we shall use Diophantine conditions. Indeed, for a given $\tau > n-1$ and $\gamma > 0$ we define the set of Diophantine frequency vectors as follows.

$$\mathbb{R}^n_{\tau,\gamma} = \{\omega \in \mathbb{R} \mid |\langle \omega, k\rangle| \geq \gamma |k|^{-\tau},\ \forall k \in \mathbb{Z}^n \backslash \{0\}\}. \tag{3.3}$$

Let us describe its structure, compare with Figure 3.1. First of all it directly follows that $\mathbb{R}^n_{\tau,\gamma}$ is a closed set. Secondly, note that if $\omega \in \mathbb{R}^n_{\tau,\gamma}$ then also $c\omega \in \mathbb{R}^n_{\tau,\gamma}$, for any $c \geq 1$. Therefore $\mathbb{R}^n_{\tau,\gamma}$ is a union of closed half lines. Thirdly, if $\mathbb{S}^{n-1} \subset \mathbb{R}^n$ is the unit $(n-1)$-sphere, then the intersection $\mathbb{R}^n_{\tau,\gamma} \cap \mathbb{S}^{n-1}$ is another closed set, which again is the union of a Cantor set and a countable set, compare the arguments of Section 2.2.2. For this $\gamma > 0$ has to be sufficiently small. Finally, the complement of this Cantor set in $\mathbb{S}^{n-1}$ has a measure of order γ as $\gamma \downarrow 0$. Note that the resonance hyperplanes with equations $\langle \omega, k\rangle = 0$, $k \in \mathbb{Z}^n \setminus \{0\}$, densely fill the complement of $\mathbb{R}^n_{\tau,\gamma}$.

If $\Gamma \subset P$ is any open subset then we define $\Gamma_{\tau,\gamma} = \Gamma \cap \omega^{-1}\left(\mathbb{R}^n_{\tau,\gamma}\right)$. If the restriction of the map ω to Γ is a submersion, then $\Gamma_{\tau,\gamma}$ is a Whitney-smooth

[7] The 'topology of uniform convergence on compact sets'.

foliation of smooth manifolds (with boundary) parameterized over a Cantor set. We colloquially call such a foliation a 'Cantor set'. According to the Inverse Function Theorem, these considerations apply for a sufficiently small neighborhood Γ of μ_0 in P whenever the family X is nondegenerate at the torus $\mathbb{T}^n \times \{\mu_0\}$. We now are ready to formulate the Main Theorem of these notes.

Theorem 3.1 [24, 23] *Let $n \geq 2$. Consider the integrable real analytic family $X = X_\mu(x)$ of vector fields* (3.1)*, $x \in \mathbb{T}^n$, $\mu \in P$. For $\mu_0 \in P$, let X be nondegenerate at the torus $\mathbb{T}^n \times \{\mu_0\}$. Then, for $\gamma > 0$ sufficiently small, there exist a neighborhood Γ of μ_0 in P and a neighborhood $\mathcal{O}$ of X in the compact-open topology, such that for any perturbed family $\tilde{X} \in \mathcal{O}$ as in* (3.2)*, there exists a mapping $\Phi : \mathbb{T}^n \times \Gamma \to \mathbb{T}^n \times P$ with the following properties.*

(i) *Φ is a C^∞ diffeomorphism onto its image which is C^∞-close to the identity map. Also, Φ preserves the projection to P and is real analytic in x;*
(ii) *The restriction of Φ to $\mathbb{T}^n \times \Gamma_{\tau,\gamma}$ conjugates X to $\tilde{X}$.*

This is the second KAM Theorem mentioned in this course and its content forms a paradigm for a great many similar results; also the circle map KAM Theorem 2.2 is just a small variation of this. Many remarks following Theorem 2.2 also hold here. In particular we recall the notion of quasi-periodic stability for the integrable family X and the Corollary that typically quasi-periodicity occurs with positive measure in parameter space. Also we like to recall the remarks in Section 2.2.1 regarding C^k-versions of Theorems 2.2 and 3.1, that can be obtained in a straightforward way [73, 24]. Here it is sufficient that $k > 2\tau + 2 > 2n$. For details, compare with Exercise 15 for the case $n = 1$ and [23] for the multi-frequency case.

Remark For $n \geq 3$ the situation in the gaps of the 'Cantor sets' in between the quasi-periodic tori can be a little different from that in Section 2.2, where we mostly dealt with 2-tori. One reason is that for flows 3-tori can contain strange attractors [70].

Exercise 4 (On Diophantine conditions) In the literature there exist many versions of the Diophantine conditions. Our present interest is in howfar these are equivalent. The Euclidean inner product of two vectors $u = (u_1, \ldots, u_n)$ and $\omega = (\omega_1, \ldots, \omega_n)$ in $\mathbb{R}^n$ is denoted by

$$\langle u, v \rangle = \sum_{i=1}^{n} u_i v_i.$$

For such vectors, we sometimes use the maximum norm

$$\|\omega\|_\infty = \max_{i=1,\dots n} |\omega_i| ,$$

and for integer vectors $k \in \mathbb{Z}^n$ the norm (or length)[8]

$$\|k\|_1 = \sum_{i=1}^{n} |k_i| .$$

Given $\gamma > 0$ and $\tau > 1$, consider the sets

$$\begin{aligned}
D_{\tau,\gamma} &= \{\beta \in \mathbb{R}^n \mid |\langle k, \beta\rangle - \ell| \geq \frac{\gamma}{\|k\|_1^\tau}, \ \forall k \in \mathbb{Z}^n \backslash \{0\}, \ \ell \in \mathbb{Z}\} \\
E_{\tau,\gamma} &= \{\beta \in \mathbb{R}^n \mid \left| e^{2\pi i \langle k,\beta\rangle} - 1 \right| \geq \frac{\gamma}{\|k\|_1^\tau}, \ \forall k \in \mathbb{Z}^n \backslash \{0\}\} \\
F_{\tau,\gamma} &= \{\beta \in \mathbb{R}^n \mid |\langle (k, -\ell), (\beta, 1)\rangle| \geq \frac{\gamma}{\|(k,\ell)\|_1^\tau}, \ \forall k \in \mathbb{Z}^n \backslash \{0\}, \ \ell \in \mathbb{Z}\} \\
G_{\tau,\gamma} &= \{\omega \in \mathbb{R}^{n+1} \mid |\langle h, \omega\rangle| \geq \frac{\gamma}{\|h\|_1^\tau}, \ \forall h \in \mathbb{Z}^{n+1} \backslash \{0\}\}
\end{aligned}$$

Show that:

(i) Given $\gamma > 0$ there exists $\tilde{\gamma} > 0$ such that $D_{\tau,\gamma} \subset E_{\tau,\tilde{\gamma}}$;
(ii) Given $\gamma > 0$ there exists $\tilde{\gamma} > 0$ such that $E_{\tau,\gamma} \subset D_{\tau,\tilde{\gamma}}$;
(iii) $D_{\tau,\gamma} \subset F_{\tau,\gamma}$;
(iv) Given $\beta \in F_{\tau,\gamma}$ there exists $\tilde{\gamma} > 0$ such that $\beta \in D_{\tau,\tilde{\gamma}}$;
(v) If $\beta \in F_{\tau,\gamma}$ then there exists $\tilde{\gamma} > 0$ such that $\omega = (\beta, 1) \in \mathbb{R}^{n+1}$ belongs to $G_{\tau,\tilde{\gamma}}$;
(vi) If $\omega = (\omega_1, \dots, \omega_{n+1}) \in G_{\tau,\tilde{\gamma}}$ then $\tilde{\gamma} > 0$ exists such that

$$\beta = \left(\frac{\omega_1}{\omega_{n+1}}, \dots, \frac{\omega_n}{\omega_{n+1}} \right)$$

belongs to $F_{\tau,\tilde{\gamma}}$.
(vii) Fix $n = 1$. Sketch a geometrical picture of the set $G_{\tau,\gamma}$. For inspiration see Figure 3.1. How can you interpret (iii) in terms of this picture? What is the relation between $D_{\tau,\gamma}$ and the set

$$H_{\sigma,\gamma} = \left\{ \beta \in \mathbb{R} \text{ s.t. } \left| \beta - \frac{p}{q} \right| \geq \frac{\gamma}{|q|^\sigma}, \quad \forall q \in \mathbb{Z} \backslash \{0\}, \ p \in \mathbb{Z} \right\}?$$

[8] In the text the subscripts ∞ and 1 usually are omitted.

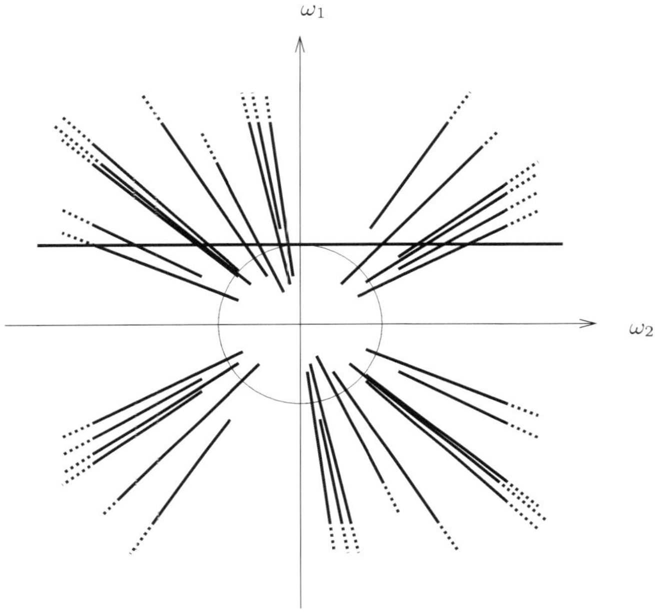

Fig. 3.1. Sketch of the set $\mathbb{R}^2_{\tau,\gamma}$ (and a horizontal line ...).

3.2 On the proof of the Main Theorem

The map Φ, conjugating the unperturbed family X with its perturbation $\tilde{X}$, will be obtained from a nonlinear conjugacy equation, compare with Section 2.2. Here we present the set-up of a proof based on a Newtonian iteration procedure that solves this nonlinear equation. At iteration step number j, $j \in \mathbb{Z}_+$, the map Φ is approximated by an analytic map Φ_j, where $\Phi_{j+1} = \Phi_j \circ \Psi_j$ and where Ψ_j is determined by a linearized conjugacy equation, also called homological equation. This is similar to Section 2.2. The limit $\Phi = \lim_{j\to\infty} \Phi_j$ is taken by the Inverse Approximation Lemma, see the Appendix, in such a way that Φ is a Whitney-C^∞ map. Here the domains of the Φ_j have shrunk to the nowhere dense union ('Cantor set') of Diophantine tori $\mathbb{T}^n \times \Gamma_{\tau,\gamma}$ in an appropriate way.

3.2.1 Introductory remarks

We need some preparations for the set-up.

Reparameterization. By the Inverse Function Theorem, near $\mu_0 \in P$ there exists an analytic diffeomorphism $\mu \mapsto (\omega(\mu), \nu(\mu))$, such that in the new parameters (ω, ν) we get the simplification

$$X_{\omega,\nu}(x) = \omega\partial_x.$$

So now the frequency vector parametrizes the X-invariant tori. Observe that the parameter ν does not show up in the unperturbed system X. We shall drop ν from now on, since it turns out that any parameter that occurs in $\tilde{X}$ in an analytic way, can be directly carried through the whole proof, ending up analytically in the map Φ. In this way the space P is replaced by an open domain of $\mathbb{R}^n$.

A compact-open neighborhood. First of all we note that the compact-open topology on holomorphic extensions corresponds to uniform convergence on compact complex domains. We specify the form of a compact-open neighborhood $\mathcal{A}$ of the family X. For given $S \subseteq \mathbb{R}^k$ and ρ denote

$$S + \rho = \{z \in \mathbb{C}^k \mid \exists s \in S \text{ such that } |z_j - s_j| \leq \rho \text{ for } 1 \leq j \leq k\}.$$

Let Γ be a compact neighborhood of $\omega(\mu_0)$ in $\mathbb{R}^n$ and $\mathcal{N}$ a compact neighborhood of $\mathbb{T}^n \times \Gamma$ in $(\mathbb{C}/(2\pi\mathbb{Z}))^n \times \mathbb{C}^n$. Without loss of generality we choose $\mathcal{N}$ of the form

$$\mathcal{N} = \overline{(\mathbb{T}^n + \kappa) \times (\Gamma + \rho)} \tag{3.4}$$

with constants $\kappa > 0$ and $0 < \rho \leq 1$. For sufficiently small Γ, κ and ρ, the unperturbed family X has a holomorphic extension to $\mathcal{N}$. So much for the 'compact' part of the compact-open neighborhood. Next we come to the 'open' part. A family $\tilde{X}$ belongs to $\mathcal{A}$ if it has the form

$$\tilde{X}_\omega = [\omega + f(x, \omega)]\partial_x$$

with real analytic f that can be extended holomorphically to $\mathcal{N}$ and such that in the supremum norm on $\mathcal{N}$

$$|f|_{\mathcal{N}} < \gamma\delta. \tag{3.5}$$

A more technical reformulation of the Main Theorem 3.1 now claims the existence of a constant δ, such that for $\tilde{X} \in \mathcal{N}$ the conclusions of the theorem hold true. It turns out that δ is independent of Γ, γ and ρ. For later use we introduce the set

$$\Gamma' = \{\omega \in \Gamma \mid \operatorname{dist}(\omega, \partial\Gamma) \geq \gamma\}. \tag{3.6}$$

In any case we need that $\gamma > 0$ is sufficiently small to let $\Gamma'_{\tau,\gamma}$ contain a 'Cantor set' of positive measure.

Remark As can be seen from (3.5), the Diophantine constant γ also enters in the smallness condition of the perturbation f. Regarding the measure of the Diophantine 'Cantor set' of invariant tori, we like $\gamma > 0$ to be as small as possible, which gives a certain conflict of interest with the tolerance for perturbations. In the case where a perturbation parameter ε is used and the perturbation is denoted $\varepsilon f(x, \mu)$, we take γ in dependence of ε, in particular $\gamma(\varepsilon) = O(\varepsilon)$ as $\varepsilon \to 0$. Compare with [23] for further discussion.

3.2.2 Idea behind the proof

The nonlinear conjugacy equation. Our goal is to find a map $\Phi : \mathbb{T}^n \times \Gamma'_{\tau,\gamma} \to \mathbb{T}^n \times P$, preserving the projection to P, which conjugates X to $\tilde{X}$, i.e., such that

$$\Phi_* X = \tilde{X}. \tag{3.7}$$

Taking Φ of the form $\Phi(x, \omega) = \left(x + \tilde{U}(x, \omega), \omega + \tilde{\Lambda}(\omega)\right)$, the conjugacy equation (3.7) translates to

$$\frac{\partial \tilde{U}(x, \omega)}{\partial x} \omega = \tilde{\Lambda}(\omega) + f\left(x + \tilde{U}(x, \omega), \omega + \tilde{\Lambda}(\omega)\right). \tag{3.8}$$

This is a nonlinear equation to be solved in $\tilde{U}$ and $\tilde{\Lambda}$, as far as possible, by the Newtonian iteration process mentioned earlier. The corresponding linearizations look like

$$\frac{\partial \tilde{U}(x, \omega)}{\partial x} \omega = \tilde{\Lambda}(\omega) + f(x, \omega), \tag{3.9}$$

which can be solved directly by Fourier series, compare with Section 2.2.2.[9] In this linearization intuitively we think of $|f|$ as an error that has to be diminuished in the iteration, in the limit yielding the expression (3.7).

A Whitney-smooth limit. As said earlier, given a perturbation $\tilde{X}$ of X, the map Φ solving the conjugacy equation $\Phi_*(X) = \tilde{X}$, see (3.7), will be obtained as a Whitney-C^∞ limit of a sequence $\{\Phi_j\}_{j \geq 0}$ of real analytic diffeomorphisms. Here Φ_j is a near-identity map, defined on a complex neighborhood D_j of $\mathbb{T}^n \times \Gamma'_\gamma$, $j \in \mathbb{Z}_+$. We ensure that $D_{j+1} \subset D_j$ for all $j \in \mathbb{Z}_+$, which directly follows from the following specification.

$$D_j = \left(\mathbb{T}^n + \tfrac{1}{2}\kappa + s_j\right) \times \left(\Gamma'_\gamma + r_j\right), \tag{3.10}$$

where $\{s_j\}_{j \geq 0}$ is any geometric sequence with ratio less than $\frac{1}{2}$ and where $r_j = \frac{1}{2} s_j^{2\tau+2}$, $j \in \mathbb{Z}_+$. The sequence $\{s_j\}_{j \geq 0}$ will be fixed later in such a way

[9] Again we speak of a '1-bite small divisor problem'.

that the iteration process converges in the sense of the Inverse Approximation Lemma A.2.

In this process the Φ_j are constructed inductively, starting with $\Phi_0 = \mathrm{Id}$. For $j > 0$, whenever Φ_j is defined, by (x_j, ω_j) we denote the components of the inverse Φ_j^{-1} and define $\tilde{X}_j = \left(\Phi_j^{-1}\right)_* (\tilde{X})$. Subsequently we introduce

$$\begin{aligned} \Phi_j(x_j, \omega_j) &= \left(x_j + \tilde{U}^j(x_j, \omega_j), \omega_j + \tilde{\Lambda}^j(\omega_j)\right) \quad \text{and} \\ \tilde{X}_{j,\omega_j}(x_j) &= [\omega_j + f^j(x_j, \omega_j)]\partial_{x_j}. \end{aligned}$$

Assuming that both X and $\tilde{X}$ have holomorphic extensions to a set $\mathcal{N}$ (see (3.4)), in the induction process we have to ensure that both Φ_j and $\tilde{X}_j$ have holomorphic extensions to the complex domain $D_j \subseteq \mathcal{N}$. Also it follows for the 'error' $|f^j|$ that $|f^j| \to 0$ as $j \to \infty$ in a 'rapid' way, as this is suitable for a Newtonian iteration process.

3.2.3 The iteration step

In order to explain how the induction works, we assume that Φ_j and $\tilde{X}_j$ are known while we want to construct Φ_{j+1} and therefore $\tilde{X}_{j+1}$ from this. Here we take $j \in \mathbb{Z}_+$, where for $j = 0$ we take $\Phi_0 = \mathrm{Id}$ and hence have $\tilde{X}_0 = \tilde{X}$. Putting $\Phi_{j+1} = \Phi_j \circ \Psi_j$ we so have to construct the map $\Psi_j : D_{j+1} \to D_j$ and then have $\tilde{X}_{j+1} = \left(\Psi_j^{-1}\right)_* (\tilde{X}_j)$. The last expression, in another tensorial shorthand, can be rewritten as $\Psi^* \tilde{X}_j = \tilde{X}_{j+1}$, where $\Psi^* = (\Psi_j^{-1})_*$. Summarizing, for all $j \in \mathbb{Z}_+$ we have

$$\Phi_{j+1} : (x_{j+1}, \omega_{j+1}) \overset{\Psi_j}{\mapsto} (x_j, \omega_j) \overset{\Phi_j}{\mapsto} (x, \omega),$$

which means that

$$\Phi_{j+1} = \Psi_0 \circ \cdots \circ \Psi_j.$$

To simplify things a bit, we introduce the plus-notation. This means that we suppress the index j and write (x, ω) and (ξ, σ) instead of (x_j, ω_j) and (x_{j+1}, ω_{j+1}) respectively. Also we replace f^j by f and f^{j+1} by f^+, D_j by D and D_{j+1} by D_+, etc. The map Ψ will be taken of the form

$$(\xi, \sigma) \mapsto (\xi + U(\xi, \sigma), \sigma + \Lambda(\sigma)),$$

where the parameter shift $\sigma \mapsto \sigma + \Lambda(\sigma)$ is needed to keep track of the frequency vectors that are invariant under near-identity conjugacy. Compare with Section 2.2.2. The unknown functions U and Λ are obtained as solutions of a homological equation

$$\frac{\partial U(\xi, \sigma)}{\partial \xi}\sigma = \Lambda(\sigma) + f(\xi, \sigma)_d, \tag{3.11}$$

where d denotes the Fourier truncation of f at order d. The integer $d = d_j$ has to be determined in the final book keeping. Compare with the earlier linearization (3.9) of the conjugacy equation (3.7). The equation (3.11) once more can be solved directly by Fourier series,[10] yielding

$$U(\xi,\sigma) = U_0(\sigma) + \sum_{0<|k|\leq d} \frac{f_k(\sigma)}{i\langle\sigma,k\rangle} e^{i\langle k,\xi\rangle}, \quad \text{and} \quad \Lambda(\sigma) = f_0(\sigma), \qquad (3.12)$$

where $U_0(\sigma)$ is arbitrary. Notice that by truncation of the Fourier series we only need finitely many Diophantine conditions on σ and we obtain U as a trigonometric polynomial in x.

The conjugacy relation $\Psi^*\tilde{X} = \tilde{X}^+$ now translates to

$$f^+(\xi,\sigma) + \frac{\partial U(\xi,\sigma)}{\partial \xi}(\sigma + f^+(\xi,\sigma)) = \Lambda(\sigma) + f(\xi + U(\xi,\sigma), \sigma + \Lambda(\sigma)),$$

which allows for 'error'-estimates of $|f^+|$ on D_+ in terms of $|f|$ on D, etc. As said before, as $j \to \infty$ we want these 'errors' $|f^j|$ on D_j to decay 'rapidly'.

The rest of the proof consists of thorough analytic book keeping, where the sequence $\{s_j\}_{j\in\mathbb{Z}_+}$, the truncation orders $d_j, j \in \mathbb{Z}_+$ and the final constant δ have to be chosen appropriately. The Paley-Wiener estimate, see the Appendix, is essential for controlling the 'tails' of the ever longer trigonometric polynomials. We refer to [23] (pp. 146-154) for further details of this convergence proof.

Remarks

- In the solution (3.12) we need only finitely many Diophantine conditions of the form $|\langle\sigma,k\rangle| \geq c|k|^{-\tau}$, namely only for $0 < |k| \leq d$. A crucial Lemma [23] (p. 147) ensures that this holds for all $\sigma \in (\Gamma_\gamma + r_j)$, compare the Exercises 5 and 6.
- The analytic book keeping mentioned above includes many applications of the Mean Value Theorem and the Cauchy Integral Formula. The latter serves to express the derivatives of a (real) analytic function in terms of this functions, leading to useful estimates of the C^1-norm in terms of the C^0-norm.
- As said earlier, the above proof is a simplification of the Lie algebra proof of [24] and thereby its set-up is characteristic for many other contexts, like for KAM Theorems in the Hamiltonian, the reversible context, etc. Compare [73, 22] and many references in [23]. Compare with Pöschel [74] for a simple version of the proof in the Hamiltonian case. For a review of several KAM proofs also see De la Llave [62].

[10] Another 1-bite small divisor problem.

Exercise 5 (Homothetic role of γ) By a scaling of the time t and of ω show that Theorem 3.1 only has to be proven for the case $\gamma = 1$. What happens to the bounded domain Γ as γ gets small?

Exercise 6 (Order of truncation) Following Exercise 5 we restrict to the case where $\gamma = 1$, taking the set Γ sufficiently large to contain a nontrivial 'Cantor set' of parameter values corresponding to $(\tau, 1)$-Diophantine frequencies. Maintaining the plus-notation consider the complexified domain

$$D = (\mathbb{T}^n + \tfrac{1}{2}\kappa + s) \times (\Gamma'_{\tau,1} + r),$$

see (3.10), assuming that $r = \frac{1}{2}s^{2\tau+2}$. For the order of truncation d take

$$d = \text{Entier}\left(s^{-2}\right).$$

(i) Show that for all integer vectors $k \in \mathbb{Z}^n$ with $0 < |k| < d$ one has $|k|^{\tau+1} \leq (2r)^{-1}$;

(ii) Next show that for all $\sigma \in \Gamma'_{\tau,1} + r$ and all k with $0 < |k| < d$ one has $|\langle \sigma, k \rangle| \geq \frac{1}{2}|k|^{-\tau}$;

(iii) As an example take $s_j = \left(\frac{1}{4}\right)^j, j \in \mathbb{Z}_+$, and express the order of truncation d_j as a function of j.

Exercise 7 (A normal form for families of circle maps) Given a 1-parameter family of circle maps $P_\lambda : \mathbb{T}^1 \to \mathbb{T}^1$ of the form

$$P_\lambda : x \mapsto x + 2\pi\beta + f(x, \lambda),$$

where x is counted mod 2π and where $f(x, 0) \equiv 0$. One has to show that by successive transformations of the form

$$H_\lambda : x \mapsto x + h(x, \lambda)$$

the x-dependence of P can be pushed away to higher and higher order in λ. For this appropriate conditions on β will be needed. Carry out the corresponding inductive process. What do you think the first step is? Then, concerning the Nth step, consider

$$P_{N,\lambda}(x) = x + 2\pi\beta + g(\lambda) + f_N(x, \lambda) + O(|\lambda|^{N+1}),$$

with $f_N(x, \lambda) = \tilde{f}(x)\lambda^N$ and look for a transformation $H = \text{Id} + h$, with $h(x, \lambda) = \tilde{h}(x, \lambda)\lambda^N$, such that in $H^{-1} \circ P_N \circ H$ the Nth order part in λ is x-independent. Formulate sufficient conditions on β, ensuring that the corresponding equation can be formally solved, in terms of Fourier series. Finally give conditions on β, such that in the real analytic case we obtain real analytic solutions h. Explain your arguments.

Exercise 8 (A problem of Sternberg) On $\mathbb{T}^2$, with coordinates (x_1, x_2) (mod 2π) a vector field X is given, with the following property. If C_1 denotes the circle $C_1 = \{x_1 = 0\}$, then the Poincaré return map $P : C_1 \to C_1$ with respect to X is a rigid rotation $x_2 \mapsto P(x_2) = x_2 + 2\pi\beta$, everything counted mod 2π. From now on we abbreviate $x = x_2$. Let $f(x)$ be the return time of the integral curve connecting the points x and $P(x)$ in C_1. A priori, f does not have to be constant. The problem now is to construct a(nother) circle C_2, that does have a constant return time. Let Φ^t denote the flow of X and express P in terms of Φ^t and f. Let us look for a circle C_2 of the form

$$C_2 = \{\Phi^{u(x)}(0, x) \mid x \in C_1\}.$$

So the search is for a (periodic) function u and a constant c, such that

$$\Phi^c(C_2) = C_2.$$

Rewrite this equation in terms of u and c. Solve this equation formally in terms of Fourier series. What condition on β in general will be needed? Give conditions on β, such that for a real analytic function f a real analytic solution u exists.

4 The normal linear part of quasi-periodic tori

Until this moment we focussed on a family of quasi-periodic attractors. The ambient dynamics was of less interest to us, since the attracting tori were normally hyperbolic and therefore persistent as invariant manifolds. For persistence of the dynamics we restricted to the n-tori as center manifolds. A central theme in the theory of dynamical systems is formed by bifurcations of attractors. In the last part of this course we address elements of the bifurcation theory regarding quasi-periodic attractors. The bifurcations at hand are all related to the loss of normal hyperbolicity.

4.1 Setting of the problem

In this section we consider certain classes of integrable and nearly integrable vector fields with invariant tori and their normal linear part. A more thorough discussion involving the normal bundle of the torus can be found in [24].

For simplicity we assume that a vector field X has $\mathbb{T}^n \times \mathbb{R}^m$ as its phase space where the n-torus $\mathbb{T}^n \times \{0\} \subset \mathbb{T}^n \times \mathbb{R}^m$ is invariant. As before $\mathbb{T}^n$ has angular coordinates $x = (x_1, x_2, \ldots, x_n)$ mod 2π. The coordinates on $\mathbb{R}^m$ are $y = (y_1, y_2, \ldots, y_m)$. As before, and again for simplicity, we assume all dependences to be real analytic. Expanding $X = X(x, y)$ in powers of y we get the following general expression,

$$\begin{aligned} \dot{x} &= \omega(x) + O(|y|) \\ \dot{y} &= \Omega(x)\, y + O(|y|^2), \end{aligned} \tag{4.1}$$

as $y \to 0$, which after truncating away the O-terms, is called the normal linear part of X at $\mathbb{T}^n \times \{0\}$. Here $\omega : \mathbb{T}^n \to \mathbb{R}^n$ and $\Omega : \mathbb{T}^n \to \mathrm{gl}(m, \mathbb{R})$ are real analytic functions. If the coordinates $x \in \mathbb{T}^n$ can be chosen in such a way that ω does not depend on x, the torus is called parallel. See Section 1.2.

The Floquet problem asks whether we can adapt the coordinates further in such a way that also Ω does not depend on x. Such an adaptation is called reduction to Floquet form. In the case where $n = 1$, i.e., the periodic case, the affirmative answer is provided by Floquet Theory [1, 39, 46]. In the multi-frequency case $n \geq 2$ the problem is not so simple, in general there are open classes of systems for which non-reducibility holds, where the obstructions may be of geometrical (topological) nature [44, 26, 27, 80, 82, 37, 54].

The following exercise shows that also for $m = 1$, i.e., for the case of codimension 1 tori, an affirmative answer can be given, provided that the frequency vector ω is Diophantine. Compare with [23] pp. 32-34.

Remark Observe that if the vector field X is integrable, which here again amounts to x-independence, automatically the normal linear part (4.1) has Floquet form.

Exercise 9 (Floquet problem on a codimension 1 torus) Consider a smooth system

$$\begin{aligned} \dot{x} &= f(x, y) \\ \dot{y} &= g(x, y), \end{aligned}$$

with $(x, y) \in \mathbb{T}^n \times \mathbb{R}^m$. Assume that $f(x, y) = \omega + O(|y|)$, which implies that $y = 0$ is a invariant n-torus, with on it a constant vector field with frequency-vector ω. Hence the torus $y = 0$ is parallel. Put $g(x, y) = \Omega(x)y + O(|y|^2)$, for a map $\Omega : \mathbb{T}^n \to \mathrm{gl}(m, \mathbb{R})$. The present problem is to find a transformation $\mathbb{T}^n \times \mathbb{R}^m \to \mathbb{T}^n \times \mathbb{R}^m$, of the form $(x, y) \mapsto (x, z) = (x, A(x)y)$, for some map $A : \mathbb{T}^n \to \mathrm{GL}(m, \mathbb{R})$, with the following property: The transformed system

$$\begin{aligned} \dot{x} &= \omega + O(|z|) \\ \dot{y} &= \Lambda z + O(|z|^2), \end{aligned}$$

is on Floquet form, meaning that the matrix Λ is x-independent. From now on, we restrict to the case $m = 1$. By a computation show that

$$\Lambda = \Omega + \sum_{j=1}^{n} \omega_j \frac{\partial \log A}{\partial x_j}.$$

From this derive an equation, expressing that Λ is constant in x. Formally solve this equation in A, given Ω. Give conditions on ω ensuring a formal solution. Also explain how to obtain a real analytic solution A, assuming real analyticity of Ω.

Remark For $m \geq 2$ the expression for Λ becomes more complicated, since then the matrices do not commute. Apart from this, as said earlier, in this case there can be topological obstructions against the existence of a Floquet form.

4.2 The perturbative point of view

The discussion in Section 4.1 raises the question of persistence of reducibility under small perturbation of the system. As in the Sections 2 and 3 we need parameters to study this problem systematically. We here include an outline of the main result in [24, 23] for the present 'dissipative' situation. Therefore consider a family $X = X_\mu(x, y)$, where $(x, y) \in \mathbb{T}^n \times \mathbb{R}^m$ and where $\mu \in P$ for an open subset P of a Euclidean space. Assume that

$$X_\mu(x, y) = [\omega(\mu) + f(x, y, \mu)]\partial_x + [\Omega(\mu)y + g(x, y, \mu)]\partial_y, \tag{4.2}$$

for $\omega : \mathbb{T}^n \times P \to \mathbb{R}^n$ and $\Omega : \mathbb{T}^n \times P \to gl(m, \mathbb{R})$ and with $f = O(|y|)$ and $g = O(|y|^2)$ as $y \to 0$. We again assume real analytic dependence on all variables and parameters. Assume that all the eigenvalues of $\Omega(\mu)$ are simple and different from 0; by continuity this holds for an open, bounded subdomain $\Gamma \subseteq P$. Moreover we assume nondegeneracy in the sense that the map

$$\mu \in \Gamma \mapsto (\omega(\mu), \mathrm{spec}(\Omega(\mu))) \tag{4.3}$$

is a submersion (if necessary please take Γ smaller).[11] If the eigenvalues of Ω are given by

$$(\delta_1, \ldots, \delta_{N_1}, \alpha_1 \pm i\beta_1, \ldots, \alpha_{N_2} \pm i\beta_{N_2})$$

with $\beta_j > 0$, for $1 \leq j \leq N_2$, then we define

$$\mathrm{spec}(\Omega) = (\delta, \alpha, \beta).$$

[11] By the Inverse Function Theorem it is sufficient that the derivative at a certain point is surjective.

The β_j are called normal frequencies. In this setting we also need extended Diophantine conditions; indeed for $\tau > n - 1$ and $\gamma > 0$ we introduce

$$\begin{aligned} \Gamma^{(2)}_{\tau,\gamma} &= \{\mu \in \Gamma \mid |\langle \omega, k\rangle + \langle \beta, \ell\rangle| \geq \gamma |k|^{-\tau}, \\ &\qquad \forall k \in \mathbb{Z}^n \setminus \{0\}, \forall \ell \in \mathbb{Z}^{N_2} \text{ with } |\ell| \leq 2\}, \end{aligned} \tag{4.4}$$

which, as before, is a 'Cantor set' of positive measure, see [66, 65, 24, 23].

Theorem 4.1 [24] *Let the family X of vector fields be as in* (4.2) *where the frequency map* (4.3) *is a submersion on the bounded set $\Gamma \subseteq P$ of parameters. Then, for sufficiently small γ and for sufficiently nearby families $\tilde{X}$ (in the compact-open topology) there exists a C^∞-diffeomorphism Φ defined near $\mathbb{T}^n \times \{0\} \times \Gamma \subset \mathbb{T}^n \times \mathbb{R}^m \times P$, that is a conjugacy from X to $\tilde{X}$ when further restricting to $\mathbb{T}^n \times \{0\} \times \Gamma^{(2)}_{\tau,\gamma}$. Moreover, Φ preserves the normal linear part.*

Note that Theorem 3.1 covers the special case $m = 0$. Theorem 4.1 roughly states that the nondegenerate family X of vectorfields (4.2), restricted to $\mathbb{T}^n \times \{0\} \times \Gamma \subset \mathbb{T}^n \times \mathbb{R}^m \times P$ is quasi-periodically stable, where the conjugating near-identity diffeomorphisms not only preserve the frequency vector, but also preserve the entire normal linear part. This is of significance for the quasi-periodic bifurcation theory to be developed later. For a proof of this version of the KAM Theorem see [24], where a more general Lie algebra setting is chosen, also compare with [66].

Remarks

- It follows that the perturbed quasi-periodic tori all are of Floquet type. As a corollary we conclude that for parametrized systems, under the above conditions regarding the spectrum of Ω and nondegeneracy, reducibility to Floquet form on 'Cantor sets', is persistent under small perturbations [24, 23].
- The more general Lie algebra formulation of Theorem 4.1 implies similar quasi-periodic stability results in a great many contexts, like in classes of volume preserving, Hamiltonian, reversible or equivariant systems. In the Hamiltonian setting the theorem deals with the persistence of lower dimensional isotropic tori [24, 22, 23, 51].
- The condition that $\Omega(0)$ should only have simple eigenvalues can be relaxed. Generalizations exist where only the eigenvalue 0 has to be avoided and where nondegeneracy needs versality of $\Omega(\mu)$ within the appropriate Lie algebra of matrices [1]. Compare with [22, 12, 20].

5 Elements of quasi-periodic bifurcation theory

In this section we deal with aspects of the bifurcation theory of invariant tori, just considering quasi-periodic attractors that lose their normal hyperbolicity. As phase space we again take $\mathbb{T}^n \times \mathbb{R}^m = \{x \pmod{2\pi}, y\}$, where we are dealing with the invariant torus $\mathbb{T}^n \times \{0\}$, which is assumed parallel.

Observe that in the integrable case, which is $\mathbb{T}^n$-symmetric, (i.e., x-independent), dividing out the toroidal symmetry, we reduce to $\mathbb{R}^m = \{y\}$, where we study the (relative) equilibrium $y = 0$. Therefore local bifurcation theory is a corner stones of the present approach. Since our interest is with persistent results, we also have to consider the effect of non-integrable perturbations.

It turns out that persistent models for torus bifurcation exist for nearly integrable systems. By a scaling [66, 24] it turns out to be 'equivalent' to assume that the unperturbed family of systems has the Floquet format at $y = 0$. Due to the abundance of resonances, we have to use KAM Theory as another corner stone of the theory, and thus speak of quasi-periodic bifurcations. It turns out that quasi-periodic analogues exist in the case of saddle-node, period doubling and Hopf bifurcation [4, 5] also see [23, 9]. The present treatment takes Theorem 4.1 as a starting point.

Bifurcations of equilibria are well understood, at least for low codimension. The same holds for bifurcations of periodic orbits and for fixed points of diffeomorphisms: two theories that are intimately connected by the Poincaré return map. For background we refer to textbooks like [35, 38, 56, 72], also see [1, 14, 53, 85]. To fix thoughts, we briefly revisit certain elements of the corresponding bifurcation theory, however, without claiming completeness.

5.1 Bifurcations of equilibria and fixed points revisited

In the present dissipative setting local bifurcations are all due to the loss of hyperbolicity. For equilibria this occurs when eigenvalues of the linear part cross the imaginary axis upon variation of parameters. For diffeomorphisms the imaginary axis is replaced by the complex unit circle. We briefly describe the codimension 1 bifurcations that occur here.

5.1.1 The saddle-node bifurcation

The simplest bifurcation for equilibria of vector fields is the saddle-node, which takes place as one (real) simple eigenvalue crosses the imaginary axis through 0. The same holds for fixed points of diffeomorphisms when a simple eigenvalue crosses the unit circle through 1. This bifurcation has codimension 1, meaning that generically it occurs in 1-parameter families for isolated values of the parameter. The bifurcation already happens with $\mathbb{R} = \{y\}$ as phase space, in which case topological normal forms, for vector fields and diffeomorphisms respectively, are given by

$$\begin{aligned} \dot{y} &= y^2 - \mu \text{ and} \qquad (5.1) \\ y &\mapsto y + y^2 - \mu, \qquad (5.2) \end{aligned}$$

$y \in \mathbb{R}$, $\mu \in \mathbb{R}$, both varying near 0. In general each of these bifurcation takes place in a 2-dimensional center manifold inside the product of phase space and parameter space [69].

Exercise 10 (What's in a name?) On $\mathbb{R}^2 = \{y_1, y_2\}$ consider the system

$$\begin{aligned} \dot{y}_1 &= y_1^2 - \mu \\ \dot{y}_2 &= -y_2. \end{aligned}$$

Draw the phase portraits of this planar vector field for μ negative, zero and positive. Can you explain the name of the bifurcation?

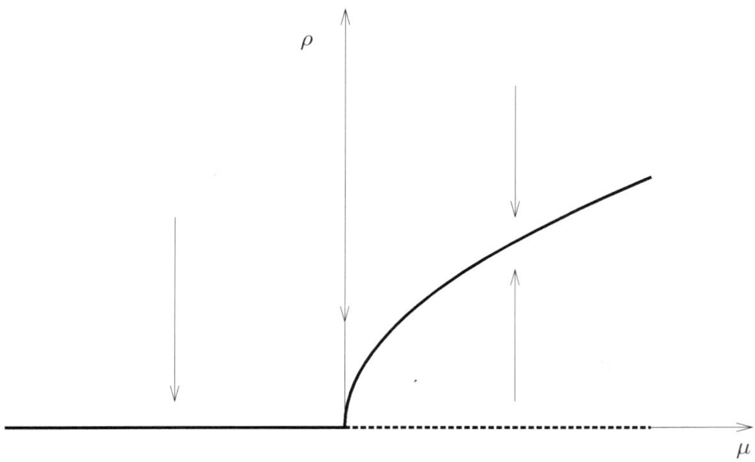

Fig. 5.1. Bifurcation diagram of the Hopf bifurcation.

5.1.2 The Hopf bifurcation

The other generic codimension 1 bifurcation for equilibria of vector fields is the Hopf bifurcation, occurring when a simple complex conjugate pair crosses the imaginary axis. A topological normal form is given by

$$\begin{pmatrix} \dot{y}_1 \\ \dot{y}_2 \end{pmatrix} = \begin{pmatrix} \alpha & -\beta \\ \beta & \alpha \end{pmatrix} \begin{pmatrix} y_1 \\ y_2 \end{pmatrix} - (y_1^2 + y_2^2) \begin{pmatrix} y_1 \\ y_2 \end{pmatrix} \tag{5.3}$$

where $y = (y_1, y_2) \in \mathbb{R}^2$, ranging near $(0, 0)$. In this representation usually one fixes $\beta = 1$ and lets $\alpha = \mu$ (near 0) serve as a (bifurcation) parameter, classifying modulo topological equivalence. In polar coordinates (5.3) so gets the form

$$\begin{aligned} \dot{\varphi} &= 1, \\ \dot{r} &= \mu r - r^3. \end{aligned}$$

Figure 5.1 shows an amplitude response diagram (often called bifurcation diagram). Observe the occurrence of the attracting periodic solution for $\mu > 0$ of amplitude $\sqrt{\mu}$.

Exercise 11 (Floquet exponents in the Hopf bifurcation) Give the Floquet exponents of the periodic solution in the Hopf bifurcation (5.3).

Let us briefly consider the Hopf bifurcation for fixed points of diffeomorphisms. A simple example has the form

$$P(y) = e^{2\pi(\alpha - i\beta)} y + O(|y|^2), \tag{5.4}$$

$y \in \mathbb{C} \cong \mathbb{R}^2$, near 0. To start with β is considered a constant, such that β is not rational with denominator less than 5, see [1, 79], and where $O(|y|^2)$ should contain generic third order terms. As before, we let $\alpha = \mu$ serve as a bifurcation parameter, varying near 0. On one side of the bifurcation value $\mu = 0$ this system by normal hyperbolicity and the Center Manifold Theorem [45] has an invariant circle. Here, due to the invariance of the rotation numbers of the invariant circles, no topological stability can be obtained [69]. Still this bifurcation can be characterized by many persistent properties. Indeed, in a generic 2-parameter family (5.4), say with both α and β as parameters, the periodicity in the parameter plane is organized in resonance tongues [1, 15, 56].[12] If the diffeomorphism is the return map of a periodic orbit for flows, this bifurcation produces an invariant 2-torus. Usually this counterpart for flows is called Neĭmark-Sacker bifurcation. The periodicity as it occurs in the resonance tongues, for the vector field is related to phase lock. The tongues are contained in gaps of a 'Cantor set' of quasi-periodic tori with Diophantine frequencies. Compare the discussion in the Sections 2 and 3 and again compare with [71].

[12] The tongue structure is hardly visible when only one parameter, like α, is used.

Remark A related object is the Arnold family $x \mapsto x + 2\pi\beta + \varepsilon \sin x$ of circle maps [1, 35, 25] as mentioned at the end of Section 2. As discussed there, quasi-periodic and periodic dynamics coexist. Periodicity in the (β, ε)-plane is organized in resonance tongues. Also here in the complement of the union of tongues, which is open and dense, there is a 'Cantor set' of hairs having positive measure.

Exercise 12 (The main tongue of the Arnold family) For the Arnold family

$$P_{\beta,\varepsilon}(x) = x + 2\pi\beta + \varepsilon \sin x$$

of circle maps, consider the region in the (β, ε)-plane where the family has a fixed point. Compute its boundaries and describe (also sketch) the dynamics on both sides of and on a boundary curve. What kind of bifurcation occurs here?

5.1.3 Period doubling

Another famous codimension 1 bifurcation is period doubling, which does not occur for equilibria of vector fields. However, it does occur for fixed points of maps, where a topological model is given by

$$P_\mu(y) = -(1 + \mu)y \pm y^3, \tag{5.5}$$

where $y \in \mathbb{R}$, $\mu \in \mathbb{R}$. It also occurs for periodic orbits of vector fields, in which case the map (5.5) occurs as a return map inside a center manifold.

Remark In all cases, the linear part of the fixed point of the return map corresponds to the normal linear part of the periodic orbit. As remarked in Section 4, the Floquet Theory of periodic orbits ensures that always the normal linear part in appropriate coordinates is constant along the orbit. If this normal linear part has the Floquet form $\omega\partial_x + \Omega y \partial_y$, $x \in \mathbb{T}^1, y \in \mathbb{R}^m$, then the bifurcations correspond to eigenvalues of Ω that cross the imaginary axis, i.e., to nonhyperbolicity of the matrix Ω.

5.2 Bifurcations of quasi-periodic tori

Quasi-periodic versions exist of the saddle-node, the period doubling and the Hopf bifurcation. Returning to the setting with $\mathbb{T}^n \times \mathbb{R}^m$ as the phase space, we remark that the quasi-periodic saddle-node and period doubling already occur for $m = 1$, or in an analogous center manifold. The quasi-periodic Hopf bifurcation needs $m = 2$. We shall illustrate our results on the latter of these cases, compare with [23, 10, 8].

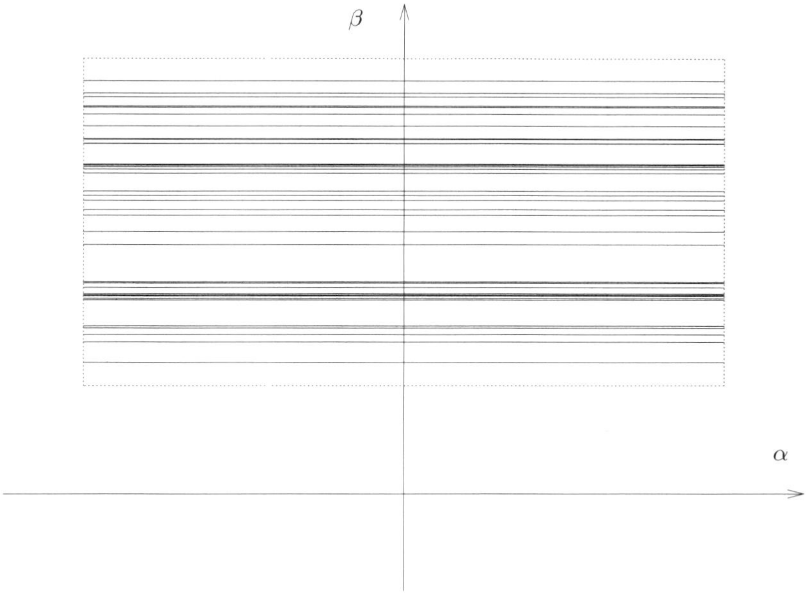

Fig. 5.2. Projection of the 'Cantor set' $\Gamma^{(2)}_{\tau,\gamma}$ on the (α, β)-plane.

5.2.1 Preliminaries

Recall that our phase space is $\mathbb{T}^n \times \mathbb{R}^m = \{x \pmod{2\pi}, y\}$, where we are dealing with the parallel invariant torus $\mathbb{T}^n \times \{0\}$. Also recall that in the integrable case, by $\mathbb{T}^n$-symmetry we can reduce to $\mathbb{R}^m = \{y\}$ and consider the bifurcations of relative equilibria. The present interest is with small non-integrable perturbations of such integrable models. It turns out that the quasi-periodic period-doubling and Hopf bifurcation can be based on Theorem 4.1, while the quasi-periodic saddle-node bifurcation is more involved [5].

5.2.2 Quasi-periodic Hopf bifurcation

The unperturbed, integrable family $X = X_\mu(x, y)$ on $\mathbb{T}^n \times \mathbb{R}^2$ has the form (4.2)

$$X_\mu(x, y) = [\omega(\mu) + f(y, \mu)]\partial_x + [\Omega(\mu)y + g(y, \mu)]\partial_y, \tag{5.6}$$

were $f = O(|y|)$ and $g = O(|y|^2)$ as before. Moreover $\mu \in P$ is a multi-parameter and $\omega : P \to \mathbb{R}^n$ and $\Omega : P \to \mathrm{gl}(2, \mathbb{R})$ are smooth maps. Here we take

$$\Omega(\mu) = \begin{pmatrix} \alpha & -\beta \\ \beta & \alpha \end{pmatrix},$$

which makes the ∂_y component of (5.6) compatible with the planar Hopf family (5.3). The nondegeneracy condition of Theorem 4.1 now requires that there is a subset $\Gamma \subseteq P$ on which the map

$$\mu \in P \mapsto (\omega(\mu), \Omega(\mu)) \in \mathbb{R}^n \times \mathrm{gl}(2, \mathbb{R})$$

is a submersion. For simplicity we even assume that μ is replaced by

$$(\omega, (\alpha, \beta)) \in \mathbb{R}^n \times \mathbb{R}^2,$$

compare with similar considerations in Section 3.2.1.

Observe that if the nonlinearity g satisfies the well-known Hopf nondegeneracy conditions, e.g., compare [38, 56], then the relative equilibrium $y = 0$ undergoes a standard planar Hopf bifurcation as described in Section 5.1.2. Here α again plays the role of bifurcation parameter and a closed orbit branches off at $\alpha = 0$. Without loss of generality we assume that $y = 0$ is attracting for $\alpha < 0$, and that the closed orbit occurs for $\alpha > 0$, and is attracting as well. For the integrable family X, qualitatively we have to multiply this planar scenario with $\mathbb{T}^n$, by which all equilibria turn into invariant attracting or repelling n-tori and the periodic attractor into an attracting invariant $(n + 1)$-torus. Presently the question is what happens to both the n- and the $(n+1)$-tori, when we apply a small near-integrable perturbation.

Persistent quasi-periodic n-tori. We start answering the question of persisting invariant n-tori, by applying Theorem 4.1 in the present setting. Therefore, for $\tau > n - 1$ and $\gamma > 0$, the Diophantine conditions (4.4) where again we restrict to a bounded set $\Gamma \subset \mathbb{R}^n \times \mathbb{R}^2 = \{(\omega, (\alpha, \beta))\}$. To be precise we reconsider the 'Cantor set'

$$\Gamma^{(2)}_{\tau,\gamma} = \{(\omega, (\alpha, \beta)) \in \Gamma \mid |\langle \omega, k\rangle + \beta\ell| \geq \gamma |k|^{-\tau}, \\ \forall k \in \mathbb{Z}^n \setminus \{0\}, \forall \ell \in \mathbb{Z} \text{ with } |\ell| \leq 2\},$$

compare (4.4); for a sketch see Figure 5.2. As a consequence of Theorem 4.1, for any family $\tilde{X}$ on $\mathbb{T}^n \times \mathbb{R}^2 \times P$, sufficiently near X in the compact-open topology, a near-identity C^∞-diffeomorphism $\Phi : \mathbb{T}^n \times \mathbb{R}^2 \times \Gamma \to \mathbb{T}^n \times \mathbb{R}^2 \times \Gamma$ exists, defined near $\mathbb{T}^n \times \{0\} \times \Gamma$, that conjugates X to $\tilde{X}$ when further restricting to $\mathbb{T}^n \times \{0\} \times \Gamma^{(2)}_{\tau,\gamma}$.

Now consider the perturbed family $\tilde{X}$ in the coordinates provided by the inverse Φ^{-1}. In other words, we study the pull-back vector field $\Phi^*\tilde{X}$, that on

$\mathbb{T}^n \times \{0\} \times \Gamma^{(2)}_{\tau,\gamma}$ coincides with the integrable family X. We directly conclude that $\Phi^* \tilde{X}$ on the 'Cantor set' $\Gamma^{(2)}_{\tau,\gamma}$ has $\mathbb{T}^n \times \{0\}$ as an quasi-periodic invariant n-torus, by the above assumptions, attracting for $\alpha < 0$ and repelling for $\alpha > 0$. Moreover, for $\tilde{X}$ we have the normal form decomposition

$$(\Phi^* \tilde{X} - X)_{\omega,\alpha,\beta}\,(x, y) = O(|y|)\partial_x + O(|y|^2)\partial_y + Q_{\omega,\alpha,\beta}(x, y), \quad (5.7)$$

as $y \to 0$. The estimates are uniform in x and ω, α, β. The C^∞-family of vector fields Q is uniformly flat on $\mathbb{T}^n \times \Delta \times \Gamma^{(2)}_{\tau,\gamma} \subset \mathbb{T}^n \times \mathbb{R}^2 \times \Gamma$, where Δ is a small neighborhood of 0 in $\mathbb{R}^m$. This means that its Taylor series completely vanishes. Indeed, for Δ small we can arrange that Q vanishes identically on the 'Cantor set' $\mathbb{T}^n \times \Delta \times \Gamma^{(2)}_{\tau,\gamma}$, whence by perfectness of Cantor sets, we conclude that all derivatives vanish.[13]

Fattening the parameter domain of invariant n-tori. What further conclusions can we draw for $\Phi^* \tilde{X}$, given that $\tilde{X}$ is close to X in the compact-open topology and that Φ is C^∞-near the identity map? For $\alpha \neq 0$, the invariant n-tori are normally hyperbolic. By [45, 81] we conclude that the parameter domain inside Γ where invariant n-tori exist is open. This means that the nowhere dense 'Cantor set', for $\alpha \neq 0$, can be fattened to an open subset of Γ. Outside the 'Cantor set' the invariant n-tori do not have to be quasi-periodic.

The fattening by hyperbolicity can be carried out using a 'standard' contraction principle, e.g., see [33], for a detailed construction using a variation of constants operator see [4]. We here restrict to describing the result of the fattening operation. To this purpose we proceed as follows.

(i) Take $\Gamma = \Gamma_\omega \times \Gamma_\alpha \times \Gamma_\beta$, i.e., of product form, compare with Figure 5.2.
(ii) In the frequency space $\mathbb{R}^n \setminus \{0\} = \{\omega\}$ define $\overline{\omega} = \omega/|\omega| \in \mathbb{S}^{n-1} \subset \mathbb{R}^n$. Also, let $d : \mathbb{S}^{n-1} \times \mathbb{S}^{n-1} \to \mathbb{R}_+$ be the metric $\mathbb{S}^{n-1}$ inherits from $\mathbb{R}^n$.
(iii) A monotonically increasing C^∞-function $p : \mathbb{R}_+ \to \mathbb{R}_+$ that is (infinitely) flat at 0.

For any fixed $\omega_0 = |\omega_0|\,\overline{\omega}_0 \in \Gamma_\omega$ and $\beta_0 \in \Gamma_\beta$, such that $(\omega_0, \alpha, \beta_0) \in \Gamma^{(2)}_{\tau,\gamma}$ for all $\alpha \in \Gamma_\alpha$, consider sets of the form

$$\{(\omega, \alpha, \beta) \in \Gamma \mid 0 < |\alpha| < C \text{ and } p(d(\overline{\omega}, \overline{\omega}_0) + |\beta - \beta_0|) < D|\alpha|^K\}, \quad (5.8)$$

where C, D and K are positive constants. Notice that this is the union of two

13 At least on a full measure subset of $\mathbb{T}^n \times \Delta \times \Gamma^{(2)}_{\tau,\gamma}$, compare with [13].

open discs with a piecewise smooth boundary, that has infinite order of contact with the bifurcation hyperplane $\alpha = 0$, see Figure 5.3.[14]

Theorem 5.1 [5] *In the above situation, given $r \in \mathbb{N}$, there exist positive constants C and D such that for all (ω_0, β_0) such that $(\omega_0, \alpha, \beta_0) \in \Gamma^{(2)}_{\tau,\gamma}$ for all $\alpha \in \Gamma_\alpha$, the corresponding discs* (5.8) *with $K = 3$, are contained in the parameter domain with normally hyperbolic $\Phi^* \tilde{X}$-invariant n-tori of class C^r. These tori are attracting for $\alpha < 0$ and repelling for $\alpha > 0$.*

Remarks

- Since Φ is a near-identity diffeomorphism, this result translates directly to the perturbed family $\tilde{X}$ of vector fields. Note that the discs grow larger as the degree of differentiability r decreases.
- The union of discs is uncountable, leaving open a countable number of holes centered around the pure resonances $(\omega, 0, \beta) \in \Gamma$ with $\langle \omega, k\rangle + \langle \beta, \ell \rangle = 0$ for some $k \in \mathbb{Z}^n \setminus \{0\}$ and $\ell = -2, -1, 0, 1, 2$.
- Such resonance holes also occur in the other quasi-periodic bifurcations at hand, and by Chenciner [29, 30, 31] were called 'bubbles' in the quasi-periodic saddle-node case.

The parameter domain of invariant $(n+1)$-tori. In order to find invariant $(n+1)$-tori we first develop a $\mathbb{T}^{n+1}$-symmetric normal form, related to both the planar normal form (5.3) and the quasi-periodic normal form (5.7). To this purpose, given $N \in \mathbb{N}$, consider the subset of $\Gamma^{(N)}_{\tau,\gamma} \subset \Gamma$, obtained by a further extension of the Diophantine conditions (4.4) to all $\ell \in \mathbb{Z}$ with $|\ell| \leq N$. Again $\Gamma^{(N)}_{\tau,\gamma}$ is a 'Cantor set' of positive measure. In these circumstances, for $|\alpha|$ sufficiently small, there exists a near-identity C^∞-diffeomorphism Φ defined near $\mathbb{T}^n \times \{0\} \times \Gamma \subset \mathbb{T}^n \times \mathbb{R}^m \times P$, such that the following normal form decomposition holds:

$$
\begin{aligned}
(\Phi^* \tilde{X})_{\omega,\alpha,\beta}(x, y) = & \\
& [\omega + |y|^2 f(|y|^2, \omega, \alpha, \beta) + O(|y|^N)]\, \partial_x \\
& \quad + [\beta + |y|^2 g(|y|^2, \omega, \alpha, \beta) + O(|y|^{N+1})]\, [-y_2 \partial_{y_1} + y_1 \partial_{y_2}] \\
& \quad + [\alpha + |y|^2 h(|y|^2, \omega, \alpha, \beta) + O(|y|^{N+1})]\, [y_1 \partial_{y_1} + y_2 \partial_{y_2}] + \\
& \qquad + Q(x, y, \omega, \alpha, \beta),
\end{aligned}
\tag{5.9}
$$

where the family Q of vector fields is uniformly flat on $\mathbb{T}^n \times \{0\} \times \Gamma^{(N)}_{\tau,\gamma}$. Note that for $N = 1$ we recover (5.7). The normal form (5.9) for $N \geq 2$ is a small

[14] In [4, 5], for historical reasons, instead of 'disc' the term 'blunt cusp' was used.

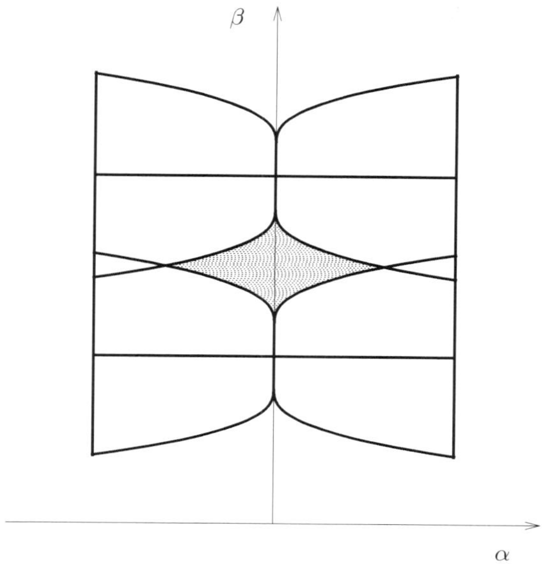

Fig. 5.3. Bubble in between discs.

variation on Theorem 4.1. Indeed, after application of Theorem 4.1 one carries out a formal normal form procedure as developed in [78, 6, 34, 4, 5]. Thus, the $\mathbb{T}^{n+1}$-symmetry of the normal linear part for $\alpha = 0$ is pushed over the formal series in y by solving successive 1-bite small divisor problems. Compare with Exercise 7.

In our application we take $N = 7$. Also the ∂_y-component of (5.9) is close to (5.3). The invariant $(n+1)$-tori now can be found by applying Center Manifold Theory [33, 45].

Theorem 5.2 [5] *In the above situation, given $r \in \mathbb{N}$, there exist positive constants C and D such that for all (ω_0, β_0) such that $(\omega_0, \alpha, \beta_0) \in \Gamma_{\tau,\gamma}^{(N)}$ for all $\alpha \in \Gamma_\alpha$, the corresponding disc* (5.8) *with $\alpha > 0$ and with $K = 7/2$ is contained in the parameter domain with normally hyperbolic $\Phi^*\tilde{X}$-invariant $(n+1)$-tori of class C^r. These tori are attracting.*

Mutatis mutandis, we have the same remarks as following Theorem 5.1. A countable number of bubbles is left out from the half plane $\alpha > 0$, 'centered' around the pure resonances $(\omega, 0, \beta) \in \Gamma$ such that for some $k \in \mathbb{Z}^n$ and some $\ell = -N, -(N-1), \ldots, 0, \ldots N-1, N$ one has $\langle \omega, k \rangle + \langle \beta, \ell \rangle = 0$. We refer to [23, 10, 8] for an overview of the quasi-periodic Hopf bifurcation. The quasi-periodic saddle-node and period doubling have a similar structure [5]. For an early treatment of such torus-bifurcations with only one parameter, see [32].

6 Concluding remarks

We summarize the above results as follows. The quasi-periodic bifurcations due to loss of hyperbolicity to some extent are similar to their periodic analogues. However, for a good description one needs to include sufficiently many parameters to keep track of the internal and normal frequencies, since resonances between these have to be avoided. The main difference with the periodic theory is that these resonance densely fill the parameter space. These resonances are avoided by introducing appropriate Diophantine conditions, giving rise to 'Cantor sets' of positive measure in the parameter space. By hyperbolicity the domains with invariant tori can be fattened, leaving over resonance 'bubbles' inside the gaps of the 'Cantor sets' and centered around the pure resonances. Restricted to the 'Cantor sets' the theory closely resembles the periodic case or the case where one only considers integrable systems, i.e., systems that are $\mathbb{T}^n$-equivariant. Notice that in the latter two cases, the subsets of the parameter space corresponding to non-hyperbolicity, are smooth manifolds. In the nearly integrable context at hand, the strands of bubbles cause a fraying of these smooth boundaries. One also could say that the Implicit Function Theorem, used to find bifurcation sets in the periodic case, is replaced by the KAM Theorem in the quasi-periodic case [92, 93]. For a description of strands of bubbles in the context of coupled oscillators, see [29, 30, 31, 4, 3, 49, 77].

6.1 Non-parallel dynamics

What is the dynamics in the complement of the parameter domains with quasi-periodic tori? Inside the normally hyperbolic n- or $(n+1)$-tori obtained by the fattening several types of dynamics can occur. One type is periodicity, also called phase lock dynamics. Another possibility is chaos. Indeed in tori of dimension 3 or higher, strange attractors can occur [70].

As said before, this fact has a certain interest for the onset of turbulence as described by the theories of Landau-Hopf-Lifschitz and of Ruelle-Takens [57, 47, 58, 75, 76]. For a discussion also see [23].

Inside a bubble we are close to a pure resonance $\langle \omega, k \rangle + \langle \beta, \ell \rangle = 0$. For $\ell = 0$ this is an internal resonance of the torus, while for $\ell \neq 0$ the resonance is normal-internal. We like to mention [3, 30, 31, 80] as examples of research in this direction. For related work on non-parallel dynamics in the Hamiltonian setting see [59, 60, 61].

6.2 Developments in quasi-periodic bifurcation theory

The quasi-periodic bifurcation theory sketched above has been extended in various directions. A direct generalization to the cusp and higher order degenerate bifurcations is performed in [84]. Here a combination of KAM Theory and Singularity Theory is being used. For a detailed study of a skew Hopf-bifurcation, see [26, 82, 27, 28, 80]; to our knowledge, this is the first contribution to KAM Theory for non-reducible systems.

Related reseach programmes are being carried out in contexts with preservation of structure. In the Hamiltonian case quasi-periodic analogues of cuspoid bifurcations of equilibria are developed in [40, 18, 19, 20, 16], also see [41]. Also here KAM Theory has to be combined with Singularity Theory and Catastrophe Theory. We refer to [12] for quasi-periodic bifurcations in the reversible context. Cases of resonant bifurcations are treated in [50, 20]. For a study of normal-internal resonances see [17].

Remark From Singularity Theory the notion of stratified set is known, often as a subset of the product of phase space and parameter space. These subsets indicate the positions of equilibria, fixed points, periodic or quasi-periodic orbits, etc. In the combination with KAM Theory, nowhere dense subsets of positive measure of such stratifications re-enter, defined by Diophantine conditions. Also Whitney-smooth images of such subsets play an important role. Colloquially the term ‘Cantor stratification’ is used for this, which generalizes the term ‘Cantor set’ as used in the present paper.

6.3 Finally . . .

Of course there is a lot more to say on this rich subject, which moreover is rapidly developing in various directions. We restrict ourselves by adding a few recommendations for further reading. Regarding the ‘classical’ KAM Theory of Hamiltonian systems, we refer the reader to [11, 62, 74] and references therein. Recently the Whitney smooth conjugacies of KAM Theory are extended globally to torus bundles of Lagrangean tori, that need not be trivial [13]. This involves notions like monodromy, which turn out to have a deep meaning in semi-classical quantum mechanics. Another area, that we did not touch at all, concerns the KAM Theory of infinite dimensional systems. Here we like to refer to, e.g., [55, 52].

Bibliography

[1] V.I. Arnol′d, *Geometrical Methods in the Theory of Ordinary Differential Equations.* Springer-Verlag 1983.

[2] V.I. Arnol′d, *Mathematical Methods in Classical Mechanics.* Springer–Verlag 1978

[3] C. Baesens, J. Guckenheimer, S. Kim and R.S. MacKay, Three coupled oscillators: Mode–locking, global bifurcation and toroidal chaos, *Physica D* **49**(3) (1991), 387–475.

[4] B.L.J. Braaksma and H.W. Broer, On a quasi–periodic Hopf bifurcation, *Ann. Institut Henri Poincaré, Analyse non linéaire* **4**(2) (1987), 115–168.

[5] B.L.J. Braaksma, H.W. Broer and G.B. Huitema, Towards a Quasi–periodic Bifurcation Theory, *Mem. AMS* **83**(421) (1990), 83–175.

[6] H.W. Broer, Formal normal forms for vector fields and some consequences for bifurcations in the volume preserving case. In: D. Rand and L.S. Young (Eds.), *Dynamical Systems and Turbulence, Warwick 1980* LNM **898** (1981), 75–89, Springer–Verlag.

[7] H.W. Broer, KAM-Theory: Multi-Periodicity in conservative and dissipative systems, *Nieuw Arch. Wisk.* **14**(1) (1996), 1–15.

[8] H.W. Broer, Coupled Hopf-bifurcations: Persistent examples of n-quasiperiodicity determined by families of 3–jets, *Astérisque* **286** (2003), 223–229.

[9] H.W. Broer, Quasi-periodic bifurcations, applications, *Proceedings of the XI CEDYA 1989*, Universidad de Málaga (1990), 3–22.

[10] H.W. Broer, Quasi-periodicity in dissipative systems, *MIHMI (Journ. Indonesian Math. Soc.)* **7**(3) (2001), 7–33.

[11] H.W. Broer, KAM theory: the legacy of Kolmogorov's 1954 paper. To appear *Bull. AMS*, 2004.

[12] H.W. Broer, M.C. Ciocci and A. Vanderbauwhede. Normal 1:1 resonance of invariant tori in reversible systems: a study in quasiperiodic bifurcation theory. *Preprint* 2004.

[13] H.W. Broer, R.H. Cushman, F. Fassò and F. Takens, Geometry of KAM tori for nearly integrable Hamiltonian systems. *Preprint* 2003.

[14] H.W. Broer, F. Dumortier, S.J. van Strien and F. Takens, Structures in dynamics,

finite dimensional deterministic studies, In E.M. de Jager and E.W.C. van Groesen (Eds.) *Studies in Mathematical Physics II* North-Holland 1991.

[15] H.W. Broer, M. Golubitsky and G. Vegter, The geometry of resonance tongues: A Singularity Theory approach, *Nonlinearity* **16** (2003), 1511–1538.

[16] H.W. Broer, H. Hanßmann and J. Hoo, The Lagrange top with quasi-periodic forcing: combination tones in gyroscopic stabilization. In preparation.

[17] H.W. Broer, H. Hanßmann, A. Jorba, J. Villanueva and F.O.O. Wagener, Normal-internal resonances in quasiperiodically forced oscillators: a conservative approach, *Nonlinearity* **16** (2003), 1751–1791.

[18] H.W. Broer, H. Hanßmann and J. You, Bifurcations of normally parabolic tori in Hamiltonian systems. Submitted.

[19] H.W. Broer, H. Hanßmann and J. You, Umbilical torus bifurcations in Hamiltonian systems. Submitted.

[20] H.W. Broer, J. Hoo and V. Naudot, Normal linear stability of quasi-periodic tori. In preparation.

[21] H.W. Broer and G.B. Huitema, A proof of the iso-energetic KAM-theorem from the ordinary one, *J. Diff. Eq.* **90**(1) (1991), 52–60.

[22] H.W. Broer and G.B. Huitema, Unfoldings of quasi–periodic tori in reversible systems, *Journ. Dynamics and Differential Equations* **7**(1) (1995) 191–212.

[23] H.W. Broer, G.B. Huitema and M.B. Sevryuk, Quasi–periodic tori in families of dynamical systems: order amidst chaos, *LNM* **1645** Springer–Verlag 1996.

[24] H.W. Broer, G.B. Huitema and F. Takens, Unfoldings of quasi–periodic tori, *Mem. AMS*, **83**(421) (1990), 1–82.

[25] H.W. Broer, C. Simó and J.C. Tatjer, Towards global models near homoclinic tangencies of dissipative diffeomorphisms, *Nonlinearity* **11**(3) (1998), 667–770.

[26] H.W. Broer and F. Takens, Mixed spectrum and rotational symmetry, *Arch. Rational Mech. An.* **124** (1993), 13–42.

[27] H.W. Broer, F. Takens and F.O.O. Wagener, Integrable and non–integrable deformations of the skew Hopf bifurcation, *Reg. and Chaot. Dyn.* **4** (1999), 17–43.

[28] H.W. Broer and F.O.O. Wagener, Quasi–periodic stability of subfamilies of an unfolded skew–Hopf bifurcation, *Arch. Rational Mech. An.* **152** (2000), 283–326.

[29] A. Chenciner, Bifurcations de points fixes elliptiques, I. Courbes invariantes, *Publ. Math. IHES* **61** (1985), 67–127.

[30] A. Chenciner, Bifurcations de points fixes elliptiques, II. Orbites périodiques et ensembles de Cantor invariants. *Invent. Math.* **80**(1) (1985), 81–106.

[31] A. Chenciner, Bifurcations de points fixes elliptiques, III. Orbites périodiques de "petites" périodes et elimination résonnantes des couples de courbes invariantes, *Publ. Math. IHES* **66** (1988), 5–91.

[32] A. Chenciner and G. Iooss, Bifurcations de tores invariants, *Arch. Rational Mech. Anal.* **69**(2) (1979), 109–198; **71**(4) (1979), 301–306.

[33] S-N. Chow and J.K. Hale, *Methods of Bifurcation Theory.* Grundlehre der mathematischen Wissenschaften **251**, Springer–Verlag 1982.

[34] S-N. Chow, C. Li and D. Wang, *Normal Forms and Bifurcation of Planar Vector Fields.* Cambridge University Press 1994.

[35] R.L. Devaney, *An Introduction to Chaotic Dynamical Systems.* Addison Wesley 1989.

[36] F. Diacu and P. Holmes, *Celestial Encounters, the origins of chaos and stability.* Princeton University Press 1996.

[37] H. Eliasson, Ergodic skew systems on $\mathbb{T}^d \times SO(3, \mathbb{R})$, *Ergod. Th. Dynam. Sys.* **22** (2002), 1429–1451.

[38] J. Guckenheimer and P. Holmes, Nonlinear Oscillations, Dynamical Systems and Bifurcations of Vector Fields, *Appl. Math. Sciences* **42** Springer–Verlag 1983.

[39] J.K. Hale, *Ordinary Differential Equations.* Wiley 1969; Krieger 1980.

[40] H. Hanßmann, The quasi-periodic centre-saddle bifurcation, *J. Diff. Eq.* **142**(2) (1998), 305–370.

[41] H. Hanßmann, A survey on bifurcations of invariant tori. *Preprint* Princeton University 2001.

[42] F. Hausdorff, *Set Theory* (2nd edition), Chelsea 1962. [German original: Veit (Leipzig) 1914.]

[43] M.R. Herman, Sur la conjugaison différentiable des difféomorphismes du cercle à des rotations, *Publ. Math. IHES* **49** (1979), 5–234.

[44] M.R. Herman, Une méthode pour minorer les exposants de Lyapounov et quelques examples montrant le caractère local d'un théroème d'Arnold et de Moser sur le tore de dimension 2, *Comment. Math. Helvetici* **58** (1983), 453–502.

[45] M.W. Hirsch, C.C. Pugh and M. Shub, Invariant Manifolds, *LNM* **583** Springer–Verlag 1977.

[46] M.W. Hirsh and S. Smale, *Differential Equations, Dynamical Systems, and Linear Algebra.* Academic Press 1974.

[47] E. Hopf, A mathematical example displaying features of turbulence. *Commun. Appl. Math.* **1** (1948), 303–322.

[48] G.B. Huitema. Unfoldings of quasi-periodic tori. *PhD thesis* Groningen 1988.

[49] G. Iooss and J.E. Los, Quasi-genericity of bifurcations to high dimensional invariant tori for maps, *Commun. Math. Phys.* **119**(3) 1988, 453–500.

[50] H.H. de Jong, Quasiperiodic breathers in systems of weakly coupled pendulums: Applications of KAM theory to classical and statistical mechanics. *PhD Thesis* Groningen 1999.

[51] A. Jorba and J. Villanueva, On the normal behaviour of partially elliptic lower-dimensional tori of Hamiltonian systems, *Nonlinearity* **10** (1997), 783–822.

[52] Th. Kappeler and J. Pöschel, KdV & KAM, Ergebnisse der Mathematik und ihrer Grenzgebiete **45**(3), Springer–Verlag 2003.

[53] A. Katok and B. Hasselblatt, *Introduction to the Modern Theory of Dynamical Systems*, Encyclopedia of Mathematics and its Applications **54**, Cambridge University Press 1995.

[54] R. Krikorian, C^0-densité globale des systèmes produits-croisés sur le cercle réductibles, *Ergod. Th. Dynam. Sys.* **19** (1999), 61–100.

[55] S.E. Kuksin, Nearly integrable infinite-dimensional Hamiltonian dynamics, *LNM* **1556** Springer–Verlag 1993.

[56] Y.A. Kuznetsov, *Elements of Applied Bifurcation Theory*, Applied Mathematial Sciences **112** Springer–Verlag 1995.

[57] L.D. Landau, On the problem of turbulence, *Akad. Nauk.* **44** (1944), 339.

[58] L.D. Landau and E.M. Lifschitz, *Fluid Mechanics.* Pergamon, Oxford 1959.

[59] A. Litvak-Hinenzon and V. Rom-Kedar, Parabolic resonances in 3 degree of freedom near integrable Hamiltonian systems, *Phys. D* **164**(3-4) (2002), 213–250.

[60] A. Litvak-Hinenzon and V. Rom-Kedar, Resonant tori and instabilities in Hamiltonian systems, *Nonlinearity* **15**(4) (2002), 1149–1177.

[61] A. Litvak-Hinenzon and V. Rom-Kedar, On Energy surfaces and the resonance web. In preparation.

[62] R. de la Llave, A tutorial on KAM Theory. *Proceedings of Symposia in Pure Mathematics* **69** (2001), 175–292.

[63] S. Marmi, *An introduction to small divisor problems*. Dipartimento di Matematica dell'Università di Pisa 2000.

[64] J.K. Moser, On invariant curves of area-preserving mappings of an annulus, *Nachr. Akad. Wiss. Göttingen, Math.-Phys. Kl. II.* **1** (1962), 1–20.

[65] J.K. Moser, On the theory of quasiperiodic motions, *SIAM Review* **8**(2) (1966), 145–172.

[66] J.K. Moser, Convergent series expansions for quasi–periodic motions, *Math. Ann.* **169** (1967), 136–176.

[67] J.K. Moser, Lectures on Hamiltonian systems, *Mem. AMS* **81** (1968), 1–60.

[68] J.K. Moser, Stable and random motions in dynamical systems, with special emphasis to celestial mechanics, *Ann. Math. Studies* **77** Princeton University Press 1973.

[69] S.E. Newhouse, J. Palis and F. Takens, Bifurcations and stability of families of diffeomorphisms. *Publ. Math. IHES* **57** (1983), 5–71.

[70] S.E. Newhouse, D. Ruelle and F. Takens, Occurrence of strange Axiom A attractors near quasi–periodic flows on $\mathbb{T}^m$, $m \leq 3$, *Commun. Math. Phys.* **64** (1978), 35–40.

[71] J. Oxtoby, *Measure and Category*. Springer–Verlag 1971.

[72] J. Palis and W.C. de Melo, *Geometric Theory of Dynamical Systems*. Springer–Verlag 1982.

[73] J. Pöschel, Integrability of Hamiltonian systems on Cantor sets. *Comm. Pure Appl. Math.* **35**(5) (1982), 653–696.

[74] J. Pöschel. A lecture on the classical KAM Theorem, *Proceedings of Symposia in Pure Mathematics* **69** (2001), 707–732.

[75] D. Ruelle, *Elements of Differentiable Dynamics and Bifurcation Theory*. Academic Press 1989.

[76] D. Ruelle and F. Takens, On the nature of turbulence, *Commun. Math. Phys.* **20** (1971), 167–192; **23** (1971), 343–4.

[77] C. Simó, Perturbations of translations in the two–dimensional torus: the case near resonance, *Proceedings VI CEDYA*, Universidad de Zaragoza 1984.

[78] F. Takens, Singularities of Vector Fields, *Publ. Math. IHES* **43** (1974), 47–100.

[79] F. Takens, Forced oscillations and bifurcations. In: Applications of Global Analysis I, *Comm. of the Math. Inst. Rijksuniversiteit Utrecht* (1974). In: H.W. Broer, B. Krauskopf and G. Vegter (Eds.), *Global Analysis of Dynamical Systems, Festschrift dedicated to Floris Takens for his 60th birthday* (2001), 1–62 IOP (ISBN 0 7503 0803 6).

[80] F. Takens and F.O.O. Wagener, Resonances in skew and reducible quasi–periodic Hopf bifurcations, *Nonlinearity* **13** (2000), 377–396.

[81] A. Vanderbauwhede, Centre Manifolds, Normal Forms and Elementary Bifurcations, *Dynamics Reported* **2** (1989), 89–170.

[82] F.O.O. Wagener, On the skew Hopf bifurcation. *PhD thesis* Groningen, 1998.

[83] F.O.O. Wagener, A Gevrey regular KAM Theorem and the Inverse Approximation Lemma. To appear *Dynamical Systems, an International Journal.*

[84] F.O.O. Wagener, On the quasi–periodic d–fold degenerate bifurcation. Submitted.

[85] S. Wiggins, *Introduction to Applied Nonlinear Dynamical Systems and Chaos.* Springer–verlag 1990.

[86] H. Whitney, Analytic extensions of diffrentiable functions defined in closed sets. *Trans. Amer. Math. Soc* **36**(1) (1934), 63–89.

[87] H. Whitney, Differentiable functions defined in closed sets. *Trans. Amer. Math. Soc.* **36**(2) (1934), 369–387.

[88] J.-C. Yoccoz, C^1-conjugaisons des difféomorphismes du cercle. In: J. Palis (Ed), *Geometric Dynamics,* Proceedings, Rio de Janeiro 1981, LNM **1007** (1983), 814–827.

[89] J.-C. Yoccoz, Travaux de Herman sur les tores invariants. In: Séminaire Bourbaki Vol 754, 1991-1992, *Astérisque* **206** (1992), 311–344

[90] J.-C. Yoccoz, Théorème de Siegel, nombres de Bruno et polynômes quadratiques, *Astérisque* **231** (1995), 3–88.

[91] J.-C. Yoccoz, Analytic linearization of circle diffeomorphisms. In: S. Marmi, J.-C. Yoccoz (Eds.) *Dynamical Systems and Small Divisors,* LNM **1784** (2002), 125–174, Springer–Verlag.

[92] E. Zehnder. An implicit function theorem for small divisor problems, *Bull. AMS* **80**(1) (1974), 174-179.

[93] E. Zehnder. Generalized implicit function theorems with applications to some small divisor problems, I and II, *Comm. Pure Appl. Math.*, **28**(1) (1975), 91-140; **29**(1) (1976), 49-111.

A
Appendix

A.1 Conservative examples: Reduction to KAM Theory of twist maps

The classical formulations of KAM Theory in the mid 20th century, are all in the world of conservative systems, see the introduction. This section is meant as a connection of this culture with the dissipative examples of Section 2.1.

As a conservative analogue of (2.3) we now consider the frictionless pendulum with time-periodic forcing

$$\ddot{y} + \omega^2 \sin y = \varepsilon \cos t.$$

In the phase space $\mathbb{R} \times \mathbb{R} \times \mathbb{T}^1$ this leads to a 3-dimensional vector field, as before,

$$\begin{aligned} \dot{y} &= z \\ \dot{z} &= -\omega^2 \sin y + \varepsilon \cos t \\ \dot{t} &= 1, \end{aligned} \tag{A.1}$$

which is now (time-dependent) Hamiltonian [2].

For $\varepsilon = 0$ most of the phase space is foliated by invariant 2-tori corresponding to the 'ordinary' oscillations of the free pendulum, with energy $H(y, z) = \frac{1}{2}z^2 - \omega^2 \cos y$, which for $\varepsilon = 0$ is a conserved quantity. As in the dissipative analogue, this motion is a combination of autonomous periodic oscillations in the (y, z)-plane and the periodic motion in the t-direction. For 'many' values of the energy $0 < |H| < \omega^2$, the corresponding motions are quasi-periodic and, as before, the question is whether they are persistent under small perturbations.

One way to treat this problem is by looking at the Poincaré map $P = P_\varepsilon(y, z)$ with respect to the section $t = 0 \,(\mathrm{mod}\, 2\pi)$, also called stroboscopic map. The map P is area preserving of the plane with coordinates y and z, see Exercise 13 below. Fore $\varepsilon = 0$ the map P_0 just is the time 2π map of the autonomous pendulum, and the foliation of invariant 2-tori above, for P_0 gives a corresponding foliation of invariant circles.

The best way to describe the dynamics of P_0 is by action-angle variables (I, φ) in the region Σ of oscilatory pendulum motions, see [2] and Exercise 14 below. Note that the unperturbed 2-tori are parametrized by the pair (φ, t), thereby showing that the tori are parallel. We so get an expression

$$P_0(I, \varphi) = (I, \varphi + 2\pi\alpha(I)) \tag{A.2}$$

for the unperturbed Poincaré map. It can be shown that the function $I \mapsto \alpha(I)$ is monotonous, although this involves manipulating an elliptic integral. Because of this monotonicity, which is the present form of non-degeneracy, the map (A.2) is called a (pure) twist map.

Moser's Twist Map Theorem [64] then guarantees that the Diophantine circles survive sufficiently small perturbations. There exist formulations of the Twist Map Theorem in the same spirit as Theorem 2.2, so in terms of quasi-periodic stability and Whitney smoothness, compare [89, 24]. As before there is a direct translation between invariant circles for the Poincaré map and invariant 2-tori for the corresponding vector field (A.1).

Observe that also this occurrence of quasi-periodicity meets with the visibility requirements formulated in Section 2.2. First, it occurs on a set of positive measure and second this phenomenon is persistent for small perturbations of the system.

Remarks

- Compare this set-up with its dissipative analogue in Section 2.2.2. Notice the fact that the parameter α from dissipative setting here is replaced by the action variable I.
- Regarding the examples with coupled oscillators presently we have a direct conservative analogue by weakly coupling two pendula. This yields a 4-dimensional (i.e., 2 degrees of freedom) Hamiltonian vector field. Restricting to a 3-dimensional energy level and taking an appropriate transversal section, gives rise to an area preserving Poincaré map as before, which again is a twist map. Compare with [67, 68, 7].

  This example can be easily generalized to n weakly coupled oscillators in the Hamiltonian setting. As before this leads to invariant n-tori and the corresponding quasi-periodic stability, [73, 21, 23, 24]. These formulations have been mimicked in Section 3.

  These examples lead us into the heart of classical Hamiltonian KAM Theory [1, 2], also see [7, 36] and many of the references in [23].

Exercise 13 (Hamiltonian vector fields give area preserving Poincaré maps) Show that the Poincaré map $P = P_\varepsilon$ of the vector field (A.1) is area preserving.

Exercise 14 (Action-angle variables for the autonomous pendulum) For $(y, z) \in \Sigma$ we let $H(y, z) = E$. In the energy level $H^{-1}(E)$ the autonomous pendulum carries out a periodic motion of period $T = T(E)$. Define

$$I(E) = \frac{1}{2\pi} \oint_{H^{-1}(E)} z\, dy$$

and show that

$$T(E) = 2\pi \frac{dI}{dE}(E).$$

Let t be the time the motion in $H^{-1}(E)$ takes to get from the line $z = 0$ to the point (y, z). Defining

$$\varphi(y, z) = \frac{2\pi}{T(E)} t,$$

show that (I, φ) is a pair of action-angle variables. Consider the Poincaré map P_0, see (A.2), and derive an integral expression for $\alpha = \alpha(I)$.

A.2 The Paley-Wiener estimate

For completeness we include the statement of the Paley-Wiener estimate, referring to [23] pp. 37-40 for a proof. Let $\Gamma \subset \mathbb{R}^n$ be a compact domain and let $\kappa, \rho \in (0, 1)$ be constant. Let the function $f : \mathbb{T}^n \times \mathbb{R}^n \to \mathbb{R}$ be real analytic in both coordinates $x \in \mathbb{T}^n$ and $\omega \in \mathbb{R}^s$. Set

$$\begin{aligned} \Gamma + \rho &= \cup_{\omega \in \Gamma} \{\omega' \in \mathbb{C}^n \mid |\omega' - \omega| < \rho\}, \\ \mathbb{T}^n + \kappa &= \cup_{x \in \mathbb{T}^n} \{x' \in (\mathbb{C}/2\pi\mathbb{Z})^n \mid |x' - x| < \kappa\}. \end{aligned}$$

Define M to be the supremum of $|f|$ over the closure $\overline{(\mathbb{T}^n + \kappa) \times (\Gamma + \rho)}$.

Theorem A.1 (Paley-Wiener) *Let $f = f(x, \omega)$ be real analytic as above, with Fourier series*

$$f(x, \omega) = \sum_{k \in \mathbb{Z}^n} f_k(\omega) e^{i<k,x>}.$$

Then, for all $k \in \mathbb{Z}^n$ and all $\omega \in \overline{\Gamma + \rho}$,

$$|f_k(\omega)| \leq M e^{-\kappa|k|}. \tag{A.3}$$

Remarks

- The converse of Theorem A.1 also holds true. Indeed, a function on $\mathbb{T}^n$ whose Fourier coefficients decay exponentially is analytic and can be extended holomorphically to a complex domain $(\mathbb{C}/2\pi\mathbb{Z})^n$ by a distance determined by the decay rate (κ in our notation).
- Theorem A.1 admits a straightforward generalization to the case where f is finitely or infinitely differentiable. For example, if $f : \mathbb{T}^1 \to \mathbb{R}$ is of class

C^r and $\omega \in \mathbb{R}$, by partial integration one easily shows that the decay rate of the Fourier coefficients is polynomial:

$$|f_k| \leq \frac{M_r}{|k|^r},$$

where $M_r = |f|_{C^r}$. Also in this case a converse result holds, compare with Exercise 15.

Exercise 15 (Loss of differentiability) Consider the linear equation

$$u(x+\omega) - u(x) = f(x),$$

where $x \in \mathbb{T}^1 = \mathbb{R}/(2\pi\mathbb{Z})$, $\omega \in \mathbb{R}$, and where $u, f : \mathbb{T}^1 \to \mathbb{R}$. Let ω satisfy a *Diophantine condition* of the form

$$\left|\omega - \frac{p}{q}\right| \geq \gamma |q|^{-\tau},$$

for all $p \in \mathbb{Z}$, $q \in \mathbb{Z}\backslash\{0\}$, with $\tau > 2$, $\gamma > 0$. By formally expanding f and u in Fourier series

$$u(x) = \sum_{k\in\mathbb{Z}} u_k \mathrm{e}^{\mathrm{i}kx}, \quad f(x) = \sum_{k\in\mathbb{Z}} f_k \mathrm{e}^{\mathrm{i}kx},$$

the linear equation transforms to an infinite system of equations in the coefficients u_k and f_k.

(i) Give a necessary condition that this system can be solved, and write down the solution.

First, assume that f is r times continuously differentiable, i.e., of class C^r.

(ii) What does this imply for the rate of decay of the coefficients f_k as $|k| \to \infty$?
(iii) Using this result and the Diophantine conditions above, estimate the rate of decay of the coefficients u_k as $|k| \to \infty$.
(iv) What is the least value r_0 of r, such that the formal Fourier series $\sum u_k \mathrm{e}^{\mathrm{i}kx}$ converges absolutely and so defines a continuous function?
(v) Given that f is C^r with $r > r_0$, what is the amount of differentiability of u?

Now, assume that f is real analytic and bounded by M on the complex strip of width $r > 0$ around $\mathbb{T}^1$. That is, for

$$x \in \mathbb{T}^1 + \sigma = \{x \in \mathbb{C}/(2\pi\mathbb{Z}) \mid |\mathrm{Im}\, x| < \sigma\},$$

assume that

$$\sup_{x \in \mathbb{T}^1 + \sigma} |f(x)| \leq M.$$

The Paley-Wiener estimate gives estimates of the kth Fourier coefficient f_k in terms of σ, M and k.

(vi) Use these and the Diophantine condition to obtain estimates of u_k.
(vii) Show that for $0 < \varrho < \sigma$, the formal Fourier series converges on $\mathbb{T}^1 + \varrho$ and defines an analytic function $u(x)$.
(viii) Derive a bound for u on $\mathbb{T}^1 + \varrho$ that depends explicitly on σ and ϱ.

A.3 On Whitney differentiability

The present version of KAM Theory uses Whitney differentiability, which is a natural way to deal with the connecting geometry inside the union of Diophantine quasi-periodic tori. The reason is that these unions are nowhere dense, closed sets in the product of phase and parameter space and for such sets the concept of Whitney differentiability is extremely suitable. Indeed, on the one hand it just means that a function is smooth on such a closed set Ω if it is the restriction of an 'ordinary' smooth function defined on an open set that contains Ω. On the other hand, there exists an intrinsic characterisation of Whitney differentiability, just in terms of the closed set [86, 87]. Also compare [23] and all the references therein. We briefly formulate the Inverse Approximation Lemma, which is used to get Whitney differentiable conjugacy results. We mainly quote from [23], but also we refer to [93, 73, 24, 62].

Let $\ell > 0$ be some order of differentiability and let $r_j = a\kappa^j$ be a fixed geometric sequence with $a = r_0 > 0$ and $0 < \kappa < 1$. Also let $\Omega \subset \mathbb{R}$ be a closed set and define

$$\Omega + r_j = \bigcup_{x \in \Omega} \{z \in \mathbb{C} \,|\, |z - x| < r_j\}.$$

For $j \in \mathbb{Z}_+$ let U^j be a real analytic function on $\Omega + r_j$. The following result states when the limit U^∞, defined on Ω of the sequence $\{U^j\}_{j=0}^\infty$ is of class C^r.

Lemma A.2 (Inverse Approximation Lemma) *Assume that $\ell \notin \mathbb{N}$. Let $\{U^j\}_{j=0}^\infty$ be as above with $U^0 \equiv 0$ and such that for $j \geq 1$*

$$|U^j - U^{j-1}|_{\Omega + r_j} \leq M r_j^\ell$$

for some constant M. Then there exists a unique function U^∞, defined on Ω, *which is of class C^ℓ and such that $|U^\infty|_\ell \leq Mc_\ell$, where the constant c_ℓ only depends on ℓ, κ. Moreover for all $s < \ell$*

$$|U^\infty - U^j|_s \to 0, \text{ as } j \to \infty.$$

Here $|-|_\ell$ and $|-|_s$ denotes the C^ℓ- and C^s-norms [15] *on Ω.*

Remarks

- In the application of this Lemma A.2 in the KAM proof of Theorem 3.1 and of all related theorems, the difference $|U^j - U^{j-1}|_{\Omega + r_j}$ decays faster than geometrical. In that case it follows that the limit U^∞ is of class C^∞.
- In most cases the decay of $|U^j - U^{j-1}|_{\Omega + r_j}$ is even exponentially fast, which implies that the limit U^∞ is Gevrey regular [83].

A.4 Hints to the exercises

Exercise 1:

Use the Poincaré (return) map of the circle $x_1 = 0$.

Exercise 4:

For (i) and (ii) use the fact that $2t/\pi \leq \sin t \leq t$ for $0 \leq t \leq \pi/2$.

As an example we give a solution of (i). Take $\beta \in D_{\gamma,\tau}$. For all $k \in \mathbb{Z}^n \backslash \{0\}$ there exists an $\ell = \ell(k) \in \mathbb{Z}$ such that $|\langle k, \beta \rangle - \ell| < \frac{1}{2}$. We then have

$$\begin{aligned}
\left| e^{2\pi i \langle k,\beta \rangle} - 1 \right| &= \left| e^{\pi i (\langle k,\beta \rangle + \ell)} \right| \left| e^{\pi i (\langle k,\beta \rangle - \ell)} - e^{-\pi i (\langle k,\beta \rangle + \ell)} \right| \\
&= \left| e^{\pi i (\langle k,\beta \rangle - \ell)} - e^{-\pi i (\langle k,\beta \rangle + \ell)} \right| \\
&= \left| e^{\pi i (\langle k,\beta \rangle - \ell)} - e^{-\pi i (\langle k,\beta \rangle - \ell)} \right| \\
&= 2 \left| \sin \pi (\langle k, \beta \rangle - \ell) \right| \geq 4 \left| \langle k, \beta \rangle - \ell \right|.
\end{aligned}$$

Exercise 5:

Set $\bar{t} = \gamma t$ and $\bar{\omega} = \gamma^{-1}\omega$.

Remarks

- Note that for $\gamma \downarrow 0$ the set $\bar{\Gamma} = \{\bar{\omega} \in \mathbb{R}^n \mid \omega \in \Gamma\}$ blows up. In the KAM proof one therefore drops the requirement that Γ should to be bounded.
- Compare with the final remark of Section 3.2.1, where also the homothetic occurrence of γ is used.

[15] This involves Hölder conditions on the $C^{[\ell]}$-th and $C^{[s]}$-th derivative, compare [23].

Exercise 6:

Regarding (i) note that by definition $d \leq s^{-2} = (2r)^{-1/(\tau+1)}$. Next, to show (ii), first observe that for $\sigma \in \Gamma'_{\tau,1} + r$ there exists $\sigma^* \in \Gamma'_{\tau,1}$ such that $|\sigma - \sigma^*| \leq r$. It then follows for all $k \in \mathbb{Z}^n$ with $0 \leq |k| \leq d$ that

$$|\langle \sigma, k \rangle| \geq |\langle \sigma^*, k \rangle| - |\sigma - \sigma^*||k| \geq |k|^{-\tau} - r|k| \geq \tfrac{1}{2}|k|^{-\tau}.$$

Exercises 7, 8 and 9:

These are all '1-bite small divisor problems', directly solvable by Fourier series. Please also consult the Paley-Wiener estimate and perhaps also Exercise 15.

For example consider Exercise 7, compare with Section 2.2.2. Here we recall the Mean Value Theorem $h(x + \varphi, \lambda) = h(x, \lambda) + h'(\xi, \lambda)\varphi$, especially in cases where $\varphi = O(|\lambda|^{N+1})$. By similar arguments it also follows that $H_\lambda^{-1}(x) = x - h(x, \lambda) + O(|\lambda|^{N+1})$. Using all this one arrives at the expression

$$\begin{aligned} &H_\lambda^{-1} \circ P_{N,\lambda} \circ H_\lambda(x) = \\ &\qquad x + 2\pi\beta + g(\lambda) + h(x, \lambda) - h(x + 2\pi\beta, \lambda) + f_N(x, \lambda) + O(|\lambda|^{N+1}), \end{aligned}$$

where we require that the Nth order part $h(x, \lambda) - h(x + 2\pi\beta, \lambda) + f_N(x, \lambda) = c(\lambda)$, for an appropriate constant $c(\lambda)$. This is a 1-bite small divisor problem where h can be solved given f_N, provided that $c = f_{N,0}$, the $\mathbb{T}^1$-average of f_N.

Regarding Exercise 8 one has that $P(x) = \Phi^{f(x)}(0, x)$. Next, using the group property for flows of vector fields (with addition in time), then keeping track of the time one finds that $f(x) - u(x) + u(P(x)) = c$. This is another 1-bite small divisor problem where u can be solved given f, provided that $c = f_0$, the $\mathbb{T}^1$-average of f.

Remark It helped us to sketch the situation as in Figure 1.2 and to draw the integral curve of X starting at $(0, x)$ that passes through $(0, P(x))$.

Exercise 13:

Use the Gauß Divergence Theorem on an appropriate flow box and the fact that the time-dependent Hamiltonian vector field has divergence zero.

Exercise 15:

To answer (i) - (vi) first prove a (generalized) Paley-Wiener estimate for C^s-functions by using integration by parts, (also see remarks in Section A.2). It follows that u is less regular than f, i.e., $s_0 < s$ in (iv); in particular $u \in C^{s-\tau}$.

VI

Symmetric Hamiltonian Bifurcations

Pietro-Luciano Buono, Frédéric Laurent-Polz & James Montaldi

Based on lectures by James Montaldi

Contents

1 Introduction

The purpose of these notes is to give a brief survey of bifurcation theory of Hamiltonian systems with symmetry; they are a slightly extended version of the five lectures given by JM on Hamiltonian Bifurcations with Symmetry. We focus our attention on bifurcation theory near equilibrium solutions and relative equilibria. The notes are composed of two parts. In the first, we review results on nonlinear normal modes in equivariant Hamiltonian systems, generic movement of eigenvalues in equivariant Hamiltonian matrices, one and two parameter bifurcation of equilibria and the Hamiltonian-Hopf Theorems with symmetry. The second part is about local dynamics near relative equilibria. Particular topics discussed are the existence, stability and persistence of relative equilibria, bifurcations from zero momentum relative equilibria and examples.

We begin with some basic facts on Lie group actions on symplectic manifolds and Hamiltonian systems with symmetry. The reader should refer to Ratiu's lectures for more details and examples.

Semisymplectic actions A Lie group G acts *semisymplectically* on a symplectic manifold $(\mathcal{P}, \omega)$ if $g^*\omega = \pm\omega$. In this case the choice of sign determines a homomorphism $\chi : G \to \mathbb{Z}_2$ called the *temporal character*, such that $g^*\omega = \chi(g)\omega$. We denote the kernel of χ by G_+; it consists of those elements acting symplectically, and if G does contain antisymplectic elements then G_+ is a subgroup of G of index 2. Some details on semisymplectic actions can be found in [MR00].

Not every semisymplectic action contains an antisymplectic element of order 2, but if it does then we can write $G = G_+ \rtimes \mathbb{Z}_2(\rho)$, where ρ is the element in question.

We write $K < G$ to mean K is a closed subgroup of G. The *fixed point set* of a subgroup $K < G$ is

$$\mathrm{Fix}(K, \mathcal{P}) = \{x \in \mathcal{P} \mid g \cdot x = x, \ \forall g \in K\};$$

it is a closed submanifold of $\mathcal{P}$. If $K < G_+$ is compact then $\mathrm{Fix}(K, \mathcal{P})$ is a symplectic submanifold. That compactness is necessary can be seen from the simple example of $t \in \mathbb{R}$ acting on $\mathbb{R}^2$ by $\begin{pmatrix} 1 & t \\ 0 & 1 \end{pmatrix}$: the fixed point space is then just the x-axis.

Throughout these lectures, we assume that G acts properly on $\mathcal{P}$. Let $\mathfrak{g}$ be the Lie algebra of G and $G_x = \{g \in G \mid g \cdot x = x\}$ the *isotropy subgroup* of $x \in \mathcal{P}$. The properness assumption implies in particular that G_x is compact.

To each element $\xi \in \mathfrak{g}$ there is an associated vector field on $\mathcal{P}$:

$$\xi_{\mathcal{P}}(x) = \frac{d}{dt} \exp(t\xi) \cdot x|_{t=0}$$

The tangent space at x to the group orbit through x is $\mathfrak{g} \cdot x = \{\xi_{\mathcal{P}}(x) \mid \xi \in \mathfrak{g}\}$.

The *adjoint action* of g on $\mathfrak{g}$ denoted $\xi \mapsto \mathrm{Ad}_g \xi$, is the tangent map of $I_g : G \to G$, $h \mapsto ghg^{-1}$ at e, $T_e I_g(\xi)$. In the case of matrix groups, this is just

$$\mathrm{Ad}_g \xi = g\xi g^{-1}.$$

Finally, dual to the adjoint action on $\mathfrak{g}$ is the coadjoint action on $\mathfrak{g}^*$:

$$\langle \mathrm{Coad}_g \mu, \eta \rangle := \langle \mu, \mathrm{Ad}_{g^{-1}} \eta \rangle .$$

In the case of matrix groups, if we identify $\mathfrak{g}^*$ with matrices via $\langle \mu, \xi \rangle = \mathrm{tr}(\mu^T \xi)$, then the coadjoint action becomes

$$\mathrm{Coad}_g \, \mu = g^{-T} \mu g^T.$$

where $g^{-T} = (g^{-1})^T = (g^T)^{-1}$. For compact groups and for semisimple groups, the adjoint and coadjoint actions are isomorphic, but in general they can be quite different—this is already the case for the 3-dimensional Euclidean group $\mathbf{SE}(2)$. See Section 5 (in Part II) for how the momentum map relates to a semisymplectic action.

Hamiltonian formalism A Hamiltonian system with symmetry is a quadruple $(\mathcal{P}, \omega, G, H)$ where:

- $(\mathcal{P}, \omega)$ is a symplectic manifold,
- G is Lie group acting smoothly and semisymplectically on $\mathcal{P}$,
- $H : \mathcal{P} \to \mathbb{R}$ is a G-invariant smooth function.

The Hamiltonian vector field X_H is defined implicitly by $\omega(-, X_H) = dH$ and of course defines a dynamical system on $\mathcal{P}$ by

$$\dot{x} = X_H(x). \tag{1.1}$$

When working in the neighbourhood of a point $x \in \mathcal{P}$, the equivariant Darboux theorem states that there exists a coordinate system such that the symplectic form is locally constant. Therefore, without loss of generality we can reduce the Hamiltonian to a vector space by identifying a neighbourhood of x in $\mathcal{P}$ with $V = T_x\mathcal{P}$ and the G action on $\mathcal{P}$ with the G action on $T_x\mathcal{P}$. Note however that for symmetric Hamiltonian systems, Montaldi *et al.* [MRS88] and Dellnitz and Melbourne [DM92] show that symplectic forms are not always locally isomorphic if the isotypic decomposition of the space contains irreducible representations of complex type.

If G is formed of symplectic elements, then X_H is G-equivariant; that is, if $x(t)$ is a solution curve of X_H then so is $g \cdot x(t)$ for all $g \in K$.

On the other hand, suppose that $\rho \in G$ is an *antisymplectic* symmetry, that is $\rho^*\omega = -\omega$ or $\omega(\rho u, \rho v) = -\omega(u, v)$, then it is time-reversing; that is, $x(t)$ is an integral curve of the X_H vector field implies $\rho \cdot x(-t)$ is also an integral curve of the vector field.

If K is compact and formed of symplectic symmetries, then $\mathrm{Fix}(K, \mathcal{P})$ is invariant under the flow of the dynamical system. If in addition K is compact, then $\mathrm{Fix}(K, \mathcal{P})$ is a Hamiltonian subsystem with Hamiltonian given by the restriction of H to $\mathrm{Fix}(K, \mathcal{P})$.

The remainder of these notes is structured as follows. There are two main parts. The first covers local dynamics near equilibria and the second local dynamics near relative equilibria.

In Section 2, we begin with the local dynamics near equilibria when the Hamiltonian has a nondegenerate quadratic part, and present the equivariant Weinstein-Moser Theorem on the existence of nonlinear normal modes in equivariant Hamiltonian systems. In Section 3 we look at bifurcations near equilibria. We start with a brief review of generic movement of eigenvalues in equivariant Hamiltonian matrices depending on parameters. Then we look at steady-state bifurcations in parameter families of Hamiltonian systems, and in particular the one-parameter case. We conclude this part with the Hamiltonian-Hopf Theorem with symmetry.

The second part deals with bifurcations from relative equilibria. In Section 5 the momentum map is defined and its equivariance is shown. Using this information, it is shown how to define reduced spaces for the dynamics using the momentum map. Then in Section 6, relative equilibria are defined and we explain how to find relative equilibria, and determine their stability and their persistence. Section 7 discusses bifurcations from zero-momentum states and in Section 8 three examples of bifurcations from zero-momentum are presented: relative equilibria of molecules, relative equilibria in point vortex models in the plane, and relative equilibria in point vortex models on the sphere.

PART I: LOCAL DYNAMICS NEAR EQUILIBRIA

2 Nonlinear normal modes

Suppose that the Hamiltonian system (1.1) has a steady-state (equilibrium) solution at some $x_0 \in \mathcal{P}$. Such solutions are critical points of H. The linearized vector field at x_0 is

$$\dot{v} = L_{x_0} v.$$

The matrix L_{x_0} is *Hamiltonian*; a matrix A is Hamiltonian if $\omega(Av, w) + \omega(v, Aw) = 0$. The set of Hamiltonian matrices on $\mathbb{R}^{2n}$ is denoted $\mathfrak{sp}(2n)$. The set $\mathfrak{sp}_G(2n) \subset \mathfrak{sp}(2n)$ is the subspace of matrices that commute with G. The eigenvalues of Hamiltonian matrices arise in quadruplets $\{\lambda, -\lambda, \overline{\lambda} - \overline{\lambda}\}$, see Lemma 4.1 of Meyer's lectures in this volume or Meyer and Hall [MH92].

Suppose that $\mathrm{Re}(\lambda) \neq 0$ for all eigenvalues λ of L_{x_0} then by the Hartman-Grobman theorem the vector field (1.1) is homeomorphic to its linear part $\dot{v} = L_{x_0} v$ in a neighbourhood of x_0; x_0 is a hyperbolic saddle point. In generic (non-hamiltonian) systems, this is usually enough to describe the local dynamics, since the eigenvalues do not (generically) lie on the imaginary axis. However, for Hamiltonian systems this is no longer true: having pure imaginary eigenvalues is a structurally stable property.

A linear Hamiltonian system with a simple nonresonant imaginary eigenvalue has a family of periodic solutions of constant period in the eigenspace of the imaginary eigenvalue. These families of periodic solutions are called *normal modes*. In nonlinear Hamiltonian systems, the search for families of periodic solutions near a steady-state or *nonlinear normal modes* has attracted a lot of interest since the seminal work of Lyapunov [L]. The Lyapunov Centre Theorem, see Meyer's lectures or [AM78] states that for each simple nonresonant eigenvalue there exists a nonlinear normal modes. A normal mode is a family of periodic orbits in a linear system, of constant period, and sweeping out the eigenspace corresponding to an imaginary eigenvalue; a nonlinear normal mode is a family of periodic orbits parametrized by energy containing a steady-state solution and tangent to the eigenspace of the imaginary eigenvalue, with period close to that of the linear system. There have been many particular extensions of this theorem, but the most general results are due to Weinstein [W73] and Moser [Mos76] who allow for multiple eigenvalues and resonance relations. Montaldi *et al* [MRS88] extend the results of Weinstein and Moser to take account of symmetry.

As we have already noted, if a compact subgroup K of G acts symplectically, then $\mathrm{Fix}(K, \mathcal{P})$ is a sub-hamiltonian system and so Lyapunov's theorem can be applied to this subsystem. The resulting periodic orbits are said to have *spatial* symmetry: the solution $\gamma(t)$ satisfies $g \cdot \gamma(t) = \gamma(t)$ for each t, and for each $g \in K$. However, using *spatio-temporal* symmetries one can go further, and we now describe this idea.

Let $v(t)$ be a 2π-periodic solution of the G-invariant Hamiltonian system (1.1) then, $g.v(t)$ is also a periodic solution of (1.1) for all $g \in G$. By uniqueness of solutions of differential equations, either $\{v(t)\} \cap \{g.v(t)\} = \{v(t)\}$ or $\{v(t)\} \cap \{g.v(t)\} = \emptyset$. In the former case, $g.v(t) = v(t-\theta)$ for some phase shift θ. We identify phase shifts with elements of the circle group $\mathbf{S}^1$ using the identification $\mathbf{S}^1 = \mathbb{R}/2\pi\mathbb{Z}$. The set

$$\Sigma_{v(t)} = \{(g, \theta) \in G \times \mathbf{S}^1 \mid g.v(t) = v(t-\theta)\} < G \times \mathbf{S}^1$$

is the (spatio-temporal) symmetry group of $v(t)$. Therefore, when searching for periodic solutions with spatio-temporal symmetries of equivariant dynamical systems we look for isotropy subgroups $\Sigma \subset G \times \mathbf{S}^1$.

We now describe the equivariant result of Montaldi, Roberts and Stewart [MRS88].

Suppose that the linearized equation $\dot{v} = L_{x_0} v$ has eigenvalue $i\alpha$ and all imaginary eigenvalues $ri\alpha$ ($r \in \mathbb{Q}$) have larger modulus. Let V_α be the set of all points that lie on $2\pi/\alpha$-periodic trajectories of the linearized equation. Then $V_{ri\alpha} \subset V_\alpha$ and V_α is called the *resonance subspace* of α. Let L_α be the restriction of L_{x_0} to V_α. Elphick *et al.* [ETBCI] show that if L_α is semisimple then $\overline{\{\exp(tL_\alpha^T)|t \in \mathbb{R}\}}$ is isomorphic to $\mathbf{S}^1$, giving rise to an action of $G \times \mathbf{S}^1$ on V_α.

Theorem 2.1 (Equivariant Weinstein-Moser Theorem) *Let $\Sigma \subset G \times \mathbf{S}^1$. Suppose that the restriction $d^2H(0)$ to $\mathrm{Fix}(\Sigma, V_\alpha)$ is definite, then on each energy level near the origin there are at least $\frac{1}{2}\dim \mathrm{Fix}(\Sigma, V_\alpha)$ periodic orbits, with period close to $2\pi/\alpha$ and symmetry at least Σ.*

Theorem 2.1 gives sufficient conditions for the existence of families of periodic solutions with spatio-temporal symmetries for H. Montaldi *et al.* [MRS90] prove a stronger theorem about the existence or nonexistence of all periodic solutions with spatio-temporal symmetries near a steady-state with imaginary eigenvalues. Moreover, their result includes the case when the Hamiltonian system also has time-reversal symmetries.

Let $\rho : \mathbb{R}^{2n} \to \mathbb{R}^{2n}$ be a time-reversing symmetry. Define

$$\tilde{\mathbf{S}}^1 = \begin{cases} \mathbf{S}^1 & \text{if } H \text{ is not time-reversible} \\ \mathbf{S}^1 \rtimes \mathbb{Z}_2^\rho \simeq \mathbf{O}(2) & \text{if } H \text{ is time-reversible} \end{cases}$$

The proof of existence of nonlinear normal modes uses a variational approach. Let $u(s) \in \mathbb{C}^1(\mathbf{S}^1, \mathbb{R}^{2n})$ be a loop in phase space of period 2π. For each real number $\alpha \neq 0$ we define a functional $S_\alpha : \mathcal{C}^1(\mathbf{S}^1, \mathbb{R}^{2n}) \times \mathbb{R} \to \mathbb{R}$ by

$$S_\alpha(u) = S(u, \alpha) = \oint u^*\beta - \alpha \int_0^{2\pi} H(u(s))ds \tag{2.1}$$

where u^* denotes the pull-back of β to $\mathbf{S}^1$ and β is a primitive of the symplectic form; $\omega = d\beta$. The functional S_α can be chosen to be $G \rtimes \tilde{\mathbf{S}}^1$-invariant.

Lemma 2.2 *$u \in \mathcal{C}^1(\mathbf{S}^1, \mathbb{R}^{2n})$ is a critical point of S_α if and only if $z(t) \equiv u(\alpha t)$ is a periodic solution of period $2\pi/\alpha$ of the Hamiltonian system defined by H.*

Theorem 2.3 *For fixed α and for sufficiently small τ the critical points of $S_{\alpha(1+\tau)}$ in a neighbourhood U of $0 \in \mathcal{C}^1(\mathbf{S}^1, \mathbb{R}^{2n})$ are in one-to-one correspondence with the critical points of a smooth finite dimensional $G \rtimes \mathbf{S}^1$-invariant mapping $F_\tau : V_\alpha \to \mathbb{R}$. Moreover this correspondence preserves symmetry groups.*

An interesting observation concerns the class of classical mechanical systems where q is position and p is momentum and the Hamiltonian is $K_q(p) + V(q)$, where $K_q(p)$—the kinetic energy—is quadratic in p. The time-reversal symmetry is $\rho.(p, q) = (-p, q)$. If a periodic solution with spatio-temporal symmetry group Σ contains a conjugate of $\mathbb{Z}_2^\rho$ then it intersects $\mathrm{Fix}(\mathbb{Z}_2^\rho) = \{(0, q)\}$ in two points. Since the velocity vanishes on $\mathrm{Fix}(\mathbb{Z}_2^\rho)$, in time-reversible systems these periodic solutions are called *brake orbits*.

Theorem 2.3 states that the search for nonlinear normal modes reduces to finding critical points of F_τ. To find the critical points it is necessary to have a convenient expression for F_τ. Such an expressions is obtained by putting the Hamiltonian function into $G \rtimes \tilde{\mathbf{S}}$-invariant Birkhoff normal form to sufficiently high order where

$$\tilde{\mathbf{S}} = \begin{cases} \overline{\{\exp(tL_{x_0}^T) : t \in \mathbb{R}\}} & \text{if } H \text{ is not time-reversible} \\ \overline{\{\exp(tL_{x_0}^T) : t \in \mathbb{R}\}} \rtimes \mathbb{Z}_2^\rho & \text{if } H \text{ is time-reversible.} \end{cases}$$

A discussion of Birkhoff normal form can be found in Section 7 of Meyer's lectures in this volume; for a complete description see Elphick *et al.* [ETBCI] or Golubitsky *et al.* [GSS88], or Cushman and Sanders [CS86] for a different approach. The next result gives the expression for F_τ in terms of the Hamiltonian function in Birkhoff normal form.

Theorem 2.4 *If H is in Birkhoff normal form to degree k then*

$$\frac{1}{2\pi} j^k F_\tau(v) = (1 + \tau) j^2 H(v) - j^k H(v)$$

for $v \in V_\alpha$ where j^k is the k-jet.

However, we do not know a priori to what order the truncation of the Birkhoff normal form that defines F_τ yields all possible nonlinear normal modes for the full system. Montaldi *et al.* [MRS90] obtain further results using singularity theory to answer the question of generic finite determinacy of the F_τ equation. We do not discuss these results here and refer the reader to the paper.

Here we present briefly one example where the equivariant Weinstein-Moser theorem gives only some nonlinear normal modes while the study of the F_τ equation yields all solutions: the time-reversible 1:1 resonance with $\mathbb{Z}_2$ symmetry.

Example 2.5 Consider Hamiltonian systems H on $\mathbb{C}^2$ where κ is the symmetry and ρ the time-reversible symmetry acting as

$$\kappa.(z_1, z_2) = (z_2, z_1) \qquad \text{and} \qquad \rho.(z_1, z_2) = (-\overline{z_1}, -\overline{z_2}).$$

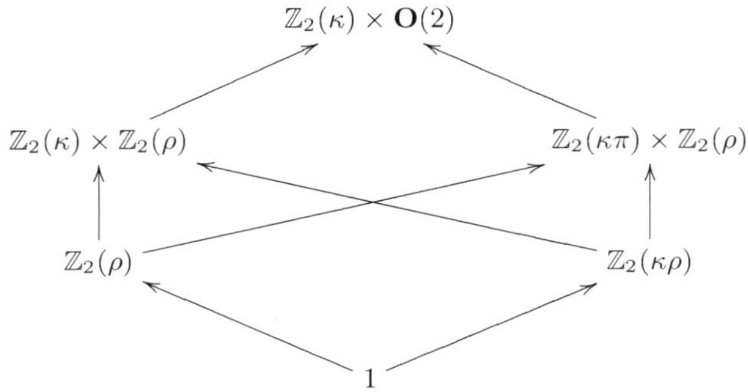

Fig. 2.1. Isotropy lattice of the $\mathbb{Z}_2(\kappa) \times \mathbf{O}(2)$ action on $\mathbb{C}^2$.

At a $1:1$ resonance the linearization $L_0 = (dX_H)_0$ has double eigenvalue $\pm i$ with positive definite quadratic part H_2. We shall see in Section 3 the $1:1$ resonance in more details. Since L_0 commutes with $\mathbb{Z}_2(\kappa)$, L_0 is semisimple and generates the $\mathbf{S}^1$ action

$$\theta.(z_1, z_2) = (e^{i\theta} z_1, e^{i\theta} z_2).$$

These actions combine to give the action of the group $\mathbb{Z}_2(\kappa) \times \mathbf{O}(2)$.

The isotropy subgroups for the $\mathbb{Z}_2(\kappa) \times \mathbf{O}(2)$ action are given in Figure 2.1. It is easy to check that

$$\begin{aligned} \mathrm{Fix}(\mathbb{Z}_2(\kappa) \times \mathbb{Z}_2(\rho)) &= \{z_1 = z_2 \in i\mathbb{R}\}, \text{ and} \\ \mathrm{Fix}(\mathbb{Z}_2(\kappa\pi) \times \mathbb{Z}_2(\rho)) &= \{z_1 = -z_2 \in i\mathbb{R}\} \end{aligned}$$

therefore in the complexification these are two-dimensional. From the equivariant Weinstein-Moser theorem, since H_2 is definite we know immediately that there are nonlinear normal modes with symmetry corresponding to the maximal isotropy subgroups $\mathbb{Z}_2(\kappa) \times \mathbb{Z}_2(\rho)$ and $\mathbb{Z}_2(\kappa\pi) \times \mathbb{Z}_2(\rho)$. Now, $\mathrm{Fix}(\mathbb{Z}_2(\kappa\rho)) = \{z_1, z_2 \in i\mathbb{R}\}$ and $\mathrm{Fix}(\mathbb{Z}_2(\rho)) = \{z_1 = -\overline{z_2}\}$ are four-dimensional, Theorem 2.1 guarantees the existence of two nonlinear normal modes with isotropy containing $\mathbb{Z}_2(\kappa\rho)$ and $\mathbb{Z}_2(\rho)$. However, these may be the solutions with maximal isotropy found above. Therefore, Theorem 2.1 cannot guarantee the existence of solutions with submaximal symmetry in this case.

We look at the F_τ equation where the system is in Birkhoff normal form to degree 4, $H = H_2 + H_4$ where

$$H_2 = N, \qquad H_4 = \alpha_1 N^2 + \alpha_2 P + \beta_1 Q^2 + \beta_2 N Q,$$

$N = |z_1|^2 + |z_2|^2$, $P = |z_1|^2 |z_2|^2$ and $Q = \mathrm{Re}(z_1 \overline{z_2})$. The critical points of

F_τ are given by

$$\phi_1(z_1, z_2, \tau) = i[(-\tau + 2\alpha_1 N + \alpha_2|z_2|^2 + \beta_2 Q)z_1 + \left(\tfrac{1}{2}\beta_2 N + \beta_1 Q\right) z_2$$
$$\phi_2(z_1, z_2, \tau) = i[(-\tau + 2\alpha_1 N + \alpha_2|z_1|^2 + \beta_2 Q)z_2 + \left(\tfrac{1}{2}\beta_2 N + \beta_1 Q\right) z_1. \tag{2.2}$$

We solve for solutions in $\mathrm{Fix}(\kappa\rho)$. Set $z_1 = ix$ and $z_2 = iy$ then (2.2) becomes

$$(-\tau + 2\alpha_1(x^2+y^2) + \alpha_2 y^2 + \beta_2 xy)x + \left(\tfrac{1}{2}\beta_2(x^2+y^2) + \beta_1 xy\right) y = 0$$
$$(-\tau + 2\alpha_1(x^2+y^2) + \alpha_2 x^2 + \beta_2 xy)y + \left(\tfrac{1}{2}\beta_2(x^2+y^2) + \beta_1 xy\right) x = 0.$$

After simplification we see that nonlinear normal modes with $x \neq 0$ and $y \neq 0$ are solutions found by setting $\tau = 2\alpha_1(x^2+y^2) + \beta_2 xy$ and solving

$$\frac{1}{2}\beta_2(x^2+y^2) + (\alpha_2+\beta_1)xy = 0. \tag{2.3}$$

We can solve (2.3) if $|\beta_2| < |\alpha_2 + \beta_1|$. Thus, for an open set of values of the coefficients $(\alpha_1, \alpha_2, \beta_1, \beta_2)$ nonlinear normal modes with isotropy exactly $\mathbb{Z}_2(\kappa\rho)$ exist in Birkhoff normal form. The same is true for nonlinear normal modes with isotropy exactly $\mathbb{Z}_2(\rho)$. In [MRS90] it is shown that for an open and dense set of values of the coefficients, the truncation of the Birkhoff normal form to degree four $H = H_2 + H_4$, is sufficient to determine all nonlinear normal modes in the full system.

Remark 2.6 Using these methods, one can show that in a 2 degree of freedom system in $1:1$ resonance, with no assumptions on the symmetry, generically there are 2, 4 or 6 pairwise transverse nonlinear normal modes, depending on the coefficients in H_4. Each of the three possibilities is obtained for an open set in the space of coefficients of H_4.

An application of these ideas to a symmetry breaking problem can be found in [Mo99], where a description is given of the nonlinear normal modes obtained after adding a magnetic term to the spherical pendulum (which breaks the reflexional symmetry).

3 Generic bifurcations near equilibria

The quadratic form $d^2H(x_0)$ is degenerate if and only if L_{x_0} has a zero eigenvalue. In this case, there exist arbitrarily small perturbations of H which remove the zero eigenvalues of L_{x_0}. However, in families of Hamiltonian matrices zero eigenvalues are inevitable; that is, they occur stably. In this section, we look at such families of Hamiltonian systems.

3.1 Generic movement of eigenvalues

In parametrized families of Hamiltonian matrices, it is typical for eigenvalues to cross the imaginary axis at the origin or for some eigenvalues to enter in resonance. In this section, we study the generic movement of eigenvalues for one-parameter families of matrices in $\mathfrak{sp}_G(2n)$. We look at the case of zero eigenvalues and the case of $1:1$ resonance. First, we need to introduce some concepts from group representation theory.

Suppose that the compact group G acts linearly on the vector space V. A subspace $W \subset V$ is *G-invariant* if $G(W) = W$. Moreover, if W does not contain any proper G-invariant subspaces, then W is an *irreducible representation* of G.

Let V be an irreducible representation of G. If the only linear mappings on V that commute with G are scalar multiples of the identity, then the representation V is *absolutely irreducible*. More precisely, if V is an irreducible representation and $\mathcal{D}$ is the space of linear mappings from V to itself commuting with G then $\mathcal{D}$ is a division algebra isomorphic to $\mathbb{R}$, $\mathbb{C}$ or $\mathbb{H}$, where $\mathbb{H}$ is the group of quaternions. Thus, if $\mathcal{D}$ is isomorphic to $\mathbb{C}$ or $\mathbb{H}$ then V is nonabsolutely irreducible.

If V is a symplectic representation, a subspace $W \subset V$ is *G-symplectic* if it is G-invariant and symplectic, and a symplectic representation is *irreducible* if it contains no proper G-symplectic subspace. Irreducible symplectic representations arise in several types depending on the underlying ordinary representation (i.e. forgetting the symplectic structure): firstly if the underlying representation is not irreducible, then $V = V_0 \oplus V_0$ where V_0 is an absolutely irreducible subspace (which is Lagrangian and one can identify $V = T^*V_0$). Secondly, if the underlying representation is irreducible, it must be either complex or quaternionic, and for a given representation of complex type there are two distinct symplectic representations, which are said to be *dual*. Any two symplectic representations whose underlying representations are equivalent quaternionic are equivalent as symplectic representations. See [MRS88] for more details.

Zero eigenvalues

Proposition 3.1 *Suppose that $L \in \mathfrak{sp}_G(2n)$ has a nonzero kernel. Then E_0, the generalized eigenspace of 0, is a G-symplectic subspace of $\mathbb{R}^{2n}$.*

The structure of the generalized eigenspace and corresponding movement of eigenvalues is described in the next result.

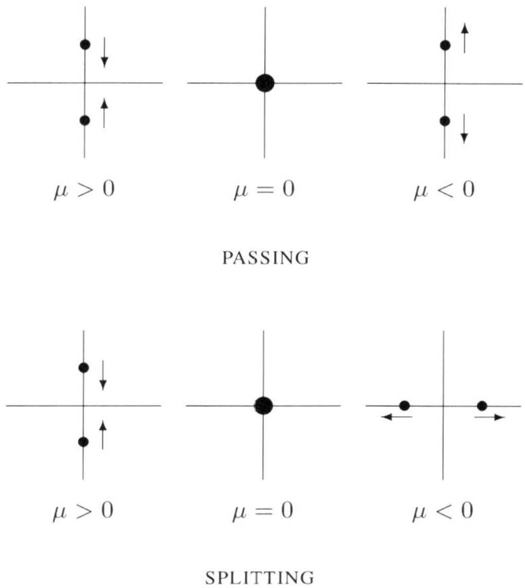

Fig. 3.1. The two scenarios for the generic movement of eigenvalues in steady-state bifurcation: see Theorem 3.2.

Theorem 3.2 (Golubitsky and Stewart [GS86]) *Let L_μ be a generic one-parameter family in $\mathfrak{sp}_G(2n)$ such that 0 is an eigenvalue of L_0. Then either*

(i) *the action of G on E_0 is nonabsolutely irreducible (in which case $L_0|_{E_0} = 0$), and the eigenvalues of L_μ lie on the imaginary axis and cross through 0 with nonzero speed ('passing' in Figure 3.1), or*
(ii) *$E_0 = V \oplus V$, where V is absolutely irreducible (in which case $L_0|_{E_0} \neq 0$ but $L_0^2|_{E_0} = 0$) and the eigenvalues cross through 0 going from purely imaginary to real or vice-versa ('splitting' in Figure 3.1).*

1:1 resonance

Suppose now that L is part of a one-parameter family and has a pair of purely imaginary eigenvalues $\pm i\eta$. By rescaling we can always assume that $\eta = 1$. The generalized eigenspace is denoted $E_{\pm i}$; it is a G-symplectic subspace. We consider two cases, either the eigenvalue i is G-simple ($E_{\pm i}$ is symplectic irreducible) or it is of G-multiplicity 2 also called a $1:1$ *resonance*. It is customary in Hamiltonian systems to distinguish two classes of $1:1$ resonance: the $1:1$ and the $1:-1$ resonances, depending on whether the Hamiltonian is definite or indefinite, respectively. The structure of the generalized eigenspace in the $1 : \pm 1$ resonance is given in the following theorem.

Theorem 3.3 (Dellnitz *et al.* [DMM92], van der Meer [vdM90]) *Let L_μ be a generic one-parameter family in $\mathfrak{sp}_G(2n)$ such that L_0 has eigenvalues $\pm i$ with G-multiplicity 2. Then $E_{\pm i} = U_1 \oplus U_2$ where for $j = 1, 2$, either*

(i) *U_j is nonabsolutely irreducible; or*
(ii) *$U_j = V \oplus V$, with V absolutely irreducible.*

Understanding how eigenvalues may move as a system passes through a $1 : \pm 1$ resonance requires a combination of group-theoretic results along with the analysis of the Hamiltonian quadratic form defined on the generalized eigenspace $E_{\pm i}$. Since $E_{\pm i}$ is a symplectic subspace of $\mathbb{R}^{2n}$, the restriction of the symplectic form ω to $E_{\pm i}$ is nondegenerate and thus $\omega_i = \omega|_{E_{\pm i}}$ is a symplectic form on $E_{\pm i}$. The Hamiltonian $Q(z) = \omega(z, Lz)$ is therefore a non-degenerate quadratic form on $E_{\pm i}$.

Recall that there are precisely two isomorphism classes of irreducible symplectic representations for a given complex underlying representation; these representations are dual to each other [MRS88].

Theorem 3.4 (Dellnitz *et al.* [DMM92]) *With the same hypotheses as in the theorem above, and with Q the Hamiltonian quadratic form induced on $E_{\pm i}$, precisely one of the following occurs:*

(i) *U_1 and U_2 are not isomorphic and the eigenvalues pass independently along the imaginary axis; Q may be indefinite or definite.*
(ii) *$U_1 = U_2 = V \oplus V$, V real, or $U_1 = U_2 = W$, W quaternionic, the eigenvalues split, and Q is indefinite.*
(iii) *U_1 and U_2 are complex of the same type, the eigenvalues pass and Q is indefinite.*
(iv) *U_1 and U_2 are complex duals and the eigenvalues pass or split depending on whether Q is definite or indefinite.*

The two splitting cases (ii) and (iv) of the theorem correspond to the Hamiltonian-Hopf bifurcation which is the Hamiltonian version of the Hopf bifurcation theorem for dissipative systems, see Section 4 below.

In the nonsymmetric case $G = 1$, $E_{\pm i} = V \oplus V \oplus V \oplus V$ where V is the one-dimensional trivial representation. The one-parameter unfolding of the

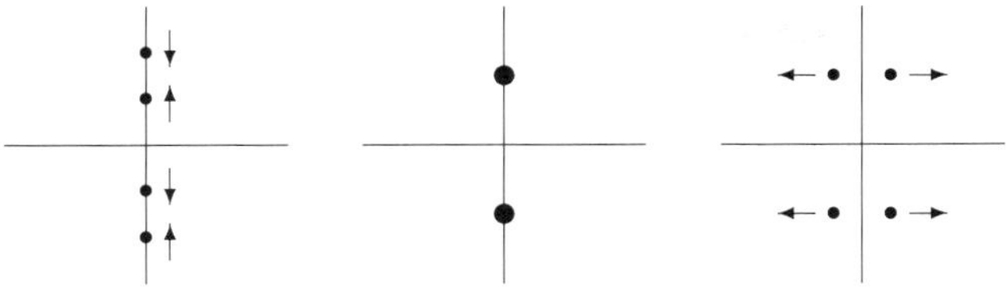

Fig. 3.2. Generic movement of eigenvalues in the $1 : -1$ resonance: the splitting case.

normal form for the $1 : -1$ resonance is given by

$$M(\mu) = \begin{bmatrix} 0 & -1 & \rho & 0 \\ 1 & 0 & 0 & \rho \\ \mu & 0 & 0 & -1 \\ 0 & \mu & 1 & 0 \end{bmatrix}. \tag{3.1}$$

where $\rho = \pm 1$. This normal form is used in the Hamiltonian-Hopf theorem without symmetry, Section 4. Note that without symmetry the $1 : 1$ resonance is of codimension 3 (and codimension 2 in time-reversible systems) and so is not usually considered.

Further information can be found in [MD93] and [Me93], where Melbourne and Dellnitz extend to symmetric systems both Williamson's results on normal forms for linear Hamiltonian systems and Galin's results on their versal deformations. For example in [Me93] one finds that the normal form for case (ii) of the theorem above is also given by (3.1), where each scalar is interpreted as scalar multiplication in V.

3.2 Bifurcation of equilibria

In multiparameter families of Hamiltonian systems, the eigenvalues of the linearization typically cross the imaginary axis leading to bifurcations of equilibria or periodic solutions.

In this section, we look at bifurcations of equilibria. We begin with bifurcations of equilibria in one-parameter families of one degree of freedom Hamiltonian systems. Then, we explain how this information is used in bifurcation diagrams for multiparameter families of one-degree of freedom Hamiltonian systems. We conclude with some comments on bifurcations in many-degrees of freedom Hamiltonians. An important tool for the study of Hamiltonian systems is the Splitting Lemma (or Morse Lemma with parameters) [BG92], which separates out a nondegenerate part of the function from the remainder.

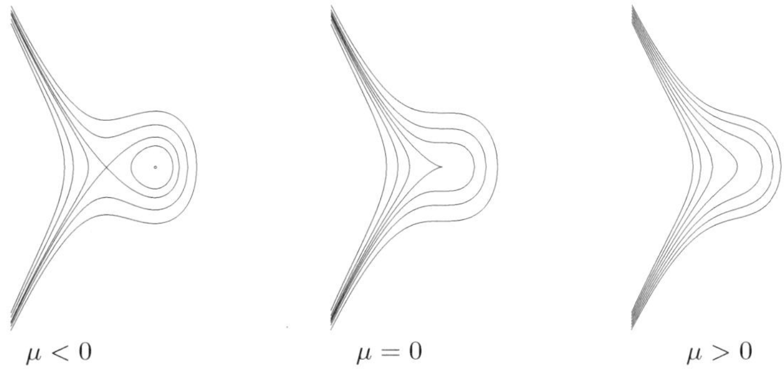

Fig. 3.3. Saddle-centre bifurcation with $G = 1$.

Theorem 3.5 (Splitting Lemma) *Let $F : \mathbb{R}^N \times \mathbb{R}^l \longrightarrow \mathbb{R}$ be a smooth function. Denote a point in $\mathbb{R}^N \times \mathbb{R}^l$ by $(x, \lambda) = (x_1, \ldots, x_N, \lambda_1, \ldots, \lambda_l)$, and suppose that $d_x F(0,0) = 0$ and that the Hessian matrix $d_x^2 F(0,0)$ is non-degenerate. Then in a neighbourhood of the origin, there is a change of coordinates of the form $\Psi(x, \lambda) = (\psi(x, \lambda), \lambda)$ with $\psi(0,0) = 0$, such that*

$$F \circ \Psi(x, \lambda) = \sum_j \varepsilon_j x_j^2 + h(\lambda),$$

where $\varepsilon_j = \pm 1$ and h is a smooth function with $h(0) = F(0,0)$.

In practice $h(\lambda)$ can be found by solving the equation $d_x F(x, \lambda) = 0$ for $x = c(\lambda)$ (by the implicit function theorem), and then $h(\lambda) = F(c(\lambda), \lambda)$. (Of course, the change of coordinates cannot in general be taken to be symplectic!)

3.2.1 One degree of freedom: one-parameter family

Let $H : \mathbb{R}^2 \times \mathbb{R} \longrightarrow \mathbb{R}$ be a smooth generic one degree of freedom Hamiltonian function depending on a single parameter μ; such a family will have just codimension-1 bifurcations. As always in bifurcation theory, the meaning of "codimension-1" depends on the context, for example on the symmetry. We illustrate this with three simple examples: $G = 1$, $G = \mathbb{Z}_2$ and $G = \mathbf{SO}(2)$. Let L_μ be the linearization of $J\nabla H$ and suppose that L_0 has a zero eigenvalue. This corresponds to Theorem 3.2(ii). We discuss the bifurcation diagram for several group actions.

$G = 1$: Saddle-centre bifurcation. The trivial group has only a one-dimensional irreducible representation which is of course absolutely irreducible.

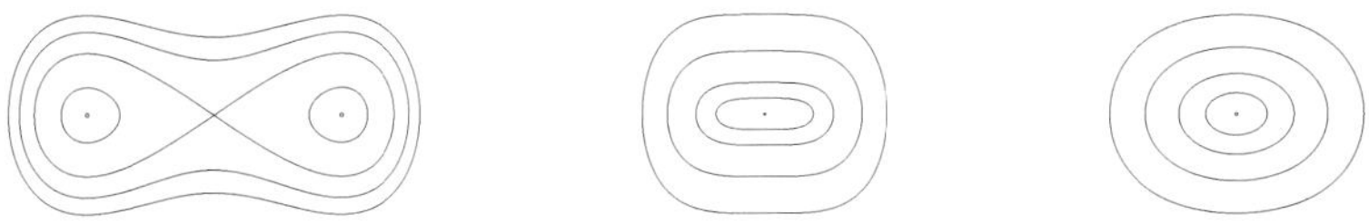

Fig. 3.4. Supercritical $\mathbb{Z}_2$-pitchfork bifurcation.

Therefore, generically $E_0 = \mathbb{R} \times \mathbb{R}$ and L_0 is nilpotent. Suppose that

$$L_0 = \begin{bmatrix} 0 & 1 \\ 0 & 0 \end{bmatrix}, \tag{3.2}$$

then $H(p, q, 0) = q^2 + \tilde{H}$ where $\tilde{H}$ contains terms of degree three and up. By the Splitting Lemma above, $H(p, q, \mu)$ is equivalent to $f(p, \mu) + q^2$. The map f has a fold catastrophe at $\mu = 0$, see Poston and Stewart [PS78]. The universal unfolding is $p^3 + \mu p$. The local dynamics for $p^3 + \mu p + q^2$ near $(0, 0, 0)$ is illustrated by Figure 3.3. This is called the *saddle-centre* bifurcation.

***G* = $\mathbb{Z}_2$: Pitchfork bifurcation.** The group $\mathbb{Z}_2$ has one nontrivial irreducible representation which is of dimension one. Therefore it is absolutely irreducible and $E_0 = \mathbb{R} \oplus \mathbb{R}$ where $\mathbb{Z}_2$ acts by -1 on each copy of $\mathbb{R}$. Thus, H is $\mathbb{Z}_2$-invariant: $H(-p, -q, \mu) = H(p, q, \mu)$. Let L_0 be given by (3.2) then $H(p, q, 0) = q^2 + \tilde{H}$ where $\tilde{H}$ contains terms of degree higher than two. Using the Splitting Lemma above, H is equivalent to $f(p, \mu) + q^2$ as in the $G = 1$ case. However, the universal unfolding must be $\mathbb{Z}_2$-invariant, thus the family of maps is $\pm p^4 + \mu p^2 + q^2$. The family $p^4 + \mu p^2 + q^2$ is a supercritical pitchfork bifurcation where a pair of stable centres for $\mu < 0$ coalesce at $\mu = 0$ into one centre at the origin, see Figure 3.2.2. The $-p^4 + \mu p^2 + q^2$ case is a subcritical pitchfork bifurcation, see Figure 3.5, where the bifurcating branch of equilibria are unstable.

Generally, a bifurcation is *subcritical* if at the instant of bifurcation (here $\mu = 0$) the equilibrium is unstable, and *supercritical* if it is stable.

***G* = *SO*(2):** The only nontrivial irreducible representation of $\mathbf{SO}(2)$ is two-dimensional of complex type. The action on $\mathbb{C}$ is $\theta.z = e^{mi\theta}z$ for $\theta \in \mathbf{SO}(2)$ and some $m \in \mathbb{Z}$. By Theorem 3.2, $E_0 = \mathbb{C} = \mathbb{R}^2$, the eigenvalues are purely imaginary and cross 0 with nonzero speed. The general $\mathbf{SO}(2)$-invariant function of two variables is $H(p, q) = f(p^2 + q^2)$ for some smooth function f. Since L_0 is identically zero at $\mu = 0$, then $H(p, q) = \mu(p^2 + q^2) + o(p^2 + q^2)$.

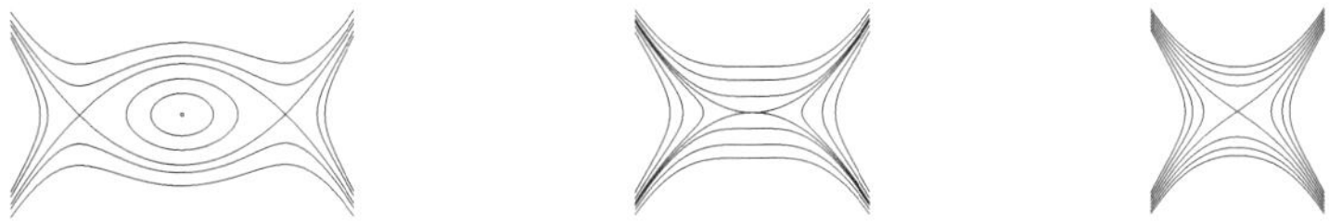

Fig. 3.5. Subcritical $\mathbb{Z}_2$-pitchfork bifurcation.

In polar coordinates, $r^2 = p^2 + q^2$, we obtain $f(r) = \mu r^2 + o(r^2)$ and so the unfolding of the generic singularity is given (up to sign) by $\mu r^2 + r^4$. For $\mu < 0$ the origin is surrounded by periodic solutions. At $\mu = 0$ the bifurcation occurs and an $\mathbf{SO}(2)$-orbit of equilibria of amplitude $\sqrt{\mu}$ for $\mu > 0$ is created.

3.2.2 One degree of freedom: two-parameter family

$G = 1$: the pitchfork revisited. In $\mathbb{Z}_2$-symmetric systems the pitchfork bifurcation is of codimension 1, while if there is no symmetry present then it is of codimension 2, as we see now. Consider a generic two-parameter family $H(p, q, \mu_1, \mu_2)$ of Hamiltonian functions. One can[1] take $H(p, q, 0, 0) = q^2 + \tilde{H}$ where $\tilde{H}$ has terms of degree three and up. By the Splitting Lemma H is equivalent to $f(p, \mu_1, \mu_2) + q^2$. To be of codimension 2, we use $f_\pm(p, 0, 0) = \pm p^4$, and then a two-parameter unfolding is $f_\pm(p, \mu_1, \mu_2) = \pm p^4 + \mu_1 p^2 + \mu_2 p$. Notice that $\mu_2 = 0$ corresponds to the $\mathbb{Z}_2$-pitchforks considered above.

Since $dH^\pm(x) = (\pm 4p^3 + 2\mu_1 p + \mu_2, q)$, equilibria are solutions of $q = 0$ and $f'(p, \mu_1, \mu_2) = \pm 4p^3 + 2\mu_1 p + \mu_2 = 0$. The Hessian is

$$\begin{bmatrix} 2 & 0 \\ 0 & \pm 12p^2 + 2\mu_1 \end{bmatrix}.$$

The Hessian is degenerate at $\mu_1 = \mp 6p^2$. Replacing in f' we obtain $\mu_2 = \pm 8p^3$, thus the cusp $\Delta \equiv 8\mu_1^3 \pm 27\mu_2^2 = 0$ is the bifurcation set. Since f' is a cubic polynomial it has at least one real root. The Hamiltonian system has a saddle-centre on the cusp curve where the number of roots of $f'(p, \mu_1, \mu_2) = 0$ (ie, of equilibria) jumps from 1 to 3 (or vice-versa).

Typical level contours of H^+ are shown in Figure 3.6 as (μ_1, μ_2) crosses the bifurcation set from the region with 3 equilibria to the region with 1. The analogous figure for H^- is left to the reader.

$G = \hat{\mathbb{Z}}_2$: the reversible umbilic. Up to now we have only considered bifurcations with *symplectic* symmetries. Hanßmann [H98] studies two-parameter

[1] If $H(p, q, 0, 0)$ has no quadratic terms then it is of codimension at least 3

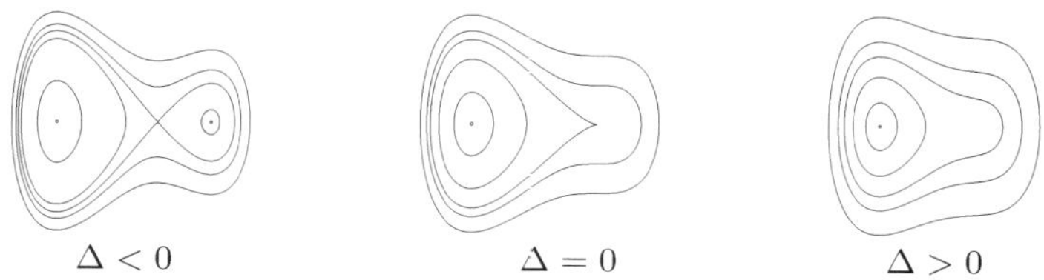

Fig. 3.6. Level contours of $H^+(p, q, \mu_1, \mu_2)$ (supercritical pitchfork) for different values of (μ_1, μ_2) as it crosses the bifurcation curve $\Delta = 0$.

families of one degree of freedom Hamiltonian with the *reversing* symmetry $\kappa.(q,p) = (q,-p)$ (whence the hat in $\hat{\mathbb{Z}}_2$). In two-parameter families the zero linearization occurs generically for reversible Hamiltonian systems. The least degenerate singularities in this case are given by $H^{\pm}(q,p) = p^2 q \mp \frac{1}{3}q^3$. Hanßmann shows that the versal unfolding in this case is a restriction of the umbilic catastrophe [PS78] given by

$$H_{\mu}^{\pm}(q,p) = p^2 q \mp \frac{1}{3}q^3 + \mu_1(p^2 \pm q^2) + \mu_2 q$$

(where $\mu = (\mu_1, \mu_2)$) and called the *reversible umbilic*. The upper sign corresponds to the elliptic umbilic and the lower one to the hyperbolic umbilic. We only consider the hyperbolic reversible umbilic in these notes.

The critical points of H_{μ}^{-} are given by solutions of $p^2 + q^2 - 2\mu_1 q + \mu_2 = p(q + \mu_1) = 0$. The Hessian is given by

$$\begin{bmatrix} 2(q-\mu_1) & 2p \\ 2p & 2(q+\mu_1) \end{bmatrix}.$$

Solving for critical points with degenerate quadratic form yields the bifurcation set given by the union of the parabolas $\{\mu_2 = \mu_1^2\}$ and $\{\mu_2 = -3\mu_1^2\}$. On the curve $\{\mu_2 = \mu_1^2\}$ the system has a unique equilibrium at $(q,p) = (\mu_1, 0)$ with degenerate Hessian

$$\begin{bmatrix} 0 & 0 \\ 0 & 1 \end{bmatrix}.$$

Now $(q,p) = (\mu_1, 0) \in \mathrm{Fix}(\kappa)$ and the kernel of the Hessian lies in $\mathrm{Fix}(\kappa)$, therefore the $\mathbb{Z}_2$-reversing symmetry is not broken and the parabola $\{\mu_2 = \mu_1^2\}$ is a saddle-centre bifurcation curve. For $\mu_1 \neq 0$, the system goes from no equilibria when $\mu_2 > \mu_1^2$ to a saddle and a centre for $\mu_2 < \mu_1^2$.

On the curve $\{\mu_2 = -3\mu_1^2\}$, the system has a unique equilibrium at $(q,p) = (-\mu_1, 0) \in \mathrm{Fix}(\kappa)$ with the kernel of the Hessian transverse to $\mathrm{Fix}(\kappa)$. Hence, the bifurcation on the parabola $\{\mu_2 = -3\mu_1^2\}$ breaks the $\mathbb{Z}_2(\kappa)$-symmetry and so is a curve of Hamiltonian pitchfork bifurcation.

The reversing symmetry forces some interesting dynamics. Since the bifurcating equilibria are related by the κ symmetry they lie on the same energy level. By reversibility, they must be connected by a heteroclinic connection. See Hanßmann [H98] for details and the full bifurcation picture.

3.2.3 Many degrees of freedom Hamiltonian

For Hamiltonian systems with symmetry having more than one degree of freedom, one can use reduction methods to determine part or all of the dynamics in the neighbourhood of an equilibrium point. A symmetry-based reduction (sometimes called discrete reduction, although the group in question need not be discrete) is given by the following result.

Proposition 3.6 *Let $H : V \to \mathbb{R}$ be a G-invariant Hamiltonian and let X_H be its associated Hamiltonian vector field. Let K be a compact subgroup of G acting symplectically. Then,* $\mathrm{Fix}(K)$ *is symplectic and X_H leaves it invariant; moreover $X_H \mid_{\mathrm{Fix}(K)}$ is a Hamiltonian vector field with Hamiltonian $H \mid_{\mathrm{Fix}(K)}$.*

In particular, if $\dim \mathrm{Fix}(K) = 2$, then the restricted system is of one degree of freedom and the dynamics/bifurcations are readily obtained as above. Note that the compactness of K is essential; for example if $K \simeq \mathbb{R}$ acts on $(\mathbb{R}^2, \omega)$ by $t \cdot (x, y) = (x + ty, y)$ then $\mathrm{Fix}(K) = \mathbb{R}$ (the x-axis).

Other reductions to one degree of freedom can be obtained for example by centre-manifold reduction.

4 Hamiltonian-Hopf bifurcation

A Hamiltonian-Hopf bifurcation occurs when two nonzero imaginary eigenvalues of an elliptic equilibrium collide in a $1 : -1$ resonance and move into the left and right half-planes, see Figure 3.2. It is named in analogy with the Hopf Bifurcation Theorem of dissipative systems where small amplitude periodic solutions bifurcate from an equilibrium that loses stability as a pair of complex eigenvalues cross the imaginary axis.

The existence of periodic solutions in the Hamiltonian-Hopf bifurcation was first established by Meyer and Schmidt [MS71], and then later by van der Meer [vdM85] who was the first to study its equivariant version [vdM90]. Recently, Chossat, Ortega and Ratiu [COR02] extended the Hamiltonian-Hopf Theorem to include relative periodic orbits.

We begin with the nonsymmetric case. For a generic one-parameter family of Hamiltonians, the generalized eigenspace $E_{\pm i}$ at a $1 : -1$ resonance is of dimension four. We restrict our study of this bifurcation to the family $(\mathbb{R}^4, \omega, 1, H_\mu)$ with $\mu \in \mathbb{R}$ (this reduction can be obtained by Lyapunov-Schmidt reduction or restriction to the centre manifold). Since the eigenvalues are far from 0, the equilibrium is non-degenerate and there is no bifurcation of equilibria in this family; we can therefore assume the origin is an equilibrium point for all μ. The following points are used implicitly in the statement of the theorem.

• Let $(x, y) \in \mathbb{R}^4$ where $x = (x_1, x_2)$ and $y = (y_1, y_2)$, with symplectic form $\omega = dx_1 \wedge y_1 + dx_2 \wedge dy_2$, and the origin is an equilibrium point of H_μ for all values of μ. Let $H_{2,\mu}$ denote the quadratic part of H_μ, then the linearization at $\mu = 0$, see matrix (3.1), implies that

$$H_{2,\mu}(x, y) = S + N + \mu P.$$

where $S = x_1y_2 - x_2y_1$ (the semisimple part of $H_{2,0}$), $N = \frac{1}{2}(x_1^2 + x_2^2)$ (the nilpotent part) and $P = \frac{1}{2}(y_1^2 + y_2^2)$. For $\mu < 0$ the linear system has two distinct pairs of imaginary eigenvalues, so the nonlinear system has 2 nonlinear normal modes, while for $\mu > 0$ the eigenvalues all have non-zero real part, so there are no nonlinear normal modes; indeed no periodic orbits in a neighbourhood of the origin. The problem is to describe this transition.

• The dynamics near equilibrium solutions is understood using Birkhoff normal form [CS86, ETBCI]. We denote by $(\mathbb{R}^4, \omega, \mathbf{S}^1, \tilde{H}_\mu)$ the Hamiltonian system of the symplectic Birkhoff normal form $\tilde{H}$ of $(\mathbb{R}^4, \omega, 1, H_\mu)$ up to some finite order k.

$$\tilde{H}(x, y, \mu) = H_2(x, y, \mu) + \tilde{H}_4(S, P) + \cdots + \tilde{H}_k(S, P), \tag{4.1}$$

where $S = S(x, y) = x_1y_2 - x_2y_1$ and $P = P(x, y) = \frac{1}{2}(y_1^2 + y_2^2)$, and $\tilde{H}_k$ is homogeneous of degree k in x, y. We also write $N = \frac{1}{2}(x_1^2 + y_1^2)$.

Theorem 4.1 (Hamiltonian-Hopf Bifurcation) *Suppose that the family $(\mathbb{R}^4, \omega, 1, H_\mu)$ of Hamiltonian systems has, at $\mu = 0$, imaginary eigenvalues in $1 : -1$ resonance. If the coefficient, a of P^2 in $\tilde{H}_4$ is nonzero then for each $k > 0$ there is a neighbourhood of the origin in $\mathbb{R}^4 \times \mathbb{R}$ in which the set of short periodic solutions of the system $(\mathbb{R}^4, \omega, 1, H_\mu)$ is C^k-diffeomorphic to the set of short periodic solutions of the system $(\mathbb{R}^4, \omega, \mathbf{S}^1, S + G_\nu)$ with*

$$G_\nu(x, y) = N + \nu P + aP^2. \tag{4.2}$$

A point z lies on a short periodic solution of $(\mathbb{R}^4, \omega, \mathbf{S}^1, S + G_\nu)$ if and only if it is a critical point of the "energy-momentum map" $(S, S + G_\nu)$, and so of

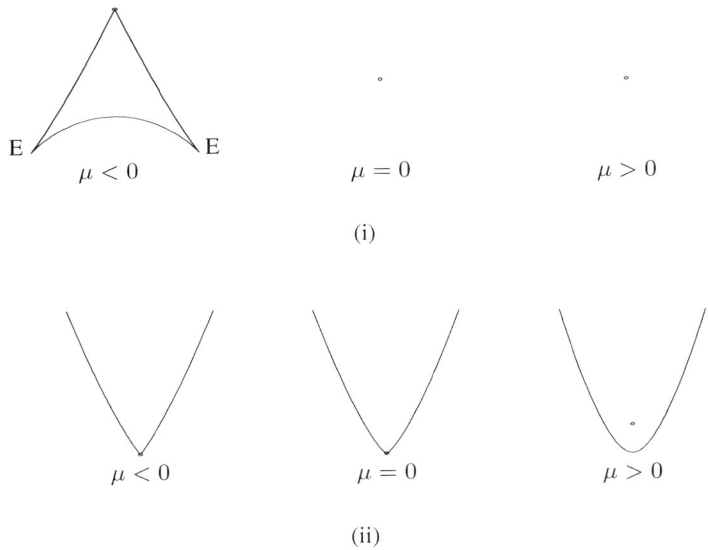

Fig. 4.1. Families of periodic orbits in the two scenarios of the Hamiltonian-Hopf bifurcation. The small dot represents the origin.

the map (S, G_ν). There are two possible scenarios for the structure of the set of periodic orbits, according to the sign of a.

Theorem 4.1 (continued) *Let a be the coefficient of P^2 in the normal form (4.1) for H_μ.*

(i) *If $a > 0$, then for $\mu < 0$ the 2 nonlinear normal modes are globally connected in a single compact family; as $\mu \to 0^-$ this family collapses to the origin and disappears.*
(ii) *If $a < 0$, then for $\mu < 0$ the two nonlinear normal modes are distinct in a neighbourhood of the origin, intersecting only at the origin. As μ passes through 0 they pull away from the origin as a single family.*

The "pulling away from the origin" in case (ii) is similar to that of a 1-sheeted hyperboloid pulling away from the origin as it deforms from a cone, though the analogy cannot be taken very far.

The two cases (i) and (ii) are illustrated in Figures 4.1 (i) and (ii) respectively and are sometimes referred to as the subcritical and supercritical Hamiltonian-Hopf bifurcations. However, the reader should beware that the nomenclature is not universally consistent. Iooss and Pérouème [IP93] refer to (i) as supercritical and (ii) as subcritical, while Hanßmann and van der Meer [HM02] have it the other way round. The lectures of Cushman in this book use Hanßmann and van der Meer's convention. The illustrations in Figure 4.1 show the images in energy-momentum space of the families of periodic orbits.

Recall that for a non-resonant elliptic equilibrium, the nonlinear normal modes are also elliptic sufficiently close to the equilibrium point. In the bifurcation of type (i) for fixed $\mu < 0$, when the equilibrium is elliptic, if one follows the compact family of periodic orbits emanating from the origin along one nonlinear normal mode and returning along the other, there is a transition from elliptic to hyperbolic periodic orbits, and then back to elliptic again; Iooss and Pérouème [IP93] refer to the transition points as *Eckhaus points*, marked E on Figure 4.1(i). There are therefore hyperbolic periodic orbits in any neighbourhood of the origin $(x, y, \mu) = (0, 0, 0)$ in $\mathbb{R}^4 \times \mathbb{R}$ (and Iooss and Pérouème also show that in the reversible setting there are orbits homoclinic to certain hyperbolic periodic orbits; presumably this would be true also in the Hamiltonian setting, but to our knowledge this has not been checked). An example of both supercritical and subcritical Hamiltonian-Hopf bifurcations are mentioned in Cushman's lectures (Section C.4). Finally, Sokol′skiı́ has shown [So74][1] that at the bifurcation point $\mu = 0$, the Hamiltonian-Hopf case (ii) scenario ($a < 0$) has an unstable equilibrium, while in the other scenario the origin is *formally stable*, meaning that the equilibrium is stable for the dynamics of the normal form approximation at any order.

Symmetric Hamiltonian-Hopf

As in the equivariant Weinstein-Moser theorem (Theorem 2.1) periodic solutions with spatio-temporal symmetries are found by considering the action of $G \times \mathbf{S}^1$, this time on the bifurcation eigenspace $E_{\pm i}$. The equivariant version of Theorem 4.1 is obtained by finding four dimensional fixed point subspaces of the action of $G \times \mathbf{S}^1$ and showing that the hypotheses of Theorem 4.1 are satisfied for the fixed point subspace. The $\mathbf{S}^1$ action is generated by the semisimple part of $H_{2,0}$ on $E_{\pm i}$.

So, we consider a one-parameter family of Hamiltonian systems $(\mathbb{R}^{2n}, \omega, G, H_\mu)$ with nontrivial symmetry G, and we will suppose for simplicity that $E_{\pm i} = \mathbb{R}^{2n}$.

Lemma 4.2 *Let* Σ *be an isotropy subgroup of* $G \times \mathbf{S}^1$, *so that* $\mathrm{Fix}(\Sigma) \neq \{0\}$. *Then* $\dim\, \mathrm{Fix}(\Sigma) \geq 4$ *and the restriction of* $X_{H_{2,0}}$ *to* $\mathrm{Fix}(\Sigma)$ *gives rise to a* $1 : -1$*-resonance.*

Lemma 4.2 guarantees that on each nonzero fixed point subspace of the

[1] the lecturer would like to thank Ken Meyer for pointing this out to him

action of $G \times \mathbf{S}^1$ the movement of eigenvalues on the subspace does correspond to a Hamiltonian-Hopf bifurcation. The result is the following.

Theorem 4.3 (Hamiltonian-Hopf Theorem with symmetry [vdM90]) *Let Σ be an isotropy subgroup of $G \times \mathbf{S}^1$ with $\dim \mathrm{Fix}(\Sigma) = 4$. Let a_Σ be the coefficient of P^2 in the normal form of H_0 on $\mathrm{Fix}(\Sigma)$. Then, provided $a_\Sigma \neq 0$, the same two scenarios occur as for the ordinary Hamiltonian-Hopf bifurcation, according to the sign of a_Σ. Moreover the resulting periodic orbits all have spatio-temporal symmetry at least Σ.*

An example of Hamiltonian-Hopf bifurcation with symmetry occurs in models of point vortices on the sphere, see Laurent-Polz [LP00].

Remark 4.4 In the Hamiltonian-Hopf bifurcation, for $\mu > 0$ the eigenvalues have non-zero real parts so there is a stable manifold and an unstable manifold for the equilibrium point, while for $\mu < 0$ the system is elliptic and there are no such manifolds. In an interesting recent paper McSwiggen and Meyer [MM03] have studied this transition and shown that there are again two scenarios that mirror quite remarkably the two scenarios for the nonlinear normal modes. One would expect to have similar behaviour in the symmetric case, but to our knowledge this has not been checked.

Hamiltonian-Hopf and Relative Periodic Orbits

Here we describe very briefly a result of Chossat, Ortega and Ratiu [COR02] which extends the results described above to finding *relative* periodic orbits. A trajectory $\gamma(t)$ of a dynamical system is periodic if $\gamma(T) = \gamma(0)$ for some $T > 0$, and it is a *relative periodic orbit* (or RPO) if there is an element $g \in G$ and $T > 0$ such that $\gamma(T) = g \cdot \gamma(0)$. As usual T is called the period, and g is called the *phase*.

The authors investigate the situation in which a point x_0 is a non-degenerate equilibrium point of a G-invariant Hamiltonian system H_0 with eigenvalues $\pm i\nu$, for which the generalized eigenspace $E_{\pm i\nu}$ is such that it decomposes as the sum of two symplectic irreducibles of complex dual type for the $G \times S^1$-action, where the S^1-action is that derived from the linearization on $E_{\pm i\nu}$; write $E_{\pm i\nu} = U_1 \oplus U_2$. As described in Theorem 3.4(iv) this hypothesis is generic for a 1-parameter family of G-invariant Hamiltonians. Furthermore H_μ is assumed to be a *generic* G-invariant deformation of H_0, which is an assumption on the movement of the eigenvalues as they collide (Figure 3.2 (p. 369)).

Let S be the unit sphere of V; it is $G \times S^1$ invariant and of odd-dimension.

For $\xi \in \mathfrak{g}$ let G_ξ denote the isotropy subgroup of ξ under the adjoint action. The main result of Chossat, Ortega and Ratiu is

Theorem 4.5 *With the setup and genericity assumptions described above then for each $\xi \in \mathfrak{g}$ sufficiently small there is a smooth $G_\xi \times S^1$-equivariant vector field on the sphere S such that on each energy level near x_0 and for each relative equilibrium of the vector field, there is a value of $\mu \approx 0$ for which there is a relative periodic point of H_μ with phase* $\exp(T\xi)$ *for some $T \approx 2\pi$.*

It is possible to use topological methods to estimate the minimal possible numbers of relative equilibria of $G_\xi \times S^1$ equivariant vector fields. In particular it is always positive, since S^1 acts freely on S, so the $G_\xi \times S^1$ equivariant vector field on S descends to a G_ξ-equivariant vector field on S/S^1. This orbit space S/S^1 has non-zero Euler characteristic so every vector field there has a zero (it is diffeomorphic to a complex projective space).

The proof is based on the reduction method of Vanderbauwhede and van der Meer [VvdM]. The reader should beware that the paper [COR02] mis-states the result by saying that each RPO exists for every value of μ rather than for some value of μ. It would be interesting to understand better the behaviour of the RPOs as μ is varied.

As we have seen, a considerable amount is known about the Hamiltonian-Hopf bifurcation (the "splitting" cases of Theorem 3.4). On the other hand, very little is known about the passing cases, probably because it is of higher codimension if there is no symmetry present: namely codimension 3, and codimension 2 in reversible systems.

PART II: LOCAL DYNAMICS NEAR RELATIVE EQUILIBRIA

5 Momentum map and reduction

5.1 Noether's theorem

Emmy Noether's theorem associates to any 1-parameter group of symmetries a conserved quantity for the dynamics. For a "several-parameter" group, there are correspondingly several conserved quantities, which together are called the momentum map.

Given a symplectic action of a group G, a map $\mathbf{J} : \mathcal{P} \to \mathfrak{g}^*$ is called a *momentum map* if $X_{\mathbf{J}^\xi} = \xi_{\mathcal{P}}$ for each $\xi \in \mathfrak{g}$, where $\mathbf{J}^\xi(x) = \langle \mathbf{J}(x), \xi \rangle$, $x \in \mathcal{P}$. The defining equation for the momentum map is

$$\langle d\mathbf{J}_x(v), \xi \rangle = \omega_x(v, \xi_{\mathcal{P}}(x))$$

for all $x \in \mathcal{P}$, $v \in T_x\mathcal{P}$ and $\xi \in \mathfrak{g}$ (the Lie algebra of G). The momentum map is thus defined up to a constant, and $\mathrm{Im}(\mathrm{d}\mathbf{J}_{\mathrm{x}}) = \mathfrak{g}_{\mathrm{x}}^\circ \subset \mathfrak{g}^*$ (where $\mathfrak{g}_x^\circ$ denotes the annihilator of $\mathfrak{g}_x$ in $\mathfrak{g}^*$). It follows that the momentum map is a submersion in a neighbourhood of any point where the action is locally free (i.e., where $\mathfrak{g}_x = 0$).

The momentum map always exists locally, but to ensure the global existence of the momentum map one needs some hypothesis such as semisimplicity of the group, or simple connectedness of the phase space (see [GS84]).

Theorem 5.1 (Noether) *Let H be a G-invariant Hamiltonian on $\mathcal{P}$ with a momentum map $\mathbf{J}$. Then $\mathbf{J}$ is conserved on the trajectories of the Hamiltonian vector field X_H.*

Proof. Differentiating the G-invariance condition, we get $dH \cdot \xi_{\mathcal{P}} = 0$. Since $dH \cdot \xi_{\mathcal{P}} = \{H, \mathbf{J}^\xi\} = -\{\mathbf{J}^\xi, H\} = -d\mathbf{J}^\xi \cdot H$, the functions $\mathbf{J}^\xi$ are conserved on the trajectories of X_H for every ξ in $\mathfrak{g}$. ■

5.2 Equivariance of the momentum map

Given a symplectic action of G on $\mathcal{P}$ and a momentum map $\mathbf{J} : \mathcal{P} \to \mathfrak{g}^*$, one can construct an action of G on $\mathfrak{g}^*$ such that the momentum map is equivariant with respect to these actions. Usually, but not always, this turns out to be the coadjoint action of G on $\mathfrak{g}^*$. The construction was found by Souriau [S70], and proceeds as follows. Let θ be the *cocycle*

$$\begin{aligned} \theta : G &\to \mathfrak{g}^* \\ g &\mapsto \mathbf{J}(g \cdot x) - \mathrm{Coad}_g \mathbf{J}(x) \end{aligned}$$

This map is well-defined if $\theta(g)$ is independent of x, which it is provided $\mathcal{P}$ is connected. We then define the *modified coadjoint action* by

$$\mathrm{Coad}_g^\theta\, \mu := \mathrm{Coad}_g\, \mu + \theta(g)$$

A short calculation shows that it is indeed an action.

Theorem 5.2 (Souriau) *Let the Lie group G act on the connected symplectic manifold $\mathcal{P}$ in such a way that there exists a momentum map $\mathbf{J} : \mathcal{P} \to \mathfrak{g}^*$. Then $\mathbf{J}$ is equivariant with respect to the modified coadjoint action on $\mathfrak{g}^*$:*

$$\mathbf{J}(g \cdot x) = \mathrm{Coad}_g^\theta\, \mathbf{J}(x)$$

Furthermore, if G is either semisimple or compact then the momentum map can be chosen such that $\theta = 0$.

For proofs see [S70], and [GS84] for semisimple groups, and [Mo97] for compact groups.

Remark 5.3 The above arguments can be extended to semisymplectic actions (see the introduction to these lectures). If $\chi : G \to \mathbb{Z}_2$ is the temporal character, then one defines the χ-twisted coadjoint action by

$$\mathrm{Coad}_g^\chi\, \mu = \chi(g)\, \mathrm{Coad}_g\, \mu, \tag{5.1}$$

and similarly $\mathrm{Coad}_g^{\chi,\theta}\, \mu$ and everything then follows as before [MR00].

5.3 Reduction

By Noether's theorem the dynamics preserve the level sets of the momentum map $\mathbf{J}$. It is then natural to study the dynamics on one level set at a time. However, these level sets are not in general symplectic manifolds and the induced dynamics are therefore not Hamiltonian. But, if one passes to the orbit space[1] of one of these level sets, then the resulting space is symplectic (provided the action is free and proper).

Let $\mu \in \mathfrak{g}^*$ and G_μ be the isotropy subgroup of the modified coadjoint action:

$$G_\mu = \{g \in G \mid \mathrm{Coad}_g^\theta\, \mu = \mu\}.$$

For example, if $G = \mathbf{SO}(3)$ and $\mu \neq 0$, G_μ is the set of rotations with axis $\langle\mu\rangle$ and so is isomorphic to $\mathbf{SO}(2)$, while $G_0 = \mathbf{SO}(3)$.

[1] that is, identify points in the level sets which lie in the same group orbit

By the equivariance of the momentum map, G_μ acts on the level set $\mathbf{J}^{-1}(\mu)$. We can then define the (Meyer-Marsden-Weinstein) *reduced space* $\mathcal{P}_\mu$ to be:

$$\mathcal{P}_\mu = \mathbf{J}^{-1}(\mu)/G_\mu.$$

Refer to Ratiu's lectures for details (and see also [MR] and [OR]). Since G acts freely, $\mathcal{P}_\mu$ is a smooth manifold. Moreover it is symplectic with symplectic form ω_μ given by $\omega_\mu(\pi(u), \pi(v)) = \omega(u, v)$, where $u, v \in T_p\mathcal{P}$ and π is the projection $T_p\mathcal{P} \to T_{\overline{p}}\mathcal{P}_\mu$.

Given an invariant Hamiltonian H, its restriction to $J^{-1}(\mu)$ is invariant under G_μ, and so determines a well-defined function on $\mathcal{P}_\mu$, the *reduced* Hamiltonian denoted H_μ. The dynamics induced on the reduced space is determined by a vector field X_μ which is defined by $dH_\mu = \omega_\mu(-, X_\mu)$.

Recall that $\mathcal{P}_\mu$ is diffeomorphic to $\mathbf{J}^{-1}(\mathcal{O}_\mu)/G$ where $\mathcal{O}_\mu$ is the coadjoint orbit through μ. One defines the *orbit momentum map* $\mathcal{J} : \mathcal{P}/G \to \mathfrak{g}^*/G$ by:

$$\begin{array}{ccc} \mathcal{P} & \xrightarrow{\quad \mathbf{J} \quad} & \mathfrak{g}^* \\ \big\downarrow & & \big\downarrow \\ \mathcal{P}/G & \xrightarrow{\quad \mathcal{J} \quad} & \mathfrak{g}^*/G \end{array}$$

where the vertical arrows are the quotient maps. Then the reduced spaces are the fibers of $\mathcal{J}$.

Now, we introduce the notion of *symplectic slice* to provide a local model for the reduced spaces. Recall that a slice to a group action at a point $p \in \mathcal{P}$ is a submanifold S through p satisfying $T_pS \oplus \mathfrak{g} \cdot p = T_p\mathcal{P}$. If G_p is compact, it can be chosen to be G_p-invariant and then S/G_p provides a local model for the orbit space $\mathcal{P}/G$. If G acts by isometries, one usually chooses $T_pS = N = \mathfrak{g}.p^\perp$ (the normal space to the group orbit).

Definition 5.4 Suppose G_p is compact. Let N be a G_p-invariant subspace satisfying $T_x\mathcal{P} = N \oplus \mathfrak{g} \cdot p$. We then define the *symplectic slice* to be

$$N_1 := N \cap \operatorname{Ker}(d\mathbf{J}(p)).$$

Again, if G acts by isometries, one usually chooses $N = (\mathfrak{g} \cdot p)^\perp$. Note that $(\mathfrak{g} \cdot p)^\perp \cap \operatorname{Ker}(d\mathbf{J}(p)) = (\mathfrak{g}_\mu \cdot p)^\perp \cap \operatorname{Ker}(d\mathbf{J}(p))$, and so one can choose the right hand space to be the symplectic slice.

6 Relative equilibria

6.1 Definition and properties of relative equilibria

A point $x_e \in \mathcal{P}$ is called a *relative equilibrium* if for all t there exists $g_t \in G$ such that $x_e(t) = g_t \cdot x_e$, where $x_e(t)$ is the dynamic orbit of X_H with $x_e(0) = x_e$. In other words, the trajectory is contained in a single group orbit. There are different ways to define relative equilibria as the following proposition shows.

Proposition 6.1 *Let* $\mathbf{J}$ *be a momentum map for the* G*-action on* $\mathcal{P}$ *and let* H *be a* G*-invariant Hamiltonian on* $\mathcal{P}$*. Let* $x_e \in \mathcal{P}$ *and* $\mu = \mathbf{J}(x_e)$*. The following assertions are equivalent:*
i) x_e *is a relative equilibrium*
ii) the group orbit $G \cdot x_e$ *is invariant under the dynamics*
iii) there is a $\xi \in \mathfrak{g}$ *such that* $x_e(t) = \exp(t\xi) \cdot x_e$
iv) there is a $\xi \in \mathfrak{g}$ *such that* x_e *is a critical point of the* ***augmented Hamiltonian****:*

$$H_\xi(x) = H(x) - \langle \mathbf{J}(x), \xi \rangle$$

v) x_e *is a critical point of the restriction of* H *to* $\mathbf{J}^{-1}(\mu)$
vi) the image $\overline{x_e} \in \mathcal{P}_\mu$ *of* x_e *is a critical point of the reduced Hamiltonian* H_μ*.*

Remarks 6.2 • The vector ξ appearing in (iii) is called a *velocity* of the relative equilibrium, it is the same as the vector appearing in (iv). Of course, for all $\eta \in \mathfrak{g}_{x_e}$, $\xi + \eta$ is also a velocity of x_e. However, with $N = N_{G_\mu}(G_{x_e})$, the normalizer of G_{x_e} in G_{μ_e}, and given a G_{x_e}-invariant inner product on $\mathfrak{n}_\mu := \mathrm{Lie(N)}$, one can define *the angular velocity* of x_e to be the component of ξ in $\mathfrak{g}_{x_e}^\perp$, the orthogonal complement of $\mathfrak{g}_{x_e}$ in $\mathfrak{n}_\mu$. With this setting, the angular velocity is unique (see [Or98]).
• Note that (iii) implies that relative equilibria cannot meander around a group orbit: they must move in a rigid fashion. For N-body problems in space, the relevant group is $\mathbf{SO}(3)$ and relative equilibria are therefore motions where the shape of the body doesn't change, and these motions are always rigid rotations about some axis.
• If $\mathbf{J}^{-1}(\mu)$ is singular, then it has a natural stratification (see [SL91]) and the condition in assertions (iv) and (v) should be interpreted as being a *stratified* critical point; that is all derivatives of H along the stratum containing x_e vanish at x_e.

Proof. The logic goes as follows:
$i) \Rightarrow ii) \Rightarrow iv) \Rightarrow iii) \Rightarrow i)$ and $iv) \Rightarrow v) \Rightarrow vi) \Rightarrow i)$.

First assume (i), let x_e be a relative equilibrium and let $x = k \cdot x_e$, $k \in G$. By G-equivariance of X_H, $x(t) = k \cdot x_e(t)$ and then $x(t) = kg_t k^{-1} x$, x is a relative equilibrium , which is (ii).

Next, assume (ii). From (ii), we have $x_e(t) \in G \cdot x_e$ for all t. So $X_H(x_e) = T_{x_e}(G \cdot x_e) = \mathfrak{g} \cdot x_e$ and there is a $\xi \in \mathfrak{g}$ such that $X_H(x_e) = \xi_{\mathcal{P}}(x_e)$. By definition of the momentum map, $X_{\mathbf{J}^\xi} = \xi_{\mathcal{P}}$ and then $X_{H-\mathbf{J}^\xi}(x_e) = 0$. Since $w(X_H, \cdot) = dH$, it turns that x_e is a critical point of H_ξ, which is (iv).

Assume (iv). Let φ_t and ψ_t^ξ be the flows of H and $\mathbf{J}^\xi$ respectively, so $\psi_t^\xi(x_e) = \exp(t\xi) \cdot x_e$. Since H is G-invariant, φ_t and ψ_t^ξ commute, it follows that $\varphi_t \circ \psi_{-t}^\xi$ is the flow of $H - \mathbf{J}^\xi$. The critical point x_e of H_ξ is therefore fixed by $\varphi_t \circ \psi_{-t}^\xi$, and so $\varphi_t(x_e) = \psi_t^\xi(x_e) = \exp(t\xi) \cdot x_e$ which is (iii).

Clearly, (iii) implies (i).

Assume (iv). If $\mathbf{J}^{-1}(\mu)$ is a manifold, (v) follows from the Lagrange multipliers theorem. If $\mathbf{J}^{-1}(\mu)$ is singular, then $\mathfrak{g}_{x_e} \neq 0$ since $\operatorname{Im}(d\mathbf{J}(x)) = \mathfrak{g}_x^\circ \subset \mathfrak{g}^*$. By the theorem of Sjamaar and Lerman [SL91], $\mathbf{J}^{-1}(\mu)$ is stratified by the subsets $\mathcal{P}^{(K)} = \{x \in \mathcal{P} \mid G_x \text{ is conjugate to } K\}$ where K is a subgroup of G. Let $\mathcal{P}_{G_x}$ be the set of points with isotropy precisely G_x, this is an open symplectic submanifold of $\operatorname{Fix}(G_x, \mathcal{P})$ containing x. Let $N(G_x)$ be the normalizer of G_x in G and $L = N(G_x)/G_x$, L acts on $\mathcal{P}_{G_x}$. The subsystem $(\mathcal{P}_{G_x}, \tilde{H}, \tilde{\omega}, L)$ is Hamiltonian. Let $\mathbf{J}_L : \mathcal{P}_{G_x} \to \text{Lie(L)}^*$ the corresponding momentum map, $\mathbf{J}_L$ is a submersion since L acts freely on $\mathcal{P}_{G_x}$, thus $\mathbf{J}_L^{-1}(\mu)$ is a manifold, and so we can apply the regular case. The result follows then from the Principle of Symmetric Criticality (see Section 6.2 for a statement).

That (v) implies (vi) follows by passing to the quotient. Finally (vi) implies that the equivalence class $\overline{x_e}$ is a fixed equilibrium of the reduced dynamics. Then $x_e(t)$ lies in $G \cdot x_e$ for all t and this is (i). ■

Proposition 6.3 *Let x_e be a relative equilibrium with angular velocity ξ and $\mu = \mathbf{J}(x_e)$. Then*

$$\operatorname{Coad}^\theta_{\exp(t\xi)} \mu = \mu.$$

Proof. This is simply because $\mathbf{J}$ is equivariant, $\exp(t\xi)$ generates the motion and μ is conserved. ■

For $G = \mathbf{SO}(3)$, this implies that ξ and the momentum vector are parallel vectors, but for the Euclidean group the corresponding relationship is more complicated.

6.2 How does one locate relative equilibria?

From Proposition 6.1, relative equilibria are critical points of the Hamiltonian restricted to the level sets of the momentum map, so results on critical point of G-invariant functions are of particular interest. Mainly, these results are due to Palais [P79] and Michel [Mi71].

Let G be a Lie group and $H : \mathcal{P} \to \mathbb{R}$ a G-invariant function. Assume that G is either compact or acts isometrically on $\mathcal{P}$ Riemannian. The *Principle of Symmetric Criticality* [P79] claims that if the directional derivatives $dH_x(u)$ vanish for all directions u at x tangent to $\mathrm{Fix}(K, \mathcal{P})$, then directional derivatives in directions transverse to $\mathrm{Fix}(K, \mathcal{P})$ also vanish. In particular, any isolated point of $\mathrm{Fix}(K, \mathcal{P})$ is a critical point of H. In our context of Hamiltonian system with symmetry, one obtains the following theorem as a corollary of the Principle of Symmetric Criticality. Recall that if G acts semisymplectically, we denote by G_+ the subgroup (of index 2) of elements acting symplectically.

Theorem 6.4 *Let G act semisymplectically on $\mathcal{P}$. Suppose $x \in \mathrm{Fix}(K, \mathcal{P})$ for some subgroup K of G, and $\mu = \mathbf{J}(x)$. If x is an isolated point in $\mathrm{Fix}(K, \mathcal{P}) \cap \mathbf{J}^{-1}(\mu)$, then x is a relative equilibrium. If in addition $K < G_+$, then x is a fixed equilibrium.*

Note that equilibria derived by this theorem do not depend on the form of the Hamiltonian, they depend only on the action of the symmetry group on the phase space. Note also that this result uses the fact that H is G-invariant, and not that G acts symplectically. In the examples of Section 8, we use this theorem with antisymplectic symmetries ($g^*\omega = -\omega$) as well as symplectic ones.

This theorem provides relative equilibria with large isotropy subgroups, and hence is not usually sufficient to determine bifurcating branches of relative equilibria since symmetry-breaking occurs at a bifurcation.

One way to find these relative equilibria with less symmetry is to determine critical points of the restriction of $H_{|\mathrm{Fix}(K,\mathcal{P})}$ to $\mathbf{J}^{-1}(\mu)$ (using Lagrange multipliers for example). Indeed by the Principle of Symmetric Criticality, we have determined critical points of the restriction of H to $\mathbf{J}^{-1}(\mu)$ which are precisely relative equilibria.

As we shall see in Section 7, one can also determine relative equilibria in a neighbourhood of a zero-momentum relative equilibrium by a bifurcation argument.

6.3 Stability

When one has found a relative equilibrium, a natural question arises: is it *stable*? We first review different definitions of stability in Hamiltonian systems of finite dimension.

Let x_0 be a fixed equilibrium of an Hamiltonian dynamical system and L_0 the matrix of the linearized system at x_0. The equilibrium x_0 is said to be *spectrally stable* if the eigenvalues of L_0 all lie on the imaginary axis. If in addition L_0 is semisimple, the equilibrium is said to be *linearly stable*. To end, an equilibrium x_0 is said to be *Lyapunov stable* if for any neighbourhood U of x_0, there is a neighbourhood V of x_0, $V \subset U$ such that any trajectory which intersects V remains in U for all time. Note that Lyapunov stability is interesting for nonlinear dynamics; for a linear system Lyapunov stability is equivalent to linear stability.

These different concepts are related:

• Linear stability implies spectral stability, but the converse is not true as resonance can generate instabilities.

• Lyapunov stability implies spectral stability. Note that spectral instability implies Lyapunov instability, and this is therefore a way to prove Lyapunov instability.

In order to prove Lyapunov stability, one has the *Lagrange-Dirichlet* criterion: *If the Hessian matrix $d^2H(x_0)$ is positive- or negative-definite, then the equilibrium x_0 is Lyapunov stable.* Indeed, a use of the Morse Lemma states that the level sets of H near the equilibrium are topologically spheres, and by conservation of energy, if a trajectory starts on one of these spheres, it remains on it. (See the lectures of Meyer.)

For relative equilibria, the previous definitions of stability are not suitable since relative equilibria which are not fixed equilibria are unstable in some directions tangent to the group orbit. The appropriate concept of stability is *stability modulo a subgroup* as follows. Let K be a subgroup of G; a relative equilibrium x_e is said to be *stable modulo* K, if for all K-invariant open neighbourhoods V of $K \cdot x_e$ there is an open neighbourhood $U \subseteq V$ of x_e such that the trajectory through any point of U is entirely contained in V.

One would like an analogue of the Lagrange-Dirichlet criterion for the stability of relative equilibria. In the case of a free and proper action with μ regular, this was obtained by the energy-momentum method of Arnold and Marsden (see [Ma92] and reference therein). This result was extended by Patrick [Pa92] to allow locally free actions μ not regular and more importantly concluding not only G-stability but G_μ-stability. More recently, the freeness assumption was dropped as the following theorem shows. The lack of freeness means that the velocity ξ of a relative equilibrium x_e is no longer unique, but only unique modulo $\mathfrak{g}_{x_e}$. Given a G_μ-invariant inner product on $\mathfrak{g}$ the *orthogonal* angular velocity is the unique angular velocity $\xi \in \mathfrak{g}_{x_e}^{\perp}$.

Theorem 6.5 (Lerman, Singer [LS98], Ortega, Ratiu [OR99])
Let G act properly on $\mathcal{P}$, x_e be a relative equilibrium with orthogonal angular velocity ξ, and $\mu = \mathbf{J}(x_e)$. Suppose further that:
i) G_μ acts properly on $\mathcal{P}$ and $\mathfrak{g}$,
ii) the restriction of the Hessian $d^2 H_\xi(x_e)$ to the symplectic slice N_1 is definite.
Then x_e is Lyapunov stable modulo G_μ.

6.4 Persistence

Given a relative equilibrium, one asks if the relative equilibrium *persists*; that is, if there are also nearby relative equilibria for nearby values of the momentum map. First, we give two definitions:

Definition 6.6 A relative equilibrium x_e is said to be *non-degenerate* if the restriction of the Hessian $d^2 H_\xi(x_e)$ to the symplectic slice N_1 is a non-degenerate quadratic form.

Definition 6.7 A point $\mu \in \mathfrak{g}^*$ is a *regular point* of the (modified) coadjoint action if in a neighbourhood of μ all the isotropy subgroups are conjugate.

For $G = \mathbf{SO}(3)$, all points of $\mathfrak{g}^* \simeq \mathbb{R}^3$ are regular except the origin. The following theorem due to Arnold [A78] was the first result on persistence. The proof is an application of the implicit function theorem.

Theorem 6.8 (Arnold) *Let x_e be a non-degenerate relative equilibrium and suppose that G acts freely in a neighbourhood of x_e, and that $\mu = J(x_e)$ is a regular point of the (modified) coadjoint action. Then in a neighbourhood of x_e there exists a smooth family of relative equilibria parametrized by $\mu \in \mathfrak{g}^*$.*

Again, this result was extended by Patrick [Pa95] who showed one need not assume that μ is regular, but only that $G_\mu \cap G_\xi$ is abelian (it always contains a maximal torus of G) together with non-degeneracy of the reduced hamiltonian. He concludes that the set of relative equilibria forms a submanifold of dimension $\dim G + \operatorname{rank} G$. This has been extended further by Lerman and Singer [LS98].

Theorem 6.9 (Lerman, Singer [LS98]) *Let G act properly on $\mathcal{P}$, x_e be a relative equilibrium with angular velocity ξ, and $\mu = \mathbf{J}(x_e)$. Suppose further that:*

i) G_μ is a compact Abelian group,

ii) the restriction of the Hessian $d^2 H_\xi(x_e)$ to $\operatorname{Ker} d\mathbf{J}(\mu)$ is of maximal rank.
Then there exists a symplectic manifold M of relative equilibria passing through x_e of dimension $\dim G + \dim G_\mu - 2\dim G_{x_e}$. Furthermore, if the restriction of the Hessian $d^2 H_\xi(x_e)$ to the symplectic slice N_1 is definite, then a neighbourhood of x_e in M consists of relative equilibria which are Lyapunov stable modulo subgroups conjugate to G_μ.

Note that if G is compact, then for a generic $\mu \in \mathfrak{g}^*$, G_μ is a torus and so hypothesis (i) of the theorem holds for generic μ.

These theorems do not provide information about the symmetry of the relative equilibria of the manifold. Such a result is given in the thesis of Ortega [Or98]. Another extension of Arnold's and Patrick's theorems has been obtained by Patrick and Roberts [PR00] which, again for free actions, describes a stratification of the set of relative equilibria near a given one, depending on G_μ and G_ξ. A statement of their theorem would take us too far afield.

For the case of extrema of H_μ, one has the following persistence result which does not rely on any regularity hypotheses; however it gives no information about the structure of the set of relative equilibria.

Theorem 6.10 (Montaldi [Mo97], Montaldi, Tokieda [MT03]) *Let G act properly on $\mathcal{P}$ and $x_e \in \mathbf{J}^{-1}(\mu)$ be a relative equilibrium, with G_μ compact. Suppose $[x_e] \in \mathcal{P}_\mu$ is an extremum of the reduced Hamiltonian H_μ. Then there is a G-invariant neighbourhood U of x_e such that, for all $\mu' \in \mathbf{J}(U)$ there is a relative equilibrium in $U \cap \mathbf{J}^{-1}(\mu')$.*

7 Bifurcation from zero-momentum state

In this section we present some methods to analyse bifurcations from a relative equilibrium with momentum zero. For details and complements, we refer to papers of Montaldi [Mo97] and Montaldi-Roberts [MR99].

Let x_e be a non-degenerate relative equilibrium with $\mathbf{J}(x_e) = 0$ and $G_{x_e} = 0$ (i.e. locally a free action). Since the momentum is zero, the symplectic slice can be identified with $\mathcal{P}_0$. From the Marle-Guillemin-Sternberg normal form ([GS84]), we have locally near x_e:

$$\mathcal{P}/G \simeq \mathcal{P}_0 \times \mathfrak{g}^*$$

The orbit momentum map becomes:

$$\begin{array}{rcl} \mathcal{J}: \mathcal{P}/G \simeq \mathcal{P}_0 \times \mathfrak{g}^* & \longrightarrow & \mathfrak{g}^*/G \\ (y, \nu) & \longmapsto & \mathcal{O}_\nu \end{array}$$

and (locally) $\mathcal{P}_\mu \simeq \mathcal{P}_0 \times \mathcal{O}_\mu$ since $\mathcal{P}_\mu = \mathcal{J}^{-1}(\mathcal{O}_\mu)$. In many applications, in this decomposition $\mathcal{P}_0$ corresponds to shape dynamics and $\mathfrak{g}^*$ to rigid body dynamics. Of course, the two are highly coupled.

By hypothesis, x_e is a non-degenerate critical point of the restriction of $H : \mathcal{P}_0 \times \mathfrak{g}^* \to \mathbb{R}$ to $\mathcal{P}_0 \times \{0\}$. Thus by the implicit function theorem, for each ν, the function $H(\cdot, \nu)$ has a critical point $y = y(\nu)$. Then define $h : \mathfrak{g}^* \to \mathbb{R}$ by

$$h(\nu) = H(y(\nu), \nu).$$

An easy exercise shows that the restriction of h to $\mathcal{O}_\mu$ has a critical point at ν if and only if the restriction of H to $\mathcal{P}_\mu$ has a critical point at $(y(\nu), \nu)$. The problem is therefore reduced to one of finding critical points of a function on coadjoint orbits. One can use Morse theory or Lyusternik-Schnirelman techniques to estimate the number of critical points, and so the number of bifurcating relative equilibria.

We assumed that G_{x_e} is trivial. However, if G_{x_e} is finite, the same argument applies, but now the resulting function $h : \mathcal{O}_\mu \to \mathbb{R}$ is G_{x_e} invariant. Here G_{x_e} is acting (semi)symplectically by the modified coadjoint action. Moreover, if ν is a critical point of h with isotropy $K < G_{x_e}$, then the corresponding relative equilibrium also has isotropy group K. Analogous to Theorem 6.4, we have the following theorem.

Theorem 7.1 (Montaldi, Roberts [MR99]) *Let x_e be a non-degenerate relative equilibrium with $\mathbf{J}(x_e) = 0$, and K be a subgroup of G_{x_e}. Suppose further that G_{x_e} is finite. Then an isolated point of* $\mathrm{Fix}(K, \mathcal{O}_\mu)$ *with μ close to zero, corresponds to a relative equilibrium with isotropy containing K.*

For an application of this theorem to point vortices on a sphere, see Section 8.3. In [MR99], the stability of the bifurcating relative equilibria is also calculated using these methods. In the case of a relative equilibrium with a non-zero momentum, one can also give a lower bound of the number of relative equilibria on the nearby reduced spaces (see [Mo97]).

Remark 7.2 Since the function h is defined on $\mathfrak{g}^*$, its differential at any point $dh(\nu) \in \mathfrak{g}$. If ν is a critical point of the restriction of h to $\mathcal{O}_\nu$, so ν is a relative equilibrium, then $dh(\nu)$ is in fact the angular velocity of the relative equilibrium in question, [Mo97].

8 Examples

In this section, we apply the previous work to three symmetric Hamiltonian systems:

- Molecules (as classical mechanical systems)
- Point vortices in the plane
- Point vortices on the sphere

where a *point vortex* is an infinitesimal region of vorticity in a 2-dimensional fluid flow.

The study of molecules is of interest in molecular spectroscopy, while point vortices are of interest in modelling concentrated region of vorticity such as hurricanes. Note that the action of the symmetry group is free and proper in these three examples (if one has more than 2 vortices on the sphere).

8.1 Molecules

We consider a molecule consisting of N atoms. We take advantage of the Born-Oppenheimer approximation, which means essentially that we ignore the movement of the electrons. We obtain a model for the nuclei alone, interacting via a potential energy function which incorporates the effects of the electrons. The configuration space is $\mathbb{R}^{3N}$ and the phase space is $\mathcal{P} = T^*\mathbb{R}^{3N} = \mathbb{R}^{6N}$. After fixing the centre of mass at the origin, the dimension of the phase space becomes $6N - 6$.

Let m_i, q_i and $p_i = m_i\dot{q}_i$ be respectively the mass, the position and the momentum of the ith nucleus. The Hamiltonian of the system is given by:

$$H = \sum_i \frac{1}{2m_i}|p_i|^2 + V(q_1, \ldots, q_N)$$

where V is the potential energy due to the electronic bonding between the nuclei. In the absence of external force, V is $\mathbf{O}(3)$-invariant, and so is the Hamiltonian. Moreover, if some nuclei are identical, then a subgroup Σ of the permutation group S_N acts on the set of the nuclei, and Σ leaves H invariant. As with any classical Hamiltonian system of the form "kinetic + potential", the system is time reversible: H is invariant under the involution $\tau : (p, q) \mapsto (-p, q)$. Finally, the Hamiltonian is $\hat{G}$-invariant where $\hat{G} = \mathbf{O}(3) \times \Sigma \times \mathbb{Z}_2^\tau$. Here we use $\hat{G}$ for a semisymplectic group action, whose symplectic part is G; the temporal character is just the projection $\chi : \mathbf{O}(3) \times \Sigma \times \mathbb{Z}_2^\tau \to \mathbb{Z}_2^\tau$.

The $\mathbf{SO}(3)$-symmetry leads to the following momentum map (see Ratiu's lectures for how to compute a momentum map when $\mathcal{P}$ is a cotangent bundle):

$$\mathbf{J} = \sum_i q_i \times p_i$$

where we identified $\mathfrak{so}(3)^*$ with $\mathbb{R}^3$ (this is of course the usual angular momentum for a collection of particles in $\mathbb{R}^3$). The momentum map is equivariant with the $\mathbf{SO}(3)$-coadjoint action (Theorem 5.2), but it is also $\hat{G}$-equivariant with the following action on $\mathfrak{so}(3)^* \simeq \mathbb{R}^3$:

$$(A, \sigma, \tau^k) \cdot \mu = (-1)^k \det(A) A\mu$$

where $(A, \sigma, \tau^k) \in \hat{G} = \mathbf{O}(3) \times \Sigma \times \mathbb{Z}_2^\tau$, see equation (5.1). The action of $\mathbf{O}(3) \times \Sigma$ is not free, however the action of $\mathbf{SO}(3)$ is free away from collinear configurations. We obtain the following theorem using Section 7. Let the *axis of reflection* be the line through the origin perpendicular to the plane fixed by the reflection.

Theorem 8.1 ([MR99]) *Consider a molecule with a non-degenerate equilibrium with symmetry group $\Gamma < \mathbf{O}(3) \times \Sigma$. There exists $\mu_0 > 0$ such that for all $\mu \in \mathbb{R}^3$ with $|\mu| < \mu_0$ there are at least 6 relative equilibria with momentum μ. Moreover, for each axis l of rotation or reflection in Γ, there are two relative equilibria rotating around the axis l with angular momentum μ, one rotating in each direction.*

The minimum of 6 relative equilibria is a consequence only of time reversal symmetry, and thee relative equilibria are similar to the six occurring for the rigid body.

For example, the methane molecule CH_4 has a tetrahedral symmetry, it has 7 axes of rotation and 6 axes of reflection. By the theorem, there are 26 families of relative equilibria bifurcating from the tetrahedral equilibrium. This result depends only on the tetrahedral symmetry, so it is also true for a molecule such as P_4 (white phosphorous).

However, the stability analysis depends on the molecule in question, this is carried out in [MR99]. For a complete investigation of the very interesting molecule H_3^+, see [KRT99].

8.2 Point vortices in the plane

The literature on planar point vortices is large ([Ar82, Ar83a, Ar83b, AV98, Sa92]), usually treated as a problem in fluid mechanics. Here we present the problem in terms of geometric mechanics.

We consider N point vortices in a planar flow of an ideal fluid. The equations governing the motion of the N point vortices are:

$$\dot{\bar{z}}_j = \frac{1}{2\pi i} \sum_{k, k \neq j} \kappa_k \frac{1}{z_j - z_k}$$

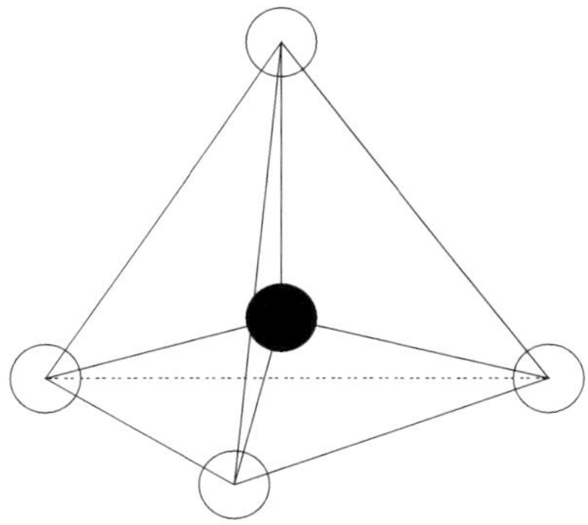

Fig. 8.1. The tetrahedral symmetry of the CH_4 molecule.

where z_j is a complex number representing the position of the j-vortex (we identified the plane with $\mathbb{C}$), and κ_j is the vorticity of the j-vortex. The Hamiltonian for this system is

$$H = -\frac{1}{4\pi}\sum_{i<j}\kappa_i\kappa_j \ln|z_i - z_j|^2$$

and the symmetry group is $\mathbf{SE}(2)$ (which is not compact). After identifying $\mathbf{SE}(2)$ with $\mathbb{C} \rtimes \mathbf{S}^1$ and so $\mathfrak{se}(2)^*$ with $\mathbb{C} \times \mathbb{R}$, the momentum map of the system is

$$\mathbf{J}(z_1, \dots, z_N) = \left(i\sum_j \kappa_j z_j,\ -\tfrac{1}{2}\textstyle\sum_j \kappa_j |z_j|^2 \right).$$

This momentum map is interesting because it fails to be equivariant with the coadjoint action on $\mathfrak{se}(2)^*$, and we must use the modified coadjoint action to make it equivariant (see Theorem 5.2):

$$\mathrm{Coad}^{\Lambda}_{(u,A)}(\nu, \psi) = \left(A\nu,\ \psi + \Im(A\nu\overline{u})\right) + \Lambda\left(iu, \tfrac{1}{2}|u|^2\right)$$

where $(u, A) \in \mathbb{C}\rtimes\mathbf{S}^1$, $(\nu, \psi) \in \mathbb{C}\times\mathbb{R}$, $\Lambda = \sum_j \kappa_j$, and $\Im(z)$ is the imaginary part of z. The term $\Lambda(iu, |u|^2/2)$ is Souriau's cocycle.

If $\Lambda = 0$, then the coadjoint orbits are of two types: either points on the ψ-axis or cylinders around that axis; if, on the other hand, $\Lambda \neq 0$, the orbits are all paraboloids, with axis equal to the ψ-axis.

2 vortices All configurations are relative equilibria. Indeed, it follows from the equations of motion that if $\Lambda = \kappa_1 + \kappa_2 \neq 0$, then the two vortices rotate about the fixed point $\frac{1}{\Lambda}(\kappa_1 z_1 + \kappa_2 z_2)$, while if $\Lambda = 0$ they translate together towards infinity in the direction orthogonal to the segment joining them.

These results can also be derived using the previous sections. Indeed, the reduced spaces are just single points, and so correspond to relative equilibria by Proposition 6.1. Moreover if $\Lambda \neq 0$, then for all μ, $G_\mu \simeq \mathbf{SO}(2)$ and the relative equilibria are Lyapunov stable modulo G_μ by Theorem 6.9 (the symplectic slice is trivial).

3 vortices A complete analysis of the motion of three planar vortices was given by Synge [Sy49]. Here we present a modern approach of the problem but restrict attention to non-collinear configurations for brevity and simplicity. Points of the orbit space $\mathcal{P}/G = \mathbb{C}^3/\mathbf{SE}(2)$ correspond to shapes of oriented triangles. Since we are away from collinear configurations, the orientation determines two isomorphic connected components, so we may ignore the orientation. A point in the orbit space is therefore determined by the three lengths l_{12}, l_{13}, l_{23} where $l_{ij} = |z_i - z_j|$. Denote $r_1 = l_{23}^2$, $r_2 = \ldots$, then the Hamiltonian on $\mathcal{P}/G$ and the orbit momentum map (Section 5.3) are respectively:

$$\begin{aligned} 4\pi H(r_1, r_2, r_3) &= -\kappa_1\kappa_2 \ln(r_3) - \kappa_2\kappa_3 \ln(r_1) - \kappa_1\kappa_3 \ln(r_2) \\ \mathcal{J}(r_1, r_2, r_3) &= p \circ \mathbf{J} = -\kappa_1\kappa_2 r_3 - \kappa_2\kappa_3 r_1 - \kappa_1\kappa_3 r_2 \end{aligned}$$

where the projection $p : \mathfrak{g}^* \simeq \mathbb{C} \times \mathbb{R} \to \mathfrak{g}^*/G \simeq \mathbb{R}$ is given by $p(\nu, \psi) = |\nu|^2 - 2\Lambda\psi$. Recall that the reduced spaces are the fibers of $\mathcal{J}$. The relative equilibria are determined by the critical points of the restriction of H to the reduced spaces, and are therefore critical points of $H - \eta\mathcal{J}$ for some η (η is a Lagrange multiplier). A short computation shows that relative equilibria are all equilateral triangle, of side r say, with $\eta = 1/(4\pi r)$. Then we rely the angular velocity ξ with η:

$$0 = d(H - \eta\mathcal{J})(x_e) = d(H - \xi\mathbf{J})(x_e).$$

Since $\mathcal{J} = p \circ \mathbf{J}$, it follows that $\xi = \eta\, dp(\mu)$ with $\mu = \mathbf{J}(x_e) = (\nu, \psi)$, so

$$\xi = \eta(\nu, -\Lambda).$$

Thus if $\Lambda = 0$, then the motion is rectilinear with constant velocity $\xi = 2\eta\nu = i\sum_j \kappa_j z_j/(2\pi r^2)$.

To determine the stability of these relative equilibria, we use the *reduced energy-momentum method* ([SLM91]); that is, we examine definiteness of the restriction of $d^2(H - \eta\mathcal{J})(x_e)$ to $T_{x_e}\mathcal{P}_\mu$. In fact, this is the energy-momentum method for the reduced Hamiltonian system $(\mathcal{P}_\mu, \omega_\mu, H_\mu)$. The tangent space $T_{x_e}\mathcal{P}_\mu$ is spanned by the two vectors $(\kappa_1, -\kappa_2, 0)$ and $(\kappa_1, 0, -\kappa_3)$, so after some calculus:

$$d^2(H - \eta\mathcal{J})(x_e)_{|T_{x_e}\mathcal{P}_\mu} = \frac{\kappa_1\kappa_2\kappa_3}{4\pi r^2}\begin{pmatrix} \kappa_1 + \kappa_2 & \kappa_1 \\ \kappa_1 & \kappa_1 + \kappa_3 \end{pmatrix}$$

This quadratic form is definite if and only if $\sigma_2(\kappa) > 0$ where $\sigma_2(\kappa) = \kappa_1\kappa_2 + \kappa_1\kappa_3 + \kappa_2\kappa_3$. It follows that the equilateral triangles are Lyapunov stable modulo G if and only if $\sigma_2(\kappa) > 0$.

The 4-vortex problem has been recently studied by Patrick [Pa00] in the case of zero-momentum configurations. Arrangements involving an arbitrary number of vortices (n-gon/kn-gon) have been considered by Lewis and Ratiu [LR96]. We suggest to check the stability of the relative equilibria of a simulation available on the web[1]. To end, the dynamics of perturbed relative equilibria is studied in [Pa99] both on the sphere and on the plane.

8.3 Point vortices on the sphere

We consider here N point vortices in a spherical layer flow of an ideal fluid. The configuration space is

$$\mathcal{P} = \{(x_1, ..., x_N) \in S^2 \times \cdots \times S^2 \mid x_i \neq x_j \text{ if } i \neq j\};$$

we do not permit collisions. Let θ_i be the co-latitude, ϕ_i be the longitude, and κ_i be the vorticity of the i-vortex. The equations governing the motion of the N point vortices on the sphere were obtained by Bogomolov [B77]:

$$\dot{\theta}_i = -\sum_{j=1, j\neq i}^{N} \kappa_j \frac{\sin\theta_j \sin(\phi_i - \phi_j)}{l_{ij}^2}, \quad i = 1...N$$

$$\sin\theta_i\, \dot{\phi}_i = \sum_{j, j\neq i} \kappa_j \frac{\sin\theta_i\cos\theta_j - \sin\theta_j\cos\theta_i\cos(\phi_i - \phi_j)}{l_{ij}^2}, \quad i = 1...N$$

where $l_{ij}^2 = 2(1 - \cos\theta_i\cos\theta_j - \sin\theta_i\sin\theta_j\cos(\phi_i - \phi_j))$ is the square of the Euclidian distance $\|x_i - x_j\|$. The system is Hamiltonian with

$$H = \sum_{i<j} \kappa_i\kappa_j \ln l_{ij}^2, \quad \omega = \bigoplus_{j=1}^{N} \kappa_j\omega_j,$$

where ω_j is the standard area form on the sphere. The dynamical system has full rotational symmetry $G = \mathbf{SO}(3)$, and hence has a 3-component conserved

[1] `http://www.mindspring.com/~brian_tvedt/java.html` and `http://www.ma.umist.ac.uk/jm/vortex.html`

quantity (see Theorem 5.1). We identify $\mathfrak{so}(3)$ with $\mathbb{R}^3$ in the usual way, the momentum map is then given by:

$$\mathbf{J}(x) = \sum_{j=1}^{N} \kappa_j\, x_j.$$

In what follows, we describe some relative equilibria.

2 vortices If the momentum μ of the configuration is non-zero, the two vortices rotate around μ at the same angular velocity, so this is a relative equilibrium. The only possible case for $\mu = 0$ is if $\kappa_1 = \kappa_2$ and $x_1 = -x_2$ which is a fixed equilibrium.

3 vortices The relative equilibria formed of 3 point vortices are completely described in the paper of Kidambi and Newton [KN98], analogous to that of Synge for the planar vortex model. There are two classes of relative equilibria, those lying on a great circle, and those which are equilateral triangle. All equilateral triangle configurations are relative equilibria. The stability of these relative equilibria is computed in [PM98]. One can also use the method described for planar equilateral triangles in the previous section. In this way, one finds that equilateral triangles which do not lie on a great circle are Lyapunov stable modulo $\mathbf{SO}(2)$ if and only if

$$\sigma_2(\kappa) := \kappa_1\kappa_2 + \kappa_1\kappa_3 + \kappa_2\kappa_3 > 0.$$

Note that the stability condition is the same as in the planar case (great circle configurations correspond to collinear configurations, the algebraic volume vanishes there).

Right-angled isosceles triangle lying on a great circle are also relative equilibria, they are Lyapunov stable modulo $\mathbf{SO}(2)$ provided

$$\kappa_2^2 + \kappa_3^2 > 2\sigma_2(\kappa)$$

where x_1 is at the right-angle [PM98].

As in the planar 2-vortex system, one can derive a stability result by a dimension count. Let x_e be a zero-momentum configuration, that is $\mu = \mathbf{J}(x_e) = 0$. Thus x_e lies on a great circle, $\mathbf{SO}(3)_\mu = \mathbf{SO}(3)$ is compact and $\dim \mathcal{P}_0 = 0$. The reduced space $\mathcal{P}_0$ consists of single points, so x_e is a relative equilibrium and is Lyapunov stable modulo $\mathbf{SO}(3)$.

Numerical simulations can be found in [MPS99], and collapse is studied in [KN98],[KN99]. To end, the effect of solid boundaries (such as continents) is taken into account in [KN00].

4 vortices A regular tetrahedron formed of four vortices of arbitrary strength is always a relative equilibrium [PM98]; the stability of these configurations has not been determined. It is easy to show from the equations of motion that a square lying on a great circle is always a relative equilibrium. Contrary to the 3-vortex case, squares not lying on a great circle are not relative equilibria unless the four vorticities are identical.

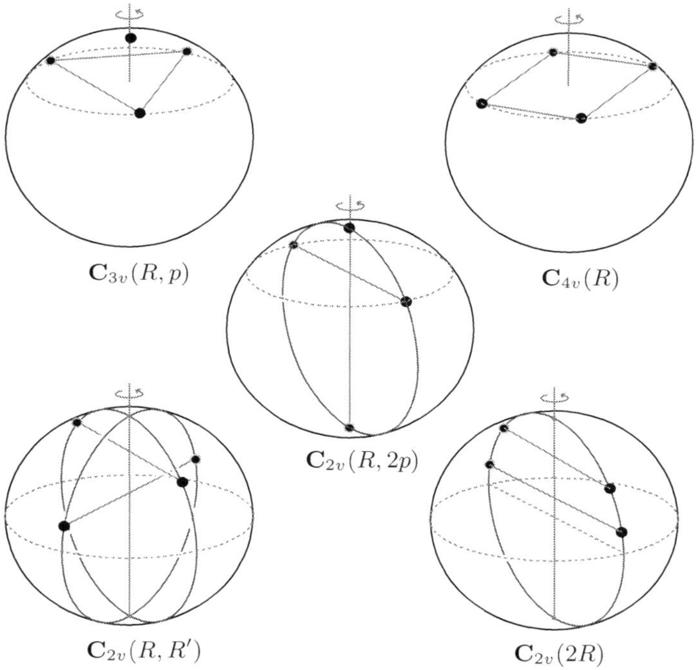

Fig. 8.2. Relative equilibria for 4 identical vortices on the sphere.

N identical vortices In the case of N identical vortices, permutation symmetries arise: the Hamiltonian is $\mathbf{O}(3) \times S_N$-invariant, whereas the vector field is only $\mathbf{SO}(3) \times S_N$-equivariant. The classification of symmetric relative equilibria is carried out in [LMR00]. For example, a regular ring (that is a regular polygon at a fixed latitude) with possibly some vortices at the poles and regular polyhedra, are all relative equilibria. As an exercise, apply Theorem 6.4 to carry out these results.

The linear stability of a regular ring of N identical vortices was studied by Dritschel and Polvani [PD93]. Recently, the results were extended in terms of *Lyapunov stability* in [LMR04], where other configurations of rings and polar vortices are also considered. The result for the single ring is as follows:

Proposition 8.2 *A regular ring of N identical vortices is Lyapunov stable if and only if one of the following assertions is satisfied:*

- $N = 2$ *or* $N = 3$
- $N = 4$ *and* $\cos^2(\theta) > 1/3$
- $N = 5$ *and* $\cos^2(\theta) > 1/2$
- $N = 6$ *and* $\cos^2(\theta) > 4/5$

where θ is the colatitude of the ring.

Note that the rings are "more" stable near the poles than near the equator; while as the number of vortices is increased so the region for which the relative equilibrium is stable diminishes.

2N vortices with opposite vorticities Here we consider N vortices with vorticity $+1$ and N vortices with vorticity -1. The Hamiltonian in this case is $\mathbf{O}(3) \times S_N \times S_N \rtimes \mathbb{Z}_2[\tau]$-invariant, where τ is a permutation of order two which exchanges the $(+1)$-vortices with the (-1)-vortices. The vector field is $\mathbf{SO}(3) \times S_N \times S_N$-equivariant. As before, the relative equilibria are determined using Section 6.2 and stability is computed using Section 6.3, the results can be found in [LP00]. For example, a regular ring formed of the $(+1)$-vortices together with a similar regular ring at the opposite latitude formed of the (-1)-vortices is a relative equilibrium if the offset between the two rings is an integer multiple of π/N.

Bifurcations Changes of stability often involve bifurcations of relative equilibria. Consider the case of a regular ring formed of 4 identical vortices ($\mathbf{C}_{4v}(R)$ in Figure 8.2). By the above proposition, this relative equilibrium is Lyapunov stable if and only if $\cos^2(\theta) > 1/3$ where θ is the colatitude of the ring. In fact, an eigenvalue vanishes when $\cos^2(\theta) = 1/3$, so a bifurcation occurs. As $\cos^2(\theta)$ decreases through $1/3$, there appears a new family of relative equilibria consisting of two rings of two vortices each, the offset between the two rings being equal to $\pi/2$ (denoted $\mathbf{C}_{2v}(R, R')$ in Figure 8.2). These bifurcating relative equilibria are Lyapunov stable: we are in the presence of a supercritical pitchfork bifurcation (see Section 3.2). Others bifurcations such as subcritical pitchfork bifurcation or Hamiltonian-Hopf bifurcation, are described in [LP00].

Bifurcations from zero-momentum state Here we apply Section 7 to our problem. At a zero-momentum configuration, bifurcations occur because the reduced spaces for $\mu = 0$ and $\mu \neq 0$ have different geometry. Consider the arrangement x_e consisting of a regular ring of N identical vortices on the equator. This configuration is a relative equilibrium (in fact a fixed equilibrium) with momentum zero, and isotropy group isomorphic to $D_{Nh} \simeq D_N \times \mathbb{Z}_2$ (we use the Schönflies-Eyring notation for subgroups of $\mathbf{O}(3)$). Nearby reduced spaces are then locally of the form $\mathcal{P}_\mu \simeq \mathcal{P}_0 \times \mathcal{O}_\mu$ where $\mathcal{O}_\mu$ is the coadjoint orbit through μ, which is a sphere in the present case. It follows from Section 7 that the relative equilibria on $\mathcal{P}_\mu$ near x_e are critical points of a D_{Nh} invariant function $h : \mathcal{O}_\mu \to \mathbb{R}$. The set $\mathrm{Fix}(\mathbf{C}_{Nv}, \mathcal{O}_\mu)$ just consists of the North and South poles, and $\mathbf{C}_{Nv}$ is a maximal isotropy subgroup for $\mu \neq 0$, so by Theorem 7.1 there exist near x_e relative equilibria with isotropy $\mathbf{C}_{Nv}$. These are rotating rings of N vortices at a fixed latitude. The subgroup $\mathbf{C}_{2v}$ with axis of rotations lying in the equatorial plane, is also a maximal isotropy subgroup for $\mu \neq 0$. The same argument holds and provides relative equilibria consisting of m 2-rings with one pole if $N = 2m + 1$, and $(m - 1)$ 2-rings with two poles if $N = 2m$.

Acknowledgements

PLB and FLP wish to thank James Montaldi for organizing the MASESS summer school in Peyresq (France) in September 2000. We also thank the staff and residents of Peyresq for their warm welcome and their kindness. FLP also wishes to thank the participants for the very friendly atmosphere (and for giving Génépi a chance...). PLB wishes to thank Mark Roberts for enabling him to attend the summer school in Peyresq. The work of PLB was partly supported by the Engineering and Physical Sciences Research Council UK (EPSRC:GR/L60029) and by a NSERC Postdoctoral Fellowship Canada (BP-230548-2000)

Bibliography

[AM78] R. Abraham and J.E. Marsden. *Foundations of Mechanics* (2nd ed) Benjamin-Cummings, Reading MA, 1978.

[Ar82] H. Aref, Point vortex motion with a center of symmetry. *Phys. Fluids* **25** (1982), 2183–2187.

[Ar83a] H. Aref, Integrable, chaotic and turbulent vortex motion in two-dimensionnal flows. *Ann. Rev. Fluid Mech.* **15** (1983), 345–389.

[Ar83b] H. Aref, The equilibrium and stability of a row of point vortices. *J. Fluid Mech.* **290** (1983), 167–181.

[AV98] H. Aref and C. Vainchtein, Asymmetric equilibrium patterns of point vortices. *Nature* **392** (1998), 769–770.

[A78] V.I. Arnold, *Mathematical Methods of Classical Mechanics.* Springer-Verlag, (1978).

[B77] V. Bogomolov, Dynamics of vorticity at a sphere. *Fluid Dyn.* **6** (1977), 863–870.

[BG92] J.W. Bruce and P.J. Giblin, *Curves and Singularities*, 2nd edition, Cambridge University Press, 1992.

[COR02] P. Chossat, J-P. Ortega and T. Ratiu. Hamiltonian-Hopf bifurcation with symmetry. *Arch. Ration. Mech. Anal.* **163** (2002), 1–33. Correction to "Hamiltonian-Hopf bifurcation with symmetry", *Arch. Ration. Mech. Anal.* **167** (2002), 83–84.

[CB97] R. Cushman and L. Bates. *Global aspects of classical integrable systems.* Birkhäuser, Basel, 1997.

[CS86] R. Cushman and J. Sanders, Nilpotent normal forms and representation theory of $\mathrm{sl}(2, R)$. *Multiparameter bifurcation theory (Arcata, Calif., 1985)*, 31–51, Contemp. Math., **56**, Amer. Math. Soc., Providence, RI, 1986.

[DM92] M. Dellnitz and I. Melbourne, The equivariant Darboux theorem. *In: Exploiting symmetry in applied and numerical analysis (Fort Collins, CO, 1992) 163–169*, Lectures in Appl. Math. **29**. Amer. Math. Soc. Providence, RI, 1993.

[DMM92] M. Dellnitz, I. Melbourne and J.E. Marsden. Generic bifurcation of Hamiltonian vector fields with symmetry. *Nonlinearity* **5** (1992) 979–996.

[ETBCI] C. Elphick, E. Tirapegui, M.E. Brachet, P. Coullet and G. Iooss. A simple global characterisation for normal forms of singular vector fields. *Physica D* **29D** (1987) 95–127.

[GS86] M. Golubitsky and I. Stewart. Generic bifurcation of Hamiltonian systems with symmetry. *Physica D* **24D** (1987) 391–405.

[GSS88] M. Golubitsky, I. Stewart and D.G. Schaeffer. *Singularities and Groups in Bifurcation Theory: Vol.II*. Appl. Math. Sci. **69**, Springer-Verlag, New York, 1988.

[GMO04] F. Grabsi, J. Montaldi and J.-P. Ortega, Bifurcation and forced symmetry breaking in Hamiltonian systems. *C.R. Acad. Sci. Paris* Sér. I **338** (2004) 565–570.

[GS84] V. Guillemin and S. Sternberg, *Symplectic Techniques in Physics*. Cambridge University Press (1984).

[H98] H. Hanßmann. The reversible umbilic bifurcation. *Physica D* **112** (1998), 81–94.

[HM02] H. Hanßmann and J.-C. van der Meer, On the Hamiltonian Hopf bifurcations in the 3D Hénon-Heiles family. *J. Dynam. Differential Equations* **14** (2002), 675–695.

[IP93] G. Iooss and M.-C. Proume, Perturbed homoclinic solutions in reversible $1:1$ resonance vector fields. *J. Differential Equations* **102** (1993), 62–88.

[KN98] R. Kidambi and P. Newton, Motion of three point vortices on a sphere. *Physica D* **116** (1998), 143–175.

[KN99] R. Kidambi and P. Newton, Collapse of three vortices on a sphere. *Il Nuovo Cimento* C **22** (1999), 779–791.

[KN00] R. Kidambi and P. Newton, Point vortex motion on a sphere with solid boundaries, *Phys. fluids* **12** (2000), 581–588.

[KRT99] I. Kozin, M. Roberts and J. Tennyson, Symmetry and structure of rotating H_3^+. *J. Chem. Phys.* **111** (1999), 140–150.

[LP00] F. Laurent-Polz, Point vortices on the sphere: a case with opposite vorticities. *Nonlinearity* **15** (2002), 143–171.

[LMR04] F. Laurent-Polz, J. Montaldi, M. Roberts, Stability of point vortices on the sphere. Preprint (2004). (ArXiv: `math.DS/0402430`.)

[LR96] D. Lewis and T. Ratiu, Rotating n-gon/kn-gon vortex configurations. *J. Nonlinear Sci.* **6** (1996), 385–414.

[LS98] E. Lerman and S. Singer, Relative equilibria at singular points of the momentum map. *Nonlinearity* **11** (1998), 1637–1649.

[LMR00] C. Lim, J. Montaldi, M. Roberts, Relative equilibria of point vortices on the sphere. *Physica D* **148** (2001) 97–135.

[L] A. Lyapunov. Problème générale de la stabilité du mouvement, *Ann. Math. Studies 17*, Princeton University Press, Princeton, NJ. (Reproduction in 1947 of the French translation).

[Ma92] J.E. Marsden, *Lectures on Mechanics*. LMS Lecture Note Series **174** Cambridge University Press 1992.

[MPS99] J. Marsden, S. Pekarsky and S. Shkoller, Stability of relative equilibria of point vortices on a sphere and symplectic integrators. *Il Nuovo Cimento C* **22** (1999), 793–802

[MR] J.E. Marsden and T.S. Ratiu, *Introduction to Mechanics and Symmetry*. (2nd ed.) Springer, 1999.

[MM03] P.D. McSwiggen and K.R. Meyer, The evolution of invariant manifolds in HamiltonianHopf bifurcations. *J. Diff. Equ.* **189** (2003), 538–555.

[Me93] I. Melbourne, Versal unfoldings of equivariant linear Hamiltonian vector fields. *Math. Proc. Camb. Phil. Soc* **114** (1993), 559–573.

[MD93] I. Melbourne and M. Dellnitz, Normal forms for linear Hamiltonian vector fields commuting with the action of a compact Lie group. *Math. Proc. Camb. Phil. Soc.* **114** (1993), 235–268.

[MH92] K. Meyer and G. Hall. *Introduction to Hamiltonian Dynamical Systems and the N-Body Problem* Appl. Math. Sci. **90**, Springer-Verlag, New-York, 1992.

[MS71] K.R. Meyer and D.S. Schmidt, Periodic orbits near L4 for mass ratios near the critical mass ratio of Routh. *Celest. Mech.* **4** (1971), 99–109.

[Mi71] L. Michel, Points critiques des fonctions invariantes sur une G-variété. *CR Acad. Sci. Paris* **272** (1971), 433–436.

[Mo97] J. Montaldi, Persistence and stability of relative equilibria. *Nonlinearity* **10** (1997), 449–466.

[Mo99] J. Montaldi, Perturbing a symmetric resonance: the magnetic spherical pendulum. In *SPT98 – Symmetry and Perturbation Theory II*. A. Degasperis and G. Gaeta eds., World Scientific (1999).

[MR99] J. Montaldi, M. Roberts, Relative equilibria of molecules. *J. Nonlinear Sci.* **9** (1999), 53–88.

[MR00] J. Montaldi, M. Roberts, A note on semisymplectic actions of Lie groups. *CR Acad. Sci. Paris* **330** (2000), 1079–1084.

[MRS88] J. Montaldi, M. Roberts and I. Stewart. Periodic solutions near equilibria of symmetric Hamiltonian systems. *Phil. Trans. R. Soc.* **325** (1988) 237–293.

[MRS90] J. Montaldi, M. Roberts and I. Stewart. Existence of nonlinear normal modes of symmetric Hamiltonian systems. *Nonlinearity* **3** (1990) 695–730.

[MT03] J. Montaldi and T. Tokieda. Openness of momentum maps and persistence of extremal relative equilibria. *Topology* **42** (2003), 833–844.

[Mos76] J. Moser. Periodic orbits near equilibria and a theorem by Alan Weinstein. *Comm. Pure. Appl. Math.* **29** (1976) 727–747.

[N01] P.K. Newton, *The N-vortex problem — Analytical techniques*. Applied Mathematical Sciences, **145**. Springer-Verlag, New York, 2001.

[Or98] J-P. Ortega, *Symmetry, Reduction, and Stability in Hamiltonian Systems*. Ph.D. Thesis. University of California, Santa Cruz (1998).

[OR99] J-P. Ortega and T. Ratiu, Stability of Hamiltonian relative equilibria. *Nonlinearity* **12** (1999), 693–720.

[OR] J.P. Ortega and T. Ratiu, *Momentum Maps and Hamiltonian Reduction*. Birkhäuser, 2003.

[P79] R. Palais, Principle of symmetric criticality. *Comm. Math. Phys.* **69** (1979), 19–30.

[Pa92] G. Patrick, Relative equilibria in Hamiltonian systems: the dynamic interpretation of nonlinear stability on a reduced phase space. *J. Geom. Phys.* **9** (1992), 111–119.

[Pa95] G. Patrick, Relative equilibria of Hamiltonian systems with symmetry: linearization, smoothness, and drift. *J. Nonlinear Sci.* **5** (1995), 373–418.

[Pa99] G. Patrick, Dynamics of perturbed relative equilibria of point vortices on the sphere or plane. *J. Nonlinear Sci.* **10** (2000) 401–415. (ArXiv: `math.SG/9906112`)

[Pa00] G. Patrick, Reduction of the planar 4-vortex system at zero-momentum. In

International Conference on Differential Equations, Vol. 2 (Berlin, 1999), 1000–1008, World Sci. Publishing, 2000. (ArXiv: `math-ph/9910012`.)

[PR00] G. Patrick and R.M. Roberts, The transversal relative equilibria of a Hamiltonian system with symmetry. *Nonlinearity* **13** (2000), 2089–2105.

[PM98] S. Pekarsky and J. Marsden, Point vortices on a sphere: Stability of relative equilibria. *J. Math. Phys.* **39** (1998), 5894–5907.

[PD93] L. Polvani and D. Dritshel, Wave and vortex dynamics on the surface of a sphere. *J. Fluid Mech.* **255** (1993), 35–64

[PS78] T. Poston and I. Stewart. *Catastrophe Theory and its Applications*. Pitman, London, 1978.

[Sa92] P. Saffman, *Vortex Dynamics*. Cambridge University Press, 1992.

[SLM91] J. Simo, D. Lewis and J. Marsden, Stability of relative equilibria. Part i: the reduced energy momentum method. *Arch. Rat. Mech. Anal.* **115** (1991), 15–59.

[SL91] R. Sjamaar and E. Lerman, Stratified symplectic spaces and reduction. *Ann. Math.* **134** (1991), 375–422.

[So74] A.G. Sokol′skiĭ, On the stability of an autonomous Hamiltonian system with two degrees of freedom in the case of equal frequencies. *J. Appl. Math. Mech.* **38** (1974), 741–749.

[S70] J-M. Souriau, *Structures des Systemes Dynamiques*. Dunod, Paris, 1970. [Engl. translation: *Structure of dynamical systems. A symplectic view of physics*. Birkhäuser, 1997.]

[Sy49] J. Synge, On the motion of three vortices. *Can. J. Math.* **1** (1949), 257–270.

[VvdM] A. Vanderbauwhede and J.C. van der Meer, A general reduction method for periodic solutions near equilibria in Hamiltonian systems. In *Normal forms and homoclinic chaos*, Fields Inst. Commun. **4** (1995), 273–294.

[vdM85] J.C Van der Meer. *The Hamiltonian-Hopf Bifurcation*. Lecture Notes in Mathematics **1160**, Springer, 1985.

[vdM90] J.C Van der Meer. Hamiltonian-Hopf Bifurcation with symmetry. *Nonlinearity* **3** (1990) 1041–1056.

[W73] A. Weinstein. Normal modes for nonlinear Hamiltonian systems. *Invent. Math.* **20** (1973) 47–57.

Made in United States
Orlando, FL
21 October 2026

89529435R00240